POST GRADUATE
ENTRANCE EXAM

POST GRADUATE
ENTRANCE EXAM

BOTANY

A Highly useful resource for PG Entrances Conducted by Different Central & State Universities

HEMA BATRA

ARIHANT PUBLICATIONS (INDIA) LTD.

Arihant Publications (India) Ltd.

卐 **ADMINISTRATIVE & PRODUCTION OFFICES**

Regd. Office
'Ramchhaya' 4577/15, Agarwal Road, Darya Ganj, New Delhi -110002
Tele: 011- 47630600, 43518550; Fax: 011- 23280316

Head Office
Kalindi, TP Nagar, Meerut (UP) - 250002
Tele: 0121-2401479, 2512970, 4004199; Fax: 0121-2401648

卐 **SALES & SUPPORT OFFICES**

Agra, Ahmedabad, Bengaluru, Bareilly, Chennai, Delhi, Guwahati, Hyderabad, Jaipur, Jhansi, Kolkata, Lucknow, Meerut, Nagpur & Pune

卐 **ISBN** : 978-93-13190-12-7

卐 **Price** : ₹ 205.00

PO No. : TXT-59-T043112-9-21

Published by Arihant Publications (India) Ltd.

For further information about the books published by Arihant log on to www.arihantbooks.com or email to info@arihantbooks.com

/arihantpub   /@arihantpub   Arihant Publications   /arihantpub

PREFACE

In the modern era of competition, every student is bound to face tough competition to join higher educational institutions. For the last few years, most of the universities and higher educational institutions have started to commence entrance tests for admission in Post Graduate courses of different subjects.

To help out the students in getting admission to PG courses of various Universities/Institutes, Arihant has prepared a Series of books for all different subjects of Post Graduate Entrance Exams.

Each book of this Series is prepared to help you out in getting higher scores in PG entrance examination and moving one step ahead towards the goal of your life.

Salient Features of the Book

- The complete syllabus is divided into chapters, each of them has sufficient theory matter.
- Theory is followed by the chapter exercise which comprises of objective type questions for practice.
- Questions are well supported by hints and solutions to clear the doubts while solving them.
- 5 scientifically designed practice sets are incorporated to assess your preparation and giving you practice to complete the exam in the giving time limit.

As the subject knowledge of students in PG entrance examinations is judged through objective type questions. Hence, inclusion of more than 1600 objective type questions on various topics of graduation is the key feature of the book.

Huge efforts have been made to make this book error free. Suggestions for further improvement and enhancement of content are welcome.

Authors

CONTENTS

1

Principles of Taxonomy

'Biology' is the science of life forms and non-living organisms. The living world is comprised of an amaging diversity of all living organisms including microorganisms, plants, animals, humans, etc. In order to facilitate this huge and vast available variety, biologists have evolved certain rules and principles in order to study, the kinds and diversity of organisms by identification, nomenclature and classification of different living beings.

Life in biological sense is defined as the characteristic that distinguishes objects having signalling and self-sustaining processes from these that do not have such functions because such functions have ceased (death) and are classified as inanimate.

Characteristic Features of Living Organisms

Living organisms shows following characteristic features:

(i) **Growth** Every living organism grows by increase in mass and in number of individuals/cells. Growth occurs continuously in plants, whereas it occurs up to certain age only in case of animals.

(ii) **Reproduction** It is a characteristic of living organisms.

It is known as the process of producing offsprings, possessing features similar to those of parents.

(iii) **Metabolism** It is an another characteristic and defining of all living things. The sum total of anabolic and catabolic reaction occurring continuously inside the body.

(iv) **Cellular organisation** The cells are building blocks of all living things whether plants, animals or humans.

(v) **Consciousness** All living organisms have excellent ability to sense their environment. They respond to various physical, chemical and biological stimuli. The various external factors to which living organisms respond are light, water, temperature, pollutants and other organisms.

(vi) **Body organisation** The body of living organisms is organised, i.e. several component and sub-components cooperate with each other for the functioning of whole body.

Types of classification

There are three types of systems of classification which are as under:

(i) **Artificial system** Only one or almost a few characters are selected arbitrarily.

(ii) **Natural system** All important characters are considered.

Phenetics It is the study of relationships among a group of organisms on the basis of the degree of similarity between them, at molecular, phenotypic or anatomical level. A tree-like network expressing phenetic relationship is called as phenogram. Phenetics is also referred as numerical taxonomy.

(iii) **Phylogenetic system** Organisms are classified according to their evolutionary and genetic tendencies.

Cladistics It is the study of the pathways of evolutionary branches which shows the connection between the branches. A tree-like network that expresses such ancestor-descendant relationship is called a **cladogram**. It is also referred as **phylogenetic classification**.

Biodiversity or Biological Diversity

It means the number and types of organisms found on the planet earth, which forms the life of the living world. Biodiversity is not only confined to the life forms.

Systematics

According to Simpson (1961), systematics is the branch of biology dealing with the unique properties of species and groups to recognise, describe, name and arrange the diverse organisms according to an organised plan.

Systematics includes following fundamental elements:

(i) **Nomenclature** Assigning names to organisms are as followed:

(a) **Common names** are the local names being assigned to an organism in any specific language into a particular region, e.g. rose, dog, cat, ginger, etc.

(b) **Scientific names** are the names been assigned by the biologists representing the particular organism in every part of the world, e.g. *Mangifera indica* (mango), *Rattus rattus* (rat), *Pisum sativum* (pea), *Felis caters* (cat).

(ii) **Classification** Mode of arranging organism into categories.

(iii) **Identification** Assigning taxonomic positions to an organism based on similarity with already existed organism.

Binomial Nomenclature

Binomial nomenclature was introduced by C Linnaeus (1753). It provides distinct and appropriate names to the organisms in which first name indicates **genus**, while the second name indicates **species**. Both of these words are written in italic. The name of the person or author who first observed and reported the plant is given in brief after the species, e.g. *Cicer arietinum* L. (L is for Linnaeus).

Trinomial nomenclature includes three words, i.e. genus, species and sub-species.

These binomial names are universally acceptable, are recognised and also remain same in all languages.

Taxonomy

Taxonomy is the science of identification, classification and nomenclature. Carolus Linnaeus is known as the father of taxonomy. In taxonomy, all living organisms are classified into different taxa, based on their special characteristics. It can further sub-divided into two main classical (old) and **modern** (new) taxonomy.

Taxonomic Hierarchy

It is known as the process of grouping of organisms into different categories that are eventually called **taxonomic categories** or **taxa**. The term 'taxon' was first introduced by ICBN during 1956.

Various taxonomic categories are given below:

(i) **Species** It was described by John Ray as a group of individual organisms with fundamental similarities as a species.

Taxonomic Categories

For Plants	For Animals
Kingdom	Kingdom
↑	↑
Division	Phylum
↑	↑
Class	Class
↑	↑
Order	Order
↑	↑
Family	Family
↑	↑
Tribe	Tribe
↑	↑
Genus	Genus
↑	↑
Species	Species

Taxonomic categories showing hierarchial arrangement in plants and animals

(ii) **Genus** A group of related species having more characters in common when compared to species of other genera is called genus.

(iii) **Tribe** In biology, a tribe is a taxonomic rank above genus, but below family and sub-family.

- In zoology, some examples include the tribes caprini (goat-antelopes), Bombini (bumble bees).
- The standard ending for the name of zoological tribe is '-ini'
- In botany, the standard ending for the name of a botanical tribe is '-eae', e.g. Hyacintheae.

(iv) **Family** A group of related genera with less number of similarities as compared to species and genus is called family.

(v) **Order** It is known as group of one or more related families that passes some similar correlated characters, lesser in number as compared to a family or genera.

(vi) **Class** It is defined as the major category including a group of related orders.

(vii) **Phylum or division** A taxonomic category higher than class and lower in rank to kingdom is called phylum or division.

(viii) **Kingdom** A highest category in the taxonomy including all organisms sharing a set of distinguished characters is called kingdom.

Out of so many classifications given for the kingdom system of organisms, five kingdom system is considered to be the most accepted system. It was founded by RH Whittaker. These kingdoms include Monera, Protista, Fungi, Plantae and Animalia.

Taxonomic Aids

These are the techniques and procedures to store the information as well as specimens for identification and classification of organisms.

Some common taxonomical aids developed by different biologists, includes

(i) **Herbarium** It is a store house of collected plant specimens, usually in dried, pressed, preserve on, sheets for future reference.

(ii) **Botanical gardens** Specialised gardens for the collection of living plants for reference.

(iii) **Museums** Place for the collection of preserved plants and animals specimens for study and reference.

(iv) **Zoological parks** Place for wild animals.

(v) **Key** Taxonomical aid used for the identification of plants and animals based on similarities and dissimilarities.

Note Flora, monograph and manuals catalogues are the other important means of keeping records of descriptions of various organisms.

Chapter Exercise

1. Phylogenetic system brings out
   (a) reproductive similarities
   (b) grouping according to morphological characters
   (c) grouping on the basis of increasing complexities
   (d) grouping according to evolutionary trends

2. A taxon is a
   (a) group of related families
   (b) group of related species
   (c) type of living organism
   (d) taxonomic group of any ranking

3. Which of the following pairs is most closely related?
   (a) Dog and tapeworm (b) Tapeworm and bacteria
   (c) Mushroom and tree (d) *Amoeba* and bacteria

4. Branches of science that deals with identification, nomenclature and classification of organisms is called
   (a) systematics (b) cytology
   (c) genetics (d) nomenclature

5. Largest herbaria of the Asia is located at
   (a) Chennai (b) Hyderabad
   (c) Sibpur (d) Xiangshan Mountain

6. Numerical taxonomy also referred to as
   (a) statistical taxonomy
   (b) phenetics
   (c) computer aided taxonomy
   (d) All of the above

7. A group of plants or animals with similar traits of any rank is
   (a) species (b) genus
   (c) order (d) taxon

8. Basic unit or smaller taxon of taxonomy is
   (a) species (b) genus
   (c) family (d) variety

9. Sequence of taxonomic categories is
   (a) Class–Phylum–Tribe–Order–Family–Genus–Species
   (b) Division–Class–Family–Tribe–Order–Genus–Species
   (c) Division–Class–Order–Family–Tribe–Genus–Species
   (d) Phylum–Order–Class–Tribe–Family–Genus–Species

10. In which kingdom would you classify the archaean and nitrogen-fixing organisms, if the five kingdom system of classification is used?
    (a) Plantae (b) Fungi
    (c) Protista (d) Monera

11. If two organisms are in the same phylum, they must also be in the same
    (a) class (b) species
    (c) family (d) kingdom

12. According to current code for botanical nomenclature, the names of class must ends with
    (a) -opsida (b) -idae
    (c) -ales (d) -aceae

13. Put these classification terms in sequence, starting with the group with the most variation
    I. Class II. Genus
    III. Phylum IV. Species
    V. Family VI. Order
    (a) III, I, VI, V, II, IV (b) IV, II, V, VI, I, III
    (c) III, I, V, VI, II, IV (d) IV, II, VI, V, I, III

14. As we go from a species to kingdom, in a taxonomic hierarchy, the number of common characteristics
    (a) will increase (b) will decrease
    (c) remain same (d) may increase or decrease

15. Which of the following suffix used for units of classification in plants indicates a taxonomic category of family?
    (a) -ales (b) -onae
    (c) -aceae (d) -ae

16. Genus represents
    (a) an individual plant or animal
    (b) a collection of plants or animals
    (c) a group of closely related species of plants or animals
    (d) None of the above

17. The taxonomic unit 'phylum' is the classification of animals is equivalent to which hierarchial level in classification of plants?
    (a) Class (b) Order
    (c) Division (d) Family

18. Which of the following is a defining characteristic of living organisms?
    (a) Growth
    (b) Ability to make sound
    (c) Reproduction
    (d) Response to external stimuli

19. Father of taxonomy is
    (a) EJ Butler (b) Theophrastus
    (c) Swanson (d) C Linnaeus

20. Family is placed between
    (a) order and genus (b) genus and family
    (c) order and class (d) species and genus

21. Term 'species' was proposed by
    (a) John Ray (b) Linnaeus
    (c) Aristotle (d) Darwin

22. A system of classification in which all important characters are considered is

(a) artificial system (b) natural system
(c) phylogenetic system (d) None of these

23. Mode of arranging organism into categories is called

(a) identification (b) nomenclature
(c) classification (d) taxonomy

24. Suffix of order ends with

(a) -ieae (b) -ales
(c) -ae (d) -ordae

25. Zoological parks are maintained by

(a) National Zoo Authority of India
(b) Central Zoo Authority of India
(c) State Government
(d) None of the above

26. The first ever monograph of a plant taxon was given by

(a) C Linnaeus (b) Ernst Mayr
(c) Robert Morison (d) Bentham

27. Basic unit of classification is

(a) genus (b) class
(c) species (d) order

28. Julian Huxley developed the concept of

(a) new systematics (b) classification
(c) old systematics (d) taxonomic hierarchy

29. The scientific name when printed should be

(a) in capital letters (b) in small letters
(c) in italics (d) underlined

30. Headquarter of Botanical Survey of India (BSI) is located at

(a) New Delhi (b) Kolkata
(c) Lucknow (d) Dehradun

31. Modern classification is based on

(a) physiology (b) fossils
(c) phylogeny (d) morphology

32. '*Systema Naturae*' was written by

(a) Wallace (b) Linnaeus
(c) Charles Darwin (d) Aristotle

33. At present time, scientists have named approximately .... species.

(a) 100,000 (b) 500,000 (c) 1,000,000 (d) 1,500,000

34. Cladistics referred to

(a) natural classification (b) artificial classification
(c) phylogenetic classification (d) None of these

35. Which of the following botanist associated with phylogenetic classification

(a) Bentham and Hooker (b) Theophrastus
(c) Linnaeus (d) Engler and Prantl

36. The Indian National Harbaria is situated is

(a) Botanical garden (NOIDA)
(b) N B R I (Lucknow)
(c) Botanical Survey of India (Howrah)
(d) Zoological Survey of India (Kolkata)

37. Five kingdom classification is given by

(a) Ernst and Haeckel (b) Bentham and Hooker
(c) R H Whittaker (d) Margulis

38. In two kingdom system of classification *Euglena* is included in

(a) Plantae (b) Protista
(c) Animalia (d) Both (a) and (b)

39. The scientist who created the group Protista for both unicellular plants and animal cell was

(a) Haeckel (b) Pasteur (c) Bentham (d) Linnaeus

40. Binomial system of nomenclature means that every organism has

(a) two names, one scientific and the other popular
(b) one name given by two scientists
(c) one scientific name consisting of generic and a specific epithet
(d) None of the above

Answers

1. (d)	**2.** (d)	**3.** (d)	**4.** (a)	**5.** (d)	**6.** (b)	**7.** (a)	**8.** (a)	**9.** (c)	**10.** (d)
11. (d)	**12.** (a)	**13.** (a)	**14.** (a)	**15.** (c)	**16.** (c)	**17.** (c)	**18.** (c)	**19.** (d)	**20.** (a)
21. (a)	**22.** (b)	**23.** (c)	**24.** (b)	**25.** (b)	**26.** (c)	**27.** (c)	**28.** (a)	**29.** (c)	**30.** (b)
31. (c)	**32.** (b)	**33.** (d)	**34.** (c)	**35.** (d)	**36.** (c)	**37.** (c)	**38.** (c)	**39.** (a)	**40.** (c)

2

Classification of Living World

Carolus Linnaeus (1707-1778) laid to the foundations for modern biological nomenclature, regulated by the nomenclature codes in 1735. He distinguishes all living organisms into two kingdoms of living things, i.e. Plantae (*Regnum vegetabile*) and Animalia (*Regnum animali*). This system of classification was accepted for long time but was failed due to the some difficulties in categorising all organisms.

Ernst Haeckel 1938, further classified organisms into three kingdoms, i.e. Plantae, Protista and Animalia. But this was also failed to be accepted due to certain limitations.

The development of electron microscope revealed important distinction between those unicellular organisms whose cells do not have a distinct nucleus (prokaryotes) and those unicellular organisms whose cells have a distinct nucleus (eukaryotes).

The four kingdom classification system was created by Herbert F Copeland (1956) as Monera, Protista, Plantae and Animalia. Clevating the protist classes of bacterial (Monera) and blue-green algae (*Phycochronacea*) to phylum the kingdom–Monera.

After that RH Whittaker (1969) recognised an additional kingdom for the fungi and divided living organisms into five main kingdoms on the basis of their cell structure (whether complex eukaryote or simple prokaryote), body structure (unicellular or multicellular), nutrition (autotrophic or heterotrophic) and ecological role (producers, consumers or decomposers). This system proposed by him had become a popular standard and with some refinement is still used in many works forming the basis for the new multi-kingdom systems. This five kingdom system may be combined with the two empire systems.

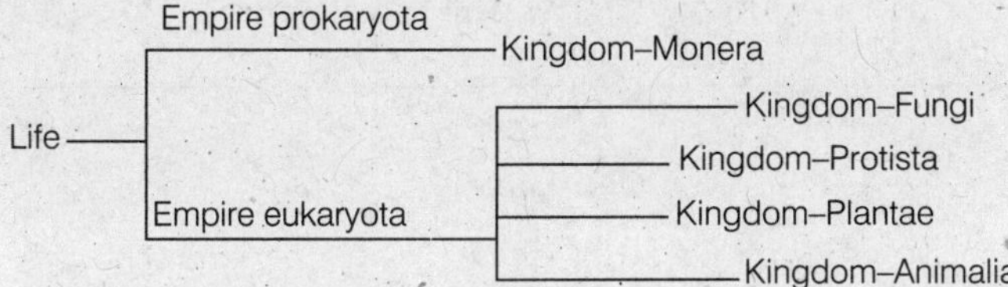

Kingdom–Monera

The main characteristics of kingdom–Monera are as follows:

- Monerans includes all prokaryotes *viz* mycoplasmas, bacteria, Actinomycetes (mycilial bacteria) and photosynthetic cyanobacteria.
- The nuclei of monerans are not organised within nuclear membrane, nucleoplasm, chromatin fibres and nucleolus and are referred as **incipient nuclei.**
- The moneran cells typically lack cell organelles such as chloroplasts, mitochondria and other membrane bound organelles like endoplasmic reticulum, Golgi bodies, etc.
- The photosynthetic pigments are present in the form of **chromatophores**.
- The respiratory enzymes which are mainly concerned with mitochondria in other organisms, however, are present along the infolding of plasma membrane called **mesosomes**.
- Mostly all monerans possess cell wall (except mycoplasmas) generally made up of **peptidoglycan** (a combination of amino acids and sugars that surround the cell like a net).
- In archaebacteria, cell wall is proteinaceous in nature.
- The monerans are unique in having specific type of ribosomes distributed in cytoplasm having 70 sedimentation coefficient (70 S).
- DNA is naked and lies inside the cytoplasm in the coiled form known as **nucleoid**.
- Flagella if present, are single stranded containing a protein called **flagellin**.
- Monerans reproduce asexually.
- The nuclear genetic material in monerans is represented by single molecule of DNA.
- The cell division does not involve any precision in the distribution of genetic material among the daughter cells. This is accomplished by a process known as **amitosis**.

Kingdom–Protista

Physiologically, the kingdom–Protista acts as a connecting link between the Monera and Fungi, Plantae and Animalia.

The important characteristics of kingdom–Protista are as follows:

- It includes all the unicellular and eukaryotic organisms like flagellates, diatoms, dinoflagellates, slime moulds, sarcodina, etc.
- It mostly includes aquatic organisms which form planktons. But some protists may also live in the bodies of animals as parasites.
- Nutrition is symbiotic, parasitic, ingestive, photosynthetic and saprophytic, etc.
- It includes producers, i.e. photosynthetic planktons like algae.
- Most of the protists are aerobic.
- It also includes another group of planktons, which is non-photosynthetic in nature and known as **zooplankton**.
- It includes *Euglena* like organism, which has dual mode of nutrition, i.e. it is photosynthetic and holozoic (in the absence of light).
- It also includes organisms that are intermediate between the wall and wall less organisms, i.e. slime moulds.
- Cellular organisation is of envelope type, i.e. plasma membrane and internal membrane which occur around certain organelles.
- It possesses certain organelles like mitochondria, endoplasmic reticulum, nucleus, food vacuoles, etc.
- They are either motile or non-motile. If motile possesses flagella or cilia (flagella with 9 + 2 organisation).

Comparative Analysis of Monera and Protista

Characters	Monera		Protista
	Eubacteria	Archaea	
Cell type	Prokaryotic	Prokaryotic	Eukaryotic
Genetic material	DNA not associated with histone protein in chromosomes	DNA associated with some proteins in chromosome	DNA associated with protein in chromosomes
Gene structure	Introns absent in most groups, present in some	Introns present	Introns present
Nuclear envelope	Absent	Absent	Double or single
Membrane bound organelles	Absent	Absent	Present
Chloroplasts	Absent	Absent	Present or absent
Cell wall	Non-cellulosic; peptidoglycan	Non-cellulosic; various types different from eubacteria	Absent or present; cellulosic or various types
Means of genetic recombination	Conjugation, transduction, transformation	Conjugation, transformation, transduction	Fertilisation and meiosis in most groups
Mode of nutrition	Autotrophic or heterotrophic by absorption	Heterotrophic or chemoautotroph (many are extremophiles)	Autotrophic or heterotrophic by absorption or phagocytosis
Flagella/Cilia	Solid (9 + 0) rotating	Solid (9 + 0) rotating	(9 + 2) microtubules
Multicellularity/Cell specialisation	Absent	Absent	Absent in most, present in algae and water moulds
Nervous system	Absent	Absent	Absent or simple
Respiration	Aerobic or anaerobic	Anaerobic or aerobic	Aerobic
Life cycle	Haploid	Haploid	Mostly haploid; some forms diploid
Unicellular spore formation	Present in some groups	Absent	Present in some groups

Kingdom–Fungi

A fungus is known to be any member of the group of eukaryotic organism including unicellular microorganism such as yeasts and moulds as well as multicellular fungi.

Fungi were the very first to be evolved soon after the evolution of monerans and protistans. It constitutes a unique kingdom of heterotrophic organisms having more than 1,00,000 species. They show a great diversity in both morphology and habitat. They are completely separate from both plants and animals.

The branch of biology dealing with the study of fungi is called **mycology**. Father of modern mycology and plant pathology is Antonde Berry, while EJ Butler is known as the father of Indian mycology and plant pathology.

Kingdom–Plantae

Plantae is the kingdom of multicellular, chlorophyll containing, photoautotrophic organisms that have the ability of growing indefinitely.

Classification of plants is the system of arranging them into suitable divisions and sub-divisions. It is mainly based upon the similarities and dissimilarities between them. The members of kingdom–Plantae have two distinct phases, the diploid saprophytic phase and the haploid gametophytic phase. This is called **alternation of generation**.

A number of workers including Eichler, Endicher, Engler, Tippo, Bessey, Smith and Alexopoulos, Bentham and Hooker and many others have given their systems of classification. Hence, plant kingdom divided into two sub-kingdom, i.e. cryptogams and phanerogams. These sub-kingdoms were further divided by Eichler (1883) as follows:

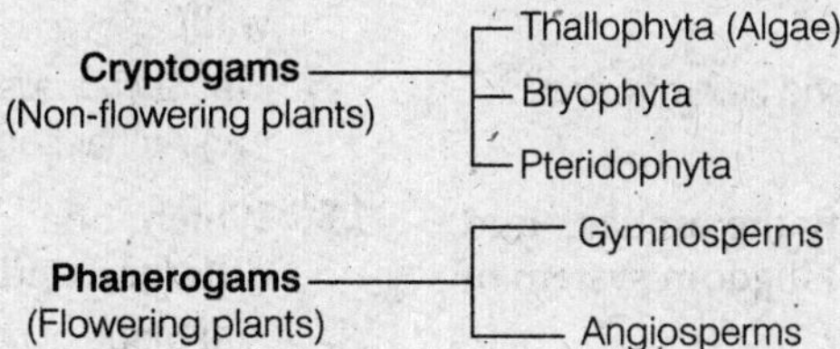

Kingdom–Animalia

The word animal is derived from the Latin word *anima* which means air or breath. Presently, animals are classified according to the broader range of characteristics including anatomy, pattern of development and their genetic makeup.

Aristotle, Father of zoology, first tried to classify the animals. Further, John Ray (1686) and Carolus Linnaeus (1758) gave their contribution in classifying animals.

Though, all the members of kingdom–Animalia are multicellular, yet all of them do not exhibit the similar pattern of cellular organisation. The complete animal kingdom is divided into 35 phyla of which 11 are considered as major phyla.

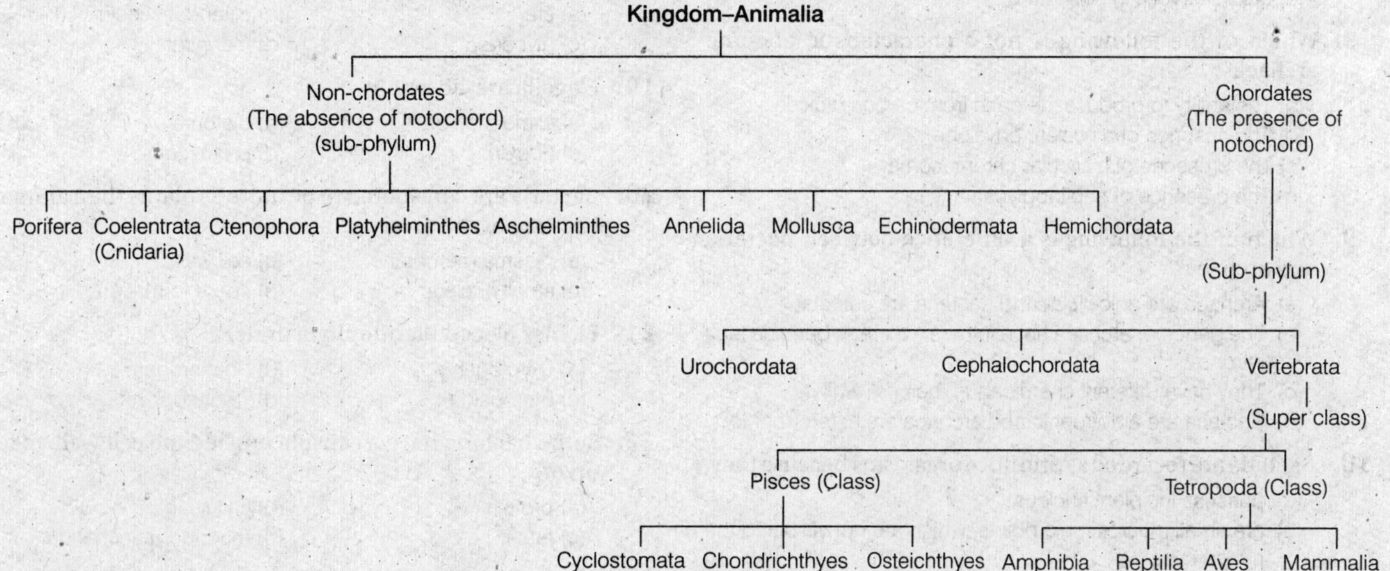

Chapter Exercise

1. In which of the following kingdoms, diatoms are placed?
(a) Plantae (b) Fungi
(c) Animalia (d) Protista

2. Organisms found in extreme temperature are
(a) cyanobacteria (b) archaebacteria
(c) fungi (d) eubacteria

3. Who proposed the five kingdom classification and named kingdoms as Monera, Protista, Fungi, Plantae and Animalia?
(a) Herbert Copeland (b) RH Whittaker
(c) Carl Woese (d) Carolus Linnaeus

4. Which of the following is not a character of Protista?
(a) Protists are prokaryotic
(b) Some protists have cell walls
(c) Mode of nutrition is both autotrophic and heterotrophic
(d) Body organisation is cellular

5. In which kingdom, would you classify the archaea and nitrogen- fixing organisms, if the five kingdom system of classification is used?
(a) Plantae (b) Fungi (c) Protista (d) Monera

6. Single-celled eukaryotes are included in
(a) Monera (b) Protista (c) Fungi (d) Archaea

7. Bacteria
(a) are sole members of kingdom–Monera
(b) have lipoproteinaceous cell wall
(c) have either DNA or RNA as genetic material
(d) are eukaryotic in cell nature

8. Which of the following is not a characteristic of some archaea?
(a) The ability to produce methane from carbon dioxide
(b) The absence of a nuclear envelope
(c) The presence of a circular chromosome
(d) The presence of peptidoglycan

9. Which of the following is a difference between bacteria and archaea?
(a) Archaea are unicellular and bacteria are colonial
(b) The genes of archaea have introns, whereas bacteria lack introns
(c) They have different chemicals in their cell wall
(d) Bacteria are autotrophic and archaea are heterotrophic

10. Bacteria are regarded as primitive organisms because they
(a) possess incipient nucleus
(b) are small, microscopic plants, which are not visible to the naked eyes
(c) cause serious diseases to human beings, domesticated animals and crop plants
(d) produce endospores, which are resistant to adverse conditions

11. Bacteria were considered more as plants than animals because of the presence of
(a) small nucleus (b) plasma membrane
(c) cell wall (d) spore formation

12. The bacteria present in the guts of several ruminant animals such as cows and buffaloes and is responsible for the production of methane is
(a) halophiles (b) methanogens
(c) thermophiles (d) acidophiles

13. Bacteria are present everywhere except in
(a) ice (b) soil
(c) sea water (d) distilled water

14. Which group of bacteria lack peptidoglycan in their cell wall?
(a) *Salmonella* (b) *E. coli*
(c) Archaebacteria (d) Vibrio

15. Which one of the following is not used to identify different kinds of bacteria?
(a) Mode of nutrition (b) Cell wall composition
(c) Number of chromosomes (d) Cell shape

16. The percentage of plankton constituted by protistans
(a) 80% (b) 90% (c) 95% (d) 100%

17. Diatoms belongs to
(a) chrysophytes (b) dinoflagellates
(c) pyrophytes (d) euglenoids

18. Storage product in chrysophytes is
(a) starch (b) oil and leucosin
(c) glycogen (d) mannitol

19. Flagella are absent in
(a) slime moulds (b) diatoms
(c) *Euglena* (d) protozoans

20. Bacteria are considered to be more as plants than animals because of the presence of
(a) plasma membrane (b) cell wall
(c) small nucleus (d) spore formation

21. Eichler placed Pteridophyta under
(a) Cryptogams (b) Phanerogams
(c) Pteropsida (d) Tracheophyta

22. Some bacteria have a capsule outside the cell wall, made up of
(a) protein (b) cellulose
(c) fats (d) mucopolysaccharide

23. Class–Chondrichthyes belong to the super-class
(a) Tetropoda (b) Pisces
(c) Both (a) and (b) (d) None of these

24. *Euglena* like organisms show dual mode of nutrition such as
(a) parasitic and holozoic
(b) photosynthetic and ingestive
(c) photosynthetic and holozoic
(d) photosynthetic and saprophytic

25. The classes of sub-kingdom–Phanerogamae are
(a) dicotyledons
(b) monocotyledons
(c) gymnosperms
(d) All of these

26. Which organism possesses characteristics of plants and animals?
(a) Bacteria (b) Monera
(c) *Euglena* (d) Mycoplasma

27. Name the phylum is characterised by the absence of true coelom.
(a) Nematoda (b) Mollusca
(c) Annelida (d) Echinodermata

28. In angiosperm characters of flowers are used in classification because
(a) flowers are large
(b) flowers are attractive
(c) characters of flowers are conservative
(d) None of the above

29. Which of the following kinds of animals is triploblastic?
(a) Corals (b) Flatworms
(c) Sponges (d) Ctenophores

30. DNA present in prokaryotes is
(a) single, circular (b) single, linear
(c) many, circular (d) many linear

31. Monerans does not include
(a) blue-green algae (b) bacteria
(c) mycoplasma (d) diatoms

32. Father of Indian mycology
(a) Birbal Sahni (b) Hargovind Khurana
(c) Antonde Barry (d) E J Butler

33. Introns are absent is
(a) *Amoeba* (b) algae
(c) fungi (d) bacteria

34. Cellular structure of an organism is studied. It was found that mitochondria, nucleus and endoplasmic reticulum present. This organism certainly do not belong to
(a) Protista (b) Monera
(c) Slime mould (d) Algae

35. Which one is not characteristic feature of kingdom fungi?
(a) Presence of hyphae (b) Cell wall-chitin
(c) Saprophytes (d) Autotrophs

36. Which protozoan is unlibely to have a contractile vacuole?
(a) *Euglena* (b) *Paramecium*
(c) *Amoeba* (d) *Plasmodium*

37. Protista does not include
(a) mycoplasma and cycanophyta
(b) pyrrophyta and xanthophyta
(c) dianoflagellates and diatoms
(d) euglenophyta and chrysophyta

38. Members of Deuteromycetes also known as imperfect fungi why
(a) Only asexual reproduct
(b) Ion present both
(c) Lack of sexual reproduction
(d) None of the above

39. Division of kingdom fungi is based on
(a) morphology of my celium
(b) mode of spore formation
(c) Both (a) and (b)
(d) None of these

40. Nitrogen fixing ability is present in members of .... and structure is known as ....
(a) Phaeophyceae, root nodules
(b) Cyanophyceae (blue green algae), heterocyst
(c) Rodophyceae, heterocyst,
(d) None of the above

Answers

1. (c)	**2.** (b)	**3.** (b)	**4.** (a)	**5.** (d)	**6.** (b)	**7.** (a)	**8.** (d)	**9.** (c)	**10.** (a)
11. (c)	**12.** (b)	**13.** (d)	**14.** (c)	**15.** (c)	**16.** (c)	**17.** (a)	**18.** (b)	**19.** (c)	**20.** (b)
21. (a)	**22.** (d)	**23.** (b)	**24.** (c)	**25.** (d)	**26.** (c)	**27.** (a)	**28.** (c)	**29.** (b)	**30.** (a)
31. (d)	**32.** (d)	**33.** (d)	**34.** (c)	**35.** (d)	**36.** (d)	**37.** (a)	**38.** (b)	**39.** (c)	**40.** (a)

3

Structure of Bacterial Cell and Bacteriophage

Cellular Structure

Bacteria are considered among the first life forms to appeared on earth. They eventually, constitutes a large domain of prokaryotic microorganisms typically a few micrometres (1-5μm) in diameter. Bacteria varies in a shape ranging from sphere, rod-shaped, to spiral. Bacteria are found in most of the habitats in habiting soil, water, acidic hot springs, radioactive wastes, etc. They are also found in the deep portions of earth's crust.

Morphology of Bacteria

Bacteria display a wide diversity of shapes and sizes. Bacterial cells are about one-tenth the size of eukaryotic cells and are typically (0.5-5.0μm) in length. Mycoplasma is considered among the smallest members of bacteria. Most bacterial species are either spherical called **cocci**, rod-shaped called **bacilli**, spiral-shaped called **spirilla**, comma-shaped called **vibrio**, etc.

Some bacterial species may exist-simply as single cells, while others associate themselves in a characteristic patterns to form diploids. Bacteria may often found attached to surfaces forming a dense aggregations called **biofilms.**

For the convenience of study, we divide the cell into following components:

(i) External (extracellular) structures to the cell wall.
(ii) The cell wall
(iii) Internal (intracellular) structures to the cell wall.

The structure of a typical bacterial cell is shown in the figure below:

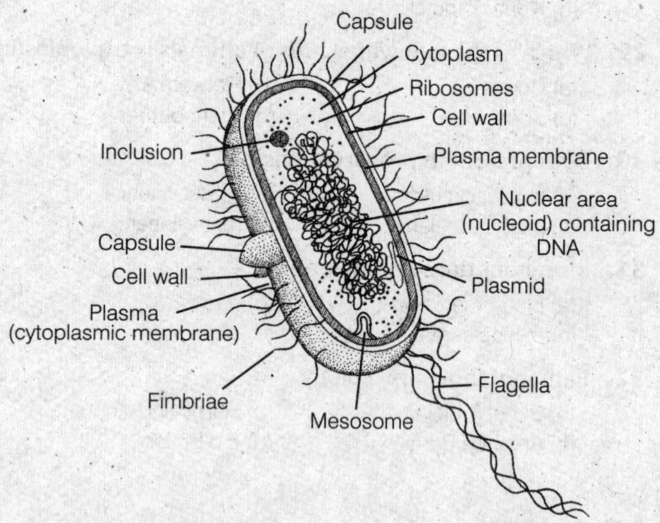

Structure of a typical prokaryotic (bacterial) cell

External (Extracellular) **Structures to the Cell Wall**

Cell wall, glycocalyx, flagella, axial filaments, fimbriae and pili are the structures found external to the cell wall.

1. Cell Wall

The bacterial cell wall is strong and rigid structure. The cell wall prevents the cell from bursting, when it absorbs water (as a result of osmosis). Tiny pores allow the passage of water, ions and small molecules. Peptidoglycan or mucopeptide or murein form the structural network of bacterial cell wall.

There are different types of cell wall in bacteria, a thick one in the **Gram-positive** and in **Gram-negative** bacteria containing thinner cell wall consisting of few layers of peptidoglycan surrounded by a second lipid membrane containing lipopolysaccharide and lipoprotein. It also contains some special amino acids, such as L-alanine, D-alanine, D-glutamic acid and either L-lysine or diaminopimelic acid.

2. **Glycocalyx** (Mucilage Sheath)

The glycocalyx (capsule or slime layer) is made up of gelatinous, polysaccharide or polypeptide. Capsule may protect pathogens from phagocytosis and enables the adherence to the surfaces which prevent desiccation and may provide nutrients. Glycocalyx differs among thickness and composition in different bacteria. It can occur in the form of loose mucilaginous sheath called slime layered, thick and tough called capsule.

3. **Flagella**

These are relatively long filamentous appendages consisting of a filament, hook and basal body.

Bacterial cell has four types of arrangement of flagella.

These are as follows:

(a) **Monotrichous** (single polar flagellum), e.g. *Vibrio cholerae*, *Thiobacillus*.

(b) **Amphitrichous** (single flagellum at each end of the cell), e.g. *Nitrosomonas.*

(c) **Lophotrichous** (two or more flagella at one end of the cell), e.g. *Pseudomonas* (fluorescence).

(d) **Peritrichous** (flagella distributed over the entire cell), e.g. *Escherichia coli and Clostridium*.

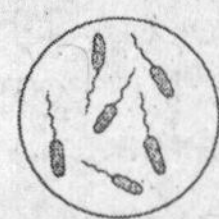
(a) Monotrichous

(b) Amphitrichous

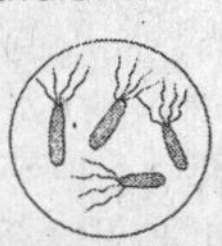
(c) Lophotrichous

(d) Peritrichous

Arrangement of flagella in a typical bacterial cell

4. **Axial Filaments**

These are the structure similar to the flagella except that they wrap around the cell. Spirochaetes are the bacteria that moves by means of axial filaments, i.e. bundles of fibrils that arises at the ends of the cell beneath the outer sheath and spiral around the cell.

5. **Fimbriae and Pili**

These are the short filamentous structures composed of protein (pilin) that extend from the surface of a cell. Pili are structurally similar to fimbriae but are typically longer. Only one or a few pili are present on the surface because they serve as receptors for certain type of viruses.

Internal Structures to the Cell Wall

There are various structures that are internal to the cell wall and perform different functions.These are as follow:

1. **Plasma** (Cytoplasmic) **Membrane or Cell Membrane**

The structure and functions of the cell surface membrane are similar to those in the eukaryotic cells. It also acts as the site of some respiratory enzymes. This membrane encloses the contents of the cell and acts as a barrier to hold nutrients, proteins and other essential components of the cytoplasm within the cell.

There is a coiled invagination arises from it which is called **mesosome.** Mesosome is analogous to mitochondria of eukaryotic cells.

2. **Cytoplasm**

It is composed of about 80% water, proteins (enzymes), carbohydrates, lipids, inorganic ions, etc. It is thick, aqueous, semi-transparent and elastic in nature. The major structures in the cytoplasm are DNA, ribosomes and reserve deposits called **inclusions**.

3. **Nucleoid**

Bacteria lack a well defined membrane bound **nucleus**. However, an amorphous lobular mass of fibrillar chromatin type material is present near the centre of the cell that occupies about 10-20% of the cell. This mass of chromatin in the prokaryotic cell is called nucleoid or **genophore.**

4. **Genetic Material** (Bacterial Chromosome)

Bacterial DNA is a single circular molecule of about 5 million base pair, i.e. much lesser than that of a eukaryotic cell. Typically, it contains several thousand genes, i.e. about 500 times fewer than a human cell.

5. **Plasmid**

A plasmid is a small, self-replicating circular extra chromosomal DNA possessing only a few genes, which generally give extra survival advantage to the cell. Some plasmids confer resistance to antibiotics, e.g. some staphylo-cocci contain a plasmid which includes a gene for the enzyme penicillinase.

6. **Ribosomes**

Bacterial cytoplasm contains numerous 70 S ribosomes, which consist of *r*RNA and protein. The 70 S ribosome consists of a small 30S sub-unit and a larger 50 S sub-unit. Several antibiotics such as streptomycin, neomycin and tetracycline work by inhibiting protein synthesis in the ribosome.

7. **Inclusions**

These are the reserve deposits which are found in both prokaryotic and eukaryotic cells. Among bacteria the inclusions found are metachromatic granules (usually glycogen or starch), lipid inclusions, sulphur granules, carboxysomes and gas vacuoles.

Metabolism in Bacteria

Bacteria exhibit an extremely wide variety of metabolic types. Bacterial metabolism is classified into nutritional groups on the basis of three major criteria.

(i) The kind of energy used for growth

(ii) Source of carbon

(iii) Electron donors used for growth.

Growth and Reproduction in Bacteria

Unlike in multicellular organisms increase in cell size (cell growth) and reproduction by cell division are tightly linked in unicellular organism. Bacteria grow to a fixed size and then reproduce through binary fission, a type of asexual reproduction. Thus, most of the bacteria reproduces by a simple form of asexual reproduction and very rarely by the sexual means.

Asexual reproduction may take place by the following methods:

1. **Vegetative Reproduction**

It occurs mainly by fission, budding and cysts formation.

2. **Sporulation**

Some bacteria produces non-motile spores such as conidia, conidiophores, etc.

3. **Sexual Reproduction**

It is proved by Lederberg and Tatum (1946). Bacteria show a primitive form of sexual reproduction as there is no gamete fusion. Exchange of genetic material takes place in bacteria, known as **genetic recombination**. Genetic recombination undergo by three forms (i) Transformation (ii) Transduction (c) Conjugation

Following three types of genetic recombinations are found:

(i) **Conjugation** in which the transfer involves cell to cell contact and a conjugative plasmid in donor cell.

(ii) **Tansformation** in which DNA released from one cell is taken by another.

(iii) **Transduction** in which donor DNA transfer is mediated by a virus.

Genetics

Most bacteria have a single circular chromosome that can range in size from only 160,000 bp in endosymbiotic bacteria to 12,200,000 bp in soil dwelling bacteria. Bacteria may often contains plasmids (an extra-chromosomal DNA that may contain genes for antibiotic resistance or virulence factors).

Virus

The term 'virus' means poisonous fluid in latin language. It was coined by Pasteur.

A virus is a small infectious agent that replicates only inside the living cells of other organisms. Viruses can infect all types of life forms, from animals and plants to microorganisms including bacteria and archea. Viruses are found in almost every ecosystem on earth and are the most abundant type of biological entity. The study of viruses is known as **virology.** Viruses do not have any cellular organisation of their own instead uses the host's mechanism or host machinery to express their genes and to replicate themselves.

Discoveries Related to Viruses

- **DJ Ivanowsky** (1892) discovered the first virus from extract of tobacco plant suffering from **TMV** (Tobacco Mosaic Virus).
- **P Rous** (1911) report that virus can cause cancer.
- **FW Twort** (1915) discovered bacteriophage and **F d' Herelle** (1917) coined the term 'Bacteriophage'.
- **W Stanley** (1933) purified the crystal form of TMV and showed that viruses could be crystallised and crystals consists largely of proteins.
- **Zinder** and **Lederberg** (1952) discovered the process of transduction.
- **TO Diener** (1971) discovered viroids.

Size

Size of viruses may range from about 10 nm (foot and mouth virus of cattle), 17 nm (alyaya mosaic virus) to 125 × 40 nm (beet yellow virus), TMV is 300 × 18.0 nm 400 nm (parrot fever virus) 1300 × 6 nm (*Pseudomonas*).

Shapes

The shape (architecture) of viruses are mainly of three types, i.e. helical, e.g. TMV, cuboidal, e.g. poliomyelitis virus, binal, e.g. bacteriophages like T_2 and T_4.

Classification of Virus

Viruses are generally classified on the basis of their host. They can either be plant virus (TMV), animal virus (chickenpox), bacterial virus or bacteriophage (λ phage), phycophage, (viruses parasitic on algae) mycophage (viruses parasitic on fungi), etc. Viroids, prions are the sub-viral agents out of which viroids are infecting *ss*RNA molecules, while the prions are infecting proteins.

Structure of Virus

A complete virus particle or a virion consists of following structural components:

(i) **Core** The central core is nucleic acid which may be either DNA or RNA but can never exists together.

(ii) **Capsid** An outer covering or a protective coat of the genetic material made up of a specific protein. It is composed of many identical sub-units called **capsomeres**.

(iii) **Envelope** It is the outermost covering present in certain viruses, hence called enveloped viruses, e.g. HIV, herpes vaccinia virus, etc, while the others are non-enveloped viruses.

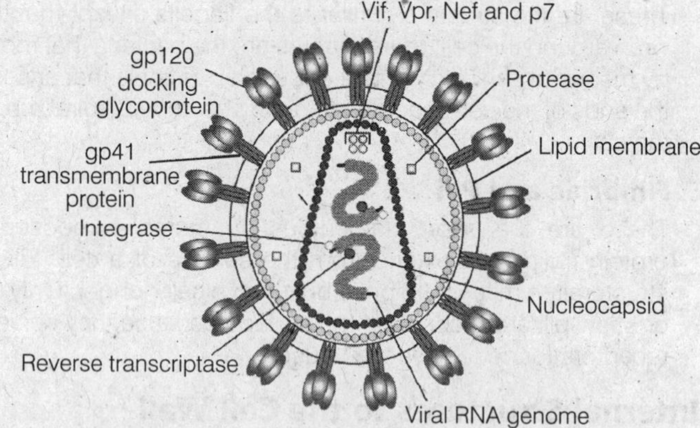

Structure of HIV

Bacterial Viruses or Bacteriophage

These are pathogenic viruses infecting bacteria are called bacteriophages or simply phages. They were discovered by Frederick Twort (1915) and Felix d' Herelle (1917). Their nucleic acid is DNA, some common bacteriophages are T_2, T_4, T_6. The T_4 bacteriophage grows on colon bacterium *Escherichia coli*. Its body contains head and tail regions. The viruses infecting *E.coli* bacteria are called **coliphages,** while the viruses infecting blue-green algae (cyanobacteria) are called **cyanophages**.

These were discovered by Schafferman and Morris (1963).

Structure of a Bacteriophage

A typical bacteriophage is composed of a head, body and a tail.

Head is icosahedron with about 2000 capsomeres and contains a 53 μm long double-stranded DNA in it, while a tail is helically symmetrical with a core tube (80 Å in diameter) surrounded by a protein tail sheath (consists of 144 sub-units arranged in 24 rings of 6 sub-units each). This sheath is connected to collar at its upper end and a base plate at its lower end.

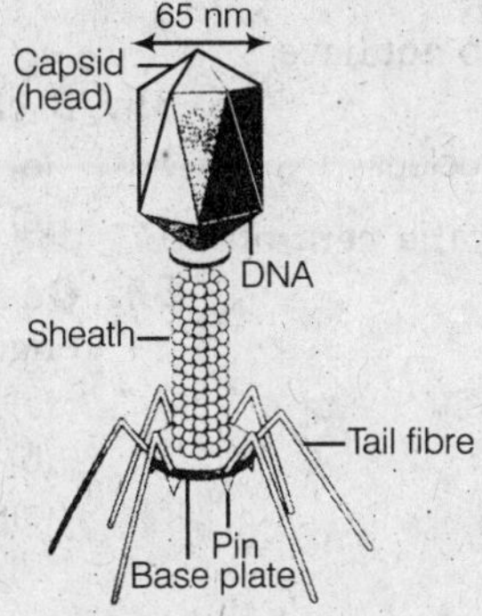

Structure of a bacteriophage

Chapter Exercise

1. Extra nuclear DNA in *E. coli* is termed as
(a) F^+ factor (b) sex factor (c) episome (d) All of these

2. Smallest cell/organism is that of
(a) vibrio (b) bacillus
(c) mycoplasma (d) *Rhizobium*

3. Bacteria which retain purple colour after staining with Gram stain is
(a) Gram +ve (b) Gram – ve
(c) trichous (d) spirillum

4. Which of the following is a rod-shaped bacterium?
(a) *Bacillus subtilis* (b) *Pneumococcus pneumoniae*
(c) *Vibrio cholerae* (d) *Streptococcus nigricans*

5. Bacteria having a tuft of flagella at both ends are called
(a) peritrichous (b) bitrichous
(c) amphitrichous (d) atrichous

6. Bacteria which are smallest in size are
(a) cocci (b) vibrio
(c) bacilli (d) spirilla

7. Bacteria, whose cells have only a single curve are known
(a) vibrio (b) cocci
(c) spirilla (d) bacilli

8. Non-photosynthetic, unicellular, cell wall bound spherical or rod-shaped microorganisms usually smaller than 5 micron in size which lack an organised nucleus are called
(a) viruses (b) bacteria
(c) cyanobacteria (d) mycoplasma

9. Organisms which have cell but no distinct nucleus are
(a) bacteria (b) cyanobacteria
(c) Both (a) and (b) (d) None of these

10. What does the bacterial cells possess in eukaryotic cells?
(a) Photosynthetic pigments
(b) A nucleoid with a circular chromosome
(c) Membrane-enclosed organelles
(d) A 9 + 2 arrangement of microtubules

11. Plasmids
(a) replicate independently of the main chromosome
(b) often contain antibiotic resistance genes
(c) are transferred from one bacterium to another by conjugation
(d) All of the above

12. Bacteria were first discovered by (bacteria were for the first time observed by)
(a) A V Leeuwenhoek (b) Robert Hooke
(c) Robert Koch (d) Louis Pasteur

13. The largest non-sulphur bacteria are
(a) vibrio (b) *Bacillus subtilis*
(c) spirillum (d) *E. coli*

14. *Staphylococcus* bacterium is a
(a) chain-like colony (b) plate-like colony
(c) bunch-like colony (d) cube-like colony

15. Which of the following is the size of bacteria?
(a) 2-10 μ (b) 10-15 μ (c) 50-100 μ (d) 100-200 μ

16. A spherical bacterium is termed as
(a) vibrio (b) coccus (c) bacillus (d) spirillum

17. What is the general shape of the *Bacilli* bacteria?
(a) Rod-shaped (b) Spherical (c) Spiral (d) Comma

18. Bacteria having a tuft of flagella at one end are called
(a) peritrichous (b) monotrichous
(c) lophotrichous (d) amphitrichous

19. The smallest bacteria measure approximately
(a) 0.5 μ (b) 0.05 μ (c) 0.005 μ (d) 0.0005 μ

20. Largest among sulphur bacteria are
(a) *Chlorobium* (b) *Beggiatoa*
(c) *Chromatium* (d) *Thiobacillus*

21. Bacteria differ from other plants in that they do not have
(a) DNA (b) RNA
(c) cell wall (d) a well-defined nucleus

22. Which one of the following does not follow the central dogma of molecular biology?
(a) HIV (b) Pea
(c) *Mucor* (d) *Chlamydomonas*

23. Infectious proteins are
(a) satellite viruses (b) gemini viruses
(c) prions (d) viroids

24. What is not true about retrovirus?
(a) DNA has no role in life cycle
(b) There is a gene for RNA dependent DNA polymerase
(c) Genetic material is RNA
(d) They are causative agents for certain cancers

25. A chemical compound that is found in all viruses is
(a) Only RNA (b) lipid
(c) RNA or DNA (d) protein

26. Bacteriophage kill
(a) fungi (b) parasite (c) bacteria (d) viruses

27. Provirus is
(a) a free virus (b) a primitive virus
(c) an integrated viral genome (d) a free DNA

28. The process which cannot take place in the absence of virus is
(a) transformation (b) conjugation
(c) translocation (d) transduction

29. Which of the following is not a viral disease?
(a) AIDS (b) Smallpox (c) Tuberculosis (d) Influenza

30. Which was the virus to be crystallised for the first time?
(a) Cauliflower mosaic virus (b) Tobacco mosaic virus
(c) Tomato mosaic virus (d) Influenza virus

31. Average size range of a virus is
(a) 10-30 nm (b) 20-14,000 nm
(c) 100-1000 nm (d) 1-10 nm

32. The main difference between Gram positive and Gram negative bacteria is
(a) cell membrane (b) cell wall
(c) ribosome (d) mitochondria

33. Helically coiled-shaped bacteria are called
(a) spirilla (b) cocci
(c) bacilli (d) vibrio

34. Extrachromosomal material of bacteria is found as
(a) circular DNA (b) extrachromosomal DNA
(c) plasmid (d) All of these

35. Bacteria pili are involved in
(a) asexual reproduction (b) sexual reproduction
(c) saprophytic nutrition (d) antibiotic resistance

36. One of the following bacteria is a good vector in genetic engineering.
(a) *Mycobacterium tuberculosis*
(b) *Agrobacterium tumefaciens*
(c) *Salmonella typhimurium*
(d) *Bacillus amyloliquefaciens*

37. From where most of the antibiotics are extracted?
(a) Algae (b) Bacteria (c) Virus (d) Plants

38. Mesosome in a bacterial cell is
(a) plasmid
(b) connection between two cells
(c) plasma membrane infolding for respiration
(d) None of the above

39. The wall of bacteria consists of
(a) N-acetyl glucosamine (b) N-acetyl muramic acid
(c) Both (a) and (b) (d) cellulose

40. Bacteria found in hot acidic aerobic conditions, are
(a) halophiles (b) thermoacidophiles
(c) methanogens (d) rickettsiae

41. Gram-negative bacteria that is not aerobic is
(a) *Escherichia* (b) *Pseudomonas*
(c) *Neisseria* (d) None of these

42. Name the enzyme which hydrolyses murein
(a) peroxidase (b) lysozyme
(c) tannase (d) None of these

43. Coliform bacteria on fermentation of carbohydrates yields
(a) acetic acid (b) lactic acid
(c) formic acid (d) All of these

44. Protein from which hook and filaments of flagella are composed is
(a) keratin (b) flagellin
(c) gelatin (d) casein

45. Germ theory of diseases was putforth by
(a) Edward Jenner (b) Weismann
(c) Louis Pasteur (d) Robert Koch

Answers

1. (d)	**2.** (c)	**3.** (a)	**4.** (a)	**5.** (c)	**6.** (a)	**7.** (a)	**8.** (b)	**9.** (c)	**10.** (b)
11. (d)	**12.** (a)	**13.** (c)	**14.** (c)	**15.** (a)	**16.** (b)	**17.** (a)	**18.** (c)	**19.** (a)	**20.** (b)
21. (d)	**22.** (a)	**23.** (c)	**24.** (d)	**25.** (c)	**26.** (c)	**27.** (c)	**28.** (d)	**29.** (c)	**30.** (b)
31. (b)	**32.** (b)	**33.** (a)	**34.** (d)	**35.** (b)	**36.** (b)	**37.** (b)	**38.** (c)	**39.** (c)	**40.** (b)
41. (a)	**42.** (c)	**43.** (d)	**44.** (b)	**45.** (c)					

4

Fungi and Lichens

Fungi

Fungi are a group of eukaryotic, non-photosynthetic, heterotrophic organisms of diverse forms, size, physiology and mode of reproduction.

They are cosmopolitan in distribution, i.e. are found to be widely distributed throughout the world.

Terrestrial fungi are considered to be more advanced, when compared to the aquatic fungi, which are primitive in origin.

General Characteristics of Fungi

Some important characteristics of fungi are:

(i) Nutrition in Fungi

Fungi are achlorophyllous, i.e. lacks chlorophyll due to which they fails to prepare their own food like plants. Hence, they are heterotrophic in nature and depend on other organism for their carbon source like animals.

Heterotrophs can be further divided into the following categories:

(a) **Parasites** Organisms that derives their nutrition from the protoplasm of another organism called host. Parasitic fungi take their food from host plants and die with the death of the host. They can not be grown artificially on a synthetic medium. Such Fungi is called **obligate parasite.**

(b) **Saprobes** Organisms that derives their nutrition or carbon source from the byproducts of organisms or dead organisms. However, under certain opportunities, saprobes may become parasitic in nature. Such organisms are called **facultative parasites.**

(c) **Symbiosis** This habitat in the strict sense is considered to the habitual 'living together' of different species. This intimate association of two dissimilar organisms in a mutually beneficial relationship is called symbiosis. e.g, 'lichens' and mycorrhizae.

(ii) Eukaryotic Fungi

Fungi are eukaryotic in nature because of having a membrane bound organelles, i.e nucleus, mitochondrion, ER, etc.

(iii) Structure of Fungi

Structurally, most fungi grows in the form of a cylindrical, thread-like structures called **hyphae** ranging 2-10 μm m diameter up to several centimeters in length.

New hyphae are usually grow at their tips by the emergence of new tips along existing hyphae by a process called branching or occasionally growing hyphal tips-tork, growing wise to two parallel hyphae. Numerous hyphae twined together to form mycelium.

The hyphae are classified into two main types:

(a) **Aseptate hyphae** In this type, cross walls or septa are not formed at the time of nuclear division.

(b) **Septate hyphae** In this type, cross walls or septa are formed after nuclear division. And cells may have one, two or many nuclei.

Fungal tissue is formed by a interweaving of fungal hyphae called as **plectenchyma** which is further divided into prosenchyma and pseudoparenchyma.

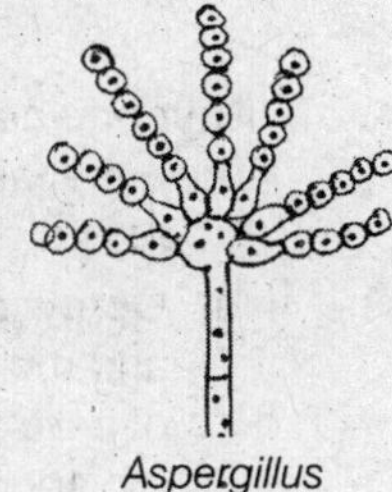

Aspergillus

Food Reserve

Food reserve is in the form of glycogen and oil glycogen is stored in plants in a different storage molecule called starch.

Reproduction in Fungi

Reproduction in fungi can be of three main types:

(i) **Vegetative** It occurs by fragmentation, budding, fission and rhizomorphs. Sclerotia are perennating bodies made up of compact masses of hyphae growing under favourable conditions to produce new mycelia. Rhizomorphs are root-like twisted masses of hyphae with a well defined apical growing point. They passes the unfavourable periods in dormant stage under favourable conditions each rhizomorph gives rise to a new mycelium. The main function of rhizomorph is absorption.

(ii) **Asexual reproduction** It occurs through the formation of spores, which are single-celled **propagules** which separates from the parent organisms. Each spore may develop into a new individual. The spores may be produced asexually or sexually and thus, named as asexual spores and sexual spores. Under asexual reproduction, only asexual spores wil be considered.

Asexual spores are innumerable and produced on the diplant mycelium in Phycomycetes and Ascomycetes.

(iii) **Sexual reproduction** A large number of fungi reproduces sexually. Usually two phases are found in the life cycle of the plants. These phases are called haploid and diploid phases respectively. The haploid phase posseses the number of chromosomes in the nucleus whereas their number because ($2n$) in the diploid phase.

The gametes are always haploid (n) and by a sexual fusion they result in diploid ($2n$) sexual spores such as zygospores, oospores, etc. To bring haploid (n) phase once again in the life cycle.

Classification of Fungi

It is based on the nature of somatic phase, kinds of asexual spores and sporangia and nature of life cycle. Based on these characters fungi are classified as under:

Fungi are often classified on the basis of morphology or form of structure and appearance of fungus, types of spores and their dispersal, life cycle and physiology of different species.

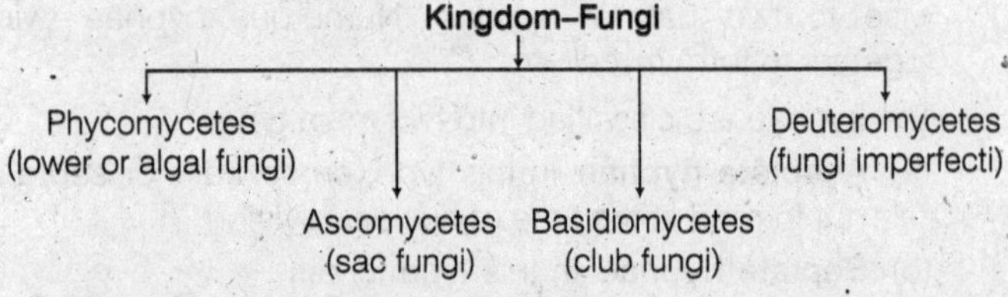

(i) **Phycomycetes** It comes under a lower group of fungi. They are aseptate and bears coenocytic hyphae. This group is divided into the following two main sub-groups:

(a) **Oomycetes** An oogamous fungi which is aseptate and multinucleate in nature. Hyphae wall is composed of cellulose and glucans. Asexual reproduction occurs by sporangia containing spores. While, the sexual reproduction is by gametangial contact.

(b) **Zygomycetes** These are fast growing fungi characterised by the primitive coenocytic hyphae. Asexual spores includes conidia, chlamydoconidia, etc. While, the sexual reproduction in isogamous which produces a thick-walled sexual resting spore called zygospore.

(ii) **Ascomycetes** These are the fungi which produces microscopic spores inside special elongated cells or sacs known as **'asci'.**

As the spore matures 'inside the ascus, increasing fluid pressure builds up inside it until the top bursts off, this eventually releases the spores rapidly desiccation. Ascomycetes can be identified from the fruiting bodies that bears asci and the way in which the asci develops.

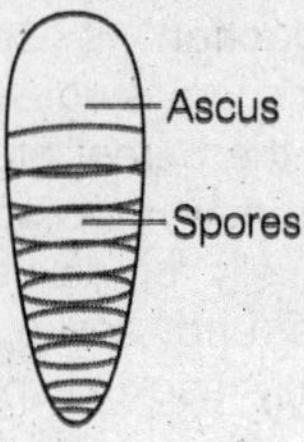

(iii) **Basidiomycetes** This class includes the members that produces their basidia and basidiospores on or in a basidiocarp. The morphology of basidium is variable until recently the morphology of the basidium was believed to be a key to determining relationship in the Basidiomycetes life cycle. Many of these fungi reproduces as sexually by spores (conidia or uidia) or by budding.

(iv) **Deuteromycetes** It is the term only closed to describe some species of fungi and has very little importance in the classification of fungi. This is the class of artificially grouped fungi having no sexual reproduction. They are called imperfecti fungi becaue of their imperfecti life cycle.

These are terrestrial, septate mycelium. No sexual reproduction takes place. Asexual reproduction by conidia.

Economic Importance of Fungi

Fungi are known to be one of the most important group of organisms on our planet due to the following reasons:

(i) Fungi, together with bacteria are responsible for the recycling of most of the dead material to the soil in a form in which it can be reused.

(ii) These are vitally important for the good growth of most plants including crops, through the development of mycorrhizal associations.

(iii) Fungi are also important directly as food for humans, e.g. mushrooms and are also used widely in the production of many food and drinks.

(iv) Fungus, is infact important for the discovery of most famous of all antibiotic drugs, e.g. penicillin, was derived from a common fungus *Penicillium*.

(v) Fungi can be extremely useful for controlling insects pests of crops.

Lichens

The term 'lichen' was coined by **Theophrastus.** The study of lichens is called **Lichenology**. It is known for symbiotic relationship between algae (phycobiont) and fungi (mycobiont) in which the fungus (mycobiont) provides space and moisture, while the algae produces food.

Internally, 95% body of the lichens constitutes fungus while, only 5% is composed by an algal component. Hence, they are also called dual organisms.

About 15,000 species of lichens are known till today. They are generally found in the places like walls, window panes, barren rocks, cooled volcanic lava, tree bark, soil and aquatic habitats.

Although lichens commonly lives under hot and humid atmosphere and exposed condition, but they bears the ability to tolerate bursts extreme desiccation. They are perennial in nature (i.e. exists for long or apparently infinite time) but have slower growth.

A lichen absorbs most of its nutrients from air and rainfall. Therefore, are considered as good **pollution indicators.** They may vary in colour from orange, dark brown, grey to yellowish, etc.

Only the algal zone contains photosynthetic partner. In about 98% lichens, the fungal partner belongs to Ascomycetes and in the rest, the fungal partner belongs to Basidiomycetes and Deuteromycetes.

Classification

On the basis of morphology lichens are of following main types:

The body of a lichen is termed as 'thallus'. Its general shape enables us to group lichens into four broad categories:

(i) **Foliose** In this, lichen's have a flat, leaf like structure.

(ii) **Fructicose** In this, lichens have erect or pendulous, bushy structure.

(iii) **Squamulose** In this, lichens have a thallus consisting of minute, scale like squamules.

(iv) **Crustose** In this, lichens produces cruston or beneath rock or free surfaces.

(v) **Filamentous** It is crust like chain of algal cell are found being wrapped around by fungal hyphae.

Structure of Lichens

Lichens vary in their degree of structural organisation. The more complex lichens have a well-defined structure often with a distinct zonation of the partners. Usually the photosynthetic cells are found in a defined band where they are closely associated with fungal hyphae for nutrient exchange.

Above this band is a **cortex** consisting of lightly packed fungal cells with a gelatinous matrix between them within and below the photosynthetic zone is a medulla of loosely packed hyphae. Often there is a thinner, **lower cortex** and several types of lichen attach themselves to a surface by a root like structure termed rhizinae.

Importance of Symbiotic Relationship

The algal partner (phycobiont) is involved in following roles:

- Nitrogen-fixation if cyanobacterial type
- Photosynthesis
- Providing vitamins and other growth substances

The fungal partner (mycobiont) is involved in following roles:

- Providing outer covering for protection
- Attachment in substratum
- Protection against harmful radiations

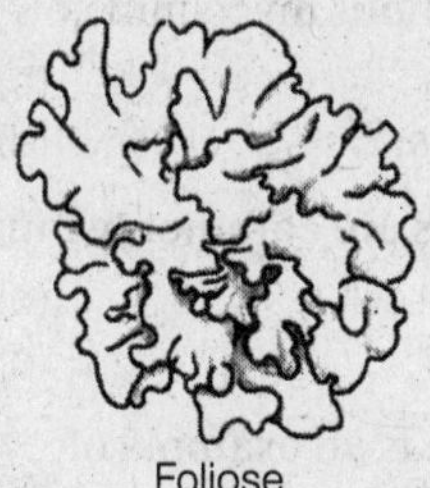

Foliose

Reproduction

Lichens may reproduce vegetatively or sexually. They reproduce vegetatively by the different ways such as fragmentation, death and decay, isidia, soredia, etc., and reproduces sexually by forming structures in such ways:

- Formation of ascomycetous fruiting body
- Formation of a basidiocarp

Chapter Exercise

1. Roots of *Pinus* seedling are associated with

(a) algae (b) fungus (c) microbes (d) bacteria

2. Fungal cell wall is composed of

(a) chitin (b) chitin and cellulose
(c) hemicellulose (d) only cellulose

3. Oomycetes is the alternate name of

(a) Zygomycetes (b) Phycomycetes
(c) Ascomycetes (d) Club fungi

4. In *Rhizopus*, we find white cottony mycelium, which shows black spots during asexual reproduction. These spots are

(a) zoospores (b) spores
(c) oospores (d) sporangia

5. One of the common fungal diseases of man is

(a) cholera (b) plague
(c) typhoid (d) ringworm

6. Sac fungi are

(a) Ascomycetes
(b) Basidiomycetes
(c) Phycomycetes
(d) Zygomycetes

7. White rust of crucifers is caused by

(a) *Albugo* (b) *Pythium*
(c) *Cercospora* (d) *Phyllactinia*

8. Yeast is abundantly found in

(a) moist bread
(b) horse dung
(c) organic substances rich in fats
(d) organic substances rich in sugar

9. The edible fungi are

(a) rusts (b) moulds
(c) mildews (d) mushrooms

10. Phycomycetes are characterised by
(a) flagellate zoospores
(b) septate mycelium
(c) oogamous reproduction
(d) imperfect sexual organ

11. The common method of reproduction in yeast is
(a) budding (b) fragmentation
(c) sexual reproduction (d) apomixis

12. Sexual reproduction is absent in
(a) Phycomycetes (b) Ascomycetes
(c) Basidiomycetes (d) Deuteromycetes

13. Fungus without mycelium is
(a) *Puccinia* (b) *Phytophthora*
(c) *Rhizopus* (d) *Saccharomyces*

14. Lomasomes are commonly found in
(a) bacterial cell (b) fungal cell
(c) algal cell (d) cyanobacterial cell

15. Mycorrhiza is an example of
(a) symbiotic relationship (b) ectoparasitism
(c) endoparasitism (d) decomposers

16. Asexual reproduction in fungi occurs by
(a) conidia (b) oospores
(c) basidiospores (d) ascospores

17. Lichens are pioneers of vegetation in a
(a) xerosere (b) hydroponics
(c) marine life (d) hydrosere

18. The common method of reproduction in yeast is
(a) fragmentation (b) fission
(c) fusion (d) budding

19. Lichen known as reindeer moss is
(a) *Lecanora* (b) *Cetraria* (c) *Cladonia* (d) *Aspicila*

20. Component secrete by lichens in high amount
(a) enzymes (b) hormones
(c) alkaloids (d) acids

21. Fungi living on dead and decaying matter are often incapable of infecting living organisms. These are known as
(a) facultative saprobes (b) obligate saprobes
(c) obligate parasities (d) None of these

22. Sexual reproduction seems to be absent in
(a) Deuteromycetes (b) *Rhizopus*
(c) Ascomycetes (d) Zygomycetes

23. Basidiospores are produced by
(a) mycelium (b) conidia
(c) basidium (d) hyphae

24. The lichens are
(a) fast growing, long lived (b) fast growing, short lived
(c) slow growing, long lived (d) slow growing, short lived

25. The loser component in a lichen is
(a) phycobiont (b) mycobiont
(c) Both (a) and (b) (d) None of these

26. The fungal partner in lichen is called mycobiont whereas, algal partner is called
(a) phycobiont
(b) mycobiont
(c) often referred as algal partner
(d) algobiont

27. Graphis is a
(a) foliose lichen (b) crustose lichen
(c) filamentous lichen (d) fruticose lichen

28. Lichens are the major pollution indicators of
(a) SO_2 (b) NO_2 (c) mercury (d) Co

29. Benefit of algae in this association is
(a) food (b) vitamins
(c) protection (d) growth substances

30. Major group of algae involved in lichen formation is
(a) red algae (b) brown algae
(c) blue-green algae (d) All of these

Answers

1. (b)	**2.** (a)	**3.** (b)	**4.** (d)	**5.** (d)	**6.** (a)	**7.** (a)	**8.** (d)	**9.** (d)	**10.** (a)
11. (a)	**12.** (d)	**13.** (d)	**14.** (b)	**15.** (a)	**16.** (a)	**17.** (a)	**18.** (d)	**19.** (c)	**20.** (d)
21. (b)	**22.** (a)	**23.** (c)	**24.** (c)	**25.** (a)	**26.** (a)	**27.** (b)	**28.** (a)	**29.** (c)	**30.** (c)

5

Cryptogams

Algae

The term 'algae' was first used by Linnaeus (1755). Algae are a group of chlorophyll bearing, photosynthetic, autotrophic, thalloid plants whose body is not differentiated into true roots, stems and leaves.

The study of algae is called **Phycology** (*Phycos*–algae; *logos*–study). MOP Ayengar is known as father of Indian Phycology. Some other Indian phycologists are Y Bhardwaj, RN Singh, TV Desikachery and GS Venkataraman.

General Characteristics of Algae

The important characteristics of algae are:

- Algae exhibits a diverse variety in thallus organisation including very small, simplest unicellular forms to the very complex forms. The complex thalli lack vascular tissue and epidermis with stomata.
- The colour of algal thallus is due to the presence of different pigments.
- Under favourable conditions, the gametophyte multiplies repeatedly by means of asexual spores called the **mitospores.**
- The sex organs are generally single celled. When multicellular, each cell is fertile and there is no jacket of sterile cells.
- There is no embryo formation after gametic union.
- Both the generations when represented in the life cycle are independent of each other.
- Except a few, all algae are aquatic in nature. The aquatic forms are either completely submerged or free floating on the surface of water. Some blue-green algae may form water blooms on the surface of water.
- Life cycle in algae is haplontic, diplontic, diplohaplontic and haplohaplontic, etc.
- An alteration of generation occurs in diplohaplontic life cycle.

For example, some economically important algae are *Gelidium, Chlorella, Anabaena, Nostoc,* etc.

Classification of Algae

Algae are divided into various classes based on the stored food, pigmentation, thallus formation, flagellation, cell wall, etc.

Salient Features of Major Classes of Algae

Characters	Chlorophyceae (green algae)	Bacillariophyceae (golden algae)	Phaeophyceae (brown algae)	Rhodophyceae (red algae)	Cyanophyceae (blue-green algae)
Habitat	Freshwater, marine or terrestrial.	Freshwater, marine or terrestrial.	Only four genera are freshwater, while rest are marine.	Mostly marine, a few are terrestrial.	Freshwater, marine or terrestrial.
Thallus form	Unicellular, colonial, coenobial, filamentous, thalloid or siphonaceous.	Unicellular or colonial.	Filamentous, branched usually heterotrichous.	Unicellular, filamentous, pseudoparenchymatous.	Unicellular, colonial or filamentous.
Cell structure	Eukaryotic, cell organised nucleus, other cell organelles also present.	Eukaryotic	Eukaryotic	Eukaryotic	Prokaryotic, membrane bound nucleus and other cell organelles like plastids, mitochondria, etc are absent.
Cell wall	Cell wall is made up of cellulose (inner layer) and pectin (outer layer) material.	Cell wall is made up of silica and pectin material.	Outer mucilaginous layer of cell wall consists of alginic and fucinic acid and the inner layer is made up of cellulose.	Outer layer is pectinaceous in nature whereas, an inner layer is cellulosic.	Mucopeptides and muramic acid are two main constituents of the cell wall.

Characters	Chlorophyceae (green algae)	Bacillariophyceae (golden algae)	Phaeophyceae (brown algae)	Rhodophyceae (red algae)	Cyanophyceae (blue-green algae)
Flagella	Motile cells with 2, 4 or more flagella. Flagella are generally anteriorly inserted equal in length and whiplash type.	Usually 1 or sometimes 2, anteriorly inserted.	Mostly two in number, laterally inserted and of unequal length, one of them is whiplash type and another is of tinsel type.	Non-flagellated, motile cells are absent.	Non-flagellated, motile cells are absent.
Dominant pigments	Chlorophyll-*a* and *b*.	Chlorophyll-*a* and *c*, β-carotene, xanthophyll (fucoxanthin).	Chlorophyll-*a* and *c*, β-carotene, xanthophyll (violaxanthin, fucoxanthin).	Chlorophyll-*a* and *d*, phycoerythrin.	Chlorophyll-*a*, β-carotene, xanthophyll and *c*-phycocyanin.
Nucleus	Well organised nucleus with chromosomes. Nucleolus is also present.	As in Chlorophyceae.	As in Chlorophyceae.	As in Chlorophyceae.	Nuclear membrane absent, DNA not associated with histones, hence, well organised chromosomes are not found.
Reserve food material	Starch and fats.	Oil, volutin and chrysose.	Complex carbohydrates such as laminarin, mannitol and sucrose.	Floridean starch.	Special type of starch (cyanophycean starch) and protein (cyanophycin granules).
Reproduction	Reproduction is by vegetative, asexual and sexual means, (isogamous, anisogamous or oogamous type) .	Reproduction is by vegetative (cell division) and sexual means, (isogamous, anisogamous or oogamous type).	Reproduction is by vegetative (fragmentation, adventitious branches), asexual (motile and aplanospores), and sexual means (isogamous, anisogamous or oogamous type).	Reproduction is by vegetative (cell division, fragmentation), asexual (monospores, neutral spores, tetraspores, carpospores) and sexual means, (complex and advanced oogamous type).	Multiplication is only through asexual methods, sexual reproduction is completely absent.
Examples	*Chlamydomonas, Volvox, Chlorella, Spirogyra, Chara, Ulothrix*, etc.	*Corethron, Pinnularia, Fragilaria, Eunotia* etc.	*Fucus, Laminaria Sargassum, Dictyota, Ectocarpus*, etc.	*Porphyra, Gelidium, Gracilaria, Ceramium, Gelidiella, Chondrus*, etc.	*Anabaena, Microcystis, Dictyosphaerium*, etc.

Bryophytes

These are known as 'amphibians of plant kingdom' because they have become completely adapted to the land habit. However, they still rely on water for sexual reproduction because the swimming habit is retained by their sperms.

Distribution

The bryophytes is a small group of primitive land dwellers having a leafy or thalloid, green plant body which is small in structure, (rarely more than five inches). They contain about 2500 species and are widely distributed throughout the world, especially in moist mountain in forests or tropics, subtropics and Antartic regions.

Morphology

The plant body lacks true roots, stem or leaves. It grows prostate on the ground attaching to the substratum by delicate, unbranched, unicellular hair-like organs called the **rhizoids.**

In the higher bryophytes (mosses), the plant body is erect which consists of a central axis bearing leaf like expansions.

General Characteristics of Bryophytes

The bryophytes lack the important characteristics of bryophytes is vascular tissue system, a characteristic of the higher plants, i.e. xylem and phloem.

- The most conspicuous phase in the life cycle is **gametophyte**. It is independent and concerned with sexual reproduction.
- Sexual reproduction is invariably highly oogamous and the sex organs are jacketed and multicellular.
- Female sex organ is in the form of an **archegonium,** which appears for the first time in the plant world and the sperms are biflagellate (both the flagella are of whiplash type). Male reproductive organs are club-shaped and called antheridium.
- Fertilisation takes place in water.
- The fertilised egg is retained within venter of the archegonium.
- The zygote undergoes repeated division to form an undifferentiated, multicellular structure called an **embryo.** The first division of the zygote is transverse and apex of the embryo develops from the outer cell.
- The embryo by further division and differentiation produces a relatively small spore producing structure (which is not independent), called **sporogonium** (sporophyte).

In all bryophytes, the sporophyte is without differentiation into the stem and leaves. It is rootless and consists of a foot, a seta and a capsule. In some cases, seta is absent (*Corsinia*) or rarely present as both foot and seta (*Riccia*).

- The sporophyte is simpler than the gametophyte and is originally attached to the parent gametophyte throughout its life and is dependent upon it partially or completely for nutrition.
- The sporogonium is concerned with the production of wind-disseminated, non-motile, cutinised spores, which belong to the category of meiospores which are morphologically are of one kind in a given species. Thus, the bryophytes in general are described as **homosporous.**
- Each spore on falling on a suitable soil germinates to give rise to the gametophyte plant either directly or indirectly as a lateral bud forming **protonema.**
- The gametophytic (haploid) phase is dominant, while sporophyte (diploid) phase is short lived. Thus, in general bryophytes show haplodiplontic pattern of alternation of generation.

Classification of Bryophytes

Schimper (1879) gave the rank of division to the group Bryophyta, while Eichler (1883) divided Bryophyta into two main classes, i.e. Hepaticae and Musci.

Smith, Campbell and Schuster further divided Bryophyta into three main classes, i.e. Hepaticae, Anthocerotae and Musci.

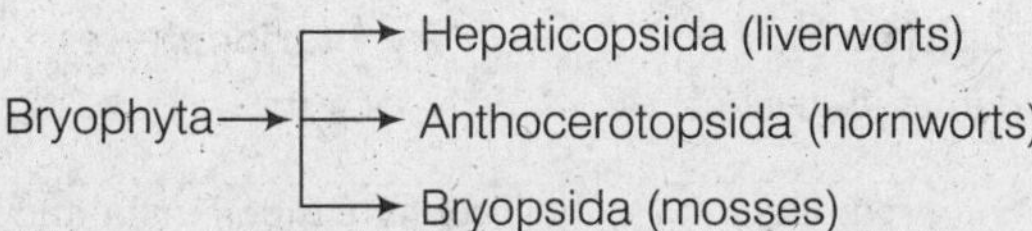

Cryptogams makes up around 84% of the world's botanical diversity they include the algae, bryophytes (mosses, liverworts and hornwort), the pteridophytes (fauns and horsetails). Cryptogams are plants that reproduces by spares without flower or seeds. Thus, cryptogams represents the non-seed bearing plants. The term cryptogam was coined by century botanists because the means of sexual reproduction in these plants was not apparent.

Hepaticopsida (Liverworts)

(i) Plant body is gametophytic, may be thallose or foliose.
(ii) Rhizoids are aseptate.
(iii) Sex organs are present on the dorsal surface of thallus.
(iv) Sporophyte or sporogonium is differentiated into foot, seta and capsule.
(v) Elaters are generally present. But are absent in some species like *Riccia*.
(vi) Sporogenous tissue develops from endothelium.
(vii) Dehisence of capsule occurs irregularly, e.g. *Riccia*, *Marchantia* and *Sphaerocarpus*,

Anthocerotopsida (Hornworts)

(i) Plant body is gametophytic, i.e. thallose.
(ii) Rhizoid are aseptate.
(iii) Sex organs are found on the dorsal surface of thallus.
(iv) Sporophyte or sporogonium is differentiated into foot, short meristematic region and capsule.
(v) Elaters in the form of pseudoelaters are found in the capsule.
(vi) Sporogenous tissue develops from amphithecium and endothecium from sterile columella, e.g. *Anthoceros*, *Megaceros* and *Notothylus*.

Bryopsida (Mosses)

(i) Gametophytic plant body is thalloid protonema and leafy gametophore.
(ii) Have obliquely septate rhizoids.
(iii) Sex organs develops from the superficial cells at the apex of leafy gametophore.
(iv) Sporophyte or sporogonium is differentiate into foot, seta and capsule.
(v) Elaters are absent.
(vi) Sporogenous tissue develops from the outer layer of endothecium, inner layer forms sterile columella.
(vii) Dehiscense of capsule is regular, e.g. *Funaria*, Sphagnum, polytrichum.

Rothmaler named these classes as Hepaticopsida, Anthoceropsida and Bryopsida and Proskauer (1957) again described Anthoceropsida-Anthocerotopsida.

Pteridophytes

Pteridophytes (*pteron*–feather; *phyton*–plant) are seedless vascular plants dominated during carboniferous period. They have vascularised root, stem and leaves, hence called **vascular cryptogams**.The term pteridophyte was coined by Haeckel (1866). They are generally found in cool, damp and shady places.

General Characteristics of Pteridophytes

- Vascular tissue consists of **xylem** (without true vessels) and **phloem** (without companion cells). Xylem transports water and minerals, while phloem conducts organic food.
- Sporangia containing leaves called **sporophylls** are present. Meiospores are formed inside the sporangia. Sporophyte represents the dominant phase and is differentiated into roots, true stem and leaves.
- In **microphyllous** leaves, the leaf trace does not disturb the vascular cylinder of the stem, e.g. lycopods and horsetails. In megasporophyllous leaves, the leaf trace produces a parenchymatous leaf gap or foliar gap in the vascular cylinder of the stem. Depending upon the development of sporangia, plants are eusporangiate or leptosporangiate.
- Small inconspicuous **gametophyte** is independent and autotrophic. Parts of gametophyte absorb water directly.
- Flagellate male gametes (antherozoids) formed in male sex organs (antheridia). Water is essential for their swimming towards the female sex organs (archegonia).
- Adventitious roots emerge from an underground, horizontally growing stem called **rhizome.** Some ferns have their stems above the ground, which may attain a height of about 20 meter, (e.g. tree ferns *Cyathea*).
- Leaves are large and called **fronds**. They may be simple with one blade or may be pinnately compound, (i.e. blade is sub-divided into leaflets or pinna). Pinnate leaves may be uni, bi or tri-pinnate. The growing point of leaves (apical meristem) stops growing at maturity. The young leaves show coiling like watch spring (circinate ptyxis).

- Mature leaflets bear sporangia in clusters called **sori**. Each sorus bears a number of stalked biconvex sporangia on a placenta (cushion-shaped) covered by a membranous sheath called **indusium**. In some cases, a false indusium (turned margin or sporophyll) is present.
- Multicellular hair or scales (ramenta) are present on the younger parts of stem, leaves and mature petiole. They are protective in function.
- Vegetative propagation takes place by bulbils, adventitious buds and fragmentation. The most unique method is seen in walking fern, i.e. *Adiantum caudatum,* which produces a new plant when its leaf tips comes in contact with the soil.
- Sporangium has a single-layered jacket and has 12 spore mother cells ($2n$). Some of the marginal cells of jacket are differentiated to form an annulus and remaining forms the stomium. Meiosis occurs in spore mother cells, which form haploid spores.
- Haploid spore germinates to form a prothallus (gametophyte) which is monoecious, e.g. both antheridia (♂) and archegonia (♀).
- Each male sex organ is hemispherical and has a three celled jacket. It contains 32-48 sperm mother cells, which produce multiciliated sperm. These swim in water to reach the female sex organs.
- Archegonia is flask-shaped organ, embedded in prothallus. Neck opens and mucilage comes out to which the sperms are attracted. The mucilage contains **malic acid**. The sperms fuses with the egg (oosphere) to form a diploid oospore.
- **Stele system** is important feature of pteridophytes. It is the core of axis which includes vascular system. Protostele, siphonostele, solenostele, dictyostele are some important type of steles found in pteridophytes.
- Ferns exhibit alternation of dominant sporophyte generation with an inconspicuous gametophyte generation (diplohaplontic).
- **Heterospory** is formation of two types of spores, i.e. microspores and megaspores. The ability to form seed is called **seed habit**. It is considered to be originated in pteridophytes.

Classification of Pteridophyte

Pteridophytes are further classified into following four classes:

(i) Psilopsida, e.g. *Psilotum*
(ii) Lycopsida, e.g. *Selaginella* and *Lycopodium*
(iii) Sphenopsida, e.g. *Equisetum*
(iv) Pteropsida, e.g. *Dryopteris, Pteris* and *Adiantum.*

Comparison of Different Characters of Algae, Bryophytes and Pteridophytes

Characteristics/Features	Algae	Bryophytes	Pteridophytes
Habitat	Freshwater, marine and terrestrial.	Mostly terrestrial (humid regions).	Mostly terrestrial.
Plant body	Unicellular, colonial, filamentous, thalloid without root, stems and leaves.	Thalloid body having root like rhizoids. No true stem and leaves.	Sporophyte is dominant and differentiated into root, stem and leaves.
Dominant phase and type of life cycle	Haploid and haplontic.	Haploid phase is dominant, but diploid phase also occurs and is short lived haplodiplontic.	Diploid body is dominant and haploid body is short lived diplohaplontic.
Vegetative reproduction	Fragmentation, cell division, etc.	Fragmentation, tuber formation, etc.	Fragmentation, bulbils adventitious buds, etc.
Asexual reproduction	Biflagellated or non-flagellated spores.	Asexual reproduction is not general in bryophytes.	Same as in bryophytes.
Sexual reproduction	Isogamous, anisogamous and oogamous.	Mainly oogamous types.	Mainly oogamous type.
Sexual organs	Main body of algae participate in sexual reproduction. No covering/ jacket on sex organs.	Male sex organ is antheridium and female is archegonium, jacketed organ.	Plant body is diploid, forms microsporangia and megasporangia that form haploid microspore and megaspore. Both spore germinate and form prothallus that bear jacketed sex organs.
Types of male and female gametes	Gametes can be flagellated or non-flagellated and similar or dissimilar in size.	Similar type of male and female gametes. Antheridia produces antherozoids, while archegonium produces egg cell.	Androcytes are formed in male prothallus and are similar type. Female prothallus bear egg cells.
Requirement of water for fertilisation	Yes	Yes	Yes
Post fertilisation development	Zygote formation occur but no embryo forms, there is no dependence on previous phase of life. Zygote on meiosis retain the haploid stage.	Zygote is first stage of sporophytic phase and dependent on gametophyte. Embryo formation occur that develop into sporophyte that consists of foot, seta and capsule. It produces spores that develop gametophytic stage.	By the formation of zygote, pteridophytes retains their diploid stage. Zygote develops embryo which develop into the main plant.

Chapter Exercise

1. FE Fritsch (1935) divided algae into classes
(a) eleven (b) thirteen
(c) twenty (d) five

2. The reserve food material of plants of class–Chlorophyceae is
(a) starch (b) laminarian
(c) glycogen (d) mannitol

3. The most abundant carotene in algae is
(a) β-carotene (b) lycopene
(c) α-carotene (d) None of these

4. Another name for class–Cyanophyceae is
(a) Myxophyceae (b) Blue-green algae
(c) Chlorophyceae (d) Both (a) and (b)

5. In which class of algae, flagella are totally absent?
(a) Phaeophyceae (b) Rhodophyceae
(c) Chlorophyceae (d) Bacillariophyceae

6. *Volvox* belong to the class
(a) Chlorophyceae (b) Cyanophyceae
(c) Rhodophyceae (d) Bacillariophyceae

7. The class of algae which has red pigment?
(a) Phaeophyceae (b) Rhodophyceae
(c) Chlorophyceae (d) Cyanophyceae

8. The *Gracilaria* algae belongs to class
(a) Chlorophyceae (b) Rhodophyceae
(c) Cyanophyceae (d) Bacillariophyceae

9. Brown algae belong to class
(a) Phaeophyceae (b) Chlorophyceae
(c) Rhodophyceae (d) Bacillariophyceae

10. Which of the following is not a class of bryophytes?
(a) Hepaticopsida (b) Anthocerotopsida
(c) Eplepsodia (d) Bryopsida

11. Another name for class–Bryopsida is
(a) Musci (b) Chlorophyceae
(c) Malvaceae (d) Lycophyta

12. The members of class–Hepaticopsida are popularly known as
(a) liverworts (b) mosses
(c) hornworts (d) ferns

13. The sporogonium of members of class–Bryopsida is differentiated into
(a) foot (b) seta
(c) capsule (d) All of these

14. The members of class–Anthocerotopsida are popularly, known as
(a) hornworts (b) mosses
(c) liverworts (d) None of these

15. Heterosporous plants produce
(a) megaspores that develop into female gametophytes and microspores that develop into male gametophytes
(b) megaspores that develop into male gametophytes and microspores that develop into female gametophytes
(c) megaspores that bear antheridia and microspores that bear archegonia
(d) spores that produce both archegonia and antheridia

16. Pteridophytes are also known as
(a) cryptogams (b) vascular cryptogams
(c) amphibious plants (d) phanerogams

17. Pteridophytes differ from bryophytes in the
(a) motility of sperms (b) vasculature
(c) archegonia (d) alternation of generations

18. When a plant produces two kind of spores the condition is known as
(a) homospory (b) heterospory
(c) apospory (d) sporogenesis

19. *Equisetum* is commonly called as
(a) club moss (b) spike moss
(c) horsetail (d) bog moss

20. Sporophyte of fern produces
(a) pollen grains (b) spores
(c) seeds (d) gametes

21. Sporangia in ferns is covered by
(a) strobilus (b) leaf margin
(c) ligule (d) indusium

22. The tallest pteridophyte is
(a) *Alsophila* (b) *Azolla*
(c) *Adiantum* (d) *Cyathea*

23. In ferns, we find
(a) sori on dorsal surface (b) circinate venation
(c) thalloid body (d) buds on leaves

24. Pteridophytes, in which sporangium develops from a single initial cell are known as
(a) leptosporangiate
(b) eusporangiate
(c) mesosporangiate
(d) metasporangiate

25. Laminarin and mannitol are the reserve food of

(a) Chlorophyceae (b) Phaeophyceae
(c) Rhodophyceae (d) Cyanophyceae

26. In ferns, young leaves are protected by the presence of

(a) indusium (b) ramenta
(c) sori (d) rhizome

27. Carrageenin, a jelly-like substance is derived from marine algae known as

(a) kelp (b) flagellates
(c) irish moss (d) diatoms

28. Which of the following is found in blue-green algae?

(a) Chrysene (b) Cyanophycean starch
(c) Starch (d) Cellulose

29. Agar-agar is obtained from

(a) *Ulva* (b) *Ulothrix*
(c) *Sargassum* (d) *Gelidium*

30. The largest algae is

(a) *Ulva* (b) *Chara*
(c) *Microcystis* (d) *Macrocystis*

31. All algae have

(a) chlorophyll-*a* and *b*
(b) chlorophyll-*b* and carotenoids
(c) chlorophyll-*a* and carotenoids
(d) phycobilins and carotenoids

32. Fucoxanthin pigment is found in

(a) blue-green algae (b) green algae
(c) red algae (d) brown algae

33. Water bloom is generally caused by

(a) *Hydrilla* (b) bacteria
(c) blue-green algae (d) green algae

34. Bryophytes are called as

(a) vascular cryptogams
(b) flowering plants
(c) amphibians of the plant kingdom
(d) None of the above

35. Which part of moss capsule help in the nutrition of spores?

(a) Columella (b) Theca
(c) Peristome teeths (d) None of these

36. One character common to all pteridophytes is

(a) protostele (b) heterospory
(c) absence of ovules (d) leptosporangiate sporangia

37. Which of the following plant group is found in freshwater, marine water and on terrestrial habitats?

(a) Bryophytes (b) Fungi
(c) Algae (d) Pteridophytes

38. Plant body of algae may be

(a) unicellular (b) colonial
(c) filamentous (d) All of these

39. In which of the following plant group, rhizoids-like roots are found?

(a) Algae (b) Pteridophytes
(c) Bryophytes (d) Cyanophyceae

40. In which of the following plant group true roots, stem and leaves are found?

(a) Algae (b) Bryophytes
(c) Fungi (d) Pteridophytes

41. Haploid plant body is dominant in which of the following plant group?

(a) Algae (b) Pteridophytes
(c) Bryophytes (d) Fungi

42. Bryophytes are showing which of the following type of alternation of generation?

(a) Haplontic (b) Diplohaplontic
(c) Haplodiplontic (d) Diplontic

43. Diplohaplontic life cycle or alternation of generation is the main characteristics of which of the following?

(a) Algae (b) Bryophytes
(c) Monera (d) Pteridophytes

44. Which of the following type of vegetative reproduction is common in algae, bryophytes and pteridophytes?

(a) Fragmentation
(b) Bulbils
(c) Budding
(d) Adventitious branching

45. Asexual reproduction by flagellated zoospore is common in which of the following?

(a) Bryophytes (b) Gymnosperms
(c) Algae (d) Pteridophytes

46. Which of the following groups/group show/sexual reproduction only by oogamous type?

(a) Algae (b) Bryophytes
(c) Pteridophytes (d) Both (b) and (c)

47. Antheridium and archegonium first occur in which of the following plant group?

(a) Algae (b) Pteridophytes
(c) Bryophytes (d) Gymnosperms

48. Heterospory is main character of which of the following plant group?

(a) Pteridophytes (b) Bryophytes
(c) Angiosperms (d) Algae

49. Jacketed sex organs are found in which of the following group/groups?

(a) Algae (b) Pteridophytes
(c) Bryophytes (d) Both (b) and (c)

50. In which of the following plant group produce prothallus after germination of spores?

(a) Algae (b) Pteridophytes
(c) Bryophytes (d) Both (a) and (b)

51. Requirement of water for sexual reproduction is necessary in which of the following group of plant?
(a) Algae
(b) Bryophytes
(c) Pteridophytes
(d) All of the above

52. In which of the following, zygote is not dependent on previous stage of plant life cycle?
(a) Algae (b) Pteridophytes
(c) Bryophytes (d) Gymnosperms

53. Sporophyte is dependent on gametophyte and divided into foot, seta and capsule. The plant group is
(a) algae (b) fungi
(c) bryophyte (d) None of these

54. By the formation of zygote, pteridophytes remains in their
(a) haploid stage
(b) haplodiploid stage
(c) diploid stage
(d) None of the above

55. Which of the plant group is economically most important?
(a) Algae (b) Pteridophytes
(c) Bryophytes (d) Both (b) and (c)

56. According to the available fossil record, the first vascular plants are
(a) horsetails (b) cycads
(c) lycopside (d) psilophytales

57. Reduction-division in pteridophytes occurs when
(a) prothallus is formed (b) spores are formed
(c) sex organs are formed (d) gametes are formed

58. Liverworts are closely related to
(a) algae (b) fungi
(c) lichen (d) mosses

59. Laminarian is an energy storage material characteristic of
(a) chlorophyta (b) chrysophyta
(c) phaeophyta (d) pyrophyta

60. Endothecium divides to form
(a) elaters (b) spores
(c) columella (d) pseudoelaters

Answers

1. (a)	2. (a)	3. (a)	4. (b)	5. (b)	6. (a)	7. (b)	8. (b)	9. (a)	10. (c)
11. (a)	12. (a)	13. (d)	14. (a)	15. (a)	16. (b)	17. (b)	18. (b)	19. (c)	20. (b)
21. (d)	22. (d)	23. (b)	24. (a)	25. (b)	26. (b)	27. (a)	28. (b)	29. (d)	30. (d)
31. (c)	32. (d)	33. (c)	34. (c)	35. (a)	36. (c)	37. (c)	38. (d)	39. (c)	40. (d)
41. (a)	42. (c)	43. (d)	44. (a)	45. (c)	46. (d)	47. (c)	48. (a)	49. (d)	50. (b)
51. (d)	52. (a)	53. (c)	54. (c)	55. (a)	56. (d)	57. (b)	58. (d)	59. (c)	60. (c)

6

Phanerogams

Phanerogams

Phanerogams are flowering or seed bearing plants. They are advanced plants. They possess well developed roots, stems, leaves and flowers. Well developed reproductive organs are visible. These are also called spermatophytes.

These are divided as gymnosperms and angiosperms.

Gymnosperms

The gymnosperms are any vascular plants in which the ovules are not enclosed by an ovary wall and remain exposed both before and after fertilisation (i.e. they are naked seeded plants). Hence, it is said that gymnosperm reproduces by the means of an expose seed or an ovule. There are small group of seed plants represented by only 900 living species.

Gymnosperms were dominant in the mesozoic era during which time some of the modern families originated. However since, the cretaceous period, gymnosperms have been gradually displaced by the more recently evolved angiosperms.

General Characteristics of Gymnosperms

The important characteristics of gymnosperms are:

Habitat

Gymnosperms are mostly found in the colder regions of Northern hemisphere, where they form extensive forests.

Morphology

Morphologically, a plant is divided into following main parts:

- These mainly includes medium sized trees or tall trees and shrubs.
- **Plant body** is sporophytic and differentiated into roots, stems and leaves.
- **Root system** is well developed with the tap root system.
- **Roots** of some gymnosperms have symbiotic relations with algae such as coralloid roots of *Cycas*.
- **Stem** is usually erect, profusely branched and woody. Leaf scars are present on stem.
- The leaves have thick cuticle layer and sunken stomata and are dimorphic, i.e.
  - (a) **Foliage leaves** which are green, simple, needle-shaped or pinnately compound.
  - (b) **Scaly leaves** which are minute and deciduous.
- Usually, leaves of gymnosperms lack lateral veins. Therefore, the lateral translocation of nutrients takes place with the help of transfusion tissue.
- The xylem contains only tracheids and xylem parenchyma, while the vessels remain absent generally but are exceptionally, present in the members of Gnetales.
- **Wood** is monoxylic (e.g. *Cycas*) or pycnoxylic (e.g. *Pinus*).
- The phloem of gymnosperms lacks companion cells.

Reproduction

Gymnosperms are heterosporous in nature, i.e. produce two types of spores, such as microspores and megaspores.

The **microsporangia** occurs on microsporophylls, while the megasporangia occurs on megasporophylls which inturn, aggregate to form compact cones or strobili.

The development of microsporangia is eusporangiate type. Usually, the cones are monosporangiate (unisexual) but in some gymnosperms such as, *Ephedra foliata*, *Ephedra intermedia*, bisporangiate cones are also present.

The **megasporangium** (i.e. ovule) is usually orthotropous and unitegmic which lies naked on the megasporophyll.

Megasporophylls aggregates to form female cones which persist for several years. Male gametes are usually non-motile with an exception of male gametes of *Cycas* and *Ginkgo* which are motile.

Each archegonium contain a single egg and a venter canal cell. Neck canal cells are absent in archegonium of gymnosperms.

Fertilisation

Pollination in gymnosperm is **anemophillous**, i.e. takes place by wind. The pollen grains are deposited in the pollen chamber where they germinate and fertilisation is **siphonogamous**, i.e. occurs through pollen tube.

Seeds

Following fertilisation, development of an embryo is meroblastic which develops only from a part of the zygote and endoscopic. There is a tendency of polyembryony in gymnosperms formed is haploid (n) in nature and provides nourishment for growth to seedling at the time of seed germination, e.g. *Cycas* is dicotyledon, i.e. containing two cotyledons, while *Pinus* contains more than two cotyledons.

Ginkgo biloba is also known as living fossil.

Classification of Gymnosperms

Chamberlian (1934) divided gymnosperms into two main classes, i.e. class–Cycadophyta and Coniferophyta.

Class–Cycadophyta

It is characterised by unbranched and stumpy stems, leaves are large and pinnately divided, male cones are large and compact with simple sporophylls, female cones are loose and consist of pinnate or simple sporophylls that bear large ovules, stems have wide cortex and monoxylic wood.

It further includes three main orders:

(i) **Cycadofilicales** includes living as well as extinct forms, e.g. *Lyginopteris*.

(ii) **Cycadeoidales** includes both living and extinct forms, e.g. *Cycadeoidea*.

(iii) **Cycadales** includes both living and extinct forms, e.g. *Cycas* and *Zamia*.

Class–Coniferophyta

The stems are profusely branched and the foliage gives cone-like appearance, leaves are simple, male and female strobili are compact and bear complex sporophylls, wood is pycnoxylic.

It further includes four main orders:

(i) **Cordaitales** includes extinct order, e.g. *Cordaites*.

(ii) **Ginkgoales** includes extinct and one, living representative, e.g. *Ginkgo biloba* (a single living species).

(iii) **Coniferales** includes both extinct and living genera, e.g. *Pinus*.

(iv) **Gnetales** includes both extinct and living genera, e.g. *Gnetum* and *Ephedra*.

Economic Importance of Gymnosperms

(i) Seeds of *Pinus gerardiana* (chilgoza) are used a food after roasting. Some other gymnosperm's edible are plant endosperm of *Ginkgo*, seed kernel of some *Cycas* and *Gnetum*, sago grains from stems of *Cycas*, etc.

(ii) Gymnosperms provide softwood for construction, plywood and paper industry.

(iii) Branches of gymnosperm trees are used as fuel food.

(iv) Saw dust of confiers in used in making plastics and linoleum.

(v) Resin is a semifluid secreted by special tubes of a number of conifers. It helps in sealing female cones after pollination, scale leaves around leaf bases and apical buds. Resin is distilled to obtain **turpentine** and **resin**. Turpentine is used for thinning paints, varnishes, etc., and resin is used for sealing joints, wheel green, preparation of writing paper, oil clothes, etc.

(vi) **Ephedrine** an antibiotic is obtained from *Ephedra*. It is used for curing respiratory problems and asthma. **Taxol** is an anticancer drug obtained from *Taxus*.

Angiosperms

The angiosperms are the seed bearing, flowering, vascular plants in which seeds are enclosed in fruits. These are the most dominant and highly evolved plants on this Earth. They appeared about 130 million years ago in Jurassic (dicot plant) and Cretaceous (monocot plants) period of the Mesozoic era, (long before the evolution of man).

They fluorished well in tertiary period of Cenozoic era which is called age of angiosperms. Monocots appeared in Oligocene epoch of tertiary period of Cenozoic era. Herbaceous angiosperms appeared in the last quaternary period of Pleistocene epoch of Cenozoic era.

General Characteristics of Angiosperms

The important characteristics of angiosperms are:

Habitat

There are about 12500 genera and 2.70 lakh species of angiosperms, out of which about 2.20 lac are dicots and 50000 are monocots. Thus, there are more than 50% of total plants on this earth. Angiosperms are found in every habitat, i.e. Antarctica, sea, plains, deserts, hot spring, at altitude of 6000 ft.

Morphology

- The plant body is divided in the form of herbs, shrubs, trees, climbers, etc.
- Primary root develops from radicle and forms the tap root system. While, the stem develops from plumule.
- **Leaves** are simple and compound bearing axillary buds which can grow into the stem branches.
- **Vascular tissues** are well developed, i.e. have vessels in xylem and phloem contains sieve tubes and companion cells.

Reproduction

Flowers are considered as the reproductive structures formed by the one or both types of sporophylls (microsporophylls or stamens and megasporophylls or carpels).

Stamens are considered as the male sex organs in angiosperms. Each stamen is composed of a slender structure called filament and an anther (at the tip) contains four microsporangia.

Carpel or **pistil** is called as the female sex organ of the angiospermic flower. It has three main parts, i.e. an ovary, style and stigma. A Megaspore Mother Cell (MMC) differentiates in the nucellus of the cell and undergoes meiosis, ultimately one functional megaspore gets enlarged and forms the female gametophyte known as **embryo sac**.

Pollination is carried out by several means such as wind, water or by other agencies in which pollen grains are dispersed to the stigma of the carpel.

During fertilisation one of the male gamete fuses with the egg cell forming a zygote, while the other male gamete fuses with the diploid secondary nucleus to produce the triploid Primary Endosperm Nucleus (PEN), because of the involvement of above mentioned two events, it is known as **double fertilisation.**

After fertilisation, synergids and antipodal cells get degenerated and the zygote develops into embryo and the primary endosperm nucleus develops into an **endosperm**.

The ovule gradually transforms into a seed and the ovary becomes the fruit.

Classification of Angiosperms

George Bentham and Joseph Dalton Hooker presented the system of classification of angiosperms published in *Genera Plantarum* (1862-1883) which appeared in three volumes.

This system of classification is used by most of the well known Herbaria of the world. Angiosperms has been divided into sub-divisions or classes, known as **monocots** (monocotyledons) and **dicots** (dicotyledons), on the basis of number of cotyledons they possess.

Class–Monocotyledonae

Embryo with one cotyledon, adventitious roots present, tap root short lived, cambium absent, vascular bundles scattered, secondary growth absent, parallel venation in leaves, floral parts are trimerous. It has 7 series such as Microspermae, Epigynae, Coronariae, Calycineae, Nudiflorae, Apocarpae and Glumaceae 34 natural orders 1495 genera and 18576 species, e.g. *Musa triticum, Vanda, Lemma*, etc.

Class–Dicotyledonae

Two cotyledons in embryo, tap root present, vascular bundles with cambium arranged in a ring, secondary growth occurs in stem and roots, reticulate venation in leaves, floral organs are pentamerous. It has 3 sub-classes, (i.e. polypetalae, gamopetalae and monochlamydeae) and 14 series, 165 natural orders, 2610 genera and 31874 species, e.g. *Citrus, Helianthus, Solanum, Petunia, Hibiscus, Euphorbia,* etc.

Economic Importance of Angiosperms

(i) The angiosperms are major source of food, fibres, spices and beverages.

(ii) They also provide valuable timber and medicines.

(iii) These also add beauty to our environments as well.

Resemblances between Gymnosperms and Angiosperms

- Root system is well developed and may be diarch, triarch, tetrarch or polyarch.
- Secondary growth is present in roots.
- The xylem is exarch in roots.
- The stem is eustelic.
- Secondary growth is present in stems.
- Secondary wood may be monoxylic or polyxylic.
- Production of seeds.
- Presence of heterospory.
- Presence of endosporic and reduced gametophytes.
- Presence of ovule.
- Presence of integument surrounding the ovule.
- Presence of pollen tube.
- Male gametophytes are highly reduced.
- Male gametes are non-motile in the majority of gymnosperms and all angiosperms.
- Germination of spores is precocious.
- The embryogeny is endosporic.

Difference between Gymnosperms and Angiosperms

- Gymnosperms are usually perennial, while angiosperms may be annual, biennial or perennial.
- In gymnosperms, vegetative reproduction is rarely found, while in angiosperms, it is commonly found.
- In gymnosperms, ovule is naked, while in angiosperms, ovule is protected by carpels.
- In gymnosperms, endosperm is haploid (n) and formed before fertilisation, whereas, in angiosperms, endosperm is triploid ($3n$) and formed after fertilisation.
- There is no double fertilisation in gymnosperms, while double fertilisation is the characteristic feature of angiosperms.
- In gymnosperms, phloem lacks companion cells, while companion cells are present in angiosperms.

Chapter Exercise

1. Phanerogamas are
   (a) seed plants in which sex organs are evident
   (b) non-seed plants
   (c) non-seed plants in which sex organs are hidden
   (d) None of the above

2. Time gap between the pollination and fertilisation in *Pinus* is
   (a) a few minutes (b) a few hours
   (c) a few weeks (d) nearly one year

3. Which of the following is lacking in gymnosperms?
   (a) Ovary (b) Embryo
   (c) Seeds (d) All of these

4. Xylem in gymnosperms is devoid of
   (a) vessels (b) tracheids
   (c) parenchyma (d) All of these

5. The plants which are fruitless but having seeds
   (a) angiosperms (b) gymnosperms
   (c) bryophytes (d) algae

6. Sexual reproduction does not require water in
   (a) algae (b) bryophytes
   (c) pteridophytes (d) gymnosperms

7. Most of the gymnosperms have
   (a) both antheridia and archegonia
   (b) archegonia but no antheridia
   (c) only antheridia
   (d) neither antheridia nor archegonia

8. Roots in gymnosperms are
   (a) adventitious roots (b) tap roots
   (c) rhizoids (d) rhizophores

9. Coralloid roots are found in
   (a) *Gnetum* (b) *Cycas*
   (c) *Pinus* (d) *Selaginella*

10. Sulphur shower is caused due to
    (a) large sulphur deposits on rocks
    (b) acid rain
    (c) as a result of SO_2, SO_3 gases in the air
    (d) pollen release during pollination of *Pinus*

11. Number of integuments in *Cycas* is
    (a) one (b) two (c) three (d) four

12. Antherozoids of *Cycas* are
    (a) uniflagellate (b) biflagellate
    (c) multicilliate (d) without flagella

13. Sunken stomata are found in
    (a) *Eucalyptus* (b) *Azadirachta*
    (c) *Cycas* (d) *Vinca*

14. Seeds are winged in
    (a) *Pinus* (b) *Cycas*
    (c) Mustard (d) All of these

15. Bentham and Hooker's system of classification is published in book
    (a) *Genera Plantarum* (b) *Species Plantarum*
    (c) *Phylum Plantarum* (d) None of these

16. A plant which produces seeds has embryo, vasculature but lacks fruit is definitely a
    (a) bryophyte (b) angiosperm
    (c) gymnosperm (d) pteridophyte

17. Chilgoza is
    (a) *Pinus gerardiana* (b) *Pinus roxburghii*
    (c) *Pinus wallichiana* (d) *Pinus merkusii*

18. A group of gymnosperms close to angiosperm is
    (a) gnetales (b) cycadales
    (c) coniferales (d) taxales

19. Which of the following is a tallest tree?
    (a) *Pinus* (b) *Sequoia*
    (c) *Metasequoia* (d) *Cedrus*

20. How many series are there in Monocotyledonae?
    (a) 8 series (b) 7 series (c) 4 series (d) 3 series

21. Gymnosperms differ from angiosperms in
    (a) presence of ovules (b) presence of pollen tube
    (c) absence of vessels (d) All of these

22. Red wood tree is
    (a) *Pinus* (b) *Cedrus* (c) *Sequoia* (d) *Dalbergia*

23. The classification given by Bentham and Hooker is
    (a) natural (b) phylogenetic
    (c) artificial (d) numerical

24. Siliqua fruit is a characteristic of the family
    (a) Leguminosae (b) Malvaceae
    (c) Liliaceae (d) Brassicaceae

25. Botanical name of cauliflower is
    (a) *Brassica oleracea* (b) *Brassica campestris*
    (c) *Mangifera indica* (d) *Sequoia gigonteum*

26. Which of the following character is shown by the members of family–Malvaceae?
    (a) Monodelphous androecium
    (b) Polyadelphous androecium
    (c) Apocarpous gynoecium
    (d) Polyandrous condition

27. Cotton belongs to the family
    (a) Malvaceae (b) Compositae
    (c) Cruciferae (d) Ranunculaceae

28. Opium plant belongs to which family?
(a) Caryophyllaceae (b) Ranunculaceae
(c) Papaveraceae (d) Rutaceae

29. The gynoecium in *Cucurbita* plant is
(a) tricarpellary, superior
(b) bicarpellary, inferior
(c) pentacarpellary, multilocular
(d) tricarpellary inferior

30. Compositae belongs to which order?
(a) Solanales (b) Asterales (c) Rosales (d) Malvales

31. The most advanced family of the dicot plant is
(a) Malvaceae (b) Cruciferae
(c) Umbelliferae (d) Compositae

32. Axile placentation is found in which of the following group of families?
(a) Liliaceae and Asteraceae
(b) Asteraceae and Fabaceae
(c) Brassicaceae and Solanaceae
(d) Liliaceae and Solanaceae

33. Oryza sativa belongs to which family
(a) Liliaceae (b) Poaceae
(c) Malvaceae (d) Brassicaceae

34. Which family have verticillaster type of inflorescence?
(a) Cucurbitaceae (b) Malvaceae
(c) Labiateae (d) Cruciferae

35. Which of the following condition of flower shows inferior ovary?
(a) Epigynous (b) Perigynous
(c) Hypogynous (d) Agynous

36. Cyathium inflorescence is found in which of the following plant?
(a) *Ficus* (b) *Dorstenia*
(c) *Morus* (d) *Euphorbia*

37. Oil glands are present in the leaves of which family?
(a) Poaceae (b) Liliaceae
(c) Rutaceae (d) Rosaceae

38. Indefinite stamens are the characteristic of
(a) Solanaceae (b) Malvaceae
(c) Gramineae (d) Asteraceae

39. *Dalbergia sisso* belongs to
(a) Leguminosae (b) Brassicaceae
(c) Malvaceae (d) Umbelliferae

40. Root nodules are present in the family
(a) Malvaceae (b) Gramineae
(c) Asteraceae (d) Papilionaceae

41. Which of the following system is well developed and same in both gymnosperms and angiosperms?
(a) Vascular system (b) Shoot system
(c) Root system (d) Both (a) and (b)

42. Secondary growth is found in
(a) gymnosperms (b) angiosperms
(c) Both (a) and (b) (d) None of these

43. Gymnosperms are usually perennial, while angiosperms are
(a) annual (b) biennial (c) perennial (d) All of these

44. Xylem is exarch in which of the following plant groups?
(a) Gymnosperms (b) Angiosperms
(c) Both (a) and (b) (d) None of these

45. Vegetative reproduction is common in angiosperms, but in gymnosperms it is
(a) most common (b) general
(c) rare (d) None of these

46. The presence of heterospory and production of seeds, both are common in
(a) gymnosperms (b) angiosperms
(c) fungi (d) Both (a) and (b)

47. In gymnosperms, ovule is naked, while in angiosperms it is protected by
(a) calyptra (b) indusium
(c) carpels (d) seed coat

48. In gymnosperms, endosperm is haploid (n) and formed before fertilisation, whereas in angiosperm it is formed after fertilisation and is
(a) n (b) $3n$
(c) $2n$ (d) All of these

49. Double fertilisation is a common character of which of the following?
(a) Angiosperms (b) Gymnosperms
(c) Both (a) and (b) (d) None of these

50. In gymnosperms, phloem lacks
(a) tracheids (b) parenchyma
(c) companion cells (d) vessels

51. The presence of endosporic and reduced gametophyte is a characteristic feature of which of the following?
(a) Angiosperms (b) Gymnosperms
(c) Both (a) and (b) (d) None of these

52. Ovules are present in both gymnosperms and angiosperms and are surrounded by
(a) indusium (b) calyptra
(c) integuments (d) All of these

53. The presence of pollen tube is a character of angiosperms besides angiosperm it is also found in
(a) algae (b) bryophytes
(c) fungi (d) None of these

54. In majority of gymnosperms and angiosperms male gametes are
(a) motile (b) non-motile
(c) Both (a) and (b) (d) None of these

55. The development of embryo in gymnosperm is endosporic, while in angiosperms it is
(a) exosporic (b) endosporic
(c) Both (a) and (b) (d) None of these

56. Male gametophyte in angiosperms and gymnosperms are
(a) highly developed
(b) profusely branched
(c) highly reduced
(d) None of the above

57. Production of fruit is the exclusive character of which of the following?
(a) Gymnosperms (b) Angiosperms
(c) Both (a) and (b) (d) None of these

58. Secondary wood may be polyxylic and monoxylic in which of the following plant group?
(a) Angiosperms (b) Gymnosperms
(c) Both (a) and (b) (d) None of these

59. Sporophyte is dominant in which of the following?
(a) Algae (b) Bryophytes
(c) Fungi (d) Gymnosperms

60. Haploid plant body is dominant in
(a) angiosperms (b) gymnosperms
(c) Both (a) and (b) (d) None of these

61. Diversification in plant life appeared
(a) due to long periods of evolutionary changes
(b) due to abrupt mutations
(c) suddenly on earth
(d) By seed dispersal

62. In gymnosperm, the endosperm is formed by
(a) fusion of two polar nuclei
(b) fertilised egg
(c) fusion of one polar nucleus
(d) germinations of one megaspore

63. The largest sperm cells are recorded in
(a) *Cycas* (b) *Pinus*
(c) *Lycopodium* (d) *Marchantia*

64. Coralloid roots of *Cycas* has
(a) *Anabaena* (b) *Mycorrhiza*
(c) *Rhizobium* (d) *Azotobacter*

65. Cleavage polyembryony occurs in
(a) *Pinus* (b) *Funaria* (c) *Cycas* (d) *Ephedra*

66. In *Cycas* pollination takes place in
(a) 3 celled stage (b) 4 celled stage
(c) 2 celled stage (d) 1 celled stage

67. *Cycas circinalis* is a source of
(a) resin (b) timber
(c) essential oil (d) starch

68. Megasporophyll of *Cycas* is equivalent to
(a) petal (b) stamen (c) carpel (d) sepal

69. Chilgoza, a gymnosperm seed, that is eaten as dry fruit is produced by
(a) *Pinus roxburghii* (b) *Pinus gerardiana*
(c) *Cimgo biloba* (d) *Cedrus deodara*

70. The pollination in *Pinus* is
(a) hydrophilous (b) entomophilous
(c) malacophilous (d) anemophilous

Answers

1. (a)	**2.** (d)	**3.** (a)	**4.** (a)	**5.** (b)	**6.** (d)	**7.** (a)	**8.** (b)	**9.** (b)	**10.** (d)
11. (a)	**12.** (c)	**13.** (c)	**14.** (a)	**15.** (a)	**16.** (c)	**17.** (a)	**18.** (a)	**19.** (b)	**20.** (b)
21. (c)	**22.** (c)	**23.** (a)	**24.** (d)	**25.** (a)	**26.** (a)	**27.** (a)	**28.** (c)	**29.** (d)	**30.** (b)
31. (d)	**32.** (d)	**33.** (b)	**34.** (c)	**35.** (c)	**36.** (d)	**37.** (c)	**38.** (b)	**39.** (a)	**40.** (d)
41. (c)	**42.** (c)	**43.** (d)	**44.** (c)	**45.** (c)	**46.** (d)	**47.** (c)	**48.** (b)	**49.** (a)	**50.** (c)
51. (c)	**52.** (c)	**53.** (d)	**54.** (a)	**55.** (b)	**56.** (c)	**57.** (b)	**58.** (c)	**59.** (d)	**60.** (d)
61. (a)	**62.** (d)	**63.** (a)	**64.** (a)	**65.** (a)	**66.** (a)	**67.** (d)	**68.** (c)	**69.** (b)	**70.** (d)

7

Morphology and Vegetative Organs (Root, Stem and Leaf)

Morphology is a branch of biology dealing with the study of the form and structure of organisms including plants and their specific structural features. It is characterised by the presence of roots, stems, leaves and flowers.

Root, stem and leaf represents morphology also provides information about the range of variations found in a species. It is also essential for recognition or identification of plants.

Root, stem and leaf are **vegetative organs** of a plant, while flowers, fruits and seeds represent the **reproductive organs** of a plant.

All these vegetative parts are involved in numerous different vegetative functions such as structural organisation, fixation, absorption, nourishment, growth, maintenance, etc.

Root

Roots are considered as the organ of a plant typically lying below the surface of the soil in case of a vascular plants. These are best defined as the non-leaf, non-nodes bearing parts of the plant's body. Root is a cylindrical structure that normally constitutes the descending or underground part of the plant axis. There is no distinction of nodes and internodes in case of roots, they give rise to endogenous branches in vascular plants.

Regions of Root

Root is divided into five different parts or regions such as

(i) **Root cap** It is the region which protects the meristematic region of root and leads the root into the soil. It is considered as a thimble-shaped or cap-like parenchymatous multicellular structures covering the root meristem.

Cells of root cap also possess starch grains. It provides the protection of the root meristem from friction of the soil particles. The role of root cap is perception of gravity was found out by Darwin in 1880.

(ii) **Meristematic zone or growing point** It is sub terminal in origin that lies below the root cap protectively. It is made up of compactly arranged meristematic cells having a dense protoplasm. It tends to divide continuously with a repeated division. It provides a new batch of young cell to the root of the plant. It helps in the growth of the root of plant.

(iii) **Region or zone of elongation or maturation** This zone lies behind the region of the growing point. It possess newly formed cells that tends to loose the power of division. It has the capacity to elongate rapidly. It helps in the absorption of water and mineral salts from the soil.

(iv) **Root hair zone** It is the region of differentiation and maturation as the differentiation of different types of primary tissues matures in the region. Root hair zone is about 1-6 cm in length. The outer cells of this zone gives rise to the lateral tubular outgrowths called root hairs. It increases the exposed surface of the root for the absorption of water.

New root hair cells are formed in the older part of the zone of elongation so that the root hairs appear in the newer parts of the soil from where water has yet been absorbed. It ensured the continuous water supply to the plant.

(v) **Zone or region of mature cells** The cells of this region undergo any further change. This region does not help in the root in the absorption of water. It allows the plant to anchors firmly in the soil.

Root Systems

The root system in plants constitutes the major part of the plant body both in the terms of function and bulk. It constitutes the network of all the roots of a plant, the system of roots and related organs that a plant develops.

The two basic types of root systems are primary root systems and the adventitious root systems.

1. Primary Root System

It is formed of root and their branches that arises from the primary root. The primary root is the first formed, main root that originates from the radicle of a seed embryo (i.e. the portion of the embryo below the cotyledon).

The two well known primary root system are:

(a) **Tap root system** which is formed by primary roots and its branches. It goes down in the deep soil and absorbs water. It is usually found in gymnosperm and most dicotyledons.

(b) **Fibrous root system** in which primary roots are replaced by the number of roots which form a bunch like structure mainly found in monocots. It has no central dominant root.

2. Adventitious Root System

It is typical of most monocotyledons in which the primary root is ephemeral in origin and soon living.

It soon arises from either a stem or a leaf tissue. While, in other plants roots arises from the structure other than radicle.

Thus, the roots and their lateral branches forms the fibrous root system.

Modifications of Roots

The modifications are the changes that occur in shape and forms of roots to carry out different and special functions other than carrying the normal functions such as absorption of water, etc. It occurs in both tap and adventitious roots.

Modifications of Tap Roots

As the tap roots are developed from radicle growing towards the soil. It is deeper in origin and found mainly in the dicot plants. These are modified to perform mainly two functions:

(i) **For the storage of food** In this type of modification, the tap root become swollen and fleshy in nature due to the storage property of food. There are mainly three types of modifications for the storage of food.

(a) **Fusiform** In this type, the root is swollen in the middle and gradually tapering towards the apex, e.g. radish (*Raphanus sativus*).

(b) **Conical form** In this type, the root is broad at the base and gradually tapers towards the apex, e.g. carrot (*Daucus carota*).

(c) **Napiform** In this type, root is considerably swollen at the upper part and becoming sharply tapering at the lower part, e.g. turnip (*Brassica rapa*).

(ii) **For respiration** Many plants grows in the swamps and marshy areas in which the oxygen is found in less quantity. This type of root modification occurs to get oxygen and help in respiration.

Pneumatophores are the special roots that are developed to help in gaseous exchange. These are aerophores or respiratory roots growing vertically upward (negatively geotropic). e.g. *Rhizophora*.

Modifications of Adventitious Roots

As these roots are not developed from radicle, the main reasons for the modifications are:

(i) **For food storage** These roots, are modified in order to store food. The main types of these roots are:

(a) **Fasciculated roots** These roots, takes the form of clusters from the base of the stem for the storage of food, e.g. *Dahlia*.

(b) **Nodulous roots** In this type, the roots become swollen near the tips, e.g. amaheldi.

(c) **Tuberous roots** These types, of roots are swollen and do not get definite shape, e.g. sweet potato.

(d) **Moniliform** These are types, of certain roots in which the swelling occurs at the regular intervals, e.g. grapes.

(e) **Annulated** In this type, of modification, roots possess a series of ring like outgrowth on its body, e.g. *Cinchona*.

(ii) **For mechanical support** These are the types, of roots that provides mechanical support to the plants.

The main type of such modifications are:

(a) **Prop roots** These are the roots that arises from the horizontal branches growing vertically downwards penetrating the soil, e.g. banyan tree.

(b) **Stilt roots** These are the roots that arises from the lower internodes of the weak stem penetrating obliquely into the soil, e.g. maize, sugarcane.

(c) **Buttress roots** These roots helps to fix the plant at its normal position. They grow as horizontal branches of the stem and vertically downward.

(d) **Climbing roots** These are the types, of roots found in the climbing plants helping them in climbing, e.g. money plant.

(e) **Hanging roots** These are the types, of roots that helps in fixing the plants to support the host, e.g. *Vanda*.

(iii) **To carry special functions** There are some roots that are modified to perform special functions.

Some of them are given below:

(a) **Absorbing roots** Some plants are known to be epiphytic in nature. In such cases these types of roots helps to absorb the atmospheric moisture, e.g. *Vanilla*.

(b) **Assimilatory roots** These are the types, of roots that have chlorophyll and prepare their own food for the plants, e.g. *Tinospora*.

(c) **Sucking roots** These are the types of roots that are found in the parasitic plants and helps in the absorption of food from the host, e.g. *Cuscuta*.

(d) **Reproductive roots** Some roots are modified to help plants in the process of reproduction, e.g. sweet potato.

(e) **Mycorrhizal roots** These types of roots are modified to help to perform the phenomenon of symbiosis (i.e. a beneficial mutual partnership between a fungus and a root) e.g. *Pinus* and *Monotropa*.

(f) **Floating roots** These are the types, of roots mainly found in the aquatic plants and thus, helps in floating, e.g. *Jussiaea*.

(g) **Contractile roots** Roots are somewhat thick and fleshy in the underground stem of some plants like onion, garlic, etc. Their apical part contracts and fixes the plant in the soil.

(h) **Nodulated roots** These roots also forms the examples of symbiosis in which the bacteria are found in the root nodules of leguminous plants, e.g. groundnut, pulses etc.

Functions of Roots

Roots in plants are tend to perform numerous functions:

(i) Major function of root system is to provide proper anchorage to plants and to absorb water, minerals, etc from the soil.

(ii) Roots also absorb food that is made in the leaves, so they also act as storage organ.

(iii) Roots are also involved in the synthesis of plant growth regulators or hormones.

(iv) In response to the concentration of nutrients roots also synthesise cytokinin which are as a signal as to how the shoot can grow.

Stem

The stem is the stalk of a plant or the main trunk of a tree. It develops from the plumule of the embryo during the time of germination. In most plants, the stem is the major vertical shoot, in some it is in conspicuous, and in others it modifies and resemble other parts of the plant. It holds up other structures of plant such as leaves and flowers.

Parts of Stem

The stem is divided into node and internodes. The region of the stem where leaves are originated is called the **node** while the gap portion between two nodes is called **internodes**.

The number of the leaves that appears at a node depends on the species of the plant, one leaf per node is common but two or more leaves may grow at the nodes of some species.

Modifications of Stem

Stems are also modified to perform function other than the conduction of water and bears leaves and flowers.

Following are some common modification:

(i) **Underground stem modifications** The underground stems are located below the surface of the soil. They protect themselves against unfavourable conditions of weather and attack of animals. They serve as the store houses for the reserve food, nature of the stem can be distinguished by the presence of nodes and internodes. The complete anatomy of underground stem resembles than that of an aerial stem.

The underground stems are of four major types:

(a) **Rhizome** It is a thick horizontally grown stem which is usually meant to store the food material. It bears nodes and internodes, scale leaves, axillary buds, adventitious roots and a terminal bud. Usually the growing points of the rhizome continues to remain underground thus causing an elongation of the rhizome. Development of roots occurs from the lower surface of the rhizome, e.g ginger, turmeric.

(b) **Tuber** It is known as the swollen end of an underground branch and arises from the axil of a lower leaf. These underground branches grows horizontally outward sin the soil. Each tuber seems to be irregular in shape due to the deposition of food material (starch) on it.

There are some leaf scars found on the surface of the each tuber. Out of which each scar encloses an axillary bud and is called an **eye**. These eyes of potato are capable of production of new plants by the method of vegetative propagation, e.g. potato.

(c) **Bulb** In this type, the stem is reduced and represented by a short disc. The lower surface of stem produces many adventitious roots, e.g. garlic and onion. In bulbs of both onion, garlic and many others, the inner leaves are fleshy while the outer ones are dry in nature. The apical bud in bulb produces the shoot while, the axillary bud tends to produce daughter bulbs.

(d) **Corm** It is a greatly swollen underground basal portion of an erect stem. This swelling occurs due to the storage of reserve food material. Corm bears scale leaves and axillary buds on their surfaces, e.g. *amorphophillus* and *Colocasia*.

(ii) **Aerial stem modification** These are the modifications to the aerial stems, vegetative buds and floral buds of plants and performs functions such as climbing, protection, synthesis of food, vegetative propagation, etc.

All the aerial stems, vegetative buds and floral buds of plants growing in different conditions undergoes modifications to perform special functions.

These includes the following:

(a) **Stem tendrils** Some weaked stemmed plants produces wiry, coiled, sensitive and delicate organs for climbing. They are called tendrils.

Their development occurs either from the axillary bud or from the terminal bud of the stem, e.g. cucumber, pumpkins, watermelons, etc.

(b) **Thorns** These are hard, woody, pointed structures meant for the protection. Thorns are often provided with the vascular tissues that may develop either from axillary or terminal buds, e.g. rose, citrus, etc.

(c) **Bulbils** These are formed when the axillary bud becomes fleshy and rounded due to the storage of food. Bulbils have the tendency of developing into a new plant if it gets detached from the plant and falls on to the ground, e.g. *Lilium, Agave, Oxalis,* etc.

(d) **Phylloclades** These are flattened, photosynthetic shoots that are usually considered to be the modified branches. These are subsets of cladodes, e.g. *Asparagus*.

Functions of Stem

(i) The major function of stem is to bear leaves, flowers and fruits and to conduct the water, minerals and all other ingredients to these parts.

(ii) They store food as reserve food materials and helps in photosynthesis and vegetative propagation.

(iii) It also helps in holding flower in position to allow the mechanism of pollination and fertilisation.

Leaves

The leaves are the lateral, flattened structures borne on the nodes of the stem of plant. They contain an axillary bud on its axil which develops into a new branch. It takes its origin mainly from the shoot apical meristem.

These are the most important parts of a plant as they are the primary photosynthetic organs of the plants.

Parts of a Leaf

Leaf is made up of three main parts:

(i) **Leaf base** (hypopodium), by which it is attached with the stem.

(ii) **Petiole** (mesopodium), a long thin flexible part of leaf which keep it towards light.

(iii) **Lamina** (leaf blade), the expanded green part of the leaf. The middle vein of the leaf is called **midrib**.

Modifications of Leaf

Leaves are also modified as per situation. Tendrils (whole leaf tendrils, leaflet tendrils, stipular tendrils, rachis tip tendrils, etc) and spines are such examples of modified leaves.

Functions of Leaf

(i) It is primarily involved in photosynthesis, transpiration, gaseous exchange, etc.

(ii) Beside this, it also performs some secondary functions.

Venation

It is the arrangement of veins and veiniet in the lamina of the leaf. Two types of venation are found in the leaves such as

(i) **Reticulate venation**, in which veinlets form a network, e.g. dicots

(ii) **Parallel venation**, in which veins run parallel to each other, e.g. dicots generally show reticulate venation, while monocots have parallel venation.

Simple and Compound Leaves

Simple leaf in which the lamina is entire, while compound leaf in which the lamina is divided into small leaflets.

Compound leaf may be pinnately compound or palmately compound. The individual units of leaf in compound leaves is called **leaflets**. In pinnately compound leaves, leaflets are present on a common rachis, while in palmately compound leaf the leaflets are attached on a common point.

Phyllotaxy

In botany, phyllotaxy is the phenomenon of arrangement of leaves, scales or bracts with flowers and the plant stem.

It is of five main types:

(i) **Spiral**, when only one leaf is borne on each node.

(ii) **Alternate**, when two leaves are on the same plane and are arranged alternatively.

(iii) **Opposite and decussate**, when opposite leaves of two successive nodes lies at the right angles.

(iv) **Opposite and superposed**, when opposite leaves of two successive nodes lies in the same plane.

(v) **Whorled or verticillate** when three or more than three leaves are borne on a single node in a circle.

Chapter Exercise

1. Prop roots are modified roots found in
(a) Banyan tree (b) *Hibiscus*
(c) *Vine* (d) *Passiflora*

2. Identify the correctly matched pair.
(a) Unifoliate compound leaf – *Citrus*
(b) Opposite phyllotaxy – *Hibiscus*
(c) Palmate convergent venation – *Musa*
(d) Bipinnate leaf – Mustard

3. What does the cells of root cap possess that helps in participating graviperception?
(a) Starch grains (b) Inulin
(c) Fat bodies (d) Protein bodies

4. The function of root cap is
(a) food-storage
(b) formation of new cells
(c) protection of root meristem
(d) water and mineral absorption

5. Absorption of water and minerals is mainly related to
(a) root hair zone (b) zone of elongation
(c) zone of mature cells (d) Both (b) and (c)

6. Zone of maturation is the name of
(a) zone of elongation (b) root hair zone
(c) zone of mature cells (d) meristematic zone

7. Fibrous roots of *Zea mays* (maize) develop from
(a) lower nodes (b) upper nodes
(c) internodes (d) leaves

8. The maximum growth in roots takes place
(a) root cap region (b) behind the root cap
(c) in zone of elongation (d) in zone of mature cells

9. Prop roots of *Ficus benghalensis* are
(a) negatively geotropic
(b) negatively hydrotropic
(c) positively heliotropic
(d) positively geotropic

10. An example of tuberous root is
(a) *Colocasia antiquorum* (b) *Ipomoea batata*
(c) *Solanum tuberosum* (d) *Raphanus sativus*

11. Roots that develop from any part of the plant but not from the radicle are called
(a) stilt roots (b) prop roots
(c) adventitious roots (d) tap roots

12. Prolonged elongation of radicle in dicot seeds gives rise to the
(a) primary root (b) secondary root
(c) tertiary root (d) adventitious root

13. Roots of *Dahlia* are
(a) tuberous and fasciculated (b) napiform
(c) fibrous (d) photosynthetic

14. Which of the following is not a modified stem?
(a) Ginger (b) Turnip
(c) *Colocasia* (d) Potato

15. The arrangement of leaves is called
(a) phyllotaxy (b) phyllotaxis
(c) ptyxis (d) venation

16. The largest leaf belongs to
(a) *Victoria amazonica* (b) *Rafflesia*
(c) *Raphia vinifera* (d) *Nerium*

17. Sucking roots are found in
(a) *Cuscuta* (b) *Tinospora*
(c) *Piper betal* (d) *Orchids*

18. Isobilateral leaves are most commonly found in
(a) monocots (b) dicots
(c) angiosperms (d) gymnosperms

19. The venation of a leaf acts as the channel for the transport of
(a) food (b) food, water and minerals
(c) water and minerals (d) food and water

20. The most common type of compound leaves occurring in nature are
(a) pinnately (b) palmately
(c) trifoliate (d) decompound

21. Bud is the modification of
(a) shoot (b) root
(c) leaf (d) flower

22. Tendrils modified from leaflets are not found in
(a) *Lathyrus aphaca* (b) *Lathyrus odoratus*
(c) *Pisum sativum* (d) Both (a) and (b)

23. Tuberous roots are thick and fleshy, and are found in
(a) carrot (b) radish
(c) turnip (d) 4'O clock plant

24. Whorled arrangement of leaves is found in
(a) Tobacco (b) *Nerium*
(c) *Calotropis* (d) *Ocimum*

25. Modification of stem for food storage is not found in
(a) potato and ginger (b) turmeric and gaurikand
(c) potato and *Colocasia* (d) *Citrus* and *Bougainvillea*

26. Prop roots are modified into adventitious roots in
(a) *Ficus* (b) *Hibiscus*
(c) *Vine* (d) *Passiflora*

27. Tuberous roots are found in
(a) *Beta vulgaris* (b) *Daucus carota*
(c) *Ipomoea batata* (d) *Raphanus sativus*

28. Stem modification in *Gladiolus* is
(a) corm (b) bulbil
(c) bulb (d) rhizome

29. The photosynthetic or assimilatory roots are observed in
(a) *Cuscuta* (b) Banyan
(c) *Vanda* (d) *Tinospora*

30. Identify the modification of leaf from the following
(a) phyllode (b) phylloclade
(c) corm (d) cladode

31. The region of root responsible for the absorption of water.
(a) Zone of mature cells (b) Meristematic zone
(c) Zone of elongation (d) Root cap

32. Type of modification of tap root responsible for respiration
(a) pneumatophores (b) fusiform
(c) conical form (d) napiform

33. Type of underground stem in which stem is reduced and represented by a short disc is
(a) rhizome (b) bulb
(c) tuber (d) corm

34. Modification of type of root that provides mechanical support to the plants by fixing the plant at its normal position.
(a) Climbing roots (b) Stilt roots
(c) Hanging roots (d) Buttress roots

35. Type of the phyllotaxy, when opposite leaves of two successive nodes lies in the same plane.
(a) Whorled and verticillate
(b) Opposite and decussate
(c) Opposite and superposed
(d) Spiral

36. Ginger multiplies vegetatively by
(a) tuber (b) bud
(c) stem (d) rhizome

37. Cotyledons and testa are edible parts in
(a) groundnut and pomegranate
(b) walnut and tamarind
(c) french bean and coconut
(d) cashewnut and litchi

38. Petiole is modified into tendril in
(a) *Gloriosa* (b) *Passiflora*
(c) *Pisum* (d) *Clematis*

39. Find the incorrect match.
(a) Still root – Turnip
(b) Taproot – Carrot
(c) Adventitius – Sweet potato
(d) Proproot – Banyan tree

40. Parallel venation is a characteristic of
(a) grasses (b) legumes
(c) parasitic plants (d) xerophytic plants

41. Wheat grains is an example of
(a) nut (b) follicle
(c) caryopsis (d) achene

42. Leaves are attached to the stem at
(a) apical meristem (b) internode
(c) node (d) axillary meristem

43. Potatoes are borne in
(a) primary roots (b) axil of scaly leaves
(c) lateral roots (d) adventitious roots

44. Root pocket does not occur in
(a) mangrove plants (b) *Trapa*
(c) *Pistia* (d) *Ipomoea*

45. Placentation in legumes is
(a) basal (b) marginal
(c) axile (d) free central

Answers

1. (a)	**2.** (a)	**3.** (a)	**4.** (c)	**5.** (a)	**6.** (c)	**7.** (a)	**8.** (b)	**9.** (d)	**10.** (b)
11. (c)	**12.** (a)	**13.** (a)	**14.** (d)	**15.** (a)	**16.** (a)	**17.** (a)	**18.** (a)	**19.** (b)	**20.** (a)
21. (a)	**22.** (a)	**23.** (d)	**24.** (b)	**25.** (d)	**26.** (b)	**27.** (c)	**28.** (d)	**29.** (d)	**30.** (a)
31. (a)	**32.** (a)	**33.** (b)	**34.** (c)	**35.** (c)	**36.** (d)	**37.** (a)	**38.** (d)	**39.** (a)	**40.** (b)
41. (b)	**42.** (c)	**43** (b)	**44.** (c)	**45.** (b)					

8

Morphology of Reproductive Organs (Plants)

Reproductive organs of plants constitute the plant reproductive morphology. It is concerned with the physical form and structure of those parts of plants that are directly or indirectly concerned with the sexual reproduction.

Among all organisms, flowers which are the reproductive structures of an angiosperm are the most varied physically and show a correspondingly great diversity in the method of reproduction.

Inflorescence

An inflorescence is a group or cluster of flowers arranged on a stem that is composed of a main branch or a complicated arrangement of branches. Morphologically, it is the part of shoot of seed plants where flowers are formed and which is accordingly modified.

Types of Inflorescence

The different types of inflorescence are

1. **Racemose**

In racemose, the main axis continues to grow and the flowers are borne laterally in **acropetal manner.**

Racemose inflorescence is indeterminate type of inflorescence and divided into following types:

(i) Raceme (mustard)
(ii) Panicle (goldmohar)
(iii) Corymb (candytuft)
(iv) Spike (bottle brush)
(v) Spikelet(wheat)
(vi) Catkin (mulberry)
(vii) Spadix (palm)
(viii) Umbel (coriander)
(ix) Capitulum (sunflower)
(x) Strobile (hop)

2. **Cymose**

In cymose, the main axis terminates into a flower and growth becomes limited. The flowers are borne in a **basipetal order.**

Cymose inflorescence is determinate type of inflorescence and divided into following types:

(i) Uniparous or monochasial cyme (potato).
(ii) Biparous or dichasial cyme (night jasmine).
(iii) Polychasial or multiparous cyme (dianthus).
(iv) Cymose head (keekar).

Besides the racemose and cymose, two other types of inflorescence are mixed and special type of inflorescence is formed.

Types of special inflorescence are as under:

(i) Cyathium (*Euphorbia*)
(ii) Verticillaster (*Ocimum*)
(iii) Hypanthodium (Peepal)
(iv) Coenanthium (*Dorstenia*)

Flower

A flower is the reproductive unit of plants. It is mainly meant for the purpose of sexual reproduction.

In a typical flower, four kinds of whorls are arranged around a swollen stalk called **thalamus**. These whorls are commonly known as calyx, corolla, androecium and gynoecium.

Gynoecium is made up from carpels and may be apocarpous or syncarpous. Androecium is composed of stamens. It may be monoadelphous, diadelphous and polyadelphous.

Calyx is also called as sepals and corolla as petals. Sepals and petals are **accessory organs**, while androecium and gynoecium are the **reproductive organs**.

Bisexual flowers have both androecium and gynoecium, while unisexual flowers either have androecium or gynoecium.

Symmetry of a Flower

Flowers have two types of symmetry, i.e. actinomorphic and zygomorphic. **Actinomorphic flowers** can be divided into two equal halves in any radial plane, while **zygomorphic flowers** can be divided into two equal halves, i.e. only in one particular plane.

Number of Floral Parts

Based on the number of floral appendages, a flower may be trimerous, tetramerous or pentamerous. **Dicotyledonous** flowers are usually di, tetra or pentamerous, while **monocotyledonous** flowers are mainly trimerous in organisation.

Insertion of Floral Parts

Flowers are described as hypogynous, perigynous and epigynous based on the position of floral appendages.

In hypogynous flower, gynoecium occupies the highest position (superior ovary), in perigynous flower the gynoecium is situated in the center, while in epigynous flowers, the gynoecium is found in the thalamus (inferior ovary).

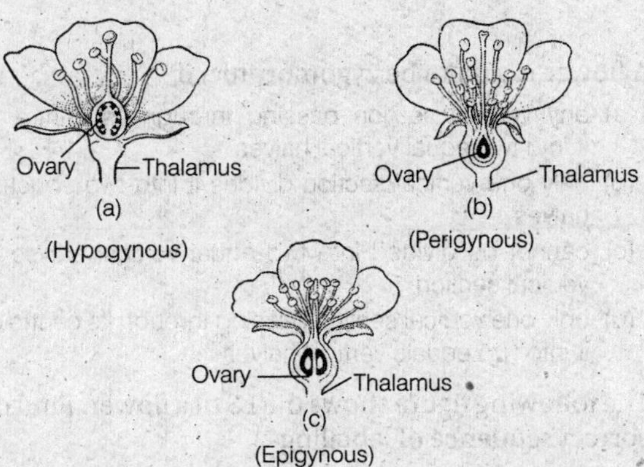

Arrangement of floral leaves on thalamus

Calyx or sepals (the outer whorl) may be gamosepalous (united sepals) or polysepalous (free sepals). Same as sepals, corolla or petals may also be gamopetalous or polypetalous.

Floral Formula

Br	Bracteate	Ebr	Ebracteate
♂	Male	♀	Female
⚥	Bisexual	⊕	Actinomorphic
%	Zygomorphic	E_p	Epicalyx
K	Calyx-free (polycepalous)	(K)	Calyx united (Gamosepalous)
C	Corolla-free (Polypetalous)	(C)	Corolla-United (Gamopetalous)
C_X	Corolla-cruciform	P	Perianth
A	Androecium-Free	(A)	Androecium-united
$\overset{\frown}{P\ A}$	Epiphyllous	$\overset{\frown}{C\ A}$	Epipetalous
G	Carpels-Free (apocarpous)	(G)	Carpels-united (syncarpous)
$\underline{G}$	Superior ovary	$\overline{G}$	Inferior ovary

Aestivation

The arrangement of whorls of flower is called aestivation. It is important in classification of plants. It may be of four types such as valvate, twisted, imbricate and vexillary.

Placentation

In biology, placentation refers to the formation, type and structure or arrangement of placentas. It occurs not only in animals, humans mainly but also in egg and flowering plants.

The function of placentation is to transfer nutrients from parent to the young embryo. In flowering plants, placentation occurs where the ovules are attached inside the ovary. The ovules inside a flower's ovary are attached *via* funiculi, the plant part equivalent to an umbilical cord in a female. The part of the ovary where the funiculus attached is referred to as placenta.

In botany, the term 'placentation' most commonly refers to the arrangement of placentas inside a flower or a fruit.

Plant placentation types includes:

(i) Basal placentation (ii) Apical placentation
(iii) Parietal placentation (iv) Axile placentation
(v) Free central placentation (vi) Marginal placentation

Fruits

These are the characteristic feature possessed only by the flowering plants. Thus, the mature, ripened ovary is called fruit. It is developed after fertilisation. When the fruit develops without fertilisation, it is called **parthenocarpic fruit**.

Parts of Fruit

Fruit is made up of an outer wall called **pericarp** and **seeds**. Pericarp is further divided into epicarp, mesocarp and endocarp. A pericarp may be dry or fleshy in nature.

Types of Fruits

Fruits may be classified as under:

(i) **Simple fruits** They have single ovary are monocarpellary or polycarpellary. These are further divided into two types on the basis of pericarp, i.e.

(a) **Fleshy fruits** involves berry, pepo, hesperidium , drupe, pome and hypanthium.

(b) **Dry fruits** can be dehiscent fruit (follicle, legume, capsule, silique) or indehiscent fruit (achene and nut).

(ii) **Aggregate fruits** A group of fruitlets which develop from a flower having polycarpellary apocarpous (free) gynoecium is called aggregate fruit and it is also called **etaerio**.

(iii) **Multiple fruits** (composite) A group of food developed from an entire inflorescence is called multiple fruits and it is composed of a number of closely associated fruits.

(iv) **Accessory fruits** These are the fruits in which some of the flesh is derived not from the ovary, but from some adjacent tissue, exterior to carpel, e.g. strawberry, pineapple, etc.

Seeds

Seeds are the structures developed from ovules after fertilisation. A seed is made up of embryo and seed coat. An **embryo** consists of radicle, plumule, embryonal axis and cotyledons.

While a seed coat is made up of outer **testa** and inner **tegmen**. Based on number of cotyledons, seeds are either monocotyledonous (single cotyledonous) or dicotyledonous (two cotyledonous).

Chapter Exercise

1. Arrangement of flowers on the floral axis is

(a) inflorescence (b) phyllotaxy
(c) placentation (d) corymb

2. The whorl of bracts present below the inflorescence of sunflower is

(a) epicalyx (b) spathe (c) involucel (d) involucre

3. Both hypanthodium and cyathium possess

(a) unisexual flowers
(b) bisexual sessile flowers
(c) bisexual pedicellate flowers
(d) sterile peduncle

4. In a racemose, the flower arrangement is

(a) basipetal (b) centripetal
(c) centrifugal (d) acropetal

5. Inflorescence consisting of unisexual sessile flowers, is

(a) spike (b) spikelet (c) catkin (d) umbel

6. Name the inflorescence, in which flowers reach the same level through which they arise from different points.

(a) Umbel (b) Catkin (c) Corymb (d) Capitulum

7. Opening of flowers in a cymose inflorescence is

(a) centrifugal (b) acropetal
(c) centripetal (d) basipetal

8. In mulberry, the inflorescence is

(a) catkin (b) capitulum (c) spadix (d) cyathium

9. When the five sepals are arranged in such a way that two are completely external and other two are completely internal, while one is partially external, then the mode of arrangement of sepals on the thalamus is

(a) valvate (b) quincuncial
(c) twisted (d) imbricate

10. The aestivation of corolla in pea (*Pisum sativum*) is

(a) quincuncial (b) vexillary
(c) ascending imbricate (d) descending imbricate

11. The condition, in which stamens are united throughout their whole length is

(a) synandrous (b) syngenesious
(c) polyandrous (d) adelphous

12. Epipetalous is the condition of

(a) adhesion of stamens with corolla
(b) cohesion of stamens with calyx
(c) aestivation of petals
(d) position of sepals with respect to petals

13. An additional floral, whorl is present in *Hibiscus rosa sinensis* called

(a) involucre (b) spathy
(c) epicalyx (d) calyx

14. A flower is said to be zygomorphic, if

(a) any vertical section passing through its centre divides it into two equal vertical halves
(b) only one vertical section divides it into two equal vertical halves
(c) cannot be divided into two equal vertical halves by any vertical section
(d) only one vertical section passing through its centre divides it into two equals vertical halves

15. The following figure shows the LS of a flower. Find out the correct sequence of labelling.

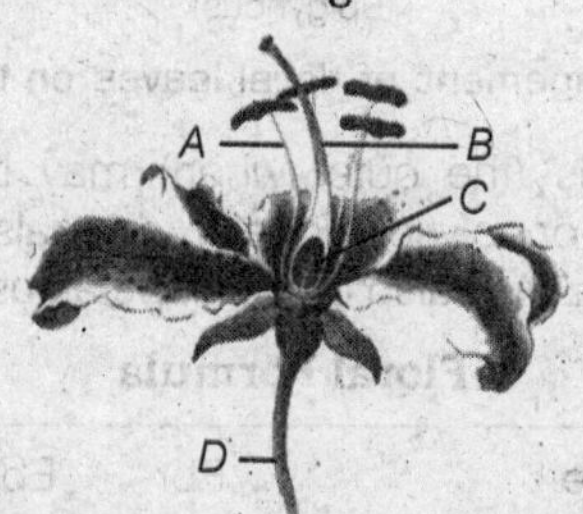

(a) A–Filament, B–Style, C–Ovules, D–Pedicel
(b) A–Style, B–Filament, C–Ovules, D–Receptacle
(c) A–Stigma, B–Style, C–Ovary, D–Pedicel
(d) A–Style, B–Stigma, C–Ovules, D–Receptacle

16. Write taxonomical terms for the various kinds of aestivation given in figures.

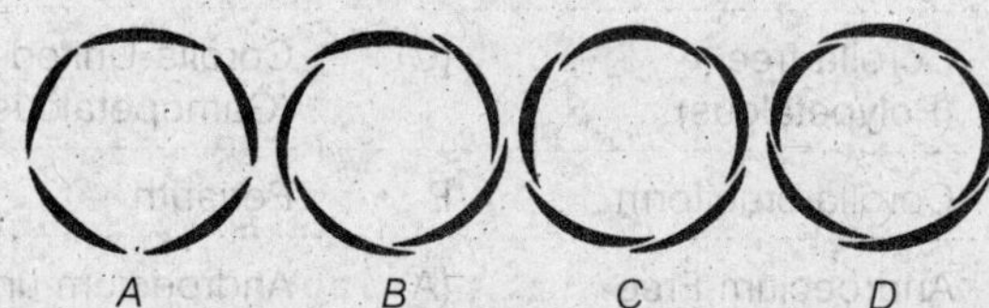

(a) A–Quincuncial, B–Twisted, C–Valvate, D–Imbricate
(b) A–Valvate, B–Twisted, C–Quincuncial, D–Imbricate
(c) A–Valvate, B–Twisted, C–Imbricate, D–Quincuncial
(d) A–Twisted, B–Valvate, C–Imbricate, D–Quincuncial

17. Identify the placentation from the given diagrams.

(a) A–Marginal, pea, B–Parietal, *Argemone*, C–Basal, wheat
(b) A–Axile, *Citrus*, B–Basal, wheat, C–Free central, silene
(c) A–Basal, sunflower, B–Superficial, *Nymphaea*, C–Free central, *Stellaria*
(d) A–Parital, radish, B–Basal, maize, C–Superficial, *Nymphaea*

18. Which one of the following is a false fruits?
(a) Cashewnut (b) Banana (c) Apple (d) Tomato

19. Pepo is a fruit of
(a) Cruciferae (b) Leguminosae
(c) Cucurbitaceae (d) Liliaceae

20. Marginal placentation is found in
(a) *Pisum* and *Lathyrus*
(b) *Citrus* and *Solanum*
(c) *Triticum* and *Ranunculus*
(d) *Zea mays* and *Nymphaea*

21. Apocarpous gynoecium is found in
(a) Ranunculaceae (b) Malvaceae
(c) Solanaceae (d) Liliaceae

22. The edible part in litchi is
(a) pericarp (b) aril
(c) mesocarp (d) None of these

23. Perianth is
(a) outer part of flower in which calyx and corolla are not always distinguishable
(b) when stamens and gynoecium are not distinguishable
(c) when only calyx is present
(d) None of the above

24. A fruit that develops from a complete condensed inflorescence, is called
(a) schizocarpic (b) composite
(c) aggregate (d) etaerio

25. From of the following contains all others structures given in gynoecium except
(a) ovary (b) stigma
(c) carpel (d) filament

26. A pome fruit is said to be false because
(a) the pericarp is in conspicuous
(b) the endocarp is cartilaginous
(c) the fruit is present in fleshy of edible thalamus
(d) the fruit is derived from inferior ovary

27. An inflorescence always forms a
(a) multiple or composite fruit (b) simple fruit
(c) dry dehiscent fruit (d) aggregate fruit

28. The correct match for edible part of fruit is
(a) Guava – Pericarp with placenta
(b) Maize – Cotyledon
(c) Tomato – Thalamus
(d) Date palm – Epicarp

29. Match the following columns.

Column I	Column II
A. Marginal placentation	1. *Petunia*
B. Axile placentation	2. *Dianthus*
C. Free central placentation	3. Mustard
D. Parietal placentation	4. Pea

Codes

	A	B	C	D		A	B	C	D
(a)	2	1	4	3	(b)	3	4	2	1
(c)	4	1	2	3	(d)	4	1	3	2

30. Identify the inflorescence

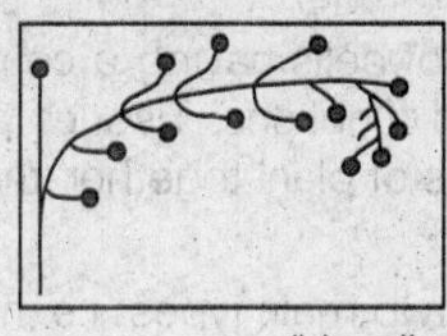

(a) raceme (b) spike
(c) helicoid (d) scorpioid

31. Labeling the given figure.

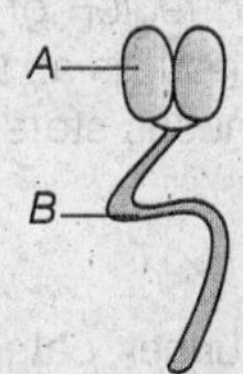

(a) A–Stigma B–style (b) A–Anther B–Filament
(c) A–Anther B–style (d) A–Stigma B–Filament

32. Which one of the following is the wrong pairing?
(a) Raceme – Mustard
(b) Spice – *Achyranthus*
(b) Compound umbel –Onion
(d) Spadix – *Musa*

33. Drupe has
(a) hard epicarp (b) hard endocarp
(d) hard mesocarp (d) no epicarp

34. Seed is a
(a) fertilised ovary (b) fertilised embryo
(c) fertilised fruit (d) fertilised ovule

35. How many plants in the list given below have composite fruits that develops from an inflorescence walnut, poppy, radish, pineapple, apple, tomato, mulberry?
(a) 5 (b) 2
(c) 3 (d) 4

Answers

1. (a)	2. (d)	3. (c)	4. (d)	5. (c)	6. (c)	7. (d)	8. (a)	9. (b)	10. (b)
11. (b)	12. (a)	13. (c)	14. (b)	15. (a)	16. (c)	17. (b)	18. (c)	19. (c)	20. (a)
21. (a)	22. (b)	23. (a)	24. (b)	25. (d)	26. (c)	27. (c)	28. (a)	29. (c)	30. (c)
31. (b)	32. (c)	33. (b)	34. (d)	35. (b)					

9

Tissue

A tissue is a group of cells having a common origin and performing common functions. Tissues are important, in joining different cells of plant together thereby, forming a complex structure.

Plant tissues are of two main types, i.e. meristematic and permanent tissue.

Meristematic Tissue

Meristem is a specialised region of cells performing cell division. A meristematic tissue is an undifferentiated mass of cells that is in a continuous state of division retaining their power of division.

Types of Meristems

Meristematic tissue is further categories into following types:

1. **Classification Based on Origin and Development**

On the basis of origin and development of initiating cells, meristems can be divided into three types:

(i) **Promeristem or primordial meristem** A group of young meristematic cells of a growing organ. It is the early embryonic meristem from which other advanced meristems are derived. In a plant, it occupies a small area at the tip of stem and root. It further divides to form primary meristem.

(ii) **Primary meristem** These are derived from promeristem. They are present below the promeristem at shoot and root apices. These cells divide and form permanent tissues.

(iii) **Secondary meristem** It is derived from primary permanent tissues which have the capacity of division, e.g. cork cambium, cambium of roots and interfascicular cambium of stem.

2. **Classification on the Basis of Position**

On the basis of their position in the plant body, meristems are of three types :

(i) **Apical meristem** These are found at the apices or growing points of root and shoot and bring about increase in length. It includes both promeristem as well as primary meristem.

Several theories have been put forward to explain the activity of apical meristem:

(a) **Apical cell theory** This theory was first proposed by Hofmeister (1857) and advanced by Nageli (1878). According to this theory, a single apical cell is the structural and functional unit of apical meristem which governs the entire process of apical growth. However, such organisation has been found only in cryptogams.

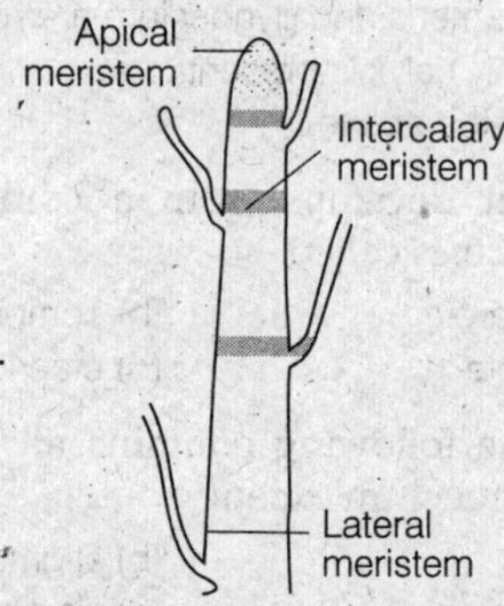

Schematic representation of position of different meristems

(b) **Histogen cell theory** This theory was given by Hanstein (1868). According to this theory root and shoot apices consists of the central or inner mass called **plerome** surrounded by the middle region composed of isodiametric cells called **periblem** and the outermost uniseriate layer of **dermatogen**. Dermatogen gives rise to epidermis, periblem to cortex and endodermis and plerome to vascular bundle and pith. These three layers were called Histogen by Hanstein.

(c) **Tunica corpus Theory** This theory was proposed by **Schmidt** (1924). According to this theory, mass of dividing cells are of two types, **tunica**, the outer consisting of one position of different meristems or more peripheral layers of cells, forming the outer region and **corpus**, the central undifferentiated multilayered mass of cell. Epidermis is derived from outer layer of tunica and the remaining tissues are derived entire from remaining layer of tunica and corpus.

(d) **Quiescent centre concept** Clowers (1959-61) in the root tip of *Zea mays* observe a central cup-like region of passive cell lying between the root cap and active root meristem. This was named as quiescent centre.

(ii) **Intercalary meristem** It lies between the region of permanent tissues and is considered as a part of primary meristem which has become detached due to formation of intermediate permanent tissues. It is found either at the base of leaf, e.g. *Pinus* or at the base of internodes, e.g. grasses.

(iii) **Lateral meristem** These are arranged parallel to the sides of origin and normally divide periclinally or radially and give rise to secondary permanent tissues. These increase the thickness of the plant part.

3. **Classification on the Basis of Function**

On the basis of their function, meristems have been classified into three types:

(i) **Protoderm meristem** It is the outermost layer of the young growing region which develops to form epidermal tissue system.

(ii) **Procambium meristem** It is composed of narrow, elongated, prosenchymatous, meristematic cells that gives rise to the vascular tissues system.

(iii) **Ground meristem** It is composed of large, thick-walled cells which develop to ground tissue system, i.e. hypodermis, cortex and pith.

4. **Classification on the Basis of Plane of Divisions**

The growth pattern and plane of division of meristematic tissue is important to govern the mode of growth.

These tissues can be divided into three types:

(i) **Mass meristem** In such meristem, cell divisions occur in all planes resulting in an increases in volume. It can be observed in meristems of cortex and pith.

(ii) **Rib or file meristem** The cells divide only on one plane, e.g., formation of filaments in algae.

(iii) **Plate meristem** These cells divide in two planes resulting to an increase in the area of an organ, e.g. leaf formation.

Permanent Tissue

It is a group of cells that become mature and do not divide further. It is formed due to the differentiation of meristematic tissues.

Permanent tissue is also categorised into following two types:

(i) **Simple Permanent Tissue**

These are the tissue which are made up of only one type of cells. Different types of simple permanent tissues found in the plant body are

(a) **Parenchyma** It is made up of isodiametric cells and the cells may be oval, round, polygonal, etc. They have thin walls and are closely packed with little intercellular spaces. Generally, it involves in photosynthesis, storage, secretion, etc.

(b) **Collenchyma** It is found beneath the epidermis and either in continuous layer or in patches. Cells are oval, spherical or polygonal with heavy deposit of **cellulose, hemicellulose** and **pectin**. Their major function is to provide mechanical support to the plant.

(c) **Sclerenchyma** In this, cells are long and narrow with thick and lignified cell wall. They have pores or pit on their cell wall. Two types of sclerenchyma cells are present in plant, out of which one is long tapering, narrow fibres and the other one is oval-shaped and sclereid.

(ii) **Complex Permanent Tissue**

These are made up of more than one type of cells and work together as a unit. Two types of complex permanent tissues are found in plants. These are two types:

(a) **Xylem** It is mainly involved in the conduction of water and minerals from roots to the stem and leaves usually made up of tracheid, vessels, xylem fibres and xylem parenchyma. Xylem is of two main types, i.e. protoxylem and metaxylem.

Protoxylem are first formed xylem and **metaxylem** are the secondly formed xylem. In stems, the protoxylem lies inside the stem and the situation is called **endarch** situation, while in roots, protoxylem lies towards periphery and this situation is called as **exarch** situation.

(b) **Phloem** It is mainly involved in the conduction of food from leaves to stem, roots and other parts of plant. It consists of sieve tube elements, companion cells, phloem parenchyma and phloem fibres.

Tissue System

On the basis of their structure and location, there are three types of tissue system, i.e., epidermal tissue system, ground or fundamental tissue system and vascular or conducting tissue system.

1. Epidermal Tissue System

It forms the outermost covering of the whole plant body. Its various components are given below:

(i) **Epidermis** It is the outermost layer of the primary plant body. The epidermal cells vary in shape and size and are compactly arranged to form a continuous layer. It is usually single-layered, but is also multilayered in the aerial root of orchids and leaves of *Nerium* and *Ficus elastica.* The cells are parenchymatous and living. Each cells has a large central vacuole and a peripheral thin cytoplasm. In roots the outermost layer called **epiblema** has tubular, unicellular, projections called root hair.

(ii) **Stomata** The stomata (sing. stoma) are openings in the epidermis of most of the aerial parts of the plants, especially the leaves. Each stomata is composed of two bean-shaped cells called as guard cells, which enclose stomatal pore. The **guard cells,** are generally much smaller in size as compared to other epidermal cells.

The guard cells are surrounded by subsidiary cells or accessory cells which differ morphologically from the other epidermal cells. The guard cells walls have special elastic properties. Which help them to stretch laterally during stomatal opening.

(iii) **Epidermal appendages** (Outgrowths) The epidermis of most plants often bear outgrowth known as epidermal appendages or epidermal outgrowths.

They protect the plant against excessive loss of water and grazing. They also helps in climbing in some plants, e.g. rose.

(a) **Trichomes** The epidermal hairs present on the stem are called trichomes. These are epidermal outgrowths present temporarily or permanently on almost all plant parts.

(b) **Emergences** (Prickles) They are multicellular, stiff and sharp epidermal outgrowths containing some inner tissues.

2. Ground Tissue System

All tissues, except epidermis and vascular bundles constitute the ground tissue system. It mainly froms the bulk of the plant body. It's various components are:

(i) **Hypodermis** This is the region situated just below the epidermis and as on outer region of cortex. It is protective and mechanical in function.

(ii) **Cortex** The cortex lies between epidermis and endodermis consisting of parenchyma, collenchyma and sclerenchyma. The cortex is distinct in dicotyledons but not in monocotyledons. The cells of cortex contain starch grains, oil, tannis and crystals.

(iii) **Endodermis** This is the innermost layer of the cortex. It is single-latered, barrel-shaped and arranged without intercellular spaces. The cells are parenchymatous. The presence of bands of **suberin** on the redial and transverse wall is the characteristic feature. These bands are called **casparian strips**.

The endodermis helps to control the movement of water and air between the cortex and xylem. It also helps to maintain the root pressure and conducts water to the protoxylem.

(iv) **Pericycle** It is made up of a single layer or many layers of cells present between endodermis and vascular tissue. The pericycle is absent in roots and stems of some aquatic plants.

(v) **Medullary rays** The medullary rays are non-vascular areas which occur between vascular boundles in dicot stems for lateral conduction.

(vi) **Pith** The central portion of root and stem is occupied by pith. It performs the function of photosynthesis. The main function of pith is storage of water and food materials.

(vii) **Ground tissue of leaves** In leaves the ground tissue of petiole is made up of parenchymatous cells with distinct intercellular spaces. These cells are thin-walled and possess chloroplasts. The main function of mesophyll is in **photosynthesis**.

3. Vascular Tissue System

A vascular bundle is a strand of conducting tissue, which is generally composed of xylem and phloem in monocots and xylem, phloem and cambium in dicots. These tissues originate from the procambium and apical meristems.

On the basis of arrangement of xylem and phloem in the vascular bundles, there are three types of bundles:

(i) **Radial** The xylem and phloem alternate with each other separated by parenchymatous cells.

(ii) **Conjoint** The xylem and phloem are present together in the same bundle on the same radius. Conjoint bundles are of two types, i.e. collateral and bicollateral.

(iii) **Concentric** A vascular bundle in which one tissue is completely surrounded by the other is called concentric. The concentric bundles are of two types, i.e. amphivasal and amphicribal.

Chapter Exercise

1. Which of the following composes the hypodermis of a dicot stem?

(a) Parenchyma (b) Collenchyma
(c) Sclerenchyma (d) Aerenchyma

2. Leguminous seeds contain elongated (rod-shaped), columnar sclereids called

(a) macrosclereids (b) brachysclereids
(c) astrosclereids (d) osteosclereids

3. Simple mechanical tissue devoid of lignin is

(a) chlorenchyma (b) parenchyma
(c) collenchyma (d) sclerenchyma

4. Collenchyma differs from sclerenchyma in

(a) being dead
(b) possessing protoplasm even at maturity
(c) possessing lumen
(d) All of the above

5. The mature sclerenchyma fibre cells are

(a) live and active (b) live but inactive
(c) dead (d) All of these

6. Lignin is a component of the secondary cell walls of

(a) collenchyma (b) parenchyma
(c) sclerenchyma (d) aerenchyma

7. The parenchyma cells that make the aquatic plants light and bouyant helping them to float are

(a) prosenchyma (b) aerenchyma
(c) assimilatory parenchyma (d) stellate parenchyma

8. The chief function of parenchymatous tissue is to

(a) conduct water and minerals
(b) synthesise food
(c) store food material
(d) provide mechanical support

9. The thread-like cytoplasmic strands joining one cell to the other are called
(a) pneumatophores (b) cytodesmata
(c) plasmodesmata (d) plasmophores

10. Which of the following makes the framework of all plant organs?
(a) Chlorenchyma (b) Collenchyma
(c) Parenchyma (d) Sclerenchyma

11. The presence of intercellular spaces is the usual conspicuous feature of
(a) parenchyma (b) sclerenchyma
(c) collenchyma (d) Both (b) and (c)

12. Storage parenchyma contains various reserve materials, which are usually either dissolved in the cell sap or present in the form of discrete bodies. Which of the following statements is correct?
(a) Sugars, starch grains and amides in dissolved state, while nitrogenous substances, oil droplets as discrete bodies
(b) Oil droplets, amides and nitrogenous substances as descrete bodies, while starch grains, sugars and protein granules in dissolved state
(c) Sugars, amides and nitrogenous substances are in dissolved state, while starch grains, protein granules and oil droplets as discrete bodies
(d) All of the above

13. Star-shaped (stellate) sclereids are present in the leaves of
(a) *Cucurbita* (b) *Pyrus*
(c) *Cinnamomum* (d) *Nymphaea*

14. Stone cells is an alternative name of
(a) cortical fibres (b) sclereids
(c) xylary fibres (d) phloem fibres

15. The tissue that brings transport of organic food is
(a) xylem (b) vascular rays
(c) phloem (d) parenchyma

16. Tracheae, tracheids, fibre and parenchyma are the constituents of
(a) phellogen (b) phelloderm
(c) phloem (d) xylem

17. Parenchymatous cells are found in
(a) pulp of fruit (b) seeds
(c) skin of fruit (d) endocarp

18. Sieve tubes differ from the sieve cells because they
(a) lack nuclei
(b) are dead
(c) are shorter
(d) have sieve pores mainly at the end walls

19. Length of the plant is mainly increased by
(a) apical meristem
(b) lateral meristem
(c) periblem
(b) dermatogen

20. Tracheids are different from vessels because they have
(a) bordered pits
(b) thick wall
(c) spiral thickening
(d) discontinuous intercalary wall

21. Which is not considered as lateral meristem?
(a) Interfascicular cambium
(b) Intrafascicular cambium
(c) Phellogen
(d) Intercalary meristem

22. Endodermis is mainly helpful in
(a) providing protection
(b) preventing loss of water from stele
(b) maintain rigidity
(d) None of the above

23. Collenchyma is
(a) living and contains protoplasm
(b) dead and hollow
(c) dead and filled with reserve food
(d) living, containing no reserve food

24. Simple mechanical tissue devoid of lignin is
(a) chlorenchyma (b) parenchyma
(c) collenchyma (d) sclerenchyma

25. Vascular bundles are found to be completely absent in
(a) pteridophytes (b) angiosperms
(c) gymnosperms (d) bryophytes

26. What is the condition called, if the protoxylem is encircled by the metaxylem?
(a) Exarch (b) Endarch
(c) Polyarch (d) Mesarch

27. Healing of wound in plants is done by the activity of
(a) apical meristem
(b) lateral meristem
(c) primordial meristem
(d) intercalary meristem

28. Root hairs are usually
(a) acellular (b) unicellular
(c) bicellular (d) multicellular

29. Ground tissue is uniform lacking differentiation which occurs in
(a) dicot stem (b) dicot root
(c) monocot stem (d) monocot root

30. Triarch vascular bundles may be found in
(a) monocot stem (b) monocot roots
(c) dicot stem (d) dicot roots

31. Tissue responsible for the length of the plant
(a) apical meristem
(b) lateral meristem
(c) intercalary meristem
(d) epidermis meristem

32. Analyse the identifying features of meristematic tissues.
(a) Thick cellulose wall, small vacuoles, dense cytoplasm, small nuclei
(b) Thin cellulose wall, almost no vacuole, dense cytoplasm, prominent nuclei
(c) Thin cellulose wall, no vacuoles, sparse cytoplasm, prominent nuclei
(d) Thick cellulose wall, large vacuole, sparse cytoplasm, small nuclei

33. Which meristem helps in increasing girth?
(a) Lateral meristem (b) Intercalary meristem
(c) Primary meristem (d) Apical meristem

34. Tunica corpus theory is connected with
(a) root apex (b) root cap
(c) shoot apex (d) secondary growth

35. Cork is formed from
(a) cork cambium (phellogen) (b) vascular cambium
(c) phloem (d) xylem

36. Pith and cortex do not differentiate in
(a) monocot stem (b) dicot stem
(c) monocot root (d) dicot root

37. Cork cambium and vascular cambium are
(a) parts of secondary xylem and phloem
(b) parts of pericycle
(c) lateral meristem
(d) apical meristem

38. Pericycle of roots produces
(a) mechanical support (b) lateral roots
(c) vascular bundles (d) adventitious buds

39. Vascular cambium produces
(a) primary xylem and primary phloem
(b) secondary xylem and secondary phloem
(c) primary xylem and secondary phloem
(d) secondary xylem and primary phloem

40. A bicollateral vascular bundle is characterised by
(a) phloem being sandwitched between xylem
(b) transverse splitting of vascular bundle
(c) longitudinal splitting of vascular bundle
(d) xylem being sandwitched between phloem

41. Bordered pits are found in
(a) sieve cells (b) vessel wall
(c) companion cells (d) sieve tube wall

42. A narrow layer of thin walled cells found between phloem/bark and wood of a dicot is
(a) cork cambium (b) vascular cambium
(c) endodermis (d) pericycle

43. Periderm is produced by
(a) vascular cambium (b) fascicular cambium
(c) phellogen (d) intrafascicular cambium

44. Which of the following meristems is responsible for extrastelar secondary growth in dicotyledonous stem?
(a) Interfascicular cambium (b) Phellogen
(c) Intercalary meristem (d) Intrafascicular cambium

45. Passage cells are thin-walled cells found in
(a) phloem elements that serve as entry points for substance for transport to other plant parts
(b) testa of seeds to enable emergence of growing embryonic axis during seed germination
(c) central region of style through which the pollen tube grows towards the ovary
(d) endodermis of roots facilitating rapid transport of water from cortex to pericycle

46. Casperian strip is made up of
(a) suberin only (b) lignin only
(c) suberin lignin and cutin (d) cutin only

47. A gymnosperm producing vessels
(a) *Pinus* (b) *Cycas*
(c) *Ephedra* (d) *Ginkgo*

48. Safranin stain used for
(a) pyrenoids (b) starch
(c) cork tissue (d) lignified calls

49. Quiescent centre is found in
(a) root tip (b) shoot tip
(c) dicot stem (d) stem and leaf both

50. The presence of vessels and companion cells are characteristics of
(a) angiosperm (b) gymnosperm
(c) ferns (d) All of these

Answers

1. (b)	**2.** (a)	**3.** (c)	**4.** (b)	**5.** (c)	**6.** (c)	**7.** (b)	**8.** (c)	**9.** (c)	**10.** (c)
11. (a)	**12.** (c)	**13.** (d)	**14.** (b)	**15.** (c)	**16.** (d)	**17.** (b)	**18.** (d)	**19.** (a)	**20.** (a)
21. (d)	**22.** (b)	**23.** (a)	**24.** (c)	**25.** (d)	**26.** (d)	**27.** (b)	**28.** (b)	**29.** (b)	**30.** (d)
31. (a)	**32.** (b)	**33.** (a)	**34.** (c)	**35.** (a)	**36.** (a)	**37.** (c)	**38.** (b)	**39.** (b)	**40.** (d)
41. (b)	**42.** (b)	**43.** (c)	**44.** (b)	**45.** (d)	**46.** (c)	**47.** (c)	**48.** (d)	**49.** (a)	**50.** (a)

10

Histological Organisation of Root and Shoot Apex

Theories of Root Apex Organisation

Various theories has been given for the organisation of root apex meristem. These are as follows:

Apical Cell Theory

According to this theory (Nageli; 1858), the apical cell was interpreted as constant structure. It is considered as the functional unit of apical meristem governing the whole process of plant growth. The vascular cryptogams (pteridophytes) have solitary apical cell, e.g. horsetail, fern, *Selaginella, Dryopteris*, while gymnosperms have two groups of initials and angiosperms have three or rarely four groups of initials.

Korper-Kappe Theory

This theory (Schuepp; 1917) states that roots change its diameter during growth and there are various points at which single longitudinal file of cell becomes a double file of cell as a result of cell division. At this point, a cell first divides transversely and thereafter one of its daughter cells divides longitudinally. This is known as **T-division**.

Histogen Theory

According to this Theory (Hanstein; 1870), the root apex is comprised of three groups of initials such as **dermatogen** which gives rise to epiblema or piliferous layer, **periblem** which gives rise to cortex including endodermis and **plerome** which gives rise to vascular tissue.

Quiescent Centre Concept

According to this concept (Clowes; 1956), root apex has an inactive region which is compared low DNA, RNA and protein content and also acts as a reservoir of the active initials.

Theories of Shoot Apex Organisation

Different theories have been put forward to explain the mode of growth of the shoot apical meristems. Brief information of some important theories are as under :

Apical Cell Theory

This theory was given by Hofmeister (1857) supported by Nageli (in 1858). According to this, in most vascular cryptogams, there is a single primordial cell in the shoot apex. Hence, this theory is proved to be good for higher algal groups, pteridophytes, bryophytes, etc., as the higher vascular cryptogams does not have a single apical cell thus, the theory not applied to gymnosperms and angiosperms.

Histogen Theory

According to this theory (Hanstein; 1870), there are three distinct meristematic zones called dermatogen, periblem and plerome, in the angiospermic shoot apices.

Tunica Corpus Theory

This theory was postulated by Schmidt (1924) in order to describe thc shoot apical meristem. He states that the apices of shoot is comprised of two main regions/halves such as:

(i) **Tunica** It is a layer of shoot apex which is been comprised of one or more than one periclinal layers of cells that divide usually by the anticlinal division (with the walls perpendicular to the surface) resulting in the surface expansion of the shoot apex. If the tunica is more than one layered, the outermost layer forms the epidermis, while the remaining later layers help in contributing to form leaf primordia and the tissue of cortex by the change of the plane of their division.

(ii) **Corpus** It is a central massive region of the shoot apex which is surrounded by the tunica layers on the external side. Unlike tunica, the cells of this zone tends to divide in almost each and every plane giving rise to the part of the cortex, vascular tissue and the pith.

Cytohistological Zonation Theory

This theory was postulated by Foster (1938), while working on a gymnosperm, *Ginkgo biloba*. He recognised that the shoot apices has zone having different cytohistological characteristics. Later on, it was assumed that these zones are found in many angiosperms. Three different zones are named as apical initial zone, central mother cell zone and peripheral meristem zone.

A brief information about shoot apex development in following plants is given below:

(i) **Pteridophytes**

More than a century ago it was demonstrated that in many pteridophytes growth takes place by the activity of a conspicuous apical cell and its derivatives. Since, then many accounts of shoot apices have appeared and on histological basis the shoot apices (Wardlaw; 1946) of pteridophytes can be referred to one of the following types:

(a) A single conspicuous apical cell, to which all cell lineages can be traced (*Equisetum*).

(b) Several definite but rather less conspicuous cells (*Lycopodium*) in it can be seen as 'central meristem'. During dichotomy the central cells (cross-marked) cease to divide and from the main meristem differentiate two new meristems (arrow-marked).

(c) Inconspicuous but definitely superficial initial cells *Dryopteris, Matteuccia*.

(ii) **Gymnosperms**

It has several interrelated growth zone derived from the group of surface initials. The apical initials divide periclinally. Peripheral zone is seat of origin of leaf primordia. The pith cell is derived at base of mother cell.

In cycads, mother cell zone is ill defined. *Pinus* and conifers have less diversified zones than cycads and less well defined than *Ginkgo*. Gnetales have definite separation between surface layer and inner core derived from its own initials. *Ephedra* and *Gnetum* have been described as having tunica corpus pattern of growth.

(iii) **Angiosperms**

The stem apex of angiosperms have tunica corpus organisation. The dicotyledons have one to five layer tunica and monocotyledons have one to three layers tunica. In *Saccharum officinarum* tunica corpus organisation is not found.

Chapter Exercise

1. **According to which theory, the root and shoot apices are composed of three histogen layers namely a central plerome, outer dermatogen and underlying periblem?**

(a) Hanstein theory (b) Histogen theory
(c) Apical cell theory (d) Tunica corpus theory

2. **The cells of tunica undergo anticlinal divisions and give rise to**

(a) epidermis (b) cortex
(c) endodermis (d) pericycle

3. **The cells of corpus undergo divisions in various plants resulting in the formation of**

(a) ground meristem (b) procambium
(c) Both (a) and (b) (d) only cortex

4. **Root apical meristem (root apex) appears cup-shaped (roughly hemispherical) due to the presence of**

(a) root cap (b) root hairs
(c) quiescent centre (d) None of these

5. **Nageli (1858) has given which of the following theory of root apex organisation?**

(a) Korper-Kappe theory (b) Quiescent centre concept
(c) Apical cell theory (d) None of these

6. **Korper-Kappe theory was given by**

(a) Nageli (b) Schuepp
(c) Clowes (d) All of these

7. **Which of the following theories was given by Hanstein?**

(a) Quiescent centre concept
(b) Apical cell theory
(c) Histogen concept
(d) None of the above

8. **Quiescent centre concept theory was given by**

(a) Nageli (b) Schuepp
(c) Clowes (d) All of these

9. **Apical cell theory of shoot apex is given by Hofmeister and supported by**

(a) Clowes (b) Nageli
(c) Both (a) and (b) (d) None of these

10. **Histogen theory of shoot apex organisation was given by Hanstein in which of the following year?**

(a) 1856 (b) 1870
(c) 1921 (d) None of these

11. **Tunica corpus theory of shoot apex organisation was given by**

(a) Clowes (b) Schmidt
(c) Nageli (d) Both (a) and (b)

12. **Korper-Kappe theory of root apex organisation was given by Schuepp in which of the following year?**

(a) 1905 (b) 1917
(c) 1921 (d) 1936

13. Root apex has three group of initials, i.e. dermatogen, periblem and plerome by which of the following theory?
(a) Korper-Kappe theory
(b) Quiescent centre concept
(c) Histogen theory
(d) None of the above

14. According to apical cell theory, gymnosperms have two groups of initials. In angiosperms, the number of initials are
(a) 2 (b) 1
(c) 3 (d) 4

15. Quiescent centre concept of root apex organisation was given by Clowes in which of the following year?
(a) 1858 (b) 1956
(c) 1917 (d) 1870

16. Schmidt gave tunica corpus theory in which year?
(a) 1924 (b) 1985
(c) 1958 (d) 1870

17. Tunica corpus theory of apex organisation is related to
(a) root apex (b) shoot apex
(c) Both (a) and (b) (d) None of these

18. In Korper-Kappe theory, a cell first divides transversely and thereafter one of its daughter cell divides longitudinally. This is known as
(a) B-division (b) C-division
(c) Y-division (d) T-division

19. According to histogen theory of root apex, periblem gives rise to
(a) cortex (b) endodermis
(c) Both (a) and (b) (d) None of these

20. Which of the following layers gives rise to vascular tissue in root apex according to histogen theory?
(a) Dermatogen (b) Periblem
(c) Plerome (d) All of these

21. The apical meristem of the root is present
(a) only in radicals
(b) only in tap roots
(c) only in adventitious root
(d) in all roots

22. In a longitudinal section of a root, starting from the tip upward, the four zones occur in the following order
(a) root cap, cell division, cell enlargement, cell maturation
(b) root cap, cell division, cell maturation, cell enlargement
(c) cell division, root cap, cell enlargement, cell maturation
(d) cell division, cell enlargement, cell maturation, root cap

23. Root cap is formed by
(a) dermatogen (b) calyptrogen
(c) vascular cambium (d) periblem

24. In a dicotyledonous stem, the sequence of tissues from the outside to the inside is
(a) phellem -pericycle-endodermis-phloem
(b) phellem-phloem- endodermis- pericycle
(c) phellem-endodermis- pericycle- phloem
(d) pericycle-endodermis-phellem-phloem

25. The quiescent centre in root meristem serves as a
(a) site for storage of food which is utilized during maturation
(b) reservoir of growth hormones
(c) reserve for replenishment of damaged cells of the meristem
(d) region for absorption of water

26. Phellogen is known as
(a) periderm (b) cork cambium
(c) vascular cambium (d) None of these

27. Which of the following is not associated with Histogen theory?
(a) Plerome (b) Phelloderm
(c) Periblem (d) dermatogen

28. The root epidermis formed from
(a) plerome (b) dermatogen
(c) periblem (d) None of these

29. The histogen layer present at the apex of the root tip is called
(a) procambium (b) calyptrogen
(c) dermatogen (d) plerome

30. In a woody dicotyledonous tree, which of the following parts will mainly consists of primary tissues
(a) all parts (b) stem and root
(c) flower, fruit and leaves (d) shoot tips and root tips

Answers

1. (b)	**2.** (a)	**3.** (c)	**4.** (c)	**5.** (c)	**6.** (b)	**7.** (c)	**8.** (c)	**9.** (b)	**10.** (b)
11. (b)	**12.** (b)	**13.** (c)	**14.** (c)	**15.** (b)	**16.** (a)	**17.** (b)	**18.** (d)	**19.** (c)	**20.** (c)
21. (d)	**22.** (a)	**23.** (b)	**24.** (c)	**25.** (c)	**26.** (b)	**27.** (b)	**28.** (b)	**29.** (b)	**30.** (c)

11

Internal Structure of Root, Stem and Leaf

Primary Structure of Root

The primary root is the one which has only primary permanent tissues that are formed from the vegetative shoot apex. The secondary tissues are absent in this. The structural differences of a primary root in dicot and monocot plants are given below:

Dicot Root

Primary structure of a dicot root in dicot plant involves the following parts:

(i) **Epiblema** terms the outermost layer in the young root made up of epidermal cells which also forms epidermis hairs. Stomata and cuticle are not found in it. Some of the epiblema cells are prolonged forming a thin-walled tubular structures called root hairs. Cells responsible for producing root hair are called **root hair cells** or **trichoblasts**. Due to the presence of root hairs, epiblema is called piliferous layer and rhizodermis.

Root hair having pectose layer on the outside, this is to help them to pass into the soil spaces for absorption of water and mineral salts. The cell of older epiblema shrivel afterward and become cutinised and subrised.

(ii) **Cortex** is the structure that lies below the epiblema and is made up of thin-walled parenchymatous cells with conspicuous intercellular spaces. This is several layered thick.

(iii) **Endodermis** is the innermost layer of cortex comprises of a single layer of barrel-shaped cells without any intercellular space. They are coated with waxy and **suberin** material. The cells of endodermis bears a band of thickening along their radial and tangential walls known as **casparian strips** or **casparian bands.** The cells of endodermis lying opposite to the protoxylem are thin-walled to permit free passage called **passage cells**.

(iv) Tissue below the endodermis, pericycle, vascular bundles and pith constitute the **stele**.

(a) **Pericycle** is the next layer to the endoderm and is made up of thick-walled parenchymatous cells.

(b) The **vascular strand** consists of a separate bundles of xylem and phloem that are arranged alternately inner to the pericycle. On the basis of the xylem bundles, the root may be diarch, triarch, tetrarch, pentarch and polyarch having two, three, four, five and more than five xylem bundles, respectively. They show **exarch** xylem condition.

The phloem and xylem are separated from each other by one or more layers of small, thin-walled cells which later becomes meristematic and forms **vascular cambium**.

(c) **Pith** is normally absent in case of dicot roots, but if present it is small and is feebly developed consisting of parenchymatous cells.

Monocot Root

Structure of monocot plant can be inferred from the following structures:

(i) **Epiblema** It is an outermost, thin-walled, compactly arranged layer of cells out of which some cells gives rise to the root hairs. The root hair is a unicellular and lie in contact with soil water. It helps in the absorption of water and minerals.

(ii) **Cortex** It is a broad zone of parenchyma cells. They store food. It provides radial movement of water and minerals from epiblema to the root interior.

(iii) **Endodermis** A single layered structure made up of barrel-shaped cells which does not enclose intercellular spaces. The young endodermal cell possesses an internal strip of suberin and lignin called casparian strip.

The unthickened endodermal cells are called passage cells or **transfusion cells**. They help in the conduction of minerals and fluids from cortex into the xylem.

(iv) **Stele** It is constituted of pericycle, vascular bundles and pith.

(a) **Pericycle** It forms the outer boundary of stele. It may be either single or multilayered. It does not form cambium.

(b) **Vascular bundles/strand** It is present in the form of several alternate and radial xylem and phloem bundles. The xylem bundles are **exarch** and **polyarch**

(c) **Pith** It is large and well developed. It is large and made up of parenchymatous cells with intercellular spaces. These cells contain starch.

Primary Structure of Stem

The structural difference of primary stem in dicot and monocot plants are given below:

Dicot Stem

Primary structure of stem of dicot plant involves the following parts

(i) **Epidermis** is the outermost protective layer of the stem. It is usually covered with the cuticle. It bears trichomes and few stomata.

(ii) **Cortex** is the layer formed below the epidermis extending till the endodermis.

Cortex is further divided into three major zones

(a) **Hypodermis** An outer layer model consisting of collenchymatous cells. It provides mechanical strength to the stem.

(b) **Cortical cells** Which are located just below the hypodermis and consists of parenchymatous cells.

(c) **Endodermis** An innermost layer of the cortex filled with starch known as **starch sheath**. The general cortex in the form of single layer of barrel-shaped cells, surrounding the stele.

(iii) **Pericycle** is present just below the endodermis and above the vascular strand. It is made up of sclerenchymatous cells, lignified cell walls.

(iv) The **vascular strand** consists of many vascular bundles, arranged in the form of a ring around a central pith and inner to pericycle. Each vascular bundle consists of phloem, xylem with a strip of cambium, between the two.

(v) In between the vascular bundle **medullary rays** of parenchymatous cells are found. Medullary ray connect pith with pericycle and cortex. It takes part in radial conduction of materials, i.e. food, water, gases, etc.

The vascular bundles are arranged in ring which can be conjoint, collateral and open. Protoxylem is **endarch**.

(vi) Below the vascular bundles, large portion of parenchymatous cells is found and is called **pith or medulla**.

The pith is the central round part of the stem consisting of rounded, oval or polygonal parenchymatous cells.

Monocot Stem

The monocot stem possesses only primary structure. The different monocot stem from outside towards inside are consist of epidermis, hypodermis, ground tissue and vascular system.

(i) **Epidermis** It is single layered, having stomata in it. The cells have a thick cuticle layer on the outside.

(ii) **Hypodermis** It is 2-3 layered having **lignified.** Sclerenchymatous cells present just below the epidermis.

(iii) **Ground tissue** It fills the whole interior of the stem containing parenchymatous cells. A number of vascular bundles are scattered in it.

(iv) **Vascular system** Each vascular bundle in vascular strand is surrounded by a sheath of sclerenchyma known as **bundle sheath cells**. The vascular bundles possess both phloem and xylem. So, these are conjoint type and also posses collateral, closed and endarch vascular bundle.

Primary Structure of a Leaf

Depending upon the internal primary structure, the leaves of angiospermic plant are generally differentiated into dorsiventral (bifacial) and isobilateral (equifacial) leaves. The structural differences of a primary leaf in dicot and monocot plants are given below.

Dicot Leaf

In dicot plant the leaves are dorsiventral and are usually horizontal with the sunlight falling on their upper surface.

Primary structure of dicot leaf involves the following parts:

(i) **Epidermis** is found on both upper (adaxial) and the lower surfaces (abaxial) of the leaf. **Cuticle** is found on epidermis and stomata are more in number at the lower surface. Both these surfaces are bounded by upper and lower epidermis.

(a) **Upper epidermis** is made up of single layered parenchymatous cells but is some times multilayered.

(b) **Lower epidermis** the stomata and chloroplast are absent in the lower epidermis except guard cells of stomata.

(ii) **Mesophyll** is the tissue found between the upper and lower epidermis. It is made up of parenchymatous cells and are of following two types:

(a) **Upper palisade parenchyma** which are made up of elongated cells and parallel to each other. It contains abundant chloroptasts.

(b) **Lower spongy parenchyma** which are found between the palisade and lower epidermis. The cells of spongy parenchyma are almost oral and irregularly arranged. So, the intercellular space formed which helps in diffusion of gases.

(iii) The **vascular system** is found in veins and veinlets. Vascular bundles are surrounded by a layer of bundles sheath. They are conjoint, collateral, endarch and closed. The vascular bundles are easily seen in the veins and midrib. The size of vascular bundles varies with the size of the veins of the leaf.

Monocot Leaf

Monocot leaves are generally isobilateral in origin but in some plants these are dorsiventral in nature. The leaves usually show parallel venation.

Various components are:

(i) **Epidermis** Monocot leaf has upper and lower epidermis made up of single layered, compactly arranged cells. The upper layer is covered with cuticle. Both upper and lower layers of epidermis contain stomata. At some places the epidermis bears a large sized, thin-walled, colourless protruding cells known as **bulliform** or **motor cells.**

(ii) **Mesophyll** It is made up of chlorenchymatous ground tissue of leaves. The mesophyll is not differentiated between palisade and spongy parenchyma. The intercellular spaces are connected with the two substomatal cavities.

(iii) **Vascular bundle** These are present in large number, out of which some are small while the other are big.

The patches of sclerenchymatous cells are present below and above the larger bundle. The vascular bundles are conjoint, collateral, endarch and closed. Xylem and phloem is found towards the upper and lower epidermis respectively.

(iv) **Midrib** It is considered as the most widest part of the leaf. The adaxial surface of leaf contains a shallow groove while, on the abaxial surface, broad ridge is found.

Chapter Exercise

1. Undifferentiated ground tissue characteristically occurs in

(a) *Zea mays* stem (b) *Cucurbita* stem
(c) *Helianthus* stem (d) *Helianthus* root

2. Which of the following characteristics differentiates monocot root from a dicot root?

(a) Radial vascular bundles
(b) Less developed large pith
(c) Scattered vascular bundles
(d) Open vascular bundles

3. The guard cells of the stomata present in the epidermis of dicot stems are

(a) dumb-bell shaped (b) elliptical
(c) rounded (d) kidney-shaped

4. Sclerenchymatous hypodermis is found in

(a) monocot roots (b) dicot roots
(c) dicot stems (d) monocot stems

5. Casparian strips are the deposition of

(a) cellulose (b) chitin (c) suberin (d) cutin

6. The oldest layer of secondary phloem in a stem lies

(a) outward to vascular cambium
(b) inward to vascular cambium
(c) inward to primary xylem
(d) outward to primary phloem

7. In an isobilateral (monocot) leaf, the stomata are

(a) usually absent or less abundant in upper epidermis
(b) usually absent or less abundant in lower epidermis
(c) almost equally distributed in upper and lower epidermis
(d) numerous in lower epidermis

8. Phloem parenchyma is absent in the phloem elements of

(a) dicots (b) monocots
(c) only dicot stem (d) only monocot stem

9. Bulliform cells occur in the epidermis of

(a) dicot leaves (b) dicot stems
(c) monocot leaves (d) monocot stems

10. Root hairs originate from

(a) pericycle (b) epidermis (c) endodermis (d) exodermis

11. Find out the correct sequence of labelling.

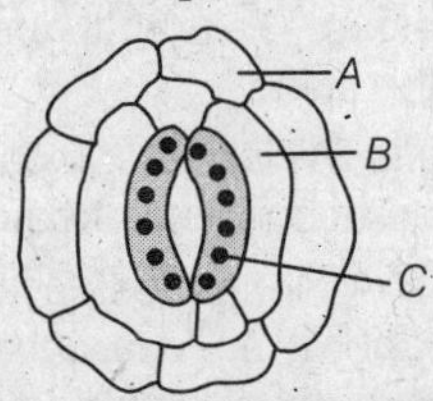

(a) A–Guard cells, B–Subsidiary cells, C–Epidermal cells
(b) A–Epidermal cells, B–Guard cells, C–Subsidiary cells
(c) A–Epidermal cells, B–Accessory cells, C–Subsidiary cells
(d) A–Epidermal cells, B–Subsidiary cells, C–Guard cells

12. Vascular bundle is enclosed within a well-developed sclerenchymatous sheath in

(a) monocot stem (b) dicot stem
(c) monocot root (d) dicot root

13. Y-shaped arrangement of xylem vessels is found in

(a) monocot stem (b) dicot stem
(c) monocot root (d) dicot root

14. When the vascular cambium is present between the xylem and phloem, the vascular bundle is called

(a) closed (b) open (c) endarch (d) exarch

15. Scattered vascular bundles are the feature of

(a) monocot stem (b) dicot stem
(c) monocot leaf (d) dicot leaf

16. Vascular bundles are completely absent in

(a) angiosperms (b) gymnosperms
(c) pteridophytes (d) bryophytes

17. Polyarch vascular bundles generally occurs in
(a) dicot stem (b) monocot stem
(c) dicot roots (d) monocot roots

18. Tissue below the endodermis in dicot root, constitute
(a) cortex (b) pith
(c) hypodermis (d) stele

19. Cortex in the stem of dicot plant made up of hypodermis, endodermis and
(a) epidermis (b) pith
(c) cortical cells (d) None of these

20. Centrally located pith is absent in
(a) monocot stem (b) dicot stem
(c) Both (a) and (b) (d) None of these

21. Cortex is known to be lying between
(a) epidermis and endodermis (b) endodermis and pith
(c) endodermis and stele (d) hypodermis and endodermis

22. The vascular bundles in dicot stems are
(a) arranged in a ring
(b) scattered
(c) amphicribral
(d) xylem and phloem situated alternately

23. When phloem and cambium are present on both sides of xylem, the vascular bundle is called
(a) collateral (b) bicollateral
(c) cocentric (d) radial

24. The oldest layer of the secondary phloem in a stem lies
(a) outward to the primary phloem
(b) outward to vascular cambium
(c) inward to primary phloem
(d) inward to vascular cambium

25. A monocot stem with secondary growth is
(a) *Lilium* (b) *Cocos*
(c) *Asparagus* (d) *Yucca*

26. Interfascicular cambium develops from the cells of
(a) xylem parenchyma (b) endodermis
(c) pericycle (d) medullary rays

27. Anatomically fairly old dicotyledonous root is distinguished from the dicotyledonous stem by
(a) the absence of secondary xylem
(b) the absence of secondary phloem
(c) the presence of cortex
(d) position of protoxylem

28. Cork tissue arises from
(a) phelloderm (b) periderm
(c) phellogen (d) phellem

29. The vascular bundles in a typical leaf are
(a) collateral and open (b) conjoint and concentric
(c) collateral and closed (d) bicollateral

30. Leaf of a bean plant does not possess
(a) lenticels (b) stomata
(c) phloem (d) guard cells

31. Which of the following statements is not correct about the arrangement of tissues in the dicot root?
(a) The specialised root tissues are seen at the zone of maturation
(b) The epidermis of the root has no cuticle on its outer surface, unlike other epidermal tissues
(c) The pericycle is the outer layer of cells in the vascular cylinder, with the ability to form secondary or branch roots
(d) The star-shaped phloem is located in the centre of the vascular cylinder, with xylem arranged between the arms of phloem

32. Of the following, which is not characteristic of a monocotyledon plant?
(a) Vascular tissue in a ring in the root
(b) Vascular bundles scattered in the stem
(c) Netted venation of leaf vascular bundles
(d) Flower parts in multiples of trees
(e) All of the above

33. Conjoint, collateral and closed vascular bundles are found in the stem of
(a) sunflower (b) grass
(c) *Cucurbita* (d) *Aristolochia*

34. An anatomical cross section with two to six exarch xylem bundles, small pith, cambium, cork cambium in the pericycle is the characteristic of
(a) dicot root (b) dicot stem
(c) monocot root (d) monocot stem

35. In monocot stems, the vascular bundles are
(a) radial (b) open collateral
(c) bicollateral (d) closed collateral

36. The hypodermis in monocot stem is
(a) parenchymatous
(b) collenchymatous
(c) sclerenchymatous
(d) collenchymatous

Answers

1. (a)	**2.** (d)	**3.** (d)	**4.** (d)	**5.** (c)	**6.** (a)	**7.** (c)	**8.** (b)	**9.** (c)	**10.** (b)
11. (d)	**12.** (a)	**13.** (a)	**14.** (b)	**15.** (a)	**16.** (d)	**17.** (d)	**18.** (d)	**19.** (c)	**20.** (a)
21. (d)	**22.** (a)	**23.** (b)	**24.** (c)	**25.** (a)	**26.** (d)	**27.** (d)	**28.** (c)	**29.** (c)	**30.** (a)
31. (d)	**32.** (c)	**33.** (d)	**34.** (a)	**35.** (d)	**36.** (c)				

12

Secondary Growth in Root and Stem of Dicot (Sunflower)

Secondary Growth

The secondary growth can be defined as the growth that increases the girth (diameter) and thickness of the plant as a result of the activities of lateral meristems, i.e. vascular cambium and cork cambium, respectively. It occurs mainly in gymnosperms and dicots such as trees and shrubs and provides strength to bear the increasing weight of the aerial growth.

Secondary Growth in Dicot Roots

The secondary growth in roots is the thickness due to the formation of secondary tissues by the lateral meristems. The secondary growth in dicot roots takes place both in **stelar region** (by vascular cambium) and in **extrastelar region** (by cork cambium), but it is clearly different in the manner of the cambium formation and tissue differentiation.

1. Formation of Vascular Cambium

The vascular cambium developed in these roots is completely secondary in origin. The pericycle cells lying above the protoxylem also become meristematic and develop another strip of cambium. The cambial ring passes below the phloem bundles and above the xylem bundles. The entire, circular ring of cambium becomes active and produces secondary xylem towards the inner side and secondary phloem towards the outer side. The primary xylem bundles remain in their original position.

The primary phloem, endodermis and cortex are normally crushed due to the enormous increase in the volume of the secondary tissue.

2. Formation of a Cork Cambium

The outer layers of pericycle become meristematic, giving rise to the thin-walled, rectangular cells, which functions as **cork cambium** or **phellogen**. The later produces secondary tissue outwardly that constitutes the **cork** or **phellem** and the **secondary cortex** or **phelloderm** towards the inner side.

The tissue of endodermis, general cortex and epiblema disorganise and are sloughed off. Thus, the cork or phellem becomes the outermost protective layer. Lenticels may also develop in the roots here and there during extrastelar secondary growth.

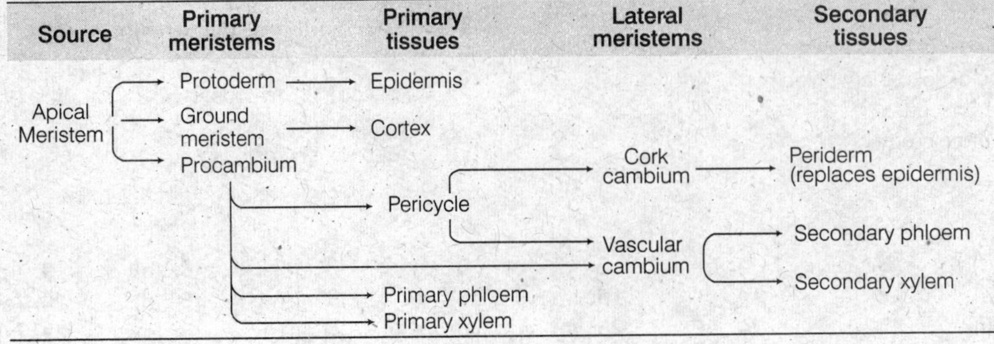

Summary of primary and secondary growth during the development of a root from a vascular plant

Secondary Growth in Dicot Stems

The secondary growth in a typical dicot stem takes place in two different regions, i.e. stellar and extrastellar region or in the stele and in the cortex region. The dicot stem in its primary state of growth contains narrow layers of intrafascicular cambium in between the xylem and phloem. The tissues involved in the secondary growth are the two lateral meristems, i.e. vascular and cork cambium.

1. Formation of Vascular Cambium

It is known as the meristematic layer being responsible for cutting off the vascular tissues such as the xylem and the phloem.

Formation of cambium ring Vascular cambium becomes meristematic and forms new strips of secondary meristems called interfascicular cambium. The interfascicular strips join the strips of fascicular cambium or both the sides to form a complete ring called **cambium ring**.

Activity of cambial ring The cambial ring becomes active and begins to form the new cell both towards the inner and the outer sides. It is made up of two types of cells, i.e. ray initials and fusiform initials.

The cells of the cambium ring divide periclinally and results in new cells both outside and inside. Cambial cells adjacent to ray initials cut off parenchyma cells extending radially through the new secondary xylem and phloem. These are called secondary the a **medullary rays**. The cambium is generally considered to be more active on the innerside than the outerside. Due to which amount of the secondary xylem produced is more as compared to the amount of secondary phloem produced soon forming a compact mass.

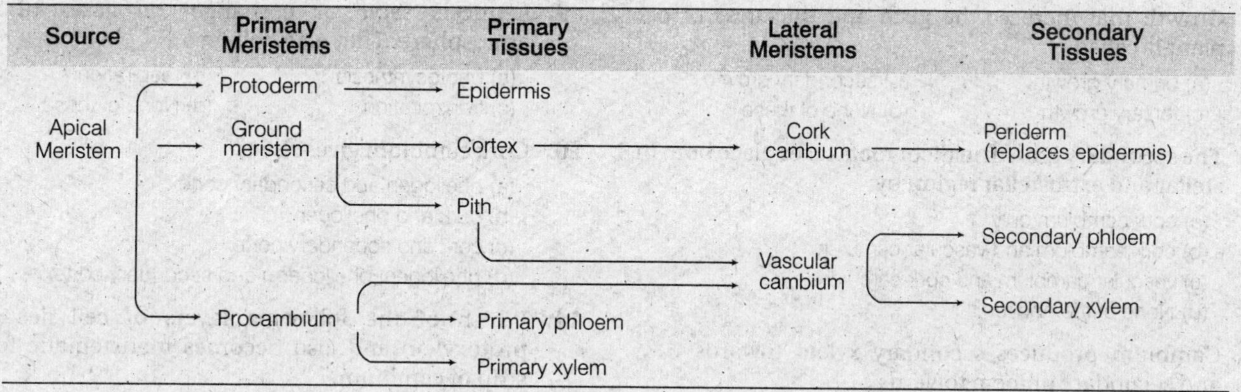

Summary of primary and secondary growth during the development of a stem from a vascular plant

2. Formation of Cork Cambium

The permanent living cells in the outer cortical region (sometimes pericycle also) become meristematic giving rise to a secondary lateral meristem called as **cork cambium** or **phellogen**.

Cork is impervious to water due to suberin and provides protection to underlying tissues. The cells cut-off from the cork cambium towards inside add to the cortex and are called secondary cortex cells or phelloderm cells. However, the phellem, phellogen and phelloderm are together called as **periderm**.

Bark

The bark includes all the tissues lying outside the xylem (wood) including phellem, phellogen, phelloderm, crushed primary cortex, secondary and primary phloem. The outer layers of the bark constantly peel off due to the addition of new secondary vascular tissues in the interior.

It is protective in nature which protects the phloem from external injuries. When bark is removed, phloem gets damaged first and root dies due to the scarcity of starch which provides energy.

Lenticels

These are the aerating pores, found on the surface of the bark as raised scars representing a small portion of periderm, where the phellogen activity is more than elsewhere. Lenticels serve the purpose of exchanging the gases between internal tissues and atmosphere and help in transpiration.

Difference between Secondary Growth in Dicot Stem and Dicot Root

Secondary growth in dicot stem	Secondary growth in dicot root
The vascular cambial ring develops as a circular ring since beginning.	The vascular cambial ring is wavy in the beginning and becomes circular later on.
The vascular cambial ring is completed by both the primary (intrafascicular) strip and secondary (interfascicular) strip.	The vascular cambial ring is secondary in origin as it is formed of conjunctive parenchyma and pericycle that become meristematic.
The vascular rays are narrow since beginning.	The vascular rays are wider in the beginning.
Annual rings or growth rings occur quite commonly.	Annual rings or growth rings are absent.
Outward layer of the cortex give birth to the cork cambium (phellogen).	Generally, the pericycle gives birth to cork cambium (phellogen).

Chapter Exercise

1. Growth that increase the girth and thickness of the plant is called

(a) primary growth (b) secondary growth
(c) tertiary growth (d) None of these

2. The secondary growth in dicot roots takes place both in stellar and extrastellar region by

(a) cork cambium only
(b) cork cambium and vascular cambium
(c) vascular cambium and cork cambium
(d) None of the above

3. Cambium produces secondary xylem towards ...... and secondary phloem towards ....

(a) outside and inside
(b) inside and outside
(c) outside only
(d) None of the above

4. Cork cambium is also called

(a) phellem (b) phellogen
(c) phelloderm (d) All of these

5. Vascular cambium becomes meristematic in dicot stem and form new secondary meristem called

(a) primary cambium (b) tertiary cambium
(c) interfascicular cambium (d) None of these

6. Cells of cambium in dicot stem divides

(a) anticlinally (b) periclinally
(c) Both (a) and (b) (d) None of these

7. Phellem, phellogen and phelloderm are together called

(a) cork (b) cork cambium
(c) vascular bundles (d) periderm

8. Aerating pores found in the bark of stem of dicot plants are called

(a) stomata (b) plasmodesmata
(c) guard cells (d) lenticels

9. Lenticels serve or exchange the gases to and from atmosphere. They also help in

(a) photosynthesis (b) respiration
(c) transpiration (d) None of these

10. Cork cambium gives rise to

(a) phellogen and secondary cortex
(b) cork and phellogen
(c) cork and secondary cortex
(d) phellogen, phelloderm and secondary cortex

11. Which of the following layers of cell lies above the protoxylem and also becomes meristematic forming the strip of cambium?

(a) Hypodermis (b) Endodermis
(c) Pericycle (d) None of these

12. During secondary growth, the tissue of endodermis, general cortex become

(a) disorganised (b) sloughed off
(c) Both (a) and (b) (d) None of these

13. The vascular cambial ring develop as a circular ring since beginning in dicot stem but in dicot root it is

(a) circular in beginning
(b) wavy in beginning and circular later on
(c) circular in beginning and wavy later on
(d) None of the above

14. The vascular rays are narrow in dicot stem, while in dicot root they are

(a) narrow in the beginning
(b) wider in the beginning
(c) narrow always
(d) None of the above

15. When bark is removed, phloem gets damaged first and root dies due to

(a) photosynthesis (b) respiration
(c) starvation (d) guttation

16. **Secondary growth occurs in**
(a) bryophytes (b) gymnosperms
(c) angiosperms (d) Both (b) and (c)

17. **Secondary growth occurs in stellar region and**
(a) non-stellar region (b) extrastellar region
(c) Both (a) and (b) (d) None of these

18. **Outward layers of the cortex give birth to cork cambium in dicot stem while in dicot roots .... gives birth to cork cambium.**
(a) endodermis (b) xylem
(c) pericycle (d) Both (a) and (b)

19. **The thickness produced by secondary growth is due to the following type of activity of cambium**
(a) more division of cambium inside
(b) more division of cambium outside
(c) equal division of cambium on both sides
(d) None of the above

20. ***Quercus suber*** **is economically important due to the extraction/production of**
(a) antibiotics (b) single cell protein
(c) commercial cork (d) All of these

21. **Abnormal secondary growth is seen in**
(a) *Triticum* (b) *Cucurbita*
(c) *Dracaena* (d) *Helianthus*

22. **The cells of cambial ring are known as**
(a) ray initials (b) interfascicular initials
(c) fusiform initials (d) Both (a) and (c)

23. **Lenticels are formed due to the rupture of**
(a) sclerenchymatous cells (b) parenchymatous cells
(c) Both (a) and (b) (d) collenchymatous cells

24. **The components of a vascular bundle develops from...... of the primary meristem.**
(a) procambial strands (b) paravascular fibres
(c) vascular cryptogams (d) monocot roots

25. **Medullary vascular bundles commonly occur in the members of the family**
(a) Brassicaceae (b) Malvaceae
(c) Amaranthaceae (d) Cucurbitaceae

26. **For a critical study of secondary growth in plants, which one of the following pairs is suitable?**
(a) Wheat and maiden hair fern
(b) Sugarcane and sunflower
(c) Teak and pine
(d) Deodar and fern

27. **Interfascicular cambium is a**
(a) type of protoderm
(b) primordial meristem
(b) primary meristematic tissue
(d) secondary meristematic tissue

28. **Identity the plant parts whose transverse sections show a clear and prominent pith.**
(a) Dicot stem and monocot stem
(b) Dicot stem and dicot root
(c) Dicot root and monocot root
(d) Dicot stem and monocot root

29. **Which one of the following is the correct sequence of tissues present in dicot stem during secondary growth?**
(a) Primary cortex, secondary cortex, phellogen, cork
(b) Cork, primary cortex, secondary cortex, phellogen
(c) Phellogen, cork, primary cortex, secondary cortex
(d) Secondary cortex, cork, phellogen primary cortex

30. **In a dicotyledonous stem, the sequence of tissues from the outside to the inside is**
(a) phellem–phloem–endodermis–phloem
(b) phellem–pericycle–endodermis–phloem
(c) phellem–endodermis–pericycle–phloem
(d) phellem–endodermis–pericycle–phloem

31. **Which of the following gives rise to procambium?**
(a) Vascular cambium (b) Cork cambium
(b) Primary vascular bundle (d) Both (a) and (c)

32. **Vascular bundle in root takes place its origin in**
(a) pericycle (b) conjunctive parenchyma
(c) Both (a) and (b) (d) None of these

33. **Secondary growth in dicot stem is the activity of**
(a) xylem (b) cork cambium
(c) vascular cambium (d) cambium ring

34. **Cork cambium and vascular cambium are the examples of**
(a) lateral meristem (b) intercalary meristem
(b) apical meristem (d) wound tissue

35. **Vascular tissue in flower develops from**
(a) periblem (b) dermatogen
(c) phellogen (d) plerome

36. **Which one of the following is the primary lateral meristem?**
(a) Interfascicular cambium (b) Intrafascicular cambium
(c) Cork cambium (d) Phellogen

37. **Old dicotyledonous root is distinguished from from dicotyledonous stem by**
(a) the absence of secondary phloem
(b) the presence of cortex
(c) position of protoxylem
(d) the absence of secondary xylem

38. **Which of the following pair is an example of lateral meristem?**
(a) Procambium and phelloderm
(b) Interfascicular cambium and phellem
(c) Phellogen and phelloderm
(d) Phellogen and fascicular cambium

39. Youngest layer of secondary xylem in wood of dicot plants is located between
(a) pith and primary xylem
(b) just outside vascular cambium
(c) just inside vascular cambium
(d) just inside cork cambium

40. Between the bark and wood in a woody stem, there is a layer of meristem called
(a) cork cambium (b) apical meristem
(c) vascular cambium (d) zone of cell division

41. Cork cambium, cork and cortex are collectively called
(a) phelloderm (b) phellogen
(c) periderm (d) phellem

42. Which of the following is not a lateral meristem?
(a) Intrafascicular cambium
(b) Interfascicular cambium
(c) Phellogen
(d) Intercalary meristem

43. During secondary growth in dicot root, phellogen form .... towards outside.
(a) periderm (b) secondary vascular tissue
(c) cork cambium (d) phellem

44. In dicot roots, cork cambium originates in
(a) pith (b) pericycle
(c) epidermis (d) hypodermis

45. Secondavy growth absent in
(a) *Funaria* (b) Sun flower
(c) *Mangifera* (d) Cycas

46. In dicot stem secondary growth occurs due to activity of
(a) apical meristem
(b) laternal meristem
(c) intercalary meristem
(d) None of the above

47. The term wood refers to
(a) primary xylem (b) secondary xylem
(c) protoxylem (d) metaxylem

48. The difference between alburnum and duramen is the presence in the former of
(a) companion cell (b) living cell
(c) race cell (d) lighnified cell

49. Heartwood is
(a) outer part of secondary xylem
(b) outer part of secondary phloem
(b) inner part of secondary xylem
(d) inner part of secondary phloem

50. Secondary phloem remains functional generally
(a) for one year
(b) for less then one year
(c) for many year
(d) as long as plant alive

Answers

1. (b)	**2.** (c)	**3.** (b)	**4.** (b)	**5.** (c)	**6.** (b)	**7.** (d)	**8.** (d)	**9.** (c)	**10.** (c)
11. (c)	**12.** (c)	**13.** (b)	**14.** (b)	**15.** (c)	**16.** (d)	**17.** (b)	**18.** (c)	**19.** (c)	**20.** (c)
21. (c)	**22.** (d)	**23.** (b)	**24.** (a)	**25.** (c)	**26.** (c)	**27.** (d)	**28.** (d)	**29.** (a)	**30.** (a)
31. (c)	**32.** (c)	**33.** (d)	**34.** (a)	**35.** (d)	**36.** (b)	**37.** (c)	**38.** (d)	**39.** (c)	**40.** (a)
41. (c)	**42.** (c)	**43.** (d)	**44.** (b)	**45.** (a)	**46.** (b)	**47.** (b)	**48.** (b)	**49.** (c)	**50.** (d)

13

Wood Anatomy

Growth Ring

The activity of the cambium ring is influenced by the climatic changes. In woody species, the period of growth and reproduction alternate with the period of inactivity during winter.

Reactivation of the cambium takes place during spring season. However, the cambium becomes more active during this season and forms plenty of xylem vessels with wider cavities known as **spring wood** or **early wood**.

In winter, however, the cambial activity slows down and gives rise to narrower xylem elements. The wood thus formed in winter is called **autumn wood** or **late wood**.

As growth continues, these two types of wood are produced year after year and appear together as a concentric ring in transverse section. This concentric ring is known as the **annual ring** or **growth ring**. It represents the total secondary xylem or wood formed in one year. Hence, by counting the number of annual rings, the age of a plant can be easily determined. Annual rings also provide the information of the climatic conditions prevailing in the past.

Dendrochronology or **tree dating** is the scientific method of dating based on the analysis of pattern of tree rings or growth rings.

Difference between Spring Wood and Autumn Wood

Spring Wood	Autumn Wood
It is also called early wood.	It is also called late wood.
It is formed during spring season.	It is formed during winter season.
It constitutes the major part of the annual ring.	It constitutes as a narrow strip in the annual ring.
Spring wood is present in the beginning of an annual ring.	Autumn wood is present at the end of an annual ring.
Forms plenty of xylem vessels with wider cavities.	The cavities of xylem vessels are narrower.
Xylem fibres are fewer in number.	Abundant xylem fibres are produced.
Wood is lighter in colour.	Wood is darker in colour.

Sapwood and Heartwood

As a result of continued secondary growth in subsequent years, the older part of secondary xylem or wood becomes non-functional as it loses the power of conduction. The cells of this wood get filled with resins or tannins produced by the adjacent functional cells. Sometimes, activities of vessels become blocked by a bladder-like in growths from the neighbouring parenchyma cells called as **tyloses**.

Due to these activities, the non-functional secondary xylem becomes hard, durable and blackish in colour known as **heartwood** or **duramen**. It fails to conduct water but provides the mechanical support to the stem.

The function of secondary xylem (water and mineral conduction from roots) in these wood can be performed by the outer younger-rings of secondary xylem, known as **sapwood** or **alburnum**. It is involved mainly in the conduction of water and minerals from root to leaf.

With the passage of time and addition of new outer rings of secondary xylem, more rings of the sapwood (alburnum) changes into heartwood (duramen). This is the reason due to which the heartwood increases in diameter year after year but the sapwood remains almost of the same thickness.

Difference between Sapwood and Heartwood

Sapwood	Heartwood
It is also known as alburnum.	It is also known as duramen.
It represents the outward wood of the plant.	It represents the central wood of the plant.
It is light in colour and lighter in weight.	It is dark in colour and heavier in weight.
It consists of living cells.	Living cells absent.
It represents functional part of the secondary xylem (wood).	It represents non-functional part of the secondary xylem (wood).
Tracheids and vessels are not plugged by tyloses.	Tracheids and vessels are plugged by tyloses.
Tannins, resins, gummy substances are not deposited in tracheary elements.	Tannins, resins, gummy substances are deposited in tracheary elements.
It is not durable.	It is durable.

Softwood and Hardwood

Softwood chiefly consists of tracheids. It is also called **non-porous wood** (as it lacks vessels) and is found in gymnosperms. The most durable softwood is that of *Cedrus deodara*. In angiosperms, however, the wood possess both tracheids and vessels is called as **hardwood**.

Difference between Softwood and Hardwood

Softwood	Hardwood
Gymnospermous wood is the softwood.	Angiospermous wood is the hardwood:
It chiefly constitutes tracheids.	It consists of both tracheids and vessels.
Since, the vessels are absent, the wood is called non-porous wood.	Due to the presence of vessels, it is called porous wood.
Tracheid percentage ranges about 90-95%.	Tracheid percentage ranges only less than 5%.
Xylem fibres fewer.	Xylem fibres are plenty.

Chapter Exercise

1. The activity of the cambium ring is influenced by
(a) transpiration (b) climatic changes
(c) respiration (d) None of these

2. Reactivation of cambium takes place during
(a) winter season (b) spring season
(c) Both (a) and (b) (d) None of these

3. The cambium becomes more active during spring season and forms plenty of xylem vessels with wide cavities called
(a) spring wood (b) autumn wood
(c) Both (a) and (b) (d) None of these

4. Spring wood is called as
(a) autumn wood (b) early wood
(c) late wood (d) None of these

5. Autumn wood is formed in winter and also called as
(a) early wood (b) late wood
(c) spring wood (d) None of these

6. Concentric rings in transverse section of stem formed is autumn and winter season are called as
(a) transverse rings (b) parallel rings
(c) growth rings (d) All of these

7. Scientific study of tree rings or growth rings are called as
(a) ornithology (b) parasitology
(c) dendrochronology (d) cytology

8. Major part of annual rings is formed by
(a) autumn wood (b) spring wood
(c) Both (a) and (b) (d) None of these

9. Autumn wood is present at the .... of annual ring.
(a) beginning (b) end
(c) middle (d) None of these

10. Xylem fibres are fewer in numbers in
(a) autumn wood (b) spring wood
(c) dark (d) All of these

11. Dark colour wood is found in
(a) autumn wood (b) spring wood
(c) Both (a) and (b) (d) None of these

12. The cells of this wood get filled with resins or tannins
(a) heartwood (b) sapwood
(c) spring wood (d) All of these

13. Heartwood is also known as
(a) sapwood (b) alburnum
(c) duramen (d) None of these

14. Alburnum is the another name of
(a) sapwood (b) duramen
(c) heartwood (d) None of these

15. Functional secondary xylem is found in
(a) duramen (b) alburnum
(c) spring wood (d) All of these

16. Which of the following is lighter in colour and weight?
(a) Autumn wood (b) Heartwood
(c) Sapwood (d) None of these

17. Tracheids and vessels are not plugged by tyloses in
(a) heart wood (b) sapwood
(c) Both (a) and (b) (d) None of these

18. Which of the following wood is durable?
(a) Heartwood (b) Softwood
(c) Sapwood (d) None of these

19. Gymnosperm wood is
(a) hardwood (b) softwood
(c) Both (a) and (b) (d) None of these

20. Tracheids percentage is 90-95% in
(a) hardwood (b) softwood
(c) Both (a) and (b) (d) None of these

21. The age of tree by counting the annual rings is called
(a) dendrochronology (b) dendrology
(c) chronology (d) countrology

22. The light coloured part in an annual ring is
(a) heartwood (b) latewood
(c) earlywood (d) sapwood

23. The sapwood is converted into heartwood by
(a) the formation of tyloses
(b) deposition of resins, oils and gums
(c) degeneration of protoplast of living cells
(d) All of the above

24. Living cells are not present in
(a) periderm (b) sapwood
(c) heartwood (d) bark

25. Cambium is most active during
(a) winters (b) rainy season
(c) snow areas (d) summers

26. What does heartwood provides to the stems?
(a) Mechanical support
(b) Conduction of water
(c) Transport of minerals
(d) Strength

27. Accumulation of organic materials like tannis resins oils and gums in the innermost layer of stem provides which colour to the secondary xylem?
(a) Red (b) Dark brown
(c) Brown (d) Orange

28. Age of plant can be determined by
(a) potometer (b) sphygmometer
(c) increment boreer (d) None of these

29. The new xylary elements have
(a) narrow vessels (b) wide vessels
(c) deep vessels (d) bread vessels

30. Which region in heartwood is light in colour?
(a) Innermost region (b) Outer most region
(c) Middle region (d) Peripheral region

31. Cork cambium results in the formation of cork which becomes impermeable to water due to the accumulation of
(a) resins (b) suberin
(c) lignins (d) tanins

32. One of the following undergoes decay when exposed to air
(a) sapwood (b) hardwood
(c) heartwood (d) light wood

33. In an old tree water is conducted through
(a) heartwood (b) hardwood
(c) sapwood (d) softwood

34. Which of the following plant product is the hardest?
(a) Lignin (b) Cutin
(c) Suberin (d) Sporopollenin

35. Which one of the following is resistant to enzyme action?
(a) Cork (b) Wood fibre
(c) Pollen exine (d) Leaf cuticle

36. Duramen is present in
(a) inner region of secondary wood
(b) part of sapwood
(c) outer region of secondary wood
(d) region of pericycle

37. Growth rings are either absent or are not sharply demorcated in
(a) temperate deciduous trees
(b) tropical evergreen
(c) temperate hot regions
(d) tropical deciduous

38. The vascular cambial ring of a dicot stem is
(a) primary in origin
(b) secondary in origin
(c) tertiary in origin
(d) partly primary and partially secondary in origin

39. Closing layers of lenticels may show deposition of
(a) cutin (b) pectin
(c) lignin (d) suberin

40. Which of the following monocot shows anamalous secondary growth?
(a) Maize (b) Wheat (c) *Dracena* (d) Rice

Answers

1. (b)	**2.** (b)	**3.** (a)	**4.** (b)	**5.** (b)	**6.** (c)	**7.** (c)	**8.** (b)	**9.** (b)	**10.** (b)
11. (a)	**12.** (a)	**13.** (c)	**14.** (a)	**15.** (b)	**16.** (c)	**17.** (b)	**18.** (a)	**19.** (b)	**20.** (b)
21. (a)	**22.** (c)	**23.** (d)	**24.** (c)	**25.** (d)	**26.** (a)	**27.** (b)	**28.** (c)	**29.** (a)	**30.** (d)
31. (b)	**32.** (a)	**33.** (c)	**34.** (d)	**35.** (c)	**36.** (a)	**37.** (b)	**38.** (d)	**39.** (d)	**40.** (c)

14

Development of Male Gametophytes

Angiosperm plants produce flowers, very beautiful and intricate structures, within which their reproductive development takes place. In flowering plants, as in other groups of plants, a diploid, spore producing generation alternates with a haploid, gamete producing generation (gametophyte), unlike, some of the more primitive plants, the male and female gametophytes of angiosperms are reduced to microscopic structures.

Male Reproductive Organ

Androecium is the third whorl of floral appendages that arises inner to corolla. It's individual appendage called stamen that represents the male reproductive organ. Each stamen consists of an anther and a filament.

A typical stamen which is the male reproductive unit of angiosperm consists of a long slender stalk known as the **filament** and the terminal bilobed structure called an **anther.**

Structure of an Anther

Usually anther is a bilobed structure whose each lobe is it self made up of two chambers or theca hence, assumed as a four-chambered structure. Sometimes an anther may also contain only one lobe and two microspqrangia. Each chamber of an anther develops a microsporangia in which microspores develops.

Structure of a Microsporangium

A typical microsporangium is surrounded by the four different walls, i.e. epidermis, endothecium, middle layers and tapetum.

Out of which the first three of the outer layers providing protection and performing other functions and the last one, i.e. tapetum, which is the innermost layer which provides nourishment to the developing pollen grains.

Based on the behaviour pattern, tapetum is of two main types, i.e. **amoeboid** (invasive or periplasmodial) which breaks down the inner radial wall of its cells, due to which the protoplast moves to the anther cavity and fuses further to form tapetal part plasmodial and **secretory** (parietal or glandular) which remains in their original position throughout the development of a microspore. The sporogenous tissue lies in the center of the sporangia and undergoes meiotic division to form the microspore tetrads.

Development of a Pollen Grain (Microsporogenesis)

The cells which are capable of developing microspore cells are called **Pollen Mother Cells** (PMCs). Microsporogenesis is the phenomenon of formation of microspore from pollen mother cells.

- After meiosis, a tetrad of four microspore is formed which gets separated when the water content of anther becomes low. Each microspore develops into pollen grains.

Pollen Grains (Male Gametophyte)

Pollen grain is regarded as the male gametophyte of angiosperms. These are generally round, oval and 10-200 micrometer in diameter. The wall of each pollen grain (called sporoderm) is two layered, i.e. outer and the inner layer.

The inner wall of pollen grain is **intine** and it is cellulosic in nature. While, its outer layer is hard and called **exine**. It is made up of **sporopollenin**, the hardest known material that can resist high temperature and acidic or basic conditions. The exine also contains pores called germ pores where sporopollenin is absent.

Mature pollen grain contains two cells, i.e. the **vegetative cell** and the **generative cell.** Out of which the vegetative cell is bigger and has an abundant food reserve and a large irregularly shaped nucleus. The generative cell is small and floats in the cytoplasm of the vegetative cell. It is spindle-shaped with dense cytoplasm and nucleus. At this two celled stage, the pollen grain shed from anther. In some angiosperm, the pollen grain shed at three-celled stage where the generated cell divides into two male gametes.

Pollen kitt is an oily layer made up of lipid and carotenoids being found inside the mature pollen grains of insect pollinated species. It is contributed by the tapetum layer of an anther.

Pollen grains are rich in nutrients. The viability of pollen grains often depends upon the temperature and humidity. It may vary from species to species.

It is possible to store pollen grains of a large number of species for many years in a liquid nitrogen (– 196°C) which can be further used as the pollen banks in several crop breeding programmes.

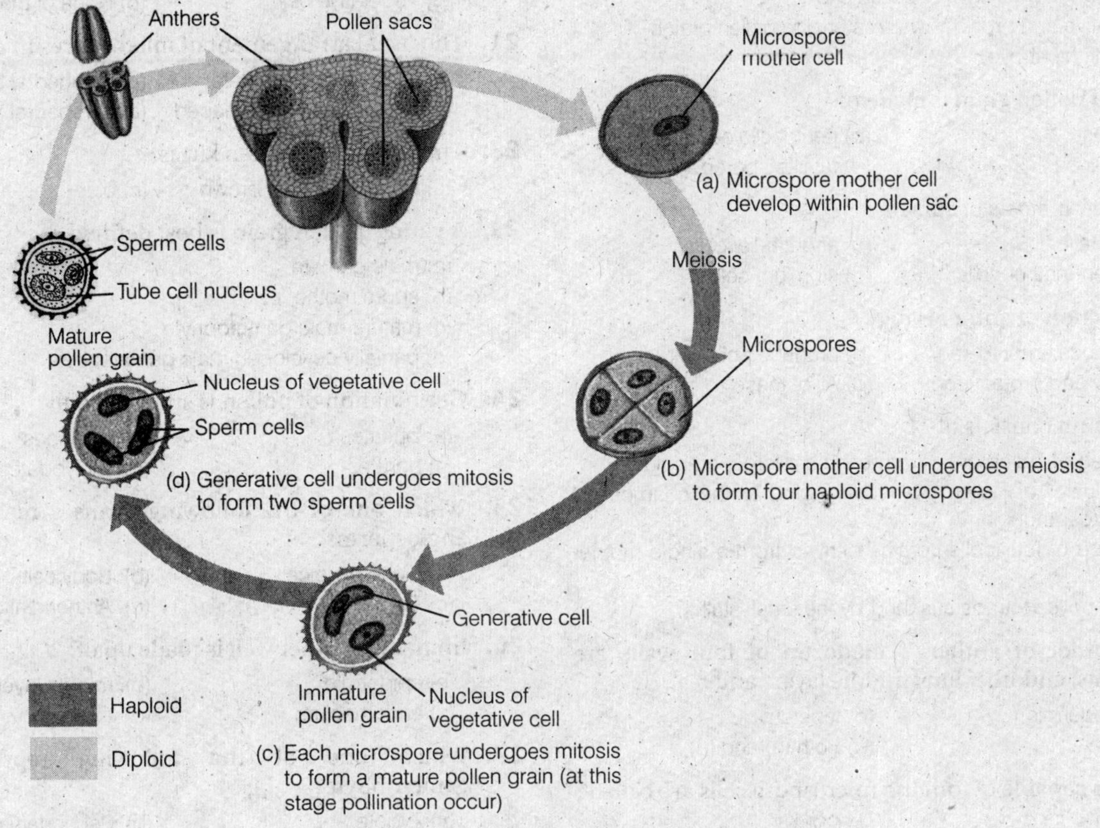

Pollen development in flowering plants

Chapter Exercise

1. Exine of pollen grain is made up of

(a) pectocellulose (b) lignocellulose
(c) sporopollenin (d) pollen kitt

2. In flower, androecium is the male reproductive organ. It is a group of stamen that consist of filament and

(a) embryo sac (b) nucellus
(c) stamen (d) None of these

3. Microscopic structure in a flower that contains the polar nuclei is

(a) polar tube (b) pollen sac
(c) embryo sac (d) male gametophyte

4. Each anther is made up of two lobes and each lobe is divided into two chambers. These chambers are also called as

(a) chalaza (b) funicle
(c) theca (d) None of these

5. The male gametes of angiosperms are

(a) uniflagellate
(b) biflagellate
(c) non-motile as there is no flagellated stage in the life cycle of angiosperms
(d) None of the above

6. Which of the following is male gametophyte in angiosperms?

(a) Embryo sac (b) Pollen grain
(c) Stamen (d) Male gamete

7. Tapetum is

(a) parietal in origin
(b) innermost wall layer of pollen sac
(c) nutritive and provides wall material to pollen grains
(d) All of the above

8. Intine of pollen grain is made of

(a) callose (b) pectocellulose
(c) cellulose (d) fat-like sporopollenin

9. In anther, meiosis occurs in

(a) tapetal cells (b) endothelial cells
(c) pollen mother cells (d) stomium cells

10. The function of anther is to

(a) produce embryo sac (b) produce pollen grains
(c) store and protect eggs (d) All of these

11. A pollinium consists of

(a) a bag of pollen grains formed in a microsporangium
(b) a cluster of pollen grains belonging to the chamber of microsporangium
(c) group of four pollen grains derived from a single mother cell
(d) two pollen tetrads attached by the small stalks

12. Each corner of anther is made up of four walls are epidermis, endothecium, middle layers and

(a) hypodermis (b) tapetum
(c) theca (d) Both (a) and (b)

13. The cells capable of forming microspore cells are called

(a) PMC (b) CMC
(c) DMC (d) None of these

14. Each microspore develops into

(a) embryo sac (b) pollen grain
(c) pollen sac (d) egg cell

15. The general diameter of pollen grains is

(a) 100-150 nm (b) 25-50 nm
(c) 10-20 μm (d) None of these

16. Mature pollen grain contains two cells, first one is the vegetative cell and the second one is

(a) tube cell (b) synergids
(c) generative cell (d) Both (b) and (c)

17. Which of the following is the hardest material?

(a) Collagen (b) Sporopollenin
(c) Cellulose (d) None of these

18. In most cases, the pollen grain shed at the stage of

(a) two cells (b) five cells
(c) four cells (d) All of these

19. Which of the following cells generates male gametes?

(a) Vegetative cell (b) Tube cell
(c) Generative cell (d) Both (a) and (b)

20. Which of the following cells is involved in the formation of pollen tube?

(a) Pollen grain (b) Generative cell
(c) Vegetative cell (d) None of these

21. The usual arrangement of microspores in a tetrad is

(a) tetrahedral and bilateral (b) tetrahedral and linear
(c) tetrahedral and T-shaped (d) tetrahedral and decussate

22. The colour of a pollen kitt is

(a) red (b) brown (c) blue (d) yellow

23. A young pollen grain is best defined as

(a) male gamete
(b) spore mother cell
(c) mature male gametophyte
(d) partially developed male gametophyte

24. Germination of pollen is stimulated by

(a) zincates (b) chlorides
(c) borates (d) molybdates

25. Which one of the following forms a male gametes in angiosperms?

(a) Generative cell (b) Body cell
(c) Tube cell (d) Antheridial cell

26. Immature anther wall is made up of

(a) epidermis (b) middle layers
(c) endothecium (d) tapetum

27. Which one of the following represents male gametophyte?

(a) Zygote (b) Secondary nucleus
(c) Pollen grain (d) Antipodals

28. What does a microspore mother cell undergoes, during the formation of pollen grains?

(a) One mitotic division
(b) One meiotic division
(c) One meiotic and one mitotic division
(d) One meiotic and two mitotic divisions

29. Outer layer of a pollen grain is made up of

(a) cellulose (b) lignin
(c) sporopollenin (d) chitin

30. The most resistant material found in the pollen grain is

(a) cuticle (b) intine
(c) pectocellulose (d) sporopollenin

31. A microspore is a

(a) male gamete
(b) first cell of male gametophyte
(c) last cell of male gametophyte
(d) diploid cell

32. In a fully developed male gametophyte, number of nuclei is
(a) one (b) two (c) three (d) four

33. Male gametophyte of an angiosperms is shed at
(a) four-celled pollen grains
(b) three-celled pollen grains
(c) microspore mother cell
(d) anther

34. Male gametophyte of angiosperms/monocots is
(a) nucellus (b) microspore
(c) microsporangium (d) stamen

35. In an angiospermic plant, how many microspore mother cells are required to produce 100 pollen grains?
(a) 75 (b) 100
(c) 25 (d) 50

36. In angiosperm, all the four microspores of tetrad are covered by a layer which is formed by
(a) pectocellulose (b) callose
(c) cellulose (d) sporopollenin

37. Male gametes in angiosperms are formed by the division of
(a) generative cell (b) microspore mother cell
(c) vegetative cell (d) microspore

38. Development of male gametophyte begins
(a) after pollination (b) before pollination
(c) on the stigma (d) in the embryo sac

39. 2 to 3 celled male gametophyte, starts its further development after pollination
(a) in the style (b) in the ovary
(c) on the stigma (d) in the ovule

40. Which part of male gametophyte, disintigrates before fertilisation?
(a) Generative nucleus (b) Tube nucleus
(c) Male gamete (d) Germ pore

41. Pollen grains preserved for long periods in fossils due
(a) cellulose (b) sporopollenin
(c) carotenoids (d) ubisch bodies

42. Ubisch bodies are provided by
(a) tapetum (b) pollen kitt
(c) exine (d) intine

43. Smallest pollens are in
(a) *Cucurbita* (b) *Nyctoanthes*
(c) *Myosotis* (d) *Hyacinthus*

44. Which forms maximum proportion in chemical composition of pollens?
(a) Proteins (b) Fats
(c) Carbohydrates (d) Water

45. 'Nemec phenomenon' relates to
(a) 7 celled female gemetophyte
(b) 8 nucleated male gemetophyte
(c) formation of pollen kitt
(d) None of the above

Answers

1. (c)	**2.** (d)	**3.** (c)	**4.** (c)	**5.** (c)	**6.** (b)	**7.** (d)	**8.** (c)	**9.** (c)	**10.** (b)
11. (a)	**12.** (b)	**13.** (a)	**14.** (b)	**15.** (b)	**16.** (c)	**17.** (b)	**18.** (a)	**19.** (c)	**20.** (c)
21. (a)	**22.** (d)	**23.** (d)	**24.** (a)	**25.** (a)	**26.** (a)	**27.** (c)	**28.** (b)	**29.** (c)	**30.** (d)
31. (b)	**32.** (c)	**33.** (b)	**34.** (b)	**35.** (c)	**36.** (a)	**37.** (a)	**38.** (b)	**39.** (c)	**40.** (b)
41. (b)	**42.** (a)	**43.** (c)	**44.** (c)	**45.** (b)					

15

Development of Female Gametophyte

Female Gametophyte

(Structure and Type of Ovule)

Gynoecium represents the female gametophyte, which is critical to many steps of the angiosperm reproductive process, including pollen grain development, the induction of seed development upon fertilisation and maternal control of seed development. The female gametophyte plays an essentially critical role in every step of the reproductive process.

The basic unit of it is **pistil**. Pistil may be either syncarpous, (i.e. more than one carpel are fused together) or may be apocarpous (carpels remains free) the pistil is made up of three main components.

These are:

(i) **Stigma**, which is the leading point of the pollen grain where the pollen grains gets lodged during pollination.

(ii) **Style**, which is a long part leading to the ovary.

(iii) **Ovary**, which is a basal swollen part being divided into various ovarian cavity or locule inside it.

In the locule, attached with placenta, inside the ovary, there lies the megasporangia which is commonly called **ovule** that may be one or more than one in number.

Structure of an Ovule (Megasporangium)

An ovule is often found attached with placenta through a small **funicle**. The region where both placenta and funicle fuses together is called **hilum** that later becomes scar on the seed.

Each and every ovule is composed of one or two **integuments** (protective layers) that encircled the ovule excepting the tip area, where no integuments are present. This end is known as **micropyle.** Opposite to the micropyle end there is another basal part of an ovule called **chalazal end.**

Within the integument, a group of cells is also found that are rich in the reserve food material and is called as **nucellus,** inside which the formation of a female gametophyte or embryo sac takes place.

Types of Ovules

On the basis of the relative position of funiculus, chalaza and micropyle, the ovules are classified into following six main types:

(i) **Atropous or Orthotropous** It is the simplest and the most primitive type of an ovule, in which micropyle, chalaza and funiculus lies in one vertical plane, e.g. all gymnosperms, *Piper nigrum, P. betel and Rumex polygonum.*

(ii) **Anatropous or Inverted** It is the most common type of ovule, occurring in more than 80% of angiospermic families. The body of the ovule is rotated by 180° and micropyle comes near to the funiculus. Chalaza and micropyle are in straight line but hilum is at 90° to the micropyle.

(iii) **Campylotropous** In this particular type, the body of an ovule is more or less at right angle to the funicle and micropyle bends downwards, e.g. Chenopodiaceae (*Chenopodium*), Capparidaceae (*Capparis*).

(iv) **Amphitropous** In this type, the body of an embryo becomes curved and looks like an anatropous ovule but the distinction lies in the embryo sac which also get curved and takes the shape of the **horse-shoe structure,** e.g. Butamaceae (*Butamus*), Loganiaceae, Alismaceae.

(v) **Hemianatropous** In this type, the body of an ovule is turned through 90°. The micropyle and chalaza in a straight line but funiculus forms 90° angle with both these two, e.g. Primulaceae and Ranunculaceae.

(vi) **Circinotropous** In this type, the length of the funiculus gets increased and covers the whole ovule, e.g. Cactaceae (*Phyllocactus, Opuntia*) and Plumbaginaceae (*Plumbago*).

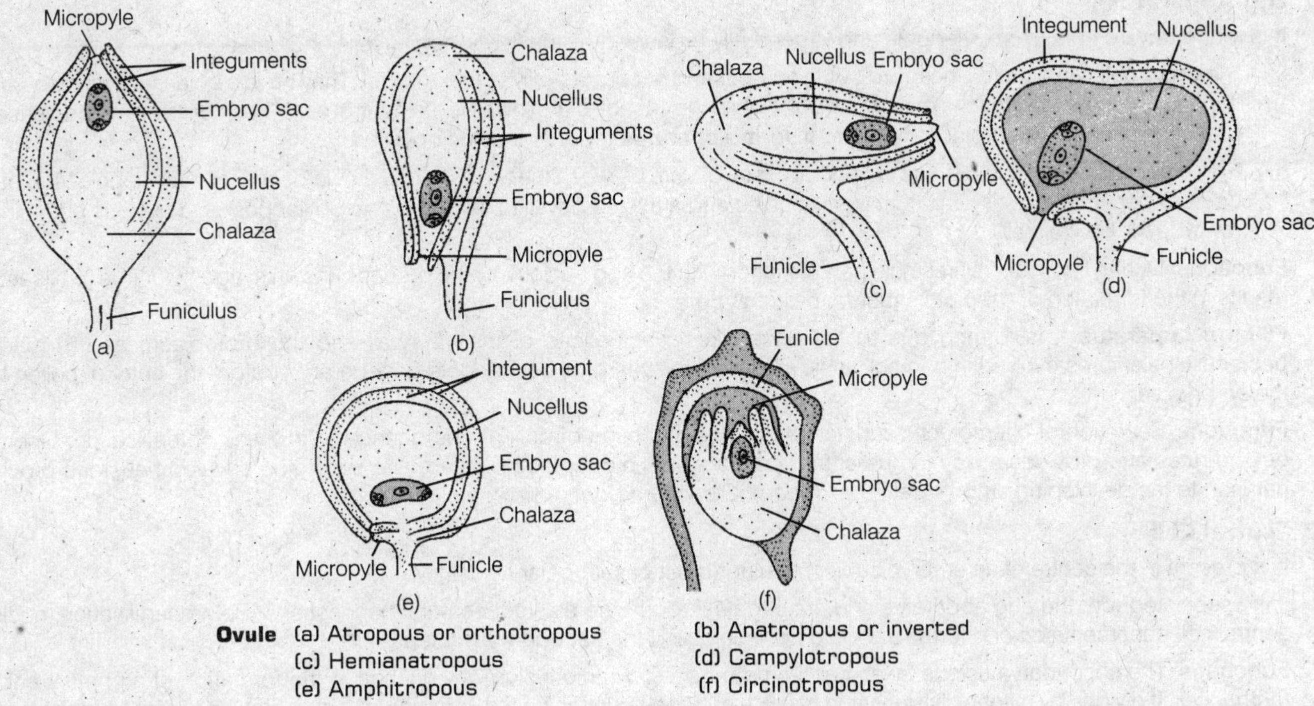

Ovule (a) Atropous or orthotropous (b) Anatropous or inverted (c) Hemianatropous (d) Campylotropous (e) Amphitropous (f) Circinotropous

Megasporogenesis and Development of Female Gametophyte

The process of formation or development of megaspore from Megaspore Mother Cell (MMC) or megasporocyte is called **megasporogenesis**. It occurs within the nucleus of the developing ovule of an angiosperms.

The development of an egg takes place inside the ovule. During this process, a single diploid megaspore mother cell matures within the ovule. Through first meiosis, it gives rise to a dyad cell after which the second meiosis occurs forming the four megaspores (haploid). In most plants, only one of these megaspore survives, however the rest are absorbed by the ovule itself.

The nucellus of the functional megaspore is called **functional megaspore**, which divides three times mitotically thereby, producing the eight haploid nucleus.

The eight nuclei are arranged in precise positions within the embryo sac such as one nuclei which is located near the opening of embryo sac and is termed as the **egg cell**.

Two nuclei are located in a single cell in the middle of the embryo sac and are called **polar nuclei**. The other two nuclei contained in cells are called **synergids** flanking the egg cell, while the other three nuclei being residing in the cells known as **antipodals** which are located at the end of the embryo sac, (i.e. opposite to the egg cell.)

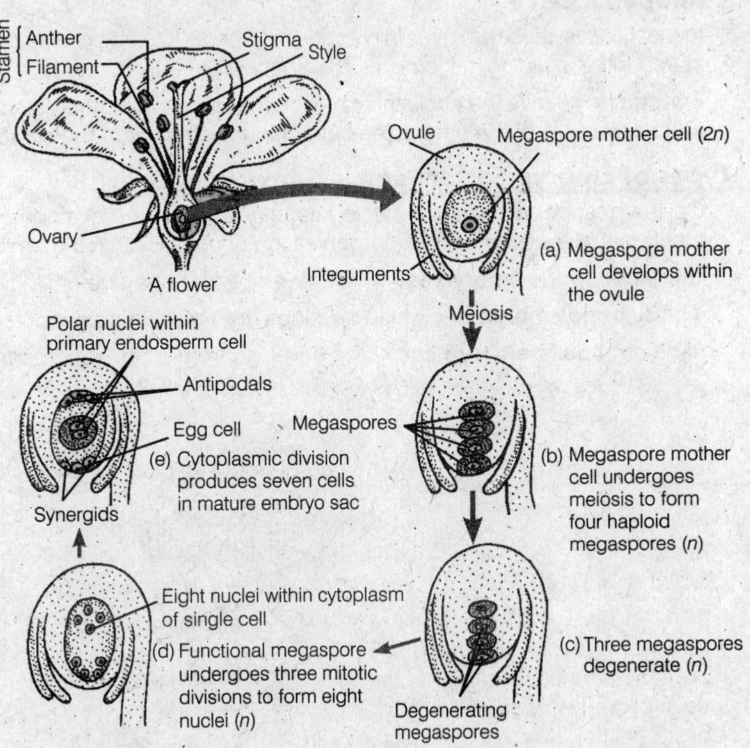

Various stages in the development of female gametophyte

Organisation of Mature Embryo Sac (Female Gametophyte)

In most cases, the mature embryo sac contains seven cells and eight nuclei. It consists of following main components:

1. **Egg Apparatus**

It is found towards the micropylar pole and is generally organised by the two synergid cells and one egg cell.

(i) **Synergid cell** In general, these are two in number and occurs laterally to the egg cell. They have a polarised protoplasm in which the nuclei occurs towards the micropylar pole, while the vacuoles are towards the central cell. These cells are characterised by the presence of finger-like projections called **filiform apparatus** towards the micropylar pole.

(ii) **Egg cell** It is one in number and in general cases occurs between the two synergids towards the micropylar pole. At maturity, the egg cell becomes polarised. It contains vacuoles and a thick cell wall towards the micropylar pole and nucleus and thin cell wall towards the central cell.

Functions During fertilisation, first male gamete fuses with the egg nucleus by the process known syngamy or true fertilisation which results in the formation of a diploid structure called **zygote**.

Filiform apparatus It is formed due to the finger-like invaginations of the cell wall into the protoplasm and is made up of pectocellulosic fibres evolved in non-cellulosic sheath. The synergids usually gets degenerated before the entry of pollen tube into the embryo sac.

Functions Secretion of chemotactic substances for the attraction of pollen tube during the process of fertilisation. One synergid degenerates and provides a way for the entry of pollen tube into the embryo sac. The synergids develop haustoria providing the nutrition to the developing embryo sac, e.g. *Sedam, Quinchamalium, Cotula,* etc.

2. **Central Cell**

It is present in the centre of an embryo sac containing (most cases) polar nuclei.

It is associated with the egg apparatus and antipodals through the plasmodesmatal connections. Before fertilisation, nuclei of the central cell become fused and forms a diploid nucleus called the **secondary nucleus**.

Functions The secondary nucleus takes part in the process of triple fusion which leads to the formation of a primary endosperm mother cell. It divides by mitotic divisions and forms a nutritive structure called endosperm which provides nutrition to the developing embryo.

3. **Antipodal Cell**

In most cases, these are three in number and occurs towards chalazal pole of the embryo sac. Their protoplasm contains a rich amount of starch, protein, lipids, ascorbic acid, oxidases, etc.

Functions It limits the chalazal expression and transfer nutrients and growth hormones to developing embryo sac. It is to be noted that all the cells of the embryo sac are uninucleate except the central cell which is binucleate.

Types of Embryo Sac or Female Gametophyte

On the basis of megaspore nuclei, contributing in the development of an embryo sac, number of mitotic divisions occurring in functional megaspore and the organisation of nuclei within the embryo sac. They are of following three main types:

(i) **Monosporic embryo sac** which develops from the one or single megaspore.

(ii) **Bisporic embryo sac** which develops from the two megaspores.

(iii) **Tetrasporic embryo sac** which develops from all the four megaspores.

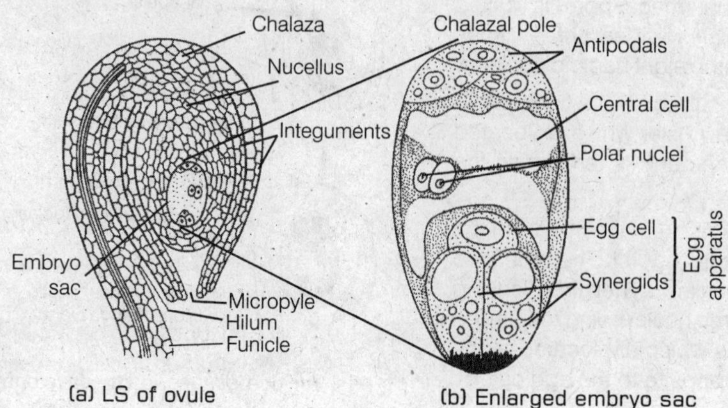

(a) LS of ovule (b) Enlarged embryo sac

Chapter Exercise

1. In most of the plants, the micropyle is formed due to absence of
   (a) outer integument
   (b) inner integument
   (c) Both (a) and (c)
   (d) None of the above

2. Which of the following is a diploid?
   (a) Egg (b) Synergids
   (c) Antipodals (d) Secondary nucleus

3. The type of megaspore tetrad most commonly formed in angiosperms is
   (a) tetrahedral (b) linear
   (c) T-shaped (d) decussate

4. Filiform apparatus is formed by
   (a) synergids (b) antipodals
   (c) egg cell (d) tapetum

5. Meiotic division in an ovule takes place in
   (a) nucellus (b) megaspore mother cell
   (c) megaspore (d) archesporium

6. The functional megaspore undergoes which of the following simultaneous divisions to form mature embryo sac?
   (a) Three meiosis (b) Two meiosis
   (c) Three mitosis (d) One mitosis

7. The embryo sac occurs within the
   (a) embryonic axis (b) embryo
   (c) ovule (d) endosperm

8. The point at which funiculus touches the ovule is
   (a) chalaza (b) hilum
   (c) raphe (d) endothelium

9. Which of the following nucleus is unlike other nuclei in the female gametophyte of angiosperms?
   (a) Egg nucleus (b) Nucleus of synergids
   (c) Secondary nucleus (d) Nuclei of antipodals

10. Opposite to micropyle, there is a basal part of ovule which is called
    (a) funicle (b) chalaza
    (c) hilum (d) None of these

11. Embryo sac in angiosperms represents
    (a) megagametophyte (b) megaspore
    (c) megasporophyll (d) megagamete

12. At the time of fertilisation, the female gametophyte is
    (a) 5-celled (b) 6-celled
    (c) 7-celled (d) 8-celled

13. Ovary is divided in various ovarian cavities called
    (a) placenta (b) locule
    (c) embryo sac (d) None of these

14. In which of the following parts of ovule reserve food is stored?
    (a) Integuments (b) Micropyle
    (c) Nucellus (d) All of these

15. The protective covering of ovule is
    (a) chalaza (b) micropyle
    (c) integuments (d) None of these

16. In atropous or orthotropous ovule, micropyle and chalaza lies
    (a) 90° to each other (b) parallel to each other
    (c) in a same verticle plane (d) None of these

17. Micropyle and funicle are parallel and very close to each other in which of the following ovule?
    (a) Amphitropous (b) Anatropous
    (c) Atropous (d) Campylotropous

18. Two cells that flanked the egg cells are called
    (a) antipodal cell (b) synergids
    (c) secondary nucleus (d) micropyle

19. Which of the following pairs of plant parts are both haploid in nature?
    (a) Nucellus and primary endosperm cell
    (b) Antipodal cells and egg cell
    (c) Nucellus and antipodal cells
    (d) Antipodal cells and megaspore mother cell

20. The endothelium is derived from
    (a) outer integument (b) inner integument
    (c) nucellus (d) funicle

21. Campylotropous ovules are found in
    (a) *Pisum* (b) *Yucca*
    (c) *Abelmoschus* (d) *Dahlia*

22. Ovules in the family Solanaceae are
    (a) orthotropous, bitegmic
    (b) anatropous, bitegmic
    (c) orthotropous, unitegmic
    (d) anatropous, unitegmic

23. In most of the plants, the micropyle is formed by
    (a) outer integument (b) inner integument
    (c) Both (a) and (b) (d) None of these

24. When body of the ovule lies at the right angles to the funicle and the micropyle lies close to the later, it is called
    (a) anatropous (b) amphitropous
    (c) campylotropous (d) hemianatropous

25. The largest cell of the embryo sac is
(a) antipodal (b) central cell
(c) egg cell (d) synergid

26. In angiosperms, the female gametophyte is
(a) egg apparatus (b) synergid
(c) embryo (d) embryo sac

27. The space between the outer and the inner integument in the region of micropyle is called
(a) exostome (b) peristome
(c) hypostome (d) endostome

28. Filiform apparatus is formed by
(a) plasma cell (b) egg cell
(c) synergids (d) antipodals

29. The ovule in a typical angiosperm is equals to
(a) megasporophyll
(b) megasporangium
(c) megaspore mother cell
(d) megaspore

30. Female gametophyte of a typical dicot plant at the time of fertilisation is
(a) 4-celled (b) 6-celled
(c) 7-celled (d) 8-celled

31. In angiosperms, embryo sac represents.
(a) female gametophyte (b) male gametophyte
(c) sporophyte (d) None of these

32. Embryo sac occurs in
(a) embryo
(b) axis part of embryo
(c) ovule
(d) endosperm

33. The cells which are found on opposite pole to egg cells are called
(a) synergids (b) antipodal cells
(c) Both (a) and (b) (d) None of these

34. Eight nucleated embryo sac is
(a) 8 celled (b) 7 celled
(c) 6 celled (d) 5 celled

35. An ovule which becomes curved so that the nucellus and embryo sac lies at the right angles to the funicle is
(a) hemitropus (b) anatropus
(c) campylotropous (d) orthotropus

36. The arrangement of nuclei in a normal embryo sac in the dicot plant is
(a) 3 + 3 + 2 (b) 2 + 4 + 2
(c) 3 + 2 + 3 (d) 2 + 3 + 3

37. An egg apparatus contains
(a) an egg + two antipodals
(b) an egg + secondary nucleus
(c) an egg + two synergid
(d) antipodal cell + two synergid

38. Embryo sac contains
(a) 2 eggs (b) 3 eggs
(c) 1 egg (d) 4 eggs

39. Out of megaspore tetrad, the functional megaspore is
(a) any megaspore (b) middle megaspore
(c) micropylar megaspore (d) chalazal megaspore

40. Ovule is straight with funicules, embryo sac, chalaza and micropyle lying on a stright line. It is called
(a) orthotropous (b) anatropous
(c) campylotropous (d) amphitropous

Answers

1. (c)	**2.** (d)	**3.** (b)	**4.** (a)	**5.** (b)	**6.** (d)	**7.** (c)	**8.** (b)	**9.** (c)	**10.** (b)
11. (b)	**12.** (c)	**13.** (b)	**14.** (c)	**15.** (c)	**16.** (c)	**17.** (b)	**18.** (b)	**19.** (b)	**20.** (b)
21. (a)	**22.** (d)	**23.** (b)	**24.** (a)	**25.** (b)	**26.** (d)	**27.** (a)	**28.** (c)	**29.** (b)	**30.** (c)
31. (a)	**32.** (c)	**33.** (b)	**34.** (b)	**35.** (a)	**36.** (c)	**37.** (c)	**38.** (c)	**39.** (d)	**40.** (a)

16

Pollination

Pollination is the phenomenon of transfer of pollen grains from the opened anther of the stamen to the receptive stigma of the carpel/pistil. Pollination is a prerequisite for fertilisation, the fusion of nuclei from the pollen grain with nuclei in the ovule.

Depending upon the source of the pollen grain, pollination is divided into two main types:

1. Self Pollination (Autogamy)

It is the kind of pollination in which the pollen from the anther of one flower is transferred to the stigma of the same flower. It is usually seen in plants like rice, wheat, pea, etc.

It is further classified as:

(i) **Chasmogamy** In this condition, flower opens at maturity, exposing the mature anther and stigma to the pollinating agents to allow the fertilisation. Flowers which show this property are called **chasmogamous flowers**.

(ii) **Cleistogamy** In this condition, flowers of some plants do not open up and the anther dehiscence inside the closed flowers only to ensure pollination. Hence, they do not possess the property of cross-pollination. Such flowers are called **cleistogamous flowers**.

2. Cross Pollination

It is the kind of pollination in which the pollen is transferred from the anther of one flower to the stigma of flower of the another plant.

It is also of following types:

(i) **Geitonogamy** in which the pollen grains is transferred between the two flowers on the same plant.

(ii) **Xenogamy** in which the pollen grains is transferred between the two flowers on different plant.

Various adaptation has been seen for the process of cross-pollination. These are given in the table below:

Adaptations for Cross Pollination

Dichogamy	Dicliny	Herkogamy	Self-incompatibility
The condition, where maturing time of stigma and anthers is such that either stigma becomes receptive before anther matures (protogyny) or the anthers are ready for the dehiscence before the stigma becomes receptive (protandry).	The presence of only one kind of reproductive whorl in a flower is called dicliny or unisexuality. A plant may be monoecious, i.e. carrying male and female flowers on the same plant. In such case, both cross and self-pollination can occur.	In some flowers, a mechanical barrier exists between the compatible pollen and stigma so, that self-pollination becomes impossible. Sometimes, a hood-like covering covers the stigma as in iris and in Calotropis.	If a pistil carrying functional female gamete fails to set seeds with viable and fertile pollen, which is capable of bringing about fertilisation in another pistil, the two are said to be incompatible, and the phenomenon is known as incompatibility.
In *Aristolochia* and *Scrophularia*, protogyny occurs and in rose, sunflower, impatiens, etc., protandry is found.	In dioecious plants, i.e. plants either with male or female flowers borne on different plants, in such a case cross-pollination is the only way of transfer of pollen.	The pollen are grouped in pollinia and are sticked to the surface till they are carried away by the insects.	Sexual incompatibility may be interspecific between the individuals of different species. The intraspecific incompatibility is also called self-incompatibility.

Agencies of Pollination

Most plants needs help moving the pollen from anther to the stigma of flowers. The agencies or agents responsible for the process of pollination are categorised into two main types:

Various Agencies of Pollination

Pollinating agency	Process		Agent
Abiotic agents	Anemophily	→	Wind
	Hydrophily	→	Water
Biotic agents	Entomophily	→	Insects
	Ornithophily	→	Birds
	Chiropterophily	→	Bats
	Malacophily	→	Snails

1. Abiotic Agents

These are agents that constitutes the non-living agents or components of pollination. Such as:

(i) **Pollination by Wind** (Anemophily)

This phenomenon was first demonstrated by **Sprengel** in 1793. It is commonly seen in flowers like maize, cotton woods, gymnosperms, grasses, etc.

The main features of wind pollinated flowers are:

- Anthers are generally versatile
- Plants produce a very large quantity of dusty pollen grains.
- Pollen grains are dry, very small sized, non-sticky and unwettable.
- Stigma may be large and feathery (grasses) or brushy (typha).
- In monoecious plants, the male flowers are more in number than female flowers.

(ii) **Pollination by Water** (Hydrophily)

It is a type of pollination being quite rare in the flowering plants and is often limited to about 30 genera, mainly monocotyledons. It is commonly seen in plants groups such as algae, bryophytes, pteridophytes, etc. The main features of water pollinated flowers are

- Inconspicuous and small flowers.
- Odour, nectar and colour is absent in flowers.
- Flowers whorls, if present are coated with wax. Stigma is unwettable but sticky.
- Pollen grains are small, light-weighted, non-sticky and coated with wax.Hydrophily is of two types:
  (a) **Hypohydrophily** It includes plants which are pollinated inside the water, e.g. *Najas*, *Zostera*.
  (b) **Epihydrophily** The flower born under water, when mature male flower detached and float on water surface when fuse with female flower pollination take place, e.g. *Vallisnaria*.

2. Biotic Agents

The agents that are living agents or components responsible for the process of pollination. These are as follows:

(i) **Pollination by Insects** (Entomophily)

Insects that are mainly involved in the pollination are bees, wasps, ants (*Hymenoptera*), hover bees (*Diptera*), moths, butterflies (*Lepidoptera*), grasshopper, locusts (*Orthoptera*), beetles (*Coleoptera*), dragon flies, etc.

The main features associated with insect-pollinated flowers are:

- Some flowers produce edible pollen grains attracting the insects to visit the flowers, e.g. rose, *Papaver*, *Clematis*.
- Flowers are fragrant and have mild scent and odour, e.g. *Jasminum*, *Cestrum*, etc.
- Flowers are bright coloured such as leaves in *Euphorbia, Words pulcherrima* and bract in *Bougainvillea*. One sepal in *Mussaenda*, bees prefer blue colour, while butterflies show attraction towards red coloured flowers.
- Stigma also secretes some exudants which makes the stigma sticky in nature.

(ii) **Pollination by Birds** (Ornithophily)

It is mainly affected by the birds like humming birds (in pendant flowers), sun birds (in erect flowers) and honey eater birds. The flowers are tubular (*Nicotiana glauca*), cup-shaped (*Callistemon*) or urn-shaped (member of Ericaceae), etc.

The main features of birds pollinated flowers are:

- Flowers are large sized, brightly coloured, odourless and produces large amount of mucilaginous nectar.
- Humming birds are the most common group of flower visiting birds in the America. These are mostly attracted to bright red coloured, odourless flowers.

(iii) **Pollination by Bats** (Chiropterophily)

Bat-pollinated flowers have parts which are usually fleshy and sweet in test, e.g. kadam.

- *Adansonia digitata* is vegetarian but pollinates the flowers of *Eidolon helvum* plant.
- *Micropteris pusillus* pollinates the flower of *Kigelia africana* (sausage tree).

(iv) **Pollination by Snails/Slugs** (Malacophily)

This type of pollination is seen in some water plants such as *Lemna* and in some members of family–Poaceae.

Benefits from Pollination

Plants benefits from pollinators because the movement of pollen allows them to reproduce by setting seeds.

However, pollinators don't know or care that the plant benefits. They pollinate to get nectar and for pollen from flowers to meet their energy requirements and to produce offspring. In the economy of nature, the pollinators provide an important service to flowering plants, while the plants pay the food for the pollinators and their offspring.

Pollinators provides a key ecosystem service vital to the maintenance of both wild and agricultural plant communities. Today pollinators are considered to be in a state of decline.

Decline in the health and population of pollinators pose what could be a significant threat to the integrity of biodiversity, to global food webs and to human health. The quality of pollinator service has declined overtime and this had led to concerns that pollination will be less resistant to extinction in the future. This is done by the source or agencies known as pollinators, which are the agents that allows such transfer effecting the pollination.

Chapter Exercise

1. Dichogamy that favours cross-pollination is a type of floral mechanism in which
   (a) anthers and stigma are placed at different levels
   (b) stamens and stigma mature at different times
   (c) structure of anther and stigma acts as barrier
   (d) pollen is unable to germinate on its own stigma
2. Pollen grains of a flower fall on the stigma of another flowers of same plant. The phenomenon is
   (a) genetically autogamy and ecologically allogamy
   (b) genetically autogamy
   (c) ecologically allogamy
   (d) genetically allogamy and ecologically autogamy
3. Cross-pollination offers certain Advantages as well as disadvantages of plants. Which one of the following statements is not true about cross-pollination?
   (a) It can fail to take place due to distance barrier
   (b) It is less economic because plants have to produce large number of pollen grains
   (c) It often gives high yield of crop
   (d) It takes place only in monoecious flowers
4. Occurrence of a fence or barrier between anthers and stigma of a flower is called
   (a) dichogamy (b) cleistogamy
   (c) herkogamy (d) heterostyly
5. What is the condition called in which a pollen of a flower falls on the stigma of another flower of the same plant?
   (a) Ecologically cross-pollination
   (b) Genetically and ecologically cross-pollination
   (c) Genetically self-pollination and ecologically cross-pollination
   (d) None of the above
6. Some flowers are usually closed. It is seen in
   (a) homogamy (b) cleistogamy
   (c) protogamy (d) allogamy
7. For the process of pollination, pollen grain must
   (a) mature and has three nuclei
   (b) land on stigma
   (c) release its sperm nuclei
   (d) release its sperm
8. Hydrophily is an excellent demonstrative property of
   (a) *Eichhornia* (b) *Nymphaea*
   (c) *Vallisneria* (d) *Nelumbium*
9. Cleistogamy is commonly seen in
   (a) *Ficus* (b) *Commelina*
   (c) *Anthocephalus* (d) *Vallisneria*
10. The *Yucca* flower is pollinated by
    (a) butterfly (b) moth
    (c) honey bee (d) bumble bee
11. Pollination in the sausages occurs by
    (a) bats (b) birds (c) insects (d) wind
12. Cleistogamous flowers
    (a) never open
    (b) open in day
    (c) always open
    (d) few open and some remain closed
13. Chalazogamy is the common features of
    (a) *Casuarina* (b) *Betula*
    (c) Both (a) and (b) (d) *Cucurbita*
14. Pollination in *Salvia* occurs through the agency of
    (a) insects (b) bats (c) ants (d) snails
15. Insect-pollinated flowers have
    (a) brightly coloured pollen
    (b) dry pollens with smooth surface
    (c) sticky pollen with rough surface
    (d) large quantities of pollen
16. Which of the following can prevent the self-pollination?
    (a) Herkogamy (b) Dichogamy
    (c) Self-sterility (d) All of these
17. Flower pollinated by humming birds is
    (a) *Bignonia* (b) Peepal
    (c) *Magnolia* (d) *Bougainvillea*
18. ……… is/are favoured by heterostyly in *Primula*.
    (a) Autogamy (b) Allogamy
    (c) Homogamy (d) All of these
19. Pollination which can occur in the same plants is
    (a) dichogamy (b) dicliny
    (c) cleistogamy (d) herkogamy
20. Plants pollinated by moths possess
    (a) strong scents (b) long corolla tube
    (c) nectaries (d) All of these
21. The cross-pollinated flowers are
    (a) brightly coloured (b) small
    (c) scented (d) Both (a) and (c)
22. Pollen of anemophilous plants are
    (a) large and heavy
    (b) small and sticky
    (c) small, dry and unwettable
    (d) large, light and hygroscopic
23. Self-pollination in potato is performed in
    (a) bud condition
    (b) cleistogamous condition
    (c) bending of filaments to bring anthers in contact with stigma
    (d) when style bends to bring stigma in contact with anthers.

24. Pollination in *Vallisneria* is
(a) hydrophilous (b) cleistogamous
(c) anemophilous (d) entomophilous

25. Maize is
(a) cleistogamous (b) anemophilous
(c) entomophilous (d) hydrophilous

26. In custard apple,
(a) self pollination is highly likely
(b) the flowers are dioecious
(c) anthers and stamens may mature at different times
(d) All of the above

27. Distance between the stigma and anthers in the same flower
(a) reduces the chances of self pollination
(b) enhances the chances of self pollination
(c) reduces the chances of cross pollination
(d) enhances the chances of cross pollination

28. In the angiosperm life cycle, which of the following is not a member of the sporophyte generation?
(a) Carpel (b) Anther (c) Pollen (d) Ovary

29. A plant that flowers in its second year of growth and then dies is
(a) a biennial (b) a semi-annual
(c) an annual (d) a perennial

30. Which of the following flowering arrangements promotes self-pollination?
(a) Monoecious (b) Dioecious
(c) Dichogamous (d) None of these

31. Early seed plants were pollinated by
(a) wind (b) bees (c) birds (d) butterflies

32. A number of megaspores degenerate after being produced by the meiosis of the megaspore mother cell. That number is
(a) 2 (b) 3 (c) 1 (d) 4

33. A trend in floral evolution is
(a) bilateral symmetry (b) lateral symmetry
(c) spiral symmetry (d) radial symmetry

34. Which of the following is not part of the innermost whorl of the flower?
(a) Stamen (b) Style
(c) Pistil (d) Ovary

35. Self-pollination can occur in which of the following?
(a) Dioecious plants (b) Dichogamous plants
(c) Pistillate plants (d) None of these

36. Which of the following is not an annual plant?
(a) Cabbage (b) Corn
(c) Sunflower (d) Wheat

37. If a flower is to be pollinated by birds, it should be
(a) blue (b) white
(c) yellow (d) red

38. Pollination is characteristic of
(a) gymnosperms (b) fungi
(c) angiosperms (d) pteridophytes

39. Wind pollinated plants differ from insect pollinated plants in having
(a) small petals and sticky pollen
(b) small coloured petals and heavy pollen
(c) coloured petal and large pollens
(d) no petals and light pollen

40. A typical example of self pollination is
(a) rice (b) wheat (c) maize (d) cotton

41. Deposition of pollen on the stigma of another flower of the same plant is
(a) homogamy (b) dichogamy
(c) xenogamy (d) geitonogamy

42. Pollen grain of flowers pollinated by insects or wind are
(a) large and showy (b) rough and sticky
(c) rough and smooth (d) rough and dry

43. Pollination by bats is called?
(a) Ornithophily (b) Entomophily
(c) Malacophily (d) Cheiropterophily

44. Insect pollinated flowers usually possess
(a) dry pollens with smooth surface
(b) sticky pollens with rough surface
(c) large quantities of pollen
(d) brightly coloured pollens

45. If a pollen of a flower falls on the stigma of another flower belonging to the same plant it is
(a) genetically self pollination and ecologically cross pollination
(b) ecologically cross pollination
(c) genetically and ecologically cross pollination
(d) None of the above

Answers

1. (b)	**2.** (a)	**3.** (d)	**4.** (c)	**5.** (c)	**6.** (b)	**7.** (a)	**8.** (c)	**9.** (b)	**10.** (b)
11. (a)	**12.** (a)	**13.** (c)	**14.** (a)	**15.** (c)	**16.** (d)	**17.** (a)	**18.** (b)	**19.** (c)	**20.** (d)
21. (d)	**22.** (c)	**23.** (d)	**24.** (a)	**25.** (b)	**26.** (c)	**27.** (a)	**28.** (c)	**29.** (a)	**30.** (d)
31. (a)	**32.** (b)	**33.** (a)	**34.** (b)	**35.** (d)	**36.** (a)	**37.** (d)	**38.** (c)	**39.** (d)	**40.** (c)
41. (d)	**42.** (b)	**43.** (d)	**44.** (b)	**45.** (d)					

17

Fertilisation and Development of an Embryo

Fertilisation

It is the process of giving rise to an embryo by the fusion of male and female gametes. In case of plants, fertilisation occurs when a compatible pollen lands on the stigma of the flower. It occurs after pollination of the carpel.

Pollen-Pistil Interaction

Through pollination, pollens land on the stigma of a female flower of the same species of plant and germinate by absorbing stigmatic secretion, which contains sugar, gum, lipids, resins and water. Now, pollen tube cell elongates and grows down the style towards the ovule in the ovary. Meanwhile, the generative cell divides by second microgametogenesis and forms two **sperm cells** or **male gametes.**

All cytoplasmic contents of the pollen grain comes out into the terminal portion below the cap block region of the growing pollen tube. These contents are endoplasmic reticulum, Golgi bodies, a tube nucleus, two male sperms, mitochondria,lipid bodies, amyloplasts, enzymes such as cellulase, pectinase, etc., hormones, like auxin and gibberellin.

After entry into stigma, pollen tube grows through the stylar tissues adhering to the stigma. It finally reaches to the ovule. Pollen tube may grow through the **solid style** or a **hollow style**. In case of solid style which generally occurs in dicot plants, the pectin rich cell wall layer of transmitting tissue or conduction tissue provides the way for the pollen tube growth.

In case of monocot plants having hollow style, pollen tube grows superficially (ectotrophic), i.e. it creeps over the surface of stylar canal cells which lines the stylar canal entry of pollen tube into an ovule. The pollen tube eventually reaches the micropyle of an ovule.

Types of Fertilisation

Depending upon the place of entry of the pollen tube into the ovule, three types of fertilisation have been recognised. These are given below:

(i) **Porogamy** occurs mostly in angiosperms in which the pollen tube enters the ovule through the **micropyle**.

(ii) **Mesogamy** occurs in those which the pollen tube enters through the **integuments** (*Cucurbita*) or through **funiculus** (*Pistacia*).

(iii) **Chalazogamy** in which the pollen tube enters through the **chalazal tissues**, e.g. *Casuarina*, *Beta vulgaris, Pterocarya, Ostrya* and *Jaglans regia*.

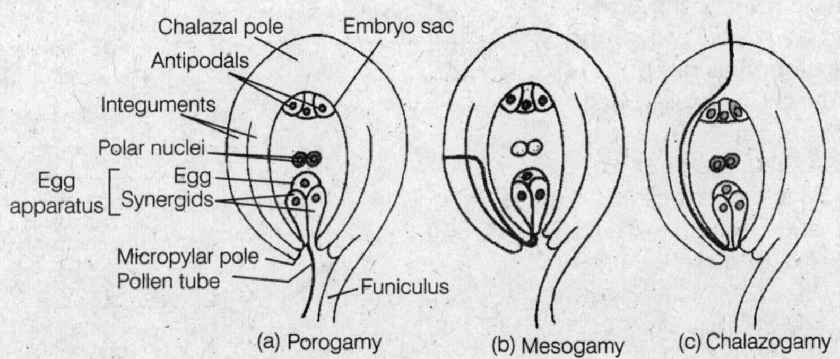

Various routes of pollen tube entry into the ovule

The tip of the pollen tube bursts by absorbing hydrolytic substances secreted by degenerating synergids and releases two male gametes, one vegetative or tube nucleus and a fair amount of cytoplasm.

Now, male gametes moves towards the chalazal pole of degenerating synergids first. After the release of male gametes into the embryo sac, one male gamete (called first male gamete) which is too near to the egg fuses with the egg nucleus, this phenomenon is known as **fertilisation** or **syngamy.**

During syngamy, random mixing of chromatin material of male and female nuclei occurs, it is known as **random assortment** of **genes.** The other male gamete (called second male gamete) migrates towards the secondary nucleus and fuses with it. This phenomenon is known as **triple fusion**, which results in the formation of a **triploid primary endosperm** mother cell.

True fertilisation together with the triple fusion is known as **double fertilisation** which is a unique phenomenon found only in angiosperms (absent in gymnosperms). It was demonstrated for the first time by Nawaschin (1898) in *Fritillaria* and *Lilium*.

Significance of Double Fertilisation

(i) Female gametophyte stops its growth at eight nucleate stage, further growth continues after double fertilisation.

(ii) The endosperm provides nutrition to the developing embryo. Thus, double fertilisation is necessary for the formation of viable seeds.

(iii) Secondary nucleus stops its division before fertilisation. So, triple fusion induces this dormant nucleus to regain its division power.

(iv) The formation of an endosperm is mainly related to its need which arises after fertilisation, i.e. particularly for the developing embryo.

The changes which take place in a fertilised flower are as follows:

Part before fertilisation	Part after fertilisation	Part before fertilisation	Part after fertilisation
Flower	Simple or aggregate fruit	Inner integument	Tagmen (delicate)
Calyx	Fall off or remain attached as fleshy structure	Nucellus	Perisperm
Corolla	Fall off	Endosperm nucleus	Endosperm
Stamens	Fall off	Egg cell	Embryo
Ovary wall	Pericarp of fruit	Funiculus	Stalk
Ovary	Simple fruit	Antipodals	Degenerate
Ovule	Seed	Synergids	Degenerate
Outer integument	Testa (tough)		

Development of an Embryo

The entire series of events that occurs between fertilisation and formation of mature embryo is called **embryogenesis.** During this process, the cells become progressively more specialised or differentiated.

In the majority of angiosperms, the zygote (fertilised egg) divides by an asymmetric mitotic division and generates two cells with two different fates. The smaller daughter cell with dense cytoplasm is situated towards the chalazal pole, called terminal cell (apical cell) or embryonal cell, which eventually gives rise to most of the embryo.

Another, comparatively larger daughter cell is situated towards the micropylar cell, called **basal cell** which continues to divide transversally and produces a thread of cells called the **suspensor**. The term **pre-embryo** is given to the structure that continues from this two-celled stage until the initiation of organs. The initial development in this two-celled pre-embryo is similar in both dicots and monocots before the formation of cotyledon and plumule occurs. Embryogeny can occur by five kinds on the basis of the following:

(i) Plane of division of the terminal daughter cell in the two celled pre-embryo.

(ii) The contribution of the basal cell and terminal daughter cell in the formation of embryo proper.

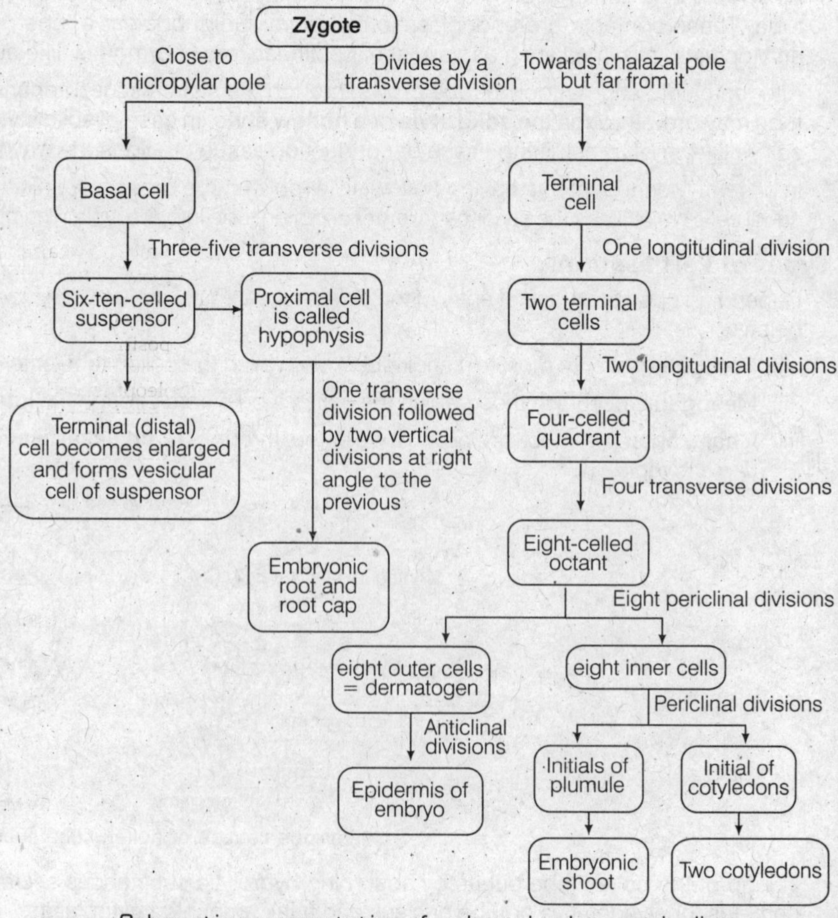

Schematic representation of embryo development in dicots

Structure of a Mature Dicot Embryo

A typical mature dicot embryo contains cotyledons that are attached laterally to the embryonal axis. The portion of this embryonal axis above the level of cotyledons forms the epicotyl, and the portion below the same tends to form the hypocotyl. The terminal portion of epicotyl forms the **plumule**, while the terminal portion of hypocotyl forms the **radicle**. When a dicot seed germinates the plumule gives rise to the shoot system and the radicle rises into the root system.

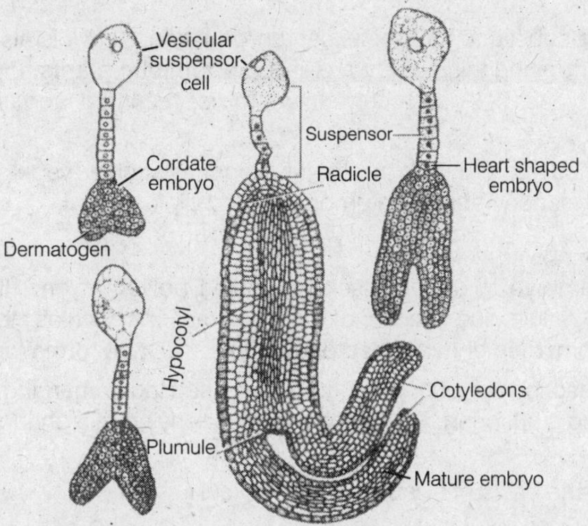

Various stages in the development of a dicot embryo

Structure of a Mature Monocot Embryo

In monocot embryo, there is not any essential difference in the early stages of development. From dicot embryo, the lateral position is occupied by the shoot apex. The embryo possesses only one cotyledon at one end known as **scutellum**. At the another end, the embryonal axis is composed of radicle and root cap. Epicotyl in monocot embryo is terminal in position. It has a shoot apex and few leaf primordia which are enclosed in a hollow foliar structure called **coleoptile**. Suspensor is joined to the embryo through the hypocotyl.

Difference between Monocot Embryo and Dicot Embryo

Characters	Monocot Embryo	Dicot Embryo
Number of cotyledons	Only one cotyledon attached to the embryonal axis.	Two cotyledons attached to an embryonal axis.
Position of plumule	Occurs laterally.	Occurs distally.
Position of cotyledons	Single cotyledon occupies terminal position.	Occurs laterally.
Coleoptile	The envelope of plumule is called coleoptile.	Absent.
Coleorhiza	This is a protective sheath of radicle.	Absent.
Scutellum	A single cotyledon.	Absent.
Suspensor	Comparatively small.	Large.

Parthenocarpy

It is defined as the mechanism of formation of an embryo **from an unfertilised egg** or a female gamete. It occurs in some plants which develop fruits without undergoing the fertilisation. Such fruits are relatively known as **parthenocarpic fruits**.

The phenomenon of parthenogenesis or parthenocarpy may be induced, genetic or environmental.

(i) In **induced parthenocarpy**, the development of homozygous haploids is achieved through some chemical agents.

(ii) In **genetic parthenocarpy**, the development of seedless fruits is done by the hybridisation or mutation.

(iii) In **environmental parthenocarpy**, several environmental conditions such as frost, fog, temperature alteration, etc., are responsible for the production of seedless fruits.

Special Mechanisms of Reproduction

Apart from sexual reproduction which is the arbitary process in almost all flowering plants. Some plants have evolved a special mechanism of producing seeds or to develop more than one embryo from single fertilised egg.

Such special mechanisms are discussed below:

1. Apomixis

It is the mechanism which acts as the substitute for the sexual reproduction, thus it is also considered as an asexual reproduction without nuclear fusion. The plants performing this activity are called **apomictic plants.** Phenomenon of apomixis was classified by **Maheshwari** (1950) into four main types, i.e. non-recurrent apomixis, recurrent apomixis, adventive embryony and vegetative propagules.

Before this **Winkler** (1908) had classified apomixis into two main categories such as vegetative reproduction and agamospermy. The agamospermy is further divided as apospory, adventive embryony and diplospory.

2. Polyembryony

The occurrence of more than one embryo in a single seed is called polyembryony. It was first demonstrated by **Antoni von Leeuwenhoek** (1917). Polyembryony results due to either of the many reasons such as cleavage in the proembryo, formation of one or more embryo in the embryo sac, formation of more than one embryo sac and formation of an embryo from sporophytic cell.

The term is also used in botany to describe the phenomenon of two seedlings emerging from one seed. In plants, polyembryony, often gives rise to the enigma of a single offspring. The mechanism underlying this phenomenon is **programmed cell death** which removes all but one embryo.

It is classified into two main types on the basis of treatment of chemical

(i) **Spontaneous** (naturally occurring) **polyembryony** which is further divided into true and false embryony.

(ii) **Induced polyembryony**.

Chapter Exercise

1. What is meant by double fertilisation?

(a) Fusion between egg and male gamete
(b) Fusion between secondary nucleus and male gamete
(c) Both (a) and (b)
(d) Formation of two zygotes

2. Of the two cells of zygote, the cell near to micropyle is termed as

(a) terminal cell (b) basal cell
(c) embryo cell (d) zygote cell

3. When pollen tube enters through the funiculus or integument it is called

(a) porogamy (b) chalazogamy
(c) mesogamy (d) isogamy

4. In the post-fertilisation changes within the ovule, the oospore forms the ...., while the fertilised fusion nucleus (product of triple fusion) develops the .... respectively.

(a) endosperm, embryo
(b) embryo, endosperm
(c) nucellus, embryo
(d) embryo, nucellus

5. In which of the following plants, pollen tube enters through integuments?

(a) *Casuarina* (b) *Malva neglecta*
(c) *Cucurbita* (d) *Beta vulgaris*

6. Double fertilisation occurs in

(a) *Pinus* (b) *Selaginella*
(c) *Funaria* (d) *Dalbergia*

7. Which of the following forms the embryo?

(a) Egg apparatus (b) Antipodal cell
(c) Fertilised ovum (d) Fertilised synergid

8. A mature dicot embryo has

(a) one cotyledon (b) two cotyledons
(c) three cotyledons (d) four cotyledons

9. Pollen tube enters through chalazal tissue is called chalazogamy and is found in

(a) *Cucurbita* (b) *Pistacia*
(c) *Casuarina* (d) None of these

10. Fertilisation of the egg takes place inside the

(a) anther (b) stigma
(c) pollen tube (d) embryo sac

11. Which of the following cells secrete enzymes to dissolve?

(a) Synergid cell (b) Egg cell
(c) Central cell (d) Both (a) and (b)

12. In porogamy, the pollen tube enters into the ovule *via*

(a) chalaza (b) micropyle
(c) integuments (d) None to these

13. Ovule converts into which of the following after fertilisation?
(a) Fruit (b) Embryo
(c) Seed (d) Both (a) and (c)

14. Fruit is formed after fertilisation by which of this following part?
(a) Ovule (b) Ovary
(c) Calyx (d) Synergids

15. Zygote divides first into basal cell and terminal cell which of these cell is present towards chalazal pole?
(a) Basal cell (b) Terminal cell
(c) Antipodal cell (d) Suspensor cell

16. The portion of the embryonal axis was above the level of cotyledons forms
(a) hypocotyl (b) epicotyl
(c) Both (a) and (b) (d) None of these

17. Cotyledons in embryo is formed from
(a) basal cell (b) terminal cell
(c) Both (a) and (b) (d) None of these

18. Calyx after fertilisation
(a) develop into fruit (b) develop into seed
(c) develop into ovary (d) fall off

19. The portion below the epicotyl is
(a) suspensor (b) cotyledons
(c) hypocotyl (d) None of these

20. In *Beta vulgaris*, the pollen tube reaches the embryo sac *via*
(a) micropyle (b) integuments
(c) chalaza (d) None of these

21. Longest pollen tube occurs in
(a) *Cucurbita* (b) *Dahlia*
(c) grass (d) maize

22. What is the another name of vegetative fertilisation?
(a) Triple fusion (b) Double fertilisation
(c) Somatogamy (d) Syngamy

23. In a flowering plant, the pollen tube first arrives in
(a) an antipodal cell (b) a synergid
(c) a central cell (d) egg

24. Which out of these mechanism does not lead to the formation of clones?
(a) Vegetative reproduction (b) Apomixis
(c) Double fertilisaion (d) Tissue culture

25. What does the phenomenon called if the pollen tube enters the embryo sac through base of a funiculus or the integument?
(a) Mesogamy (b) Isogamy
(c) Porogamy (d) Chalazogamy

26. Agamospermy is
(a) diplospory (b) apospory
(c) adventitious embryony (d) All of these

27. Formation of an embryo from egg of an embryo sac developed directly from a cell nucellus is
(a) diplospory (b) apospory
(c) adventitious embryony (d) None of these

28. Structure whose outgrowth helps in fertilisation is
(a) stigma (b) placenta or funiculus
(c) pollen tube (d) pollen grains

29. Chalazogamy is found in
(a) *Pisum* (b) *Pistacia*
(c) *Casuarina* (d) *Cucurbita*

30. Embryo from an unfertilised egg is
(a) allogamy (b) apogamy
(c) parthenocarpy (d) parthenogenesis

31. Double fertilisation in angiosperm was first explained by
(a) N. Borlaug (b) Nawaschin
(c) Maheshwari (d) Stewart

32. Apomixis is the type of reproduction that result in the development of
(a) embryo from nucellus
(b) embryo from endosperm
(c) new organism without fusion of gametes
(d) None of the above

33. A plant has 25 chromosomes in 'microspore mother cell'. The number of chromosomes in its endosperm will be
(a) 36 (b) 24
(c) 48 (d) 12

34. Process in which fertilisation occurs without fusion of gametes
(a) amphimixis (b) parthenogenesis
(c) apomixis (d) autogamy

35. During the process of fertilisation, when one nucleus fuses with egg and the second with the secondary nucleus, it is called
(a) double fertilisation (b) triple fusion
(c) single fertilisation (d) triple fertilisation

Answers

1. (c)	**2.** (b)	**3.** (c)	**4.** (b)	**5.** (c)	**6.** (d)	**7.** (c)	**8.** (b)	**9.** (c)	**10.** (d)
11. (a)	**12.** (b)	**13.** (c)	**14.** (b)	**15.** (b)	**16.** (b)	**17.** (b)	**18.** (d)	**19.** (c)	**20.** (c)
21. (d)	**22.** (a)	**23.** (b)	**24.** (c)	**25.** (a)	**26.** (d)	**27.** (b)	**28.** (b)	**29.** (c)	**30.** (d)
31. (b)	**32.** (c)	**33.** (a)	**34.** (c)	**35.** (a)					

18

Structure, Development and Types of Endosperm

Endosperm

In angiosperms as the consequence of triple fusion, when a male gamete fuses with the secondary nucleus, a triploid structure called **Primary Endosperm Mother Cell** (PEMC) is formed that divides by mitotic division and forms a mass of nutritive cells called endosperm. It is generally triploid meant for nourishing the embryo. Formation of endosperm starts prior to the embryo formation. It starts with the degeneration of nuclear tissue. It is not formed in the members of families like Podostemaceae, Trapaceae, Orchidaceae, etc.

The ploidy level of endosperm varies according to the type of embryo sac because different embryo sacs contain different ploidy level at secondary nucleus, e.g. *Oenothera* type of embryo sac contains haploid secondary nucleus (n). Thus, the ploidy level of endosperm is diploid ($2n$) in *Oenothera* type embryo sac.

Kinds of Endosperm

On the basis of its development, endosperm is classified into three main groups, i.e. cellular, nuclear and helobial endosperm.

Cellular Endosperm

It is found in 25% angiospermic families particularly dicotyledons. During cellular type of endosperm development, the primary endosperm mother cell divides by repeated mitotic divisions and each division is followed by a cell wall formation, e.g. *Adoxa*, *Verbascum*, *Cetranthus*, *Scutellaria*, *Datura*, *Myastis arvensis*, *Impatiens*, *Magnolia*, etc.

Nuclear Endosperm

It is a more common type of endosperm and occurs in about 56% of those angiospermic families.

During the development of this type of endosperm, the Primary Endosperm Mother Cell (PEMC) divides by free nuclear divisions which are not followed by the cell wall formation. Thus, free nuclei remain in the cytoplasm of embryo sac.

Later on, these nuclei arranged themselves towards the periphery of the embryo sac. Now, cell wall formation takes place from periphery to the centre (centripetal) that finally leads to the formation of a cellular type of embryo sac, e.g. *Acer, Arachis, Gravillea, Lomatia, Phoenix, Calotropis, Primula, Citrus, Malva, Acalypha*, etc.

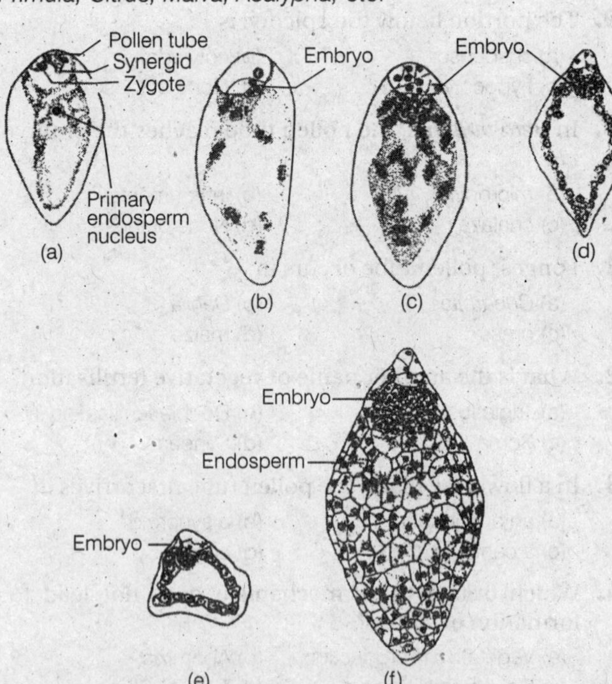

Endosperm development in *Acalypha indica* (a) Embryo sac after fertilisation, the primary endosperm nucleus and the zygote have not yet divided. (b-c) Embryo sac, showing synchronous divisions of the endosperm nuclei. (d) The endosperm nuclei have moved to the periphery. (e) The peripheral part of the embryo sac has become cellular. (f) Completely cellular endosperm.

Helobial Endosperm

It is found in 19% angiospermic families particularly in **monocotyledons**. It is an intermediate type between cellular and nuclear endosperms and is formed due to the formation of a large micropylar and a small chalazal chamber by a mitotic division in the primary endosperm mother cell.

Nucleus of chalazal chamber remains undivided or divides by a few mitotic divisions, while, micropylar nucleus divides by the free nuclear divisions which are followed by centripetal cell wall formation, e.g. *Asphodelus*.

Internal Structure of an Endosperm

Internally, the endosperm is made up of isodiamorphical, triploid (in most cases) cells that stores reserve food material in the form of starch, carbohydrates, proteins or fats. The cells of endosperm are thin-walled and do not have pits but in the presence of hemicellulose as a reserve food the cell wall becomes thick and pitted.

For the transfer of stored material, the plasmodesmata also occur between the cells of endosperm.

Aleurone Tissue

In cereals, the cells of the outermost layers of the endosperm become morphologically and physiologically specialised to form of a layer of cells called **aleurone layer** or **aleurone tissue**. Cells of this layer contain aleurone grains, each of which is covered by a single unit membrane. The main components of aleurone grains are phospholipids, proteins, phytin and some carbohydrates.

Each aleurone grain contains a globoid cavity, surrounded by a protein carbohydrate body, which inturn is covered by numerous spherosomes. The main function of aleurone cells is to secrete hydrolytic enzymes, protease (protein digesting enzymes) and amylase (starch digesting enzymes) in the germinating seeds.

Functions of Endosperm

- It forms sufficient food for the developing embryo.
- In the absence of endosperm, the growth of embryo will be poor and suppressed which results in the formation of non-viable seeds.
- It stores carbohydrates, fats and proteins, etc and provides to the seed during germination for the growth of seedlings.
- Many experiments have proved that the contents of endosperm are very necessary for the growth of embryo.

Chapter Exercise

1. The ploidy level of endosperm is
(a) haploid (b) diploid
(c) triploid (d) None of these

2. Major types of endosperms are
(a) 2 (b) 3
(c) 6 (d) 8

3. Which of the following types of endosperm is found in 25% of angiosperms?
(a) Cellular (b) Nuclear
(c) Helobial (d) All of these

4. During nuclear type of endosperm development, primary endosperm mother cell divide by
(a) mitosis
(b) meiosis
(c) free nuclear division
(d) Both (a) and (b)

5. Which of the following types of endosperm is found in *Datura*?
(a) Nuclear (b) Cellular
(c) Helobial (d) None of these

6. *Acer, Lomatia* and *Calotropis* have which of the following?
(a) Cellular (b) Nuclear
(c) Helobial (d) Both (b) and (c)

7. Which of the following type of endosperm is common in monocots?
(a) Helobial (b) Cellular
(c) Nuclear (d) None of these

8. In cereals, the cells of the outermost layer of the endosperm becomes morphologically and physiologically specialised and form a layer of cells called
(a) perisperm (b) tapetum
(c) aleurone layer (d) All of these

9. *Asphodelus* shows which type of endosperm?
(a) Cellular (b) Nuclear
(c) Helobial (d) None of these

10. In nuclear division, the nuclei after free division arranged themselves towards .... of embryo sac.
(a) center (b) periphery
(c) Both (a) and (b) (d) None of these

11. Milky water of green coconut is liquid
(a) nucellus (b) of female gametophyte
(c) endosperm (d) chalaza

12. The aleurone layer of endosperm in monocot seed is related to
(a) growth of endosperm
(b) digestion of reserve food of endosperm
(c) store the food of endosperm
(d) formation of endosperm

13. Nuclear endosperm is found in which of the following?
(a) *Adoxa* (b) *Datura*
(c) *Citrus* (d) *Asphodelus*

14. Endosperm of angiosperms is produced after the fertilisation of male gamete with
(a) synergids (b) secondary nucleus
(c) oospore (d) antipodals

15. In *Oenothera*, the ploidy level of endosperm is
(a) $2n$ (b) $3n$
(c) $4n$ (d) n

16. Double fertilisation in angiosperms results in the formation of
(a) embryo (b) endosperm
(c) Both (a) and (b) (d) None of these

17. Double fertilisation in angiosperm was discovered by whom?
(a) Nawaschin (b) Nageli
(c) Darwin (d) None of these

18. Development of an endosperm occurs in
(a) pollen sac (b) ovule
(c) embryo sac (d) locule

19. Three types of endosperm are cellular, nuclear and
(a) celluloid (b) helobial
(c) Both (a) and (b) (d) All of these

20. Which of the following is an intermediate type of endosperm?
(a) Cellular
(b) Nuclear
(c) Helobial
(d) None of the above

21. An endosperm having irregular boundaries is termed as
(a) nuclear (b) cellular
(c) helobial (d) ruminate

22. Endosperm is generally
(a) diploid (b) triploid
(c) haploid (d) polyploid

23. The maximum ploidy seen in the secondary nucleus is
(a) $7n$ (b) $14n$
(c) $15n$ (d) $10n$

24. The term *Xenia* denotes the effect of pollen on the
(a) endosperm (b) egg
(c) nucellus (d) seed coat

25. Endosperm in angiosperms results after fertilisation form
(a) zygote (b) secondary nucleus
(c) antipodals (d) synergids

26. Aleurone layer is present in
(a) the peripheral part of endosperm
(b) the peripheral part of scutellum
(c) the peripheral part of coleoptile
(d) the peripheral part of cotyledons

27. The aleurone layer of endosperm in monocot seed is related to
(a) growth of endosperm
(b) digestion of reserve food of embryo
(c) store the food of endosperm
(d) formation of endosperm

28. In which of the following helobial endosperm occurs?
(a) *Acalypha* (b) *Cocos*
(c) *Cannabis* (d) *Adoxa*

29. The structure which can show the effect of traits brought by the male gamete immediately after its formation is
(a) embryo (b) cotyledons
(c) endosperm (d) plumule

30. Free nuclear divisions are characteristic of
(a) cellular endosperm (b) nuclear endosperm
(c) helobial endosperm (d) Both (b) and (c)

Answers

1. (c)	**2.** (b)	**3.** (a)	**4.** (c)	**5.** (c)	**6.** (b)	**7.** (a)	**8.** (c)	**9.** (c)	**10.** (b)
11. (c)	**12.** (b)	**13.** (c)	**14.** (b)	**15.** (a)	**16.** (c)	**17.** (a)	**18.** (c)	**19.** (b)	**20.** (c)
21. (d)	**22.** (b)	**23.** (b)	**24.** (a)	**25.** (b)	**26.** (a)	**27.** (b)	**28.** (b)	**29.** (c)	**30.** (d)

19

Structure of Dicot and Monocot Seed

After fertilisation in plants, the ovule develops into **seeds**. A typical seed is an embryonic plant enclosed in a protective outer covering known as a **seed coat** and an **embryo**, which is a characteristic of spermatocytes (gymnosperm and angiosperm plants) and the product of the ripened ovule which occurs after fertilisation and some growth within the mother plant. The embryo is made up of radicle, embryonal axis and cotyledons. Hence, the seed leads to germination of plants.

Development of Seed

Angiosperm (flowering plants) seeds consists of three genetically distinct constituents, i.e. embryo formed from the zygote, endosperm; which is normally triploid, seed coat from tissues derived from the maternal tissue of the ovule.

The process of development of seed begins with double fertilisation, right after fertilisation, the zygote is mainly inactive in nature, but the primary endosperm divides rapidly to form the endosperm tissue which later becomes the food in early stage that the plant will consume until the roots have developed after germination.

The ovule that develops into seeds may consist of following number of components:

(i) The funicle
(ii) The nucellus
(iii) The micropyle
(iv) The chalaza

The shape of ovules, on their development affects the final shape of the seeds. Plants usually produces ovules of four different shapes, the most common of which is **anatropous** (a curved shape), **orthotropous** (ovules are straight with all parts being lined up in a long row producing an uncurved seed), **campylotropous** (ovules having a curved megagametophyte often giving the seeds a tight 'I' shape) and **amphitropous** (ovules are inverted partly and turned back 90 degrees units stalk).

Seed Types

Seeds of angiosperms have been considered to occur in twelve different types, on the basis of mainly embryo to seed 'size' ratios. This thus, reflects the degree to which the developing cotyledons absorb the nutrients of the endosperm and thus, obliterate it.

There are about six types of seed amongst monocotyledons ten amongst dicotyledons and two in gymnosperms being known.

The classification is often based on three main characteristics, i.e. embryo morphology, amount of endosperm and position of the embryo relative to the endosperm. The following types are:

(i) Broad (ii) Capitate
(iii) Lateral (iv) Peripheral
(v) Rudimentary (vi) Dwarf
(vii) Micro (viii) Linear
(ix) Spatulate (x) Investing
(xi) Bent (xii) Folded

Structure of a Dicot Seed

In dicots, seed coat is the outermost covering of the seed. It is composed of two main layers, i.e. the outer **testa** and an inner integument forming **tegmen**. In some seeds both testa and tegmen are fused together with each other to form a pericarp.

In some seeds such as gram seed, a scar is found at which point the seed is attached with its fruit. This scar is called **hilum.** Micropyle is a small pore found above the hilum.

Embryo of a dicot seed is enclosed in a seed coat and is differentiated into a **radicle, plumule** (found on two sides of embryonal axis) and two **cotyledons**. Out of which the radicle gives rise to root on development, while the plumule develops into the shoot. Fleshy cotyledons are often found full of reserve food.

Hypocotyl is known as the point of attachment of cotyledon and radicle, while the **epicotyl** is the point of attachment of cotyledons and plumule.

The endosperm (food storing tissue) of the seed which is developed after double fertilisation is also found but in some dicot seed, the endosperm is not found, such seeds are called **non-endospermous seeds.**

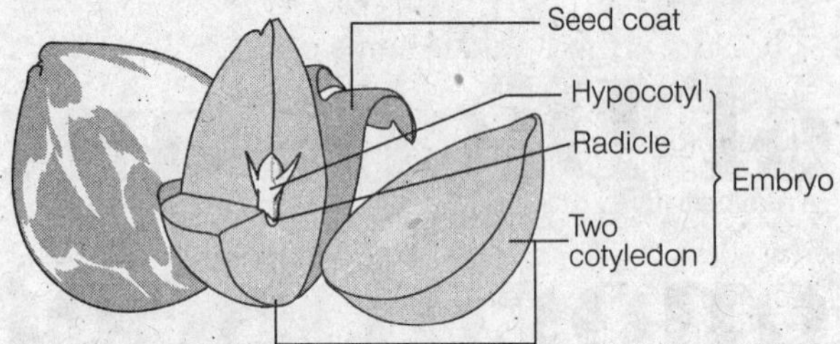

Internal structure of a gram seed (dicot seed)

Functions

Seeds serves important functions for plants that produces them.

The major functions amongst them are as follows:

(i) to provide nourishment of the embryo.
(ii) dispersal to a new location.
(iii) dormancy during unfavourable conditions.
(iv) Seeds are also the fundamental means of reproduction due to which most seeds are the product of sexual reproduction which produces a remixing of a genetic material and phenotype variability on which the natural selection acts.

Structure of Monocot Seed

Apart from the dicot seeds, the seeds of monocots are mostly endospermic but in some plants such as *Orchids,* etc., they are **non-endospermic**. The grain of monocot seed is single and thin, hard covering seed coat is membranous and fused with the wall of fruit or pericarp.

Bulky endosperm reserves the food. The proteinaceous layer of endosperm called **aleurone** layer and it separates the embryo from endosperm. The embryo is small and situated at one end on the groove of endosperm.

The single cotyledon is shield shaped and is called **scutellum.**

Plumule and radicle are found on a short axis. These structures are enclosed in sheaths which are called coleoptile and coleorhiza respectively.

Coleoptile possesses a terminal pore for the emergence of first leaf during germination. It helps the shoot of future to pass through the soil during germination.

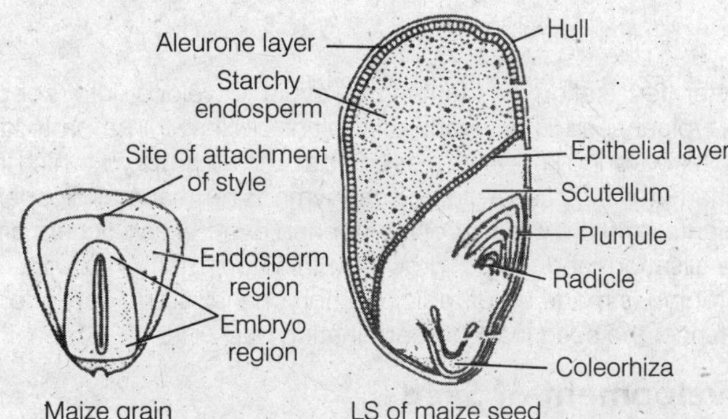

Structure of maize seed (monocot seed)

Chapter Exercise

1. A ripened/mature ovule is
(a) fruit (b) flower (c) seed (d) ovary

2. Which of the following is not found in the dicot seed?
(a) Seed coat (b) Coleoptile
(c) Endosperm (d) Radicle

3. Which of the following develops seed coat?
(a) Pericarp (b) Endosperm
(c) Ovary (d) Integuments

4. Which of the following is not found in seed?
(a) Endosperm (b) Cotyledons
(c) Style (d) All of these

5. Cotyledons are part of which of the following?
(a) Endosperm (b) Embryo
(c) Both (a) and (b) (d) None of these

6. Which of the following does not have dicotyledonous seed?
(a) Pea (b) Gram (c) Mango (d) Rice

7. In which of the following non-endospermic seeds are found?
(a) Apple (b) *Orchids* (c) Millets (d) Cereals

8. The ploidy level of endosperm is
(a) $2n$ (b) $3n$
(c) n (d) None of these

9. A small pore above the hilum in gram seed is called
(a) chalaza (b) micropyle
(c) integuments (d) None of these

10. Radicle and plumule are found on the two side of
(a) ovule (b) embryonal axis
(c) ovary (d) nucellus

11. Seed coat of maize seed is
(a) fleshy (b) membranous
(c) rough (d) None of these

12. Monocotyledonous seeds are generally
(a) endospermic (b) non-endospermic
(c) Both (a) and (b) (d) None of these

13. Which of the following proteinaceous layer separate endosperm from embryo in maize seed?
(a) Integuments (b) Hyalurone layer
(c) Aleurone layer (d) None of these

14. Embryo is small in maize and situated on one end on the groove of
(a) ovule (b) cotyledons
(c) endosperm (d) All of these

15. The cotyledon of maize seed is also called
(a) scutellum (b) coleoptile
(c) coleorhiza (d) None of these

16. Starchy endosperm is a feature of
(a) gram seed (b) mango seed
(c) pea seed (d) maize seed

17. Two cotyledonous of gram seeds are
(a) membranous (b) fleshy
(c) leathery (d) None of these

18. Number of cotyledon in maize is
(a) 2 (b) 3
(c) 1 (d) None of these

19. Mature ovary is …… and mature ovule is ……… .
(a) ovule, ovary (b) ovule, ovum
(c) fruit, seed (d) seed, fruit

20. Seed coat of gram seed has two layers called
(a) testa, integuments (b) tagmen, aleurone
(c) testa, tagmen (d) None of these

21. The outgrowth that develops along with hilum of the seed is called
(a) perisperm (b) raphe (c) plumule (d) straphiole

22. Seed in actual is a
(a) fertilised ovule (b) fertilised and ripened ovule
(c) fertilised ovary (d) ripened ovary

23. The part of *Castor* seed responsible for the extraction of oil is
(a) testa (b) endosperm
(c) hypocotyl (d) cotyledon

24. Region where food is stored in albuminous seeds is
(a) plumule (b) cotyledon
(c) testa (d) endosperm

25. Scutellum is the alternate name of
(a) leaf like cotyledon (b) caruncle
(c) aleurone layer (d) shield shaped cotyledon

26. An example of dicot plant is
(a) corn (b) bean
(c) rice (d) wheat

27. Which of the structure is unique to the seed of a monocot?
(a) Endosperm (b) Cotyledon
(c) Coleoptile (d) Radicle

28. When seeds germinates, the radicle emerges before the shoot. This allows the seedling to quickly
(a) obtain a dependable water supply
(b) mobilise stored carbohydrates
(c) protect the emerging coleoptile
(d) avoid etiolation

29. Which of the following is not matched correctly?
(a) One cotyledon – Monocot only
(b) Two cotyledons – Dicot only
(c) Coleoptile – Monocot only
(d) Endosperm – Both monocot and dicot

30. Identify the figure given below and examine the labelling marked as *A* and *B*.

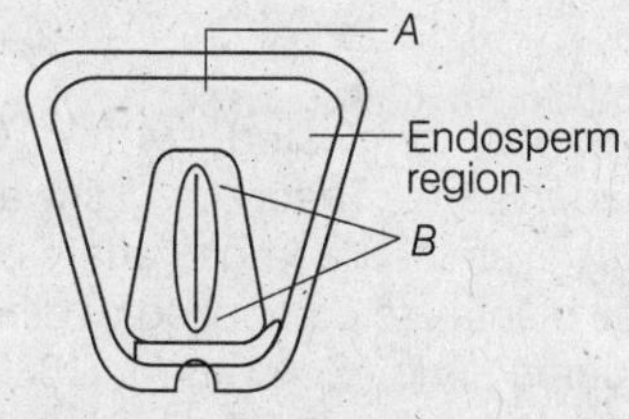

(a) Pea, *A*-Site of attachment of style, *B*-Embryo region
(b) Maize, *A*-Site of attachment of style, *B*-Coleorhiza
(c) Wheat, *A*-Radical, *B*-Embryo region
(d) Maize, *A*-Site of attachment of style, *B*-Embryo region

Answers

1. (c)	**2.** (b)	**3.** (d)	**4.** (c)	**5.** (b)	**6.** (d)	**7.** (b)	**8.** (b)	**9.** (b)	**10.** (b)
11. (b)	**12.** (a)	**13.** (c)	**14.** (c)	**15.** (a)	**16.** (d)	**17.** (b)	**18.** (c)	**19.** (c)	**20.** (c)
21. (d)	**22.** (b)	**23.** (b)	**24.** (d)	**25.** (d)	**26.** (b)	**27.** (c)	**28.** (a)	**29.** (d)	**30.** (d)

20

Ultrastructure of Cell and Chromosome

Cell

Cell is known to be the fundamental, structural and functional unit of all living organisms it may be either plants, animals or other small microorganisms. Cells may vary in shapes and sizes depending upon their origin and nature.

Robert Hooke discovered the cell, while Av Leeuwenhoek first saw the live cell. Schleiden and Schwann gave cell theory in the year 1839 which states that all living organisms are composed of cells and arise from pre-existing cells.

Functions of Cell

(i) Help to provide structure to the body.
(ii) Take in nutrients from food.
(iii) Help in the conversion of those nutrients into energy.
(iv) To carry out specialised functions.

Cell Theory

The cell theory was formulated by two German scientists, Matthias Schleiden and Theodore Schwann independently. Schleiden (1838) examined a large variety of plant tissues and observed that all plants are composed of different kinds of cells. At about the same time, Schwann (1839), closely studied different types of animal cells and found that the animal cell had a very thin outer layer known as plasma membrane.

He also concluded, from his studies based on plant tissues that animal cells differ from plant cells in lacking cell wall.

He said that all cells derived from preexisting cells *'Omnis cellula e cellula'*.

Size and Shape of a Cell

Size of a Cell

The cells exhibit an endless variation in size, life span and cellular activities, e.g. Mycoplasma (smallest cell) or PPLOs (Pleuro-Pneumonia Like Organisms) is only 0.3 μm in length and bacteria are approx. 3-5 μm in size. An ostrich egg, which is known to be the largest isolated single cell measures about 170 × 135 mm. Human Red Blood Cells (RBCs) are about 7 μm in diameter and the nerve cell of human being is the longest cell having length of 90-100 cm.

Shape of a Cell

The cells also vary in their shapes. They may be polygonal, disc-like, amoeboid, thread-like, cuboid or irregular. The cell shape is always related and vary with the function they perform.

Types of Cell

Based on the presence and absence of nucleus, cells are of two main types, i.e. eukaryotic and prokaryotic cells.

Comparison of Prokaryotes and Eukaryotes

Prokaryotic cells	Eukaryotic cells
Cells are divided by amitosis.	Cells are divided by mitosis and/or meiosis.
Cytoplasm lack membrane bound organelles.	Cytoplasm contains membrane bound organelles.
Ribosome is 70 S type in cytoplasm.	Ribosome is 80 S type in cytoplasm.
Mitochondria, Chloroplast, Golgi bodies and endoplasmic reticulum are absent.	Mitochondria, chloroplast, Golgi bodies and endoplasmic reticulum are present.
Nucleus is not bounded by nuclear membrane and is called nucleoid.	Nucleus is bounded by a nuclear membrane.
Nucleolus is absent.	Nucleolus is present.

Components of a Prokaryotic Cell

A typical prokaryotic (bacterial) cell is often composed of various components such as genetic material, cell envelope, cytoplasm, nucleoid, inclusion bodies, ribosomes, flagella, fimbrae, pilli, etc.

Cell Envelope and Its Modifications

Cell envelope is the outermost covering of protoplasm of the bacterial cell. It is known to protect the cell from mechanical shocks and injuries.

Glycocalyx (Mucilage Sheath)

It is the outermost layer, made up of macromolecules that given sticky character to the cell. Glycocalyx differs in composition and thickness among different bacteria. It could be in the form of loose mucilaginous sheath cell **slime layer** or thick and tough called **capsule**.

Components of a Eukaryotic Cell

A typical eukaryotic plant cell is composed of following components and can be divided as:

(i) Cell boundary or cell wall

(ii) Nucleus

(iii) Space between nucleus and cell boundary

Cell Wall

It is a characteristic feature of some of the cells including plants, fungi, etc.

It is a tough, rigid, non-living structure made up of cellulose, hemicellulose and pectose materials.

Cellulose is a specialised sugar been classified as a structural carbohydrate. It is not used for energy purposes. It forms the outer covering of the plasma membrane in most of the plants and fungi. It is after found to be absent in animal cells.

It is metabolically active in nature and is capable of growth. It is not only provides shape and protection to the cell from any mechanical damage and infection but also help in the cell to cell interaction and providing barrier to the unwanted and undesirable macromolecules.

Structure of Cell Wall

On the basis of the structure, cell wall is differentiated into the following three parts:

(i) **Middle lamella** It is the layer mainly made up of **calcium and magnesium pectates.** It cements the cell walls of two adjoining cells together. It is absent on the outer side of surface cells. Ripened fruits become soft due to the dissolution of middle lamella.

(ii) **Primary cell wall** It is produced inner to the middle lamella in a young and growing cell. It is capable of growth and extension. It tends to diminish gradually as the cell attain maturity.

(iii) **Secondary cell wall** The thick secondary wall is formed inner towards membrane to the primary wall.

Functions of Cell Wall

Cell wall possess the following functions:

(i) It helps in providing a definite shape to the cell and also protects protoplasm against any mechanical injury, i.e. damage and infection.

(ii) It also helps in cell to cell interaction.

(iii) It provides barrier to undesirable macro molecules and attack of pathogens.

Plasma/Cell Membrane

Beneath the cell wall, each and every living organism is covered with a thin, elastic, transparent, regenerative and a **semi-permeable** membrane called a cell membrane. It is also known as a **Plasma membrane** or a **plasmalemma**. It helps to separate the internal environment of the cell from that of the external environment.

Structure of Cell Membrance

Studies on human RBCs concluded that the cell membrane is composed of lipid which forms a bilayer with protein molecules embedded in it at places.

Lipids

The lipid molecules are amphipathic in nature and are arranged within the membrane by the help of two types of ends. These are as follows:

(i) **Polar hydrophilic end** This region is in the form of (water loving) head, which faces towards the outer sides of the cell membrane to interact with the aqueous environments on both sides.

(ii) **Non-polar hydrophobic end** This region is in the form of (water repelling) tail, both ends of which faces each other that occur towards the centre of the cell membrane.

These are formed of cholesterol (25-32%) and mainly of **phospho-glycerides** or **phospholipids** (55-75%).

The membrane proteins can be classified as:

(i) **Integral proteins** (intrinsic protein) They have stronger association and bound firmly to the membrane. These proteins are buried partially or totally in the phospholipid bilayer.

(ii) **Peripheral proteins** (extrinsic protein) They have weaker association and are bound to lipids of membrane by electrostatic interactions.

Carbohydrates

These constitute about 1-5% of chemical composition of plasma membrane. These are associated with the phospholipids or with the peripheral proteins to form glycolipids and glycoproteins respectively.

To understand the structure of plasma membrane various models are given out of which the most accepted model is **Fluid Mosaic Model**.

It was given by **Singer** and **Nicholson** (1972). According to this model, the lipid bilayer and integral proteins appear like a mosaic arrangement and the quasi-fluid nature of lipid enables the lateral movement of the proteins within the overall bilayer.

Endomembranous System

Space between the cell wall and nucleus consist of endomembranous system and cell organelles. It is called so because of their coordinated functions with each other. The endomembranous system includes :

(i) Endoplasmic Reticulum (ER)

It is the network of tubular structures. It is found in continuous with the nuclear membrane extending upto the plasma membrane.

ER helps to divide the intracellular space into the following two main compartments:

(a) Luminal (inside ER) compartment

(b) Extra luminal (cytoplasm) compartment

Types of Endoplasmic Reticulum

Endoplasmic reticulum are mainly of two types, depending upon the nature of its membranes:

(a) **Smooth Endoplasmic Reticulum** (SER) These are smooth because they do not bear ribosomes in the form of granules on their surfaces. It is present in cells where they acts as a major site for the synthesis of **lipid** and also helps in synthesis of **steroidal hormone** in animal cells.

(b) **Rough Endoplasmic Reticulum** (RER) They are found extensive and continuous with the outer membrane of nucleus. They are rough due to the presence of ribosome on the surface.

Functions of Endoplasmic Reticulum

Endoplasmic reticulum possess the following functions:

(a) It provides support to the colloidal cytoplasmic matrix so also called as **cytoskeleton** of cells.

(b) Helps in the rapid intracellular transport of the material.

(c) ER membranes contains a variety of enzymes for various metabolic processes, e.g. ATPase, phosphatases, etc.

(d) It (RER) helps in protein synthesis.

(ii) Ribosome

It is a cell organelle found on the endoplasmic reticulum. Due to the presence and absence of it, the ER is also called as **rough** and **smooth** ER. Ribosomes are involved in the vasicular transport and synthesis protein and lipid. It is also found in cytoplasm.

(iii) Golgi Complex (Golgi bodies)

These are the organelles being discovered by **Camillo Golgi** for the first time. It is made up of sac-like structure called **cisternae**. Near to the nucleus it is **convex** type, while away from the nucleus, it is **concave** type. It is mainly involved in the packaging and delivery of cellular products.

(iv) Lysosomes

These are vesicular bodies filled with the hydrolytic enzymes and are involved in the digestion of biomolecule. It is called as suicidal bag of cell.

(v) Vacuoles

These are considered as the major feature of the plant cell. These are filled with water, sap, etc. **Tonoplast** is the membrane of vacuoles that regulate the transport.

(vi) Plastids

These are the major cell organelles of the plant cell, especially chloroplast that contain chlorophyll and is involved in the mechanism of photosynthesis. A typical chloroplast is a double membranous structure that contains a stack of sac-like bodies called **granum.** The wall of granum contains a pigment called **chlorophyll** which is responsible for photosynthesis.

Leucoplasts

These are the colourless plastids of varied shapes and sizes with stored nutrients in the form of carbohydrates lipids and proteins.

Chromoplasts

These are the plastid, which are yellow or reddish in appearance because of the presence of fat soluble carotenoid pigment **carotene**.

Chloroplasts

These are the plastids which are greenish in colour containing photosynthetic pigments **chlorophyll** and **carotenoids**. These pigments are responsible for trapping the light energy, essential for the photosynthesis,

Occurrence

They occur in major number in the photosynthetic mesophyll cells of leaves and green stem.

Ultrastructure

Chloroplasts are also bounded by double membrane envelope. The inner membrane of chloroplast is less permeable than the outer one.

The inner membrane is grounded by a space known as stroma or **matrix,** a dense, colourless and a granular substance mainly formed of soluble proteins.

Thylakoids are number of membranous like flattened structures that run throughout the matrix or stroma. When several thylakoids are arranged or organised in the stack (like the piles of coins), called **grana** or the **intergranal thylakoids**.

Functions of Chloroplast

Chloroplasts possess the following functions:

(a) Helps in photosynthesis, i.e. formation of organic compounds.

(b) In consumption of CO_2 and release of O_2 in photosynthesis.

(c) May also change into chromoplast in order to provide colour to many flowers and fruits.

(d) Helps in storing fat and lipids.

(e) Functions in transduction of energy.

(f) Contains extra chromosomal DNA.

Mitochondria

Mitochondria are membrane bound cell organelles, essential for aerobic respiration of eukaryotic cells. These are also known as **power house of the cell**. Thus, they produces cellular energy in the form of ATP.

Occurrence

Mitochondria are present in all living cell except, prokaryotic cell and certain specialised eukaryotic cell such as anaerobic cells and mature RBCs.

It is revealed from the studies that mitochondria is not easily visible, unless it is specifically stained.

Shape and Size

Mitochondria vary considerably according to the shape and size. They have varying shape such as granular fibrillar, spherical, oval, discodial, etc. Average size of mitochondria is 2-6 μm in length and 0.5 μm in diameter (typical cylindrical or sausage-shaped mitochondria has diameter of 0.2-1.0 μm).

Ultrastructure

A mitochondrion contains two membranes, i.e. outer and inner. Out of which **outer membrane** is smooth and forms the continuous boundary of the organelle. The **inner membrane** is semipermeable to some metabolities. It is infolded into the matrix as incomplete partitions called **cristae** that are responsible for increasing the physiological active area or surface area. The density of cristae determines the intensity of respiration.

The outer and the inner membranes divides its lumen into two aqueous compartments separately, i.e. the outer and the inner compartment.

Inner compartment is also called **matrix**, which forms the inner core of the mitochondrion. The matrix also possesses single circular DNA molecule, a few RNA molecules, ribosomes (70S) and the components required for the synthesis of proteins.

(viii) Cilia and Flagella

These are hair-like projections of cell membrane. Both cilia and flagella are almost identical in structure but differ somewhat in length. As **cilia** are small structures, working as oars (causing the movement of either the cell or the surrounding fluid), while **flagella** are comparatively longer in size than cilia and are responsible for the movement of cell.

According to the electron microscopic studies it is predicted that the cilium or the flagellum are covered with plasma membrane. Their core called the **axoneme**, contains a number of microtubules, running parallel to the long axis. This arrangement of axonemal microtubules is referred to as the (9+2) **array**.

The central tubules are connected by bridges and is also enclosed by a **central sheath**, which is connected to one of the tubules of each peripheral doublets by a **radial spoke.**

Centrosomes is an organelle that generally have two cylindrical structures known as **centrioles**. They are basically surrounded by an amorphous pericentriolar materials.

Both the centrioles in a centrosome lie perpendicular to each other in which each has an organisation like the cartwheel.

They are usually made up of nine evenly spaced **peripheral fibrils** (triplet in nature) of tubulin protein.

The centre part of the proximal region of the centriole possess rod-shaped proteinaceous mass known as hub, which is connected with tubules of the peripheral triplets fibrils known as **radial spokes**.

Nucleus

It is the third main component of the cell. It was first described by **Robert Brown** (1831). It composed of the genetic material (DNA) in a supercoiled stage. This supercoiled stage leads to more condensed form during metaphase stage of cell division and is called **chromosome.**

Chromosomes are the structures found in the nucleus of the cell. It contains DNA and packing protein histone. A human cell has approximately 2 meter long DNA in its chromosomes.

Chromosomes have constrictions in the middle part, it is known as **centromere**. But position of centromere is not fixed at centre as it may found different position of chromosome. Based on position of centromere, chromosome are of following types:

(i) **Metacentric chromosome** in which the centromere is situated at the middle and forms V-shaped chromosome.

(ii) **Sub-metacentric chromosome** in which the centromere is situated slightly away from the centre and forms a α-shaped chromosome.

(iii) **Acrocentric chromosome** which the centromere is situated closed to one end and forms a J-shaped chromosome.

(iv) **Telocentric chromosome** in which the centromere is situated at the end.

Chapter Exercise

1. A cell biologist used ultrasonic method to disrupt the cell and found fragments in a cell organelle which served as site of ATP synthesis. These fragments actually are the parts of

(a) mitochondria (b) lysosomes
(c) liposomes (d) ribosomes

2. In which cell organelle nucleic acid is not present?

(a) Chloroplast (b) Mitochondria
(c) Golgi apparatus (d) Ribosome

3. The term 'power house' is applicable to a class of cell organelle called

(a) Golgi apparatus (b) mitochondria
(c) microsomes (d) peroxisomes

4. Elastic nature of plasma membrane is due to the presence of

(a) lipids (b) proteins
(c) carbohydrates (d) water

5. As they release hydrolase that digest old and damaged cells, the term 'suicidal bags' is used by cells biologists for the

(a) Golgi bodies (b) lysosomes
(c) glyoxysomes (d) peroxisomes

6. Which organelle help in the synthesis of lipids, cholesterol, steroids and visual pigments in epithelial cells of retina?

(a) Golgi bodies (b) RER
(c) SER (d) All of these

7. Some cellular structures are bounded by a single or double membranes, while some other without any membranes. Match the organelle in column I with the nature of membrane in column II and select the correct answer using the codes given below the columns.

Column I	Column II
A. Mitochondria	1. Without membrane
B. Lysosomes	2. Single membrane
C. Ribosomes	3. Double membrane
D Nucleus	

Codes

	A	B	C	D		A	B	C	D
(a)	1	2	3	2	(b)	3	1	1	1
(c)	3	2	1	3	(d)	2	3	1	3

8. Which of the following observations most strongly support the view that mitochondria contains electron transport enzymes that are aggregated into compact association?
   (a) Mitochondria have a highly folded inner wall
   (b) Disruption of mitochondria yields membrane fragmentation which are able to synthesise ATP
   (c) Mitochondria in animal embryos have a tendency to concentrate in cells, which are able to become locomotory structures
   (d) A contractile protein capable of utilising ATP has been obtained from the mitochondria

9. Nucleolus was discovered by
   (a) Robert Hooke (b) Robert Brown
   (c) Robert Koch (d) Robert Dixon

10. A structure found traversing most of the cells, discovered by Porter, Claude and Fullman in 1945 is the
   (a) Golgi apparatus (b) mitochondria
   (c) endoplasmic reticulum (d) lysosomes

11. The organelles, which are responsible for almost all molecular oxygen uptake of the cell as well as 90% of the ATP formed are
   (a) mitochondria (b) chloroplasts
   (c) lysosomes (d) Golgi bodies

12. Which cell structure, sub-structure acts as the site of oxidation-reduction reaction?
   (a) Glyoxysomes
   (b) Mitochondrial cristae
   (c) Lysosomes
   (d) Coated vesicles in Golgi complex

13. Which organelle of the plant cell does not contain functional DNA?
   (a) Nucleus (b) Chloroplast
   (c) Mitochondria (d) Peroxisome

14. Robert Hooke used the term cell in the year
   (a) 1665 (b) 1885
   (c) 1865 (d) 1960

15. The secondary structure of that portion of an integral protein that is buried in the lipid bilayer of a cell membrane is
   (a) β-band (b) α-helix
   (c) β-strand (d) random coil

16. A human egg is nearly a million times as large as human sperm. Most of this size difference is due to the difference in their
   (a) cytoplasmic content (b) nucleus
   (c) Both (a) and (b) (d) mitochondria

17. Cell recognition and adhesion are facilitated by the components of plasma membrane. These components are generally
   (a) protein molecules alone
   (b) lipids alone
   (c) Both (a) and (b)
   (d) glycolipids and glycoproteins

18. According to fluid mosaic model, plasma membrane is composed of
   (a) cellulose and hemicellulose
   (b) phospholipid, intrinsic and extrinsic protein
   (c) phospholipid and intrinsic protein
   (d) phospholipid and hemicellulose

19. Organelle which is responsible for the origin of lysosome is
   (a) ER (b) chloroplast
   (c) Golgi body (d) ribosome

20. Continuity of cytoplasm from cell to cell is maintained through cytoplasmic connections in plants called
   (a) ER (b) tight junction
   (c) gap junction (d) plasmodesmata

21. Lysosomes contain a rich repertoire of
   (a) hydrolases (b) isomerases
   (c) transferases (d) oxidoreductases

22. Smooth endoplasmic reticulum is well developed in the cells which synthesises
   (a) steroids (b) proteins
   (c) carbohydrates (d) All of these

23. The detoxification reaction occurs in which site of liver
   (a) free ribosome (b) RER
   (c) SER (d) hydrophobic interaction

24. Nuclear pores in the nucleus is formed by the
   (a) fusion of nuclear membranes
   (b) absence of nuclear membrane
   (c) movement of cytoplasm through nuclear membrane
   (d) breakage of nuclear membrane

25. A cellular component in bacteria resembling eukaryotic cell is
   (a) cell wall (b) plasma membrane
   (c) ribosomes (d) nucleus

26. Cell theory is not applicable to
(a) bacteria (b) fungi
(c) algae (d) viruses

27. Membrane covering vacuoles in plant cell is
(a) tonoplasm (b) cell membrane
(c) cell wall (d) tonoplast

28. Organelle found in secretory cells in large amount is
(a) chloroplast (b) lysosome
(c) mitochondria (d) vacuole

29. Diagrammatic representation of karyotype (morphological representation of chromosomes) of a species is called
(a) choloroplast (b) cladogram
(c) golgi apparatus (d) ecogram

30. At which phase, size of chromosome is measured?
(a) Metaphase (b) Anaphase
(c) Prophase (d) All of these

31. Term chromosome was coined by
(a) Sutton (b) W Flemming
(c) W Roux (d) Waldeyer

32. Cell theory states that
(a) all cells have nuclei
(b) all cells are living
(c) cells reproduces by both mitosis and meiosis
(d) cells are fundamental structural units of plants and animals

33. Organelle responsible for converting food into energy is
(a) nucleus (b) nucleolus
(c) chloroplast (d) mitochondria

34. Plasma membrane is composed of
(a) proteins (b) lipids
(c) cellulose (d) Both (a) and (b)

35. The sedimentation coefficient of ribosomes is generally 70 S. It breaks up into two subunits whose sedimentation constants are
(a) 50 S and 20 S (b) 40 S and 30 S
(c) 60 S and 10 S (d) 50 S and 30 S

36. Middle lamella of cell wall is mainly composed of
(a) hemicellulose
(b) protein
(c) carbohydrate
(d) calcium pectate

37. Keeping in view about the fluid mosaic model for the structure of cell membrane, which one of the following statements is correct with respect to the movement of lipids and proteins from one lipid monolayer to the other (described as flip flop movement)?
(a) Neither lipids, nor proteins can flip flop
(b) Both lipids and proteins can flip flop
(c) While lipids can rarely flip flop, proteins cannot
(d) While proteins can rarely flip flop, lipids cannot

38. Which of the following is not a constituent of cell membrane?
(a) Proline (b) Phospholipid
(c) Glycolipid (d) Cholesterol

39. Cell structure that along with the micro tubule forms the mitotic spindle
(a) cilia (b) centrioles (c) chromatin (d) nucleolus

40. Analyse the molecule that helps to add stability to the bilayer of the plasma membrane
(a) glycoprotein (b) cholesterol
(c) glycolipid (d) oligosaccharides

41. Cell structure that acts as the site of ribosomal RNA is
(a) cilia (b) centrioles (c) chromatin (d) nucleolus

42. Which structure of a cell serves as the site of modification, packaging and preparation for the distribution of newly formed protenis?
(a) Golgi complex (b) Mitochondria
(c) Nucleus (d) Ribosomes

43. Which of the following types of protein is synthesised in bound ribosomes?
(a) Enzyme (b) Haemoglobin
(c) Insulin (d) Chromosome

44. Which out of these has the ability to detoxing of poisons such as amphetamines, barbiturates, morphine and some pesticides in liver?
(a) Rough endoplasmic reticulum
(b) Smooth endoplasmic reticulum
(c) Golgi apparatus
(d) Lysosome

45. What is/are the functions of cell?
(a) It helps to provide structure to the body
(b) It takes in nutrients from food
(c) It helps in conversion of nutrients into energy
(d) All of the above

Answers

1. (a)	**2.** (c)	**3.** (b)	**4.** (a)	**5.** (b)	**6.** (c)	**7.** (c)	**8.** (b)	**9.** (b)	**10.** (c)
11. (a)	**12.** (b)	**13.** (d)	**14.** (a)	**15.** (b)	**16.** (a)	**17.** (d)	**18.** (b)	**19.** (c)	**20.** (d)
21. (a)	**22.** (a)	**23.** (c)	**24.** (a)	**25.** (b)	**26.** (d)	**27.** (d)	**28.** (a)	**29.** (c)	**30.** (a)
31. (d)	**32.** (d)	**33.** (d)	**34.** (d)	**35.** (d)	**36.** (d)	**37.** (c)	**38.** (a)	**39.** (b)	**40.** (b)
41. (d)	**42.** (a)	**43.** (c)	**44.** (b)	**45.** (d)					

21

Cell Cycle, Cell Division and Cell Signalling

Cell Cycle and Cell Division

Growth and reproduction are the basic characters of living organisms. Thus, **cell division** acts as an essential process in which all organism grows by dividing their cells into two daughter cells that are further divided into four and so on.

All these events by which a cell replicate its genetic material (DNA), generate the constituent material for cell division and finally divides called **cell cycle**. The complete cell cycle within any cell is divided into two main phases, i.e. **interphase** and **mitotic** or M-phase.

Interphase

It is the period or phase between the two mitotic divisions. Approximately 95% time of the cell cycle, interphase. It is also known as **resting phase** because no observable changes are seen regarding cell-division during this phase.

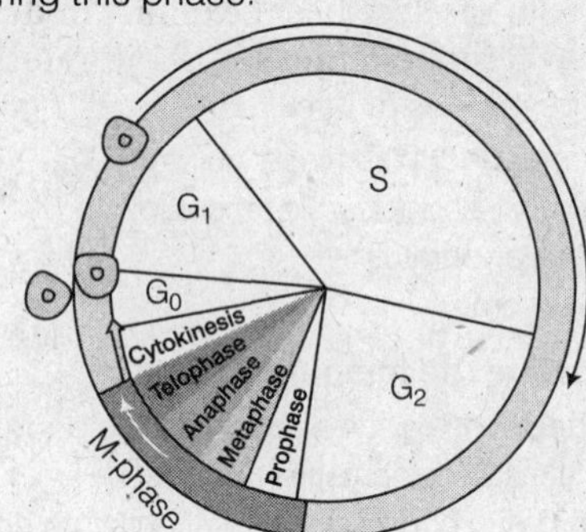

A complete interphase is divided into following three sub-stages phases:

(i) **G_1-phase** (Gap1) It corresponds to a gap phase between the mitosis and the initiation of the S-phase. In this stage, the cell grows in size but there is no duplication of genetic material.

(ii) **S-phase** (Synthesis phase) It is the most important phase of the cell cycle as it represents the duplication of DNA material, i.e. a phase where actual DNA synthesis takes place. There is no increase in the chromosome number during this phase.

(iii) **G_2-phase** This is the phase that occurs between the S-phase and M-phase. During this stage, the synthesis of DNA stops and proteins required for mitosis are synthesised.

Sometimes, the cell does not undergo division before entering into G_1-phase. These cells remain inactive and this stage of inactiveness is called **quiescent stage** (G_0) of the cell cycle.

M-phase

Following the interphase, a cell enters a stage called M-phase. **Mitosis** is a general type of cell division by which a cell only divide itself, while in **meiosis** cell not only divide itself but also divide its genetic material. Both the type of cell divisions come under the karyokinesis stage of M-phase. The other stage of M-phase is cytokinesis, i.e. division of cytoplasm.

Karyokinesis (Division of Cytoplasm)

It is further divided into following types:

Mitosis It is a type of division in which the chromosomes replicates themselves and gets divided equally into two daughter cells. It is also known as an **equational division**.

Mitosis involves the following main phases:

(i) **Prophase** It is known to be the longest and the most complex phase of cell division because it lasts for about 50 minutes nutes of total duration of mitotic phase.

This phase is known for the initiation of condensation of chromosomal material, which during the process of chromatin condensation becomes untangled and finally the **centriole** (already duplicated during S-phase of interphase) begins to move towards the opposite poles of the cell.

For the suitability in study we can categorised prophase as:

(a) **Early prophase** During this phase, condensation of chromosomal material takes place in order to form a compact mitotic chromosomes composed of two chromatids which are attached together at centromere. The most conspicuous change that takes place during prophase is the **formation of mitotic spindle**.

(b) **Late prophase** At the end of the prophase, i.e. during late prophase the **nucleolus disintegrates** gradually and the nuclear envelope disappear.

If we view cells under the microscope, during the prophase the cell will not show nucleolus, nuclear envelope, Golgi complexes, endoplasmic reticulum, etc.

(ii) **Metaphase** It is the phase that starts after the disintegration of nuclear envelope in the late prophase. The chromosomes spreadout through the cytoplasm of the cell and are seem to be slightly shortest and thickest of all. The chromosome can be easily observed under the microscope.

Mitotic Poison: Colchicine

Colchicine, an alkaloid is extracted from the corms of **autumn crocus** (*Colchicum autumnale*), acts as a poison for mitosis as it does not allow the formation of mitotic spindle to takes place by preventing the assembly of microtubules.

Each chromosome at this stage is made up of two longitudinal threads (sister chromatids) and are held together by the centromere in the centre. At the surface of each centromere a disc-shaped structures called **kinetochores** are present, which helps in the attachment of spindle fibres to the chromosomes.

The chromosomes finally arrange themselves at the equator in one equitorial plane known as **metaphase plate**.

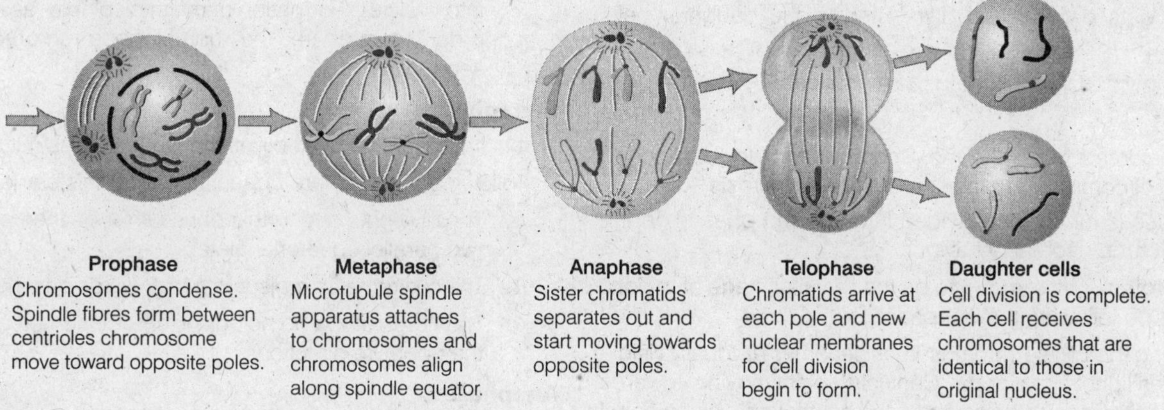

Cytokinesis

After the division of nucleus, cytoplasm gets divided by the formation of cell plate (in plant cells) between two newly formed nucleus and by a formation of cell furrow (in animal cells). Also, the cell organelles are divided equally between the two newly formed daughter cells.

Differences between Karyokinesis and Cytokinesis

	Karyokinesis	Cytokinesis
1.	It is known as division of daughter chromosomes, M-phase.	It is known as division of cytoplasm.
2.	It occurs or starts with karyokinesis itself.	It occurs at the end of M-phase.
3.	It can occur in dependent of cytokinesis.	It cannot occur without karyokinesis.

Significance of Mitosis

Mitosis has following significances:

(i) It helps in the production of diploid daughter cells with equal and identical genetic complement.

(ii) Mitosis helps in growth of multicellular organisms.

(iii) It also helps in maintaining a proper cell size by dividing somatic cell.

(iv) It is helpful in cell repair mechanism, e.g. continuous replacement of the cells like of the upper layer of the epidermis, cells of the lining of the gut and blood cells.

(v) The growth of multicellular organism is due to the mitosis as, it is known that a cell cannot grow in size beyond a certain limit without disturbing the ratio between the nucleus and the cytoplasm.

After reaching the particular size, the cell divides in order to restore the nucleocytoplasmic ratio. Therefore, the growth of cell takes place by the increase in the number of cells, rather than the increase in the size of the cell.

(vi) It is also helpful in producing new cells for healing wounds and for regeneration.

(vii) Mitosis plays an important role in a continuous growth of plants throughout their life as meristematic tissues like the apical and the lateral cambium.

Meiosis

Meiosis is another type of cell division by which the cell divides its genetic material. It is divided into two main sub-stages, i.e. meiosis-I and II.

Meiosis-I

It is known as **reductional division** because in this, the genetic material gets divided and distributed in the daughter cells equally.

It is also divided into following main phases:

Prophase-I

It is considered to be the most complicated and prolonged phase as if compared to the similar stage in mitosis.

This phase is further sub-divided into five sub-phases on the basis of chromosomal behaviour.

(i) **Leptotene** It is known to be the very first stage of meiotic division following the interphase.

During this phase chromosomes becomes gradually visible under light microscope. Centrioles start moving towards opposite ends or poles and each centriole develops astral rays. Each chromosome is attached to the nuclear envelope through the attachment plate at both of its ends.

(ii) **Zygotene** This is the next sub-stage that takes place after the completion of the previous one. This is also a short lived stage like leptotene.

During this phase homologous chromosomes pair up. This pairing is done in a such a way that the genes of the same character present on the two chromosomes lie exactly opposite to each other. This process of association is known as **synapsis**. Formation of **synaptonemal complex** takes place by a pair of homologous chromosomes that show synapsis.

(iii) **Pachytene** This is the stage which immediately follows zygotene where the pair of chromosomes become twisted spirally around each other and cannot be distinguished separately. This stage is comparatively long lived as compared to the previous two stages.

During this stage bivalent chromosomes clearly seen as tetrads. Sometimes exchange of genes or crossing over between the two non-sister chromatids of homologous chromosomes occurs at the points called **recombination nodules**. By the end of pachytene recombination gets completed leaving the chromosomes linked at the sites of crossing over.

(iv) **Diplotene** It is the stage of longest duration of all following changes are observed during this stage the **synaptonemal complex** appears to get dissolve while, the chromatids of each tetrad remain clearly visible. Recombined homologous chromosomes of the bivalents get separated and form chiasmata. Which is necessary for the separation of homologous, chromosome which have undergone the process of crossing-over.

(v) **Diakinesis** This is known to be the final stage of meiotic prophase-I. Also known as **terminalisation**, due to the shifting of chiasmata towards the end of the chromosomes.

During this stage chromosomes become fully condensed. Nucleolus degenerates. Breakdown of nuclear envelope into vesicles. Formation of meiotic spindle (as in mitosis) in order to prepare the homologous chromosomes for separation.

Metaphase-I

It is the stage followed by prophase (same as mitosis).

Following changes are observed during this stage:

(a) The bivalents during this phase arrange themselves on the two parallel equatorial plates.

(b) The centromeres project little bit towards periphery.

(c) The fibres of the homologous chromosomes are always in the opposite directions.

Anaphase-I

This is the next phase after metaphase-I in which homologous chromosomes break their connection with each other and get separated.

This process of separation of homologous chromosomes is known as **disjunction**. The separated chromosomes are univalents and are also called **dyads**.

On reaching at the end of the anaphase, the two groups of chromosomes are produced.

The homologous chromosome condensed and gets separated during all these three remaining phases i.e. **metaphase-I**, **anaphase-I** and **telophase-I.**

Meiosis-II

It is also called **homotypic** or **equational division** because it leads to the separation of chromatids and centromeres and the chromosome number remains same as produced in the meiosis-I.

During this stage of cell division, the chromatids get separated and result into the formation of four daughter cells. Like meiosis-I, it is also divided into four main phases, i.e. prophase-II, metaphase-II, anaphase-II, telophase-II. In many organisms cytokinesis occurs after mitosis-I and meiosis-II, i.e. it may not occur between these meiotic divisions.

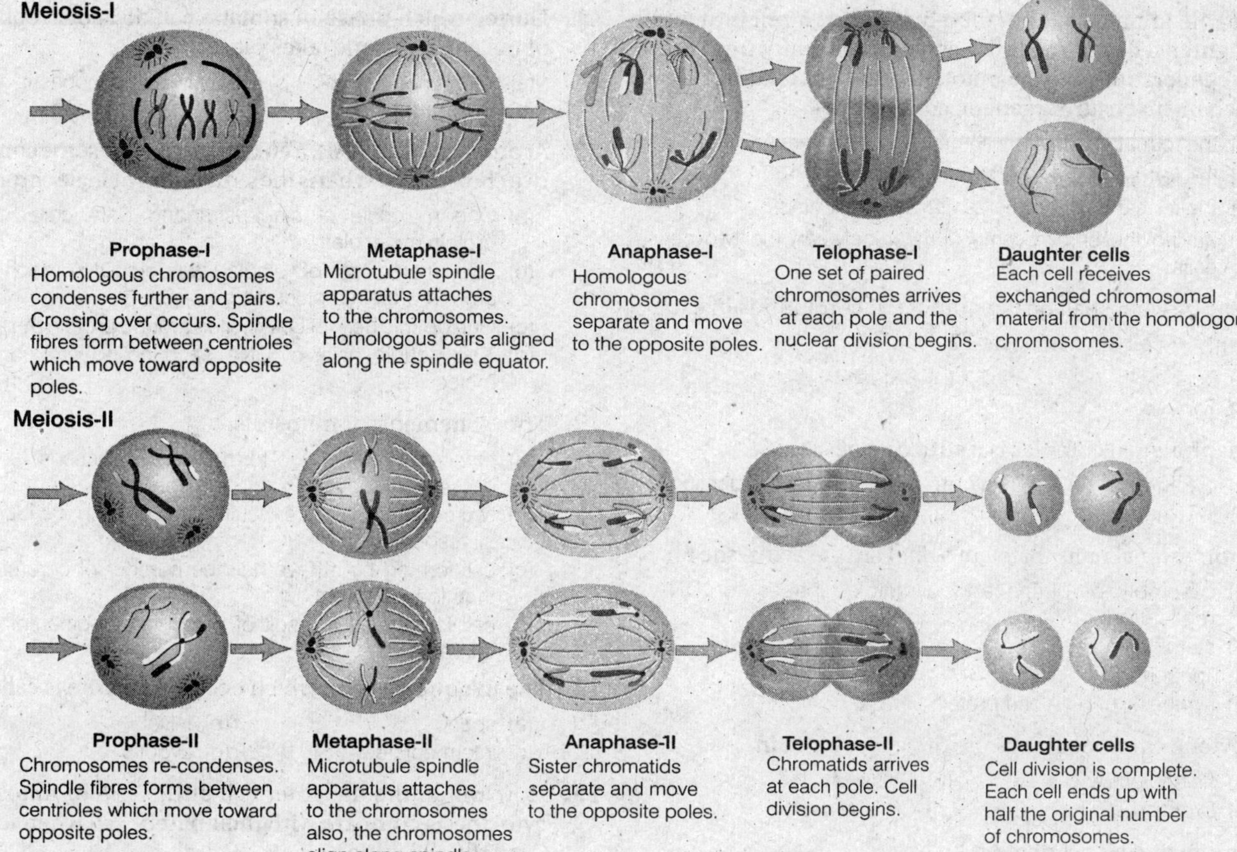

Cell Signalling

Cell signalling is the phenomenon of a cell of how one receives the message from one cell and transmit it to the other cells. It is performed with the help of different signalling molecules.

Signalling of molecules may be of following two types:

(i) Direct cell to cell signalling.

(ii) Indirect signalling involving signal molecules.

It is further divided into following three types:

(i) **Endocrine**, in which the signal molecules travel a good distance, for the target, e.g. hormone.

(ii) **Paracrine**, in which the signal molecules travel less distance, for the target, e.g. neurotransmitters.

(iii) **Autocrine**, in which the signal molecules activates cells on their own.

Signal Molecules and their Receptors

Chemically, the signal molecules may be of following types:

(i) Peptide (ii) Steroids (iii) Amino acid derivatives

On the basis of nature of signalling molecules, the receptor of these molecules are of following two types;

(i) **Membranous** or **membrane bound receptors** These are found on the membrane of the cell for proteinaceous or hydrophilic signalling molecules.

(ii) **Cytoplasmic** or **nuclear receptors** These are used for the hydrophobic signalling molecule.

Chapter Exercise

1. The role of mitosis is not merely to divide a cell into two daughter cells but to ensure genetic continuity from one cell generation to the another cell generation. The mechanism ensuring genetic continuity is
   (a) the formation of cells with new chromosomes
   (b) the formation of two daughter cells
   (c) the formation of two cells with identical DNA
   (d) having the chromosome number between the two new cells

2. If $2n = 8$, what shall be the number of chromatids in each daughter cell after meiosis-I ?
   (a) Four (b) Sixteen
   (c) Two (d) Eight

3. Synaptonemal complex is a structure of
   (a) cytokinesis (b) chromosomal disfunction
   (c) terminalisation (d) chromosomal pairing

4. Chromosomal puffs (Balbiani's ring) are the active site for
   (a) assembly of purine and pyrimidine bases for DNA replication
   (b) steroidogenesis
   (c) lipogenesis
   (d) synthesis of RNA and protein

5. Polytene chromosomes were first discovered in
   (a) *Phlebotomus* (b) *Glossina*
   (c) *Drosophila* (d) *Chironomus*

6. Identify the false statement.
   (a) Pairing of paternal and maternal chromosomes during zygotene is called synapse
   (b) Chiasmata are the site for crossing over of genetic material
   (c) Terminalisation occurs during diplotene
   (d) Prophase-I is the most complicated and longest stage of meiosis

7. Which significant changes occur(s) during the diakinesis phase of meiotic prophase-I?
   (a) Breakdown of nucleolus
   (b) Fragmentation of nuclear membrane
   (c) Terminalisation
   (d) All of the above

8. During cell division, the splitted chromosomes move towards opposite poles due to
   (a) microtubules (b) centrioles
   (c) cytoplasmic streaming (d) phragmoplast

9. Mitosis in adult animal does not occur in
   (a) skin (b) bone marrow
   (c) testes (d) nails

10. In which stage of meiotic prophase-I, chromosomes become thread-like and visible?
   (a) Diakinesis (b) Leptotene
   (c) Zygotene (d) Pachytene

11. During which phase in somatic cell division, replication of the chromosome takes place?
   (a) Interphase (b) Middle prophase
   (c) Early prophase (d) Late prophase

12. At prophase of mitosis, a eukaryote chromosome consists of two chromatids. What is the structure of a single chromatid?
   (a) One molecule of single-stranded DNA coiled around protein molecules
   (b) Two molecules of single-stranded DNA each coiled around protein molecules
   (c) One double helix of DNA coiled around protein molecules
   (d) Two double helix of DNA each coiled around protein molecules

13. True statement for mitosis is
   (a) the cell formed by it form diverse functions, i.e. show division of labour
   (b) two cells formed as a result of this division are identical in all respects
   (c) cell formed by it have half the number of chromosomes than that of parent cell
   (d) cell formed as a result of mitosis have different genetic characters

14. The meiotic division which occurs in zygote is called
   (a) sporic (b) initial
   (c) intermediate (d) zygotic

15. How would the daughter cell at the end of mitosis and cytokinesis compare with their parent cell when it was in G_1 of the cell cycle?
   (a) The daughter cells have half the amount of cytoplasm and half the amount of DNA
   (b) The daughter cells have half the number of chromosomes and half the amount of DNA
   (c) The daughter cells have the same number of chromosomes and half the amount of DNA
   (d) The daughter cells have the same number of chromosomes and same amount of DNA

16. Genetic homogeneity and exact similarity between chromosomes of various cells within the same type of tissue of a plant is largely due to
   (a) cytokinesis (b) mitosis (c) meiosis (d) fertilisation

17. Which type of chromosomes segregate when a cell undergoes meiosis?
   (a) Homologous chromosomes
   (b) Non-homologous chromosomes
   (c) Both (a) and (b)
   (d) Centric and acentric chromosomes

18. Cellular structures which always disappear during mitosis are
   (a) plastids and mitochondria
   (b) nuclear envelope and nucleolus
   (c) cell wall and plasmalemma
   (d) cell wall and nucleolus

19. **Phragmoplast is related to the**
(a) division of nucleolus
(b) cell elongation
(c) cytokinesis
(d) assemblage of chromosomes at metaphase

20. **Complete process of meiosis involves**
(a) one cytoplasmic division with only one chromosome duplication
(b) two cytoplasmic divisions with one duplication of chromosome
(c) two cytoplasmic divisions with two duplication of chromosome
(d) one cytoplasmic division with two duplication of chromosome

21. **Meiotic divisions in an angiospermic plant can be observed by examining the cells of**
(a) the apical meristem of the stem
(b) the apical dividing cells in the root tip
(c) cells of the vascular cambium when they are dividing
(d) pollen mother cells when they are undergoing the divisions of another

22. **Distribution of genetic material occurs from generation to generation and permits great genetic diversity by frequent reorganisation of genetic complement. This occurs during**
(a) mitosis (b) meiosis
(c) organogenesis (d) metamorphosis

23. **Which is the correct sequence of labelling of the given figure of interphase (G_2-phase)?**

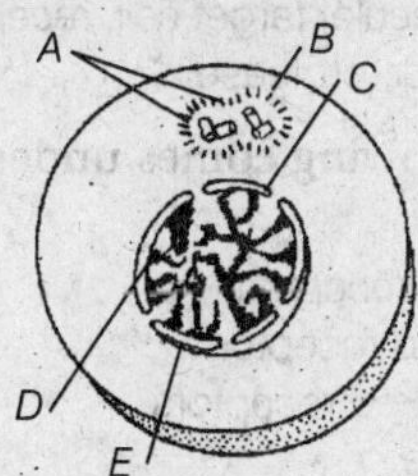

(a) A–Centrosome, B–Aster, C–Chromatin, D–Nucleus, E–Nucleolus
(b) A–Asters, B–Centriole, C–ER, D–Chromatin, E–Protoplasm
(c) A–Centrioles, B-Pre-spindle, C–Nucleus, D–ER, E–Nucleolus
(d) A–Centrosome, B–Spindle, C–Chromatin, D–Nucleolus, E–ER

24. **Centrioles plays an important role in the formation of**
(a) brush border of a cell
(b) synthesis of actin and myosin protein
(c) transduction of signals from one cell to other
(d) spindle apparatus of a dividing cell

25. **Dyad acts as a pair of**
(a) non-sister chromatids
(b) homologous chromosomes
(c) non-homologous chromosomes
(d) sister chromatids

26. **The nuclear membrane disappears at which stage of mitosis**
(a) telophase (b) prophase
(c) metaphase (d) anaphase

27. **The shortest phase of the cell cycle is**
(a) G_2 (b) S
(c) G_1 (d) M

28. **A disc-shaped structure found on the sides of centromere of chromosome is**
(a) kinetochore (b) telomere
(c) chromocentre (d) centronema

29. **The major component of the synaptonemal complex is**
(a) protein (b) lipid
(c) fat (d) nucleic acid

30. **During pachytene stage of meiosis, the chromosomes appears**
(a) four-stranded (b) single-stranded
(c) double-stranded (d) eight-stranded

31. **Formation of a chromatid takes place in**
(a) S-phase (b) Metaphase
(c) G_1-phase (d) G_2-phase

32. **When cells respond to an extracellular signal, they often converts the information from one form to the another. The process is**
(a) signal transduction (b) signal transformation
(c) signal amplification (d) signal interference

33. **Steroid hormones, estrogen and testosterone are most likely to bind to**
(a) enzyme linked membrane receptor
(b) cytoplasmic receptors
(c) membrane ion channel
(d) G-protein linked membrane receptor

34. **Which signal binds to receptors and stimulates the near by cells?**
(a) Exocrine (b) Endocrine
(c) Paracrine (d) Autocrine

35. **When could the cell normally proceed to mitosis without intruption?**
(a) It has entered the G_2-phase
(b) It has entered the S-phase
(c) Anytime during the cell division
(d) None of the above

36. **How many mitotic divisions are required to produce 128 cells from a single cell?**
(a) 7 (b) 8
(c) 16 (d) 32

37. **Hypnotic meiosis occurs in**
(a) *Puccinia* (b) *Chlamydomonas*
(c) *Pteris* (d) *Puccinia*

38. **During metaphase mitosis chromosomes**
(a) undergoes coiling (b) moves towards the poles
(c) line up at the equator (d) break and disintegrate

39. If a cell divides in every one mintute. And this rate of division it can fill a 100 mL of beaker in one hour. Calculate how much time does it take to fill a 50 mL beaker?

(a) 30 min (b) 60 min (c) 59 min (d) 65 min

40. Which aspect of mitosis is affected by Colchicine in inducing polypeptide?

(a) Spindle formation (b) Cell plate formation
(c) Chromosome doubling (d) DNA duplication

41. Cell cycle of germinal cell has

(a) 2 successive mitotic divisions
(b) 2 successive reductive divisions
(c) very short prophase in first division
(d) on reduction division followed by one mitotic division

42. During cell division, sometimes there will be failure of separation of sister chromatids. This cucuta is called

(a) coincidence (b) fusion
(c) non-disjunction (d) interference

43. During synapsis, the number of thread (chromoemata) in each chromosome is

(a) 2 (b) 4 (c) 8 (d) many

44. Synapis is a characteristic of

(a) leptotene (b) diplotene (c) zygotene (d) pachytene

45. Cell division of mitosis in a normal process in a living cell but sudden and abnormal mitosis in an organ will sometimes result in

(a) cancer (b) new organ
(c) zygote (d) gastrula

46. Estimate the number of generations of mitotic divisions that must occur in a cell of root tip to form 256 cells

(a) 8 (b) 32 (c) 64 (d) 128

47. DNA content gets doubled during which phase of cell cycle?

(a) G_1-phase (b) G_2-phase
(c) Metaphase (d) S phase

48. Which of the following in unique to mitosis and is not a part of meiosis?

(a) Homologous chromosomes pair and forms bivalents
(b) Chromatids are separated during anaphase
(c) Homologous chromosomes behaves independently
(d) Homologous chromosomes crossover

49. Arrange the following events of meiosis in the correct sequence.

I. Terminalisation II. Crossing over
III. Synapsis IV. Disjunction of genomes

(a) II, I, IV, III (b) III, II, I, IV
(c) IV, III, II, I (d) I, IV, III, II

50. In paracrine signalling, the signalling molecule affects only

(a) target cell close to the cell from which it was secreted
(b) target cells distant from its site of synthesis in cells of an endocrine organ
(c) Both (a) and (b)
(d) None of the above

51. Identify the one which is not a type of a signalling molecule.

(a) Insulin (b) Thyroxin
(c) Testosterone (d) Acetylate cycles

52. SH_2 domain specifically binds to

(a) Ca^{2+}
(b) GDP
(c) phosphorylated serine residue
(d) phosphorylated tyrosine residue

53. When a .... reaches its ...., there is a specific means of recurring it and acting on the message. This task is the responsibility of a specialised proteins called ....

(a) signalling molecule, receptor, G-protein
(b) signalling molecule, target cell, G-proteins
(c) signalling molecule, target cell, receptors
(d) kinase, receptor, protease

54. Which of the following comes under the category of cell surface receptor?

(a) Enzyme linked receptors
(b) G-protein linked receptors
(c) Ion-channel linked receptors
(d) All of the above

55. In terms of cell communication, what do the bacterial pathogens such as cholera and anthrax have in common?

(a) they prevent the production of key signalling molecules
(b) they alter the chemical structure of key signalling molecules
(c) they destroy the receptors for key signalling molecules.
(d) they block the normal functioning of signal transduction mechanisms.

Answers

1. (c)	**2.** (a)	**3.** (d)	**4.** (d)	**5.** (c)	**6.** (c)	**7.** (d)	**8.** (a)	**9.** (d)	**10.** (b)
11. (a)	**12.** (c)	**13.** (b)	**14.** (d)	**15.** (c)	**16.** (b)	**17.** (a)	**18.** (b)	**19.** (c)	**20.** (b)
21. (d)	**22.** (b)	**23.** (b)	**24.** (d)	**25.** (d)	**26.** (b)	**27.** (d)	**28.** (a)	**29.** (a)	**30.** (a)
31. (a)	**32.** (a)	**33.** (b)	**34.** (c)	**35.** (b)	**36.** (a)	**37.** (b)	**38.** (c)	**39.** (c)	**40.** (a)
41. (d)	**42.** (c)	**43.** (a)	**44.** (c)	**45.** (a)	**46.** (a)	**47.** (d)	**48.** (c)	**49.** (b)	**50.** (a)
51. (d)	**52.** (d)	**53.** (c)	**54.** (d)	**55.** (d)					

22

Basis of Genetics and Mendelian Law of Inheritance

Genetics, Heredity and Variation

Genetics is the branch of biology dealing with the study of heredity and variations of organisms in a population. The term genetics was postulated by Bateson in the year 1906.

Heredity is the phenomenon of transmission of characters from parents to the offsprings. This is the fact due to which we look like our parents.

Variation is the difference, can be physiological, morphological, cytological among the individuals of a population of a same species and offsprings of the same parents. It may be somatic or germinal affecting the somatic and germinal cells of an organism respectively.

Some Common Terms Related to Inheritance

- **Gene** is the unit of inheritance.
- **Allele** is the pair of contrasting characters of a single gene.
- **Phenotype** is observable characters of an offspring.
- **Genotype** is genetic makeup of an organism.
- **Test cross** performed between unknown dominant genotype and a recessive parent to determine the genotype of an organism.
- **Multiple allele** are multiple forms of a Mendelian character occurring on the same gene, e.g. ABO blood group.
- **Pleiotropy** is a mechanism in which a single gene product is responsible for multiple functions i.e. more than one phenotypic effect.
- **Pedigree analysis** is a method used to study the pattern of inheritance of genes in human genetics.
- **Punnett square** is a graphical representation to calculate the probability of all possible genotypes of offsprings in a genetic cross.

Mendel's Experiment

Mendel experimented on garden pea plant (*Pisum sativum*) and gave several postulates in the field of genetics. He selected pea plant because it is bisexual, self pollinating plant and is also easy to grow and took small place.

In pea plant, out of the seven characters Mendel studied the following dominant and recessive traits:

The Seven Characters Mendel Studied
(The dominant and recessive traits)

Characters	Traits	
	Dominant	Recessive
Length of stem	Tall	Dwarf
Position of flower	Axial	Terminal
Colour of flower	Red or purple	White
Shape of pod	Inflated	Constricted
Colour of pod	Green	Yellow
Shape of seed	Round (smooth)	Wrinkled
Colour of seed	Yellow	Green

On the basis of his experimental studies, Mendel postulated three basic laws known as Mendel's laws of heredity or Mendel's law of inheritance.

Mendel's Law of Inheritance

The three basic laws of inheritance are given below:

1. **Law of Dominance** (First Law)

This law has its basis from the monohybrid cross. According to this law, when a cross is made between two homozygous (pure line) individuals considering contrasting traits of simple character, then the trait that appears in F_1 hybrid is called **dominant** and the other one is called **recessive trait.**

Monohybrid Cross

In the monohybrid cross, between a **tall** and a **dwarf** plant, only tall plants appears in F_1-generation. But when the F_1 offsprings allowed to be self-fertilised, both tall and dwarf characters appear in F_2-generation. It shows that in F_1-generation hybrid, the character of tallness dominates or conceals the character of dwarfness which inturn is unable to express itself in F_1-generation.

Besides pea plant, the law of dominance has also been observed in different characters in many plants.

For example

- **Sunflower** (branching), in which the branched habit is dominant over unbranched habit.
- **Cotton** (colour of lint), in which coloured lint is dominant over white lint.
- **Maize** (type of kernel), in which round starchy kernel is dominant over wrinkled sugary kernel.
- **Snapdragon** (colour of flower), in which red-coloured flowers are dominant over non-red colour flowers.

The another very interesting example is **wheat** in which the susceptibility to rust is dominant over immunity to rust.

Deviation from Law of Dominance

The law of dominance does not occur universally. After Mendel, several cases were recorded by different scientists, where a clear cut deviation from law of dominance was seen.

Such a deviation may be seen in the form of incomplete dominance and codominance.

(i) **Incomplete dominance** When F_1 hybrids exhibited a mixture or blending of characters of two parents, the case is considered as that of **incomplete or blending inheritance**. It simply means that the two genes of allelomorphic pair are not related as dominant or recessive but each of them express itself partially, e.g. in case of 4 O' clock plant (*Mirabilis jalapa*), when plants with red flower (homozygous, dominant, RR) are crossed with the plants having white flower (homozygous, recessive, rr), the heterozygous F_1 hybrid (Rr) bears pink flower.

When these pink flowers are self-pollinated they develop red (RR), pink (Rr) and white (rr) flowers in the ratio 1 : 2 : 1, respectively.

(ii) **Codominance** In codominance, both alleles of a pair express themselves fully in F_1 hybrid (it is contrary to the situation seen in incomplete dominance, where traits express themselves only partially), e.g. It is best observed in case of cattles, in which, when a cattle with red coat (RR) is crossed with the cattle having white coat (WW) the F_1 hybrid possess roan Appearance (RW). In roan appearance, the red and white hairs occur in patches but no hairs have intermediate colour of red and white.

2. Law of Segregation or Law of Purity of Gametes (Second Law)

This law has the monohybrid cross as its basis. According to this law, in F_2 **hybrid**, the dominant and recessive characters though remain together for a long time but does not contaminate or get mixed with each other thus, get separated or segregated out at the time of gametogenesis (gamete formation).

The gamete thus, formed receives either dominant or the recessive character out of them.

This law is also known as law of purity of gametes because a gamete when formed is always pure for a particular trait, due to the fact that it always contains the factor which determines that single trait pertaining a particular character.

For proper understanding of Mendel's law of segregation, the formation of hybrid is considered from pure line homozygous parents, e.g. pure tall plant (homozygous, TT) when crossed with pure dwarf plant (homozygous, tt), hybrid tall plants (heterozygous, Tt) are produced.

This heterozygous hybrid have one factor of tallness and one factor of dwarfness.

It shows that in F_1-generation hybrid, one factor for plant length comes from one parent and the other from second parent.

Further confirmation of segregation can be done when the plants of F_1-generation are self-fertilised. The recessive dwarf trait reappears in F_2-generation in the phenotypic ratio of 3 : 1 or genotypic ratio of 1 : 2 : 1.

Reappearance of dwarfness in F_2-generation is not possible until the plants in F_2-generation are formed by the fusion of pure plant gametes of F_1-generation.

Thus, the law of segregation is universal in its applications and has been found to occur in plants and animal both.

3. Law of Independent Assortment (Third Law)

This law has its basis from the dihybrid cross. According to this law, the inheritance of one character is always independent to the inheritance of other character within the same individual. In other words, we can say that if the inheritance of more than one character is studied simultaneously, the factors or genes for each characters assort out independently to the other gene or factor.

Dihybrid Cross

The dihybrid cross of Mendel can be a very good example of independent assortment. The presence of new combinations, i.e. round yellow (RRYY) and wrinkled green (rryy) suggests that the genes for shape of seed and colour of seed are assorted independently.

When we look at the place of such an independent assortment then it looks like the gamete formation, where presence of four different combinations (RY, Ry, rY, ry) suggests that both the genes for both the characters assort independently to each other.

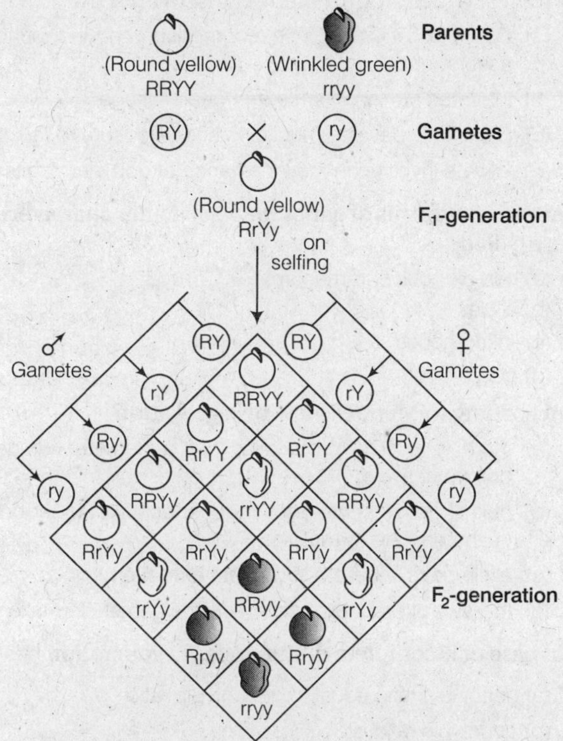

Phenotypic ratio Round yellow : Round green : Wrinkled yellow : Wrinkled green = 9 : 3: 3 : 1

Genotypic ratio 1 : 2 : 2 : 4 : 1 : 2 : 1 : 2 :1

Result of dihybrid cross where the two parents differed in two pairs of contrasting traits seed

Exceptions to Principle of Dominance and Principle of Paired Factors

Incomplete Dominance

It is a phenomenon in which phenotype of the F_1-hybrid offspring do not resemble any of the parent, but is an intermediate between the expression of two alleles in their homozygous state.

It is not similar to pre-Mendelian blended inheritance because parental characters reappear in F_2-generation. Here, the phenotypic ratio deviates from Mendel's monohybrid ratio, e.g. snapdragon (dog flower, *Antirrhinum majus*) and 4 O'clock plant (*Mirabilis jalapa*).

Codominance

It is the phenomenon in which two alleles express themselves independently when present together in an organism. In other words, it is the phenomenon in which offspring shows resemblance to both the parents, e.g. ABO blood grouping in humans. More than two alternative forms (allele) of a gene occupying the same locus on a chromosome in a population are known as multiple alleles.

The ABO blood grouping is a good example of **multiple alleles**. In this case, more than two, i.e. three alleles are present governing the same character.

Test Cross

This is a method devised by Mendel to determine the genotype of an organism. In this cross, the organism with an unknown dominant genotype is crossed with the recessive parent, instead of self-crossing, e.g. in a monohybrid cross, between violet colour flower (W) and white colour flower (w), the F_1-hybrid was a violet colour flower.

Pleiotropy

It is the phenomenon in which a single gene product may produce multiple or more than one phenotypic effect.

A pleiotropic gene often has a more evident effect on one trait called the **major effect** and less evident effects on other traits called **secondary effects**.

When a number of related changes are caused by a pleiotropic gene, they are called **syndrome**.

- **Phenylketonuria** (PKU) is a defect in which an individual lacks enzyme needed to change phenylalanine to tyrosine in liver. This results in multiple phenotypes including mental retardation, eczema and pigment defects.

Polygenic Inheritance

In this case, traits are controlled by three or more genes and the graded phenotypes are due to the additive or cumulative effect of all the different genes of the trait, e.g. human skin colour, height and intelligence.

Polygene is a gene where one dominant allele controls only a unit or partial quantitative expression of a trait.

It also takes into account the influence of environment and is called **quantitative inheritance**, as the character/phenotype can be quantified, like amount of pigment, intelligence in human beings and milk yield in animals have been found to be determined by many genes and their effects have been cumulative.

Complementary Genes

These genes complement the effect of each other to produce a phenotype so, they are called **complementary genes**, e.g. there are two varieties of sweet pea (*Lathyrus odoratus*), controlled independently by two different genes.

Chapter Exercise

1. How did Mendel's studies in genetics differ from earlier studies of breeding and inheritance?

(a) Mendel worked with plants, earlier studies used animals
(b) Mendel was able to explain the 'blending' hypothesis
(c) Mendel's work was more quantitative
(d) Mendel worked with wild species, not the domesticated ones

2. Which statement describes the work of Gregor Johnn Mendel?

(a) He developed some basic principles of heredity without having knowledge of chromosomes
(b) He developed the microscope for the study of genes in pea plants
(c) He explained the principle of dominance on the basis of the gene chromosome theory
(d) He used his knowledge of gene mutations to explain the appearance of new traits in organisms

3. One of the Mendel's pure strains of pea plants had green peas. How many different kinds of eggs could such a plant produce with regard to pea colour?

(a) One (b) Two (c) Four (d) Eight

4. A codominance cross between a homozygous red and a homozygous white snapdragons produces plants with

(a) flower with both red and white patches
(b) red flowers
(c) pink flowers
(d) sterile flowers

5. Mendel developed his basic principles of heredity by

(a) microscopic study of chromosomes and genes
(b) mathematical analysis of the offspring of pea plants
(c) breeding experiments with *Drosophila*
(d) ultracentrifugation studies of cell organelles

6. Mendel started his experiments with pure strains of peas which was developed by removing all

(a) female plants
(b) male plants
(c) typical plants in each generation
(d) weak plants

7. Why was Mendel fairly certain that he had a pure breed variety when he collected seeds from a pea plant?

(a) The flower was self-pollinating in the plants
(b) Peas bear a large number of 'offspring'
(c) Because the varieties are so distinctive
(d) He maintained extremely accurate records

8. Mendel was the first biologist

(a) to study the mechanism of inheritance
(b) to discover that chromosomes that are responsible for inheritance
(c) to use statistical methods in analysing his results
(d) All of the above

9. Using the results of his experiments with pea plant crosses, Gregor Johnn Mendel discovered the

(a) principles of dominance, segregation and independent assortment
(b) intermediate inheritance and gene linkage
(c) pea plants develop mutations after exposure to radiation
(d) DNA is involved in the inheritance of dominant traits

10. Alternative forms of genes for a particular characteristic are called

(a) homologous chromosomes
(b) alleles
(c) linked genes
(d) genotypes

11. According to Mendel's law of segregation,

(a) there is a 50% probability that a gamete will get a dominant allele
(b) pair of genes segregates independently of other genes in the gamete formation
(c) allele pairs separate in gamete formation
(d) the laws of probability determine the gamete formation

12. In case of incomplete dominance, F_2-generation has

(a) genotypic ratio equal to phenotypic ratio
(b) genotypic ratio is 3 : 1
(c) phenotypic ratio is 3 : 1
(d) None of the above

13. In Mendel garden peas, the tall allele (T) is dominant over the dwarf allele (t) and the green pod allele is dominant (Y) over the yellow pod allele (y). What is the genotype of a heterozygous tall, homozygous green pod pea plant?

(a) TtYy (b) TTYY
(c) TtYY (d) TTYy

14. Mendel made some crosses where he looked two characteristics, at once round yellow peas crossed with wrinkled green peas, e.g. he did this because he wanted to find out

(a) how new characteristics originated
(b) whether different characteristics were inherited together or separately
(c) how plants and animals adapt to their environments
(d) whether the characteristics influence each other whether the colour affects degree of roundness

15. One reason for Mendel's success was that he

(a) repeated his experiments many times
(b) carefully used controlled experiments
(c) used plants with easily observable traits
(d) All of the above

16. Mendel developed his basic principles of heredity by
(a) examining chromosomes with microscope
(b) analysing large numbers of offsprings
(c) using X-rays to induce mutations
(d) observing crossing over during meiosis

17. Segregation of genes takes place during
(a) metaphase (b) anaphase
(c) prophase (d) zygote formation

18. Mendel published the results of his experiments in
(a) 1568 (b) 1773
(c) 1866 (d) 1921

19. If Mendel had chosen more than seven traits in his plants, he might have run into some confusing results regarding the
(a) phenomenon of dominance
(b) law of segregation
(c) law of independent assortment
(d) All of the above

20. Number of characters used by Mendel in his experiment are
(a) 8 (b) 7 (c) 6 (d) 4

21. Mendel published the results of his experiments on the garden pea in the year
(a) 1856 (b) 1875 (c) 1776 (d) 1866

22. The trait phenotypic ratio of a haploid organism is
(a) 1 : 2 : 1 (b) 3 : 1
(c) 1 : 1 (d) 9 : 3 : 3 : 1

23. Multiple alleles control the character of
(a) only skin colour
(b) only blood groups
(c) sickle-cell anaemia
(d) blood group and skin colour

24. Number of gametes produced by individuals having the genotype of AaBbCc are
(a) six (b) four (c) eight (d) two

25. Phenomenon of controlling several phenotypic characters by a single gene is
(a) epistasis (b) codominance
(c) multiple allelism (d) pleiotropy

26. In Mendel's experiment, the contrasting traits
(a) showed blending at F_1 stage
(b) showed blending at F_2 stage
(c) did not showed any blending at either F_1 or F_2 stage
(d) showed blending at both F_1 and F_2 stage

27. Diagram which shows the production of gametes by the parents, the formation of zygotes, the F_1 and F_2-generation is
(a) Punnett square (b) idiogram
(c) checker board (d) pedigree

28. A genotype AaBbCcDd will produce gametes in a number of
(a) 4 (b) 16
(c) 32 (d) 64

29. Which of the following will best prove the codominance?
(a) Flower colour in snapdragon
(b) blood group in human
(c) Skin colour in humans
(d) Both (b) and (c)

30. The ratio of 1 : 2 : 1 with the pink flower in F_2-generation indicates the phenomenon of
(a) dominance (b) codominance
(c) incomplete dominance (d) segregation

31. All these obeys Mendel's laws except
(a) linkage (b) independent assortment
(c) dominance (d) purity gametes

32. The title of Mendel's paper while presenting at Burn Natural History Society in 1865 was
(a) law of inheritance
(b) law of heredity
(c) experiment on pea plants
(d) experiment in plant hybridisation

33. Both homozygosity and heterozygosity of an individual can be determined by, which of the best method?
(a) Test cross (b) Back cross
(c) Self-fertilisation (d) Inbreeding

34. Each gamete carries
(a) only recessive allele (b) only dominant allele
(c) only one of the alleles (d) All of the above

35. Cross in which parents differs in two pairs of contrasting characters is called
(a) monohybrid cross (b) dihybrid cross
(c) trihybrid cross (d) tetrahybrid cross

36. What is the pattern of inheritance called in when the phenotype is influenced by many genes?
(a) Epistasis (b) Monohybrid cross
(c) Polygenic inheritance (d) Dihybrid cross

37. In blood group typing in human being, if an allele contributed by one parent is I^A an and allele contributed by the other parent is i, the resulting blood group of the offspring will be
(a) 'A' (b) 'B' (c) 'AB' (d) 'O'

38. Out of these given below, which one of the following is true?
(a) Factors occurs in pairs
(b) Discrete unit that controls a particular character is called a factor
(c) Out of one pair of factors, one is dominant and other one is recessive
(d) Alleles do not show any blending and both the characters recovers as such in the F_2-generation

39. Mendel's law of independent assortment holds good for genes situated on the
(a) non-homologous chromosomes
(b) homologous chromosomes
(c) extra chromosomal DNA
(d) same chromosome

40. A cross between two tall plants resulted in offsprings having few dwarf plants. What would be the genotype of been the parents?
(a) TT and tt (b) TT and TT
(c) Tt and Tt (d) Tt and tt

41. Mother and father of a person with 'O' blood group have 'A' and 'B' blood group respectively. What would be the genotype of both mother and father?
(a) Mother is homozygous 'A' blood group and father is heterozygous for 'B'
(b) Mother is heterozygous for 'A' blood group and father is homozygous for 'B'
(c) Both mother and father are heterozygous for 'A' and 'B' blood group respectively
(d) Both mother and father are homozygous for 'A' and 'B' blood group respectively

42. In the F_2-generation of a Mendelian dihybrid cross the number of phenotypes and genotypes a
(a) phenotypes–4, genotypes–16
(b) phenotypes –9 genotypes–4
(c) phenotypes–9, genotypes–4
(d) phenotypes–4, genotypes–9

43. Occasionally, a single gene may express more than one effect. The phenomenon is called
(a) pleiotropy
(b) multiple allelism
(c) mosaicism Ne
(d) polygeny

44. In a dihybrid cross, it you get the ratio 9 : 3 : 1. it denotes that
(a) the alleles of two genes are interacting with each other
(b) it is a multigenic inheritance
(c) it is a case of multiple allelism
(d) the alleles of two genes are segregating independently

45. Due to the cross between TTRR × ttrr, the resultant progenies show. What percent of tall, red flowered plants
(a) 50% (b) 75% (c) 25% (d) 100%

46. On selfing a plant F_1- generation with genotype 'AABbCC', the phenotypic ratio in F_2-generation will be
(a) 3 : 1 (b) 1 : 1
(c) 9 : 3 : 3 : 1 (d) 27 : 9 : 9 : 9 : 3 : 3 : 3 : 1

47. Two crosses between the same pair of genotypes or phenotypes in which the sources of the gamete are reversed is one cross, is known as
(a) test cross (b) reciprocal cross
(c) dihybrid cross (d) reverse cross

48. Which one of the following traits of garden pea studied by Mendel was recessive feature?
(a) Axial flower position (b) Green seed colour
(c) Green pod colour (d) Round seeds

49. In a plant, red fruit (R) is dominant over yellow fruit (r) and tallness (T) is dominant over shortness (t). If a plant with RR Tt genotype is crossed with a plant that is rrtt
(a) 25% will be tall with red fruit
(b) 50% will be tall with red fruit
(c) 75% will be tall with red fruit
(d) All the offspring will be tall with red fruit

50. The genes controlling the seven pea characters studied by Mendel are now known to be located on how many different chromosomes?
(a) seven (b) six (c) five (d) four

Answers

1. (c)	**2.** (a)	**3.** (b)	**4.** (a)	**5.** (b)	**6.** (b)	**7.** (a)	**8.** (d)	**9.** (a)	**10.** (b)
11. (b)	**12.** (a)	**13.** (c)	**14.** (b)	**15.** (d)	**16.** (b)	**17.** (b)	**18.** (c)	**19.** (b)	**20.** (b)
21. (d)	**22.** (c)	**23.** (d)	**24.** (c)	**25.** (d)	**26.** (c)	**27.** (a)	**28.** (b)	**29.** (b)	**30.** (c)
31. (a)	**32.** (d)	**33.** (b)	**34.** (c)	**35.** (b)	**36.** (c)	**37.** (a)	**38.** (d)	**39.** (a)	**40.** (c)
41. (c)	**42.** (d)	**43.** (a)	**44.** (d)	**45.** (a)	**46.** (d)	**47.** (b)	**48.** (b)	**49.** (b)	**50.** (d)

23

Chromosomes and their Behaviour

Chromosomal Theory of Inheritance

Wilhelm Roux in 1883, postulated that the chromosomes within the nucleus of a cell are responsible for the transference of hereditary factors from generation to generation. This was later confirmed by T Boveri and WS Sutton in 1902, through their experiments. Although, the theory of the gene as a discrete unit of chromosome was developed by TH Morgan later on.

On various basis, a parallelism is seen between the chromosomes and genes as given below:

- Each species possesses a specific and characteristic number of chromosomes in its cells. In diploid cells (with two sets of chromosomes), these chromosomes are found in pairs called homologous pairs, which resembles in their morphology and genetic content.
- As out of two sets of chromosomes in a cell, one set belongs to paternal side and other set belongs to maternal side. Similarly, among the two chromosomes of a homologous pair one belongs to the maternal and other to the paternal side. Genes on the other hand also exist in allelomorphic pair, which are contributed equally by the two parents.
- Each chromosome is distinctly a double stranded structure consisting of two sister chromatids during cell division.
- During mitosis, the two chromatids get separated into two daughter nuclei. Thus, both the members of allelic pair find place in two different daughter nuclei during mitosis.
- During meiosis, the members of a chromosome pair are separated to go to daughter nuclei. Similarly, the members of an allele pair are also separated. Hence, each haploid cell or gamete contains only one allele out of the pair.
- When two haploid gametes get fuse together during fertilisation, not only the diploid chromosome number but also every allele pair is restored.

Thus, by accepting the fact that genes are located on the homologous pair of chromosomes, we can easily explain both the principle of segregation and independent.

With the separation of members of homologous pair, genes of allelic pair will also segregate and go to different gametes (that's how the segregation of genes belonging to an allelic pair is seen).

When gametes belonging to two different homologous pairs fuse, a new allelic pair is formed, in which each new combination is possible (formation of a new combination after fusion of segregated gametes tells us about independent assortment of segregated genes).

Experimental Proof of Chromosomal Theory

The experimental proof of chromosomal theory was given by C Bridges, a student of Morgan in 1916, through his non-disjunction experiment.

In this experiment, a cross was made between white-eyed *Drosophila* females and red-eyed males. The result of such a cross was on the expected lines with red-eyed females and white-eyed males. However, one in about every 2000 F_1 flies was an exception, i.e. either a white-eyed female or a red-eyed male.

Bridges showed that these rare exceptional offsprings results from the occasional failure of two X-chromosomes in the mother to separate from each other during meiosis, i.e. through **non-disjunction**.

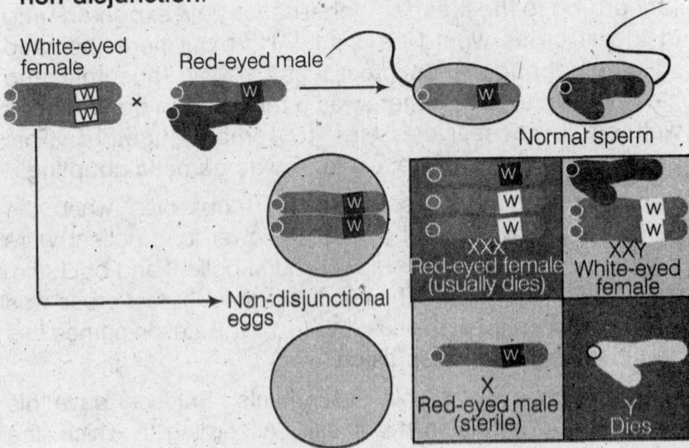

Non-disjunction experiment of bridges

Linkage

According to the 'chromosomal theory of inheritance', the chromosomes are considered as the vehicles of inheritance. Hence, the number of genes per individual far exceed the number of chromosomal pairs, i.e. each chromosome bears many genes. These genes are arranged in linear fashion over the chromosome and cannot show independent assortment.

In other words, it can be said that the genes of a particular chromosome show the tendency to get inherited together. This phenomenon of genetic inheritance in which genes of a particular chromosome show their tendency to inherit together, (i.e. tendency to retain their parental combination even in the offsprings) is known as linkage.

Linkage Groups

All those genes which are located in the single chromosome constitute a linkage group. This simply means that the total number of linkage groups in an organism are equal to the number of chromosomal pairs, e.g. *Drosophila melanogaster* has four linkage groups, similarly 23 of them are present in man, seven in sweet pea, etc.

Mechanism of Linkage

The process of linkage was fully explained as a theory by TH Morgan (1911) but its existence was predicted even before that by different workers.

The first suggestion about the tendency of linkage was given by Sutton (1903), but he was lacking in experimental verifications of his thoughts. Afterwards, it was Bateson and Punnett (1906), suggested their coupling and repulsion hypothesis on linkage.

Coupling and Repulsion Experiment

Bateson (1905), described a cross in sweet pea, (Lathyrus odoratus) where a deviation from independent assortment was observed.

Case I In this experiment, plants of sweet pea bearing blue flowers and long pollen (BBLL) were crossed with red flowers and round pollen (bbll). F_1 individuals had long pollen and blue flowers with genetic constitution (BbLl).

Normally, if independent assortment have took place (according to the Mendel's inheritance) the expected ratio in a test cross would be 1 : 1 : 1 : 1 but here the ratio obtained during the test cross was 7:1:1:7. Thus, this ratio clearly indicated that there was a tendency in dominant as well as in recessive alleles to remain together. This deviation was, therefore, explained as **gametic coupling**.

Case II The similar results were not observed when the parents of F_1 hybrid (BbLl, i.e. blue flower long pollen) were crossed with BBll (i.e. blue flower round pollen) and bbLL (i.e. red flower long pollen). The result this time for test cross was 1 : 7 : 7 : 1 instead of expected 1: 1 : 1 : 1. Bateson named this deviation as **gametic repulsion**.

Now on the basis of his experiments, Bateson gave his coupling and repulsion hypothesis. According to which, the coupling and repulsion was due to the tendency of genes to inherit together.

Types of Linkage

According to Morgan the linkage can be

(i) **Complete or perfect** It is implied for all those cases, in which genes remain together in the same chromosome for atleast two generations of genes for bent wings (bt) and shaven bristles (srn) of 4th chromosome mutant of *Drosophila*.

(ii) **Incomplete or imperfect** It is implied for all those cases, where the genes remain together within the same chromosome for less than two generations. This linkage is very common and is studied mostly in almost all the organisms. The main cause of incomplete linkage is crossing over during meiosis.

This type of linkage is normally shown by those linked genes which are widely located on the same chromosome,

e.g. in man, the incomplete linkage is commonly seen between the genes for haemophilia, colour blindness, genes for ABO blood group and papilla syndromes, etc.

In maize, the incomplete linkage is seen between the seed colour and seed shape.

Factors Affecting the Strength of Linkage

The factors which affects the strength of linkage besides intergenic distance are

(i) **Age of an animal** It supports that with the increase in age strength, linkage increases. It is just because of the fact that with the advancement of age, chances of crossing over decreases.

(ii) **Temperature** It indicates that the strength of linkage decreases with the temperature enhancement. It is because of the fact that the increase in temperature increases the frequency of chiasmata formation (i.e. incidence of crossing over).

(iii) **Exposure to X-rays** It indicates that if an organism is exposed to X-rays or any other similar radiations, the strength of linkage is decreased.

Crossing Over

The process of interchange of alleles (homologous segments) between non-sister chromatids of homologous chromosomes in F_2-generation is termed as crossing over. It was first postulated by **Morgan**.

Thus, the phenomenon of crossing over can be defined as the phenomenon of recombinations of linked genes which is the result of interchange of homologous segments between the non-sister chromatids of homologous chromosome pair.

For example, a cross between grey bodied vestigial wings (BBvv) and black bodied long wings (bbVV) of *Drosophila* produces four types of offsprings in F_2-generation on test cross as

I. Grey - vestigial — 41.5%
II. Black - long — 41.5%
} Both parental type

I. Black-vestigial – 8.5%
II. Grey-long – 8.5%
} Both new combination formed due to crossing over

Types of Cross Over

Depending upon the number of chromatids involved and chiasmata formed the crossing over may be of following types:

(i) A **single cross over** always produces two cross over chromatids and two non-cross over chromatids.

(ii) The **double cross over** can be either reciprocal (i.e. same two chromatids are involved in both the crossing overs) and can be complementary (i.e. all the four chromatids are involved in crossing over).

In complementary type of cross over two chromatids forms the first cross over and rest two forms the other cross over. The result of this type of cross overs is seen in the form of four cross over chromatids and none of them is non-cross over type.

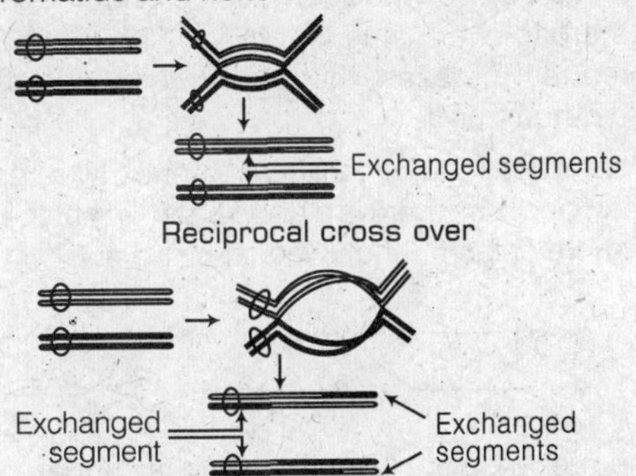

(iii) The **multiple cross** over occurs very rare when compared to single and double cross overs. In this case, more than two chiasmata are formed.

Types of Crossing Over

Kinds of crossing over according to its occurrence in the somatic or germ cells, there are two types:

1. Somatic or Mitotic Crossing Over

It is the process of very rare occurrence which occurs mainly in somatic cells of an organism. This process occurs during mitotic cell division and has no genetic significance.

It was first reported by **C Stern** in somatic cells of *Drosophila* for bristles and hairs. Later on, also reported by **G Pontecorvo** in fungus, *Aspergillus nidulans*.

The somatic cross over produces patches of cross over tissue between the normal cells, i.e. **mosaic appearance**. A mosaic of even one cell has been recognised in *Drosophila* for bristles and hairs.

2. Germinal or Meiotic Crossing Over

It is the process which occurs in germ cells or germinal cells during meiosis, i.e. during gametogenesis. It is universal in its occurrence and is of great genetic significance. Usually, it is of following three main types:

(i) **Single crossing over** is the condition in which crossing over is seen only at one point.

(ii) **Double crossing over** is the condition in which crossing over takes place at two points within the same homologous pairs.

(iii) **Multiple crossing over** is the condition in which crossing over takes place at more than two points within the same homologous pairs.

Mechanism of Crossing Over

The mechanism of crossing over involves the following steps:

(i) During the process of crossing over, first of all pairing of homologous chromosomes takes place during the **zygotene** stage of prophase-I of first meiotic division. This pairing is called **synapsis** and the resultant pairs are termed as **bivalents**.

(ii) This is followed by chromosome duplication, in which each homologous chromosome of bivalent splits longitudinally to form two identical sister chromatids, which remain held together by an unsplitted centromere. At this stage, each bivalent contains four arms or two chromatids hence, are called **bivalent** (tetrad). It occurs in pachytene substages of meiosis-I.

(iii) After the formation of tetrads crossing over occurs, during which the two non-sister chromatids exchange their segments and the chromosomal parts start repelling each other except at the point of contact.

(iv) The result will be seen in the form of cross-shaped or X-shaped structures called **chiasmata**. The crossing over that takes place at this point, results into the formation several chiasmatas. Hence, the number of chiasmata depends upon the length of chromosome and the type of cross over. Stern and Hotta (1969), have described that the breakage of chromatids is brought about by a nuclear enzyme **endonuclease**. The fusion of broken segments takes place due to action of another enzyme **ligase**. Thus, these two enzymes help in exchange of segments during crossing over.

(v) After the formation of chiasmata the chromosomes start separating progressively from each other beginning at centromere and moving towards the chiasma.

(vi) Similarly, the chiasma moves itself in a zipper fashion towards the ends of tetrads. This zipper movement of chiasma is called **terminalisation** and results to complete separation of homologous pairs.

Linkage Maps

The linkage or chromosomal map or genetic map is the graphical representation of the relative distances between the genes in a linkage group. These are expressed in terms of percentage of crossing over and given for a single group of chromosomes.

First chromosomal map was given by Strutevant and Morgan in second decade of 20th century. It was the graphic representation of five genes on the chromosome of *Drosophila*.

An arbitrary unit of measurement called **map unit** is used to describe the distances between the linked genes.

Thus, one map unit represents the linear distance, within which, 1% crossing over takes place,

i.e. 1 map unit = 1% of crossing over

One map unit is now referred to as cM (centiMorgan) in the honour of Morgan's contribution.

Construction of Genetic Map

The method of construction of genetic map includes the following steps:

(i) **Determination of linkage groups and total number of genes** By undergoing the hybridisation experiments between wild and mutant strains, we can easily determine the total number of genes of any species.

By the same technique, it can also be determined that how many phenotypic traits remain together always or are linked and consequently their determiners (genes) during the course of inheritance. Thus, different linkage groups of a species can be worked out.

(ii) **Determination of map distance** It can be done after knowing the total number of genes in each linkage group. Hence, the relative distances between each linked gene are calculated according to the percentage of crossing over.

This is done so, because cross over frequency is directly proportional to the distance between the genes. Hence, for determining the map distances, test crosses are done.

Gene Mapping

An example of the process by which the recombinant frequencies (cross over values) of different alleles of pairs of genes located on the same chromosome are determined by using breeding experiments (test crosses).

Genes *A* and *B* are found to have a cross over value of 8%, so we say that genes *A* and *B* are 8 chromosome map units apart.

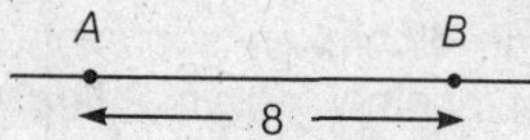

Genes *A* and *C* are then found to have a cross over value of 20%, so *A* and *C* are 20 map units apart.

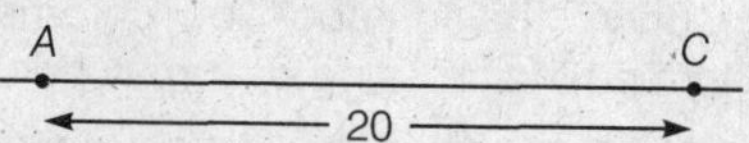

We deduce that the spacing of genes *A*, *B* and *C* is either

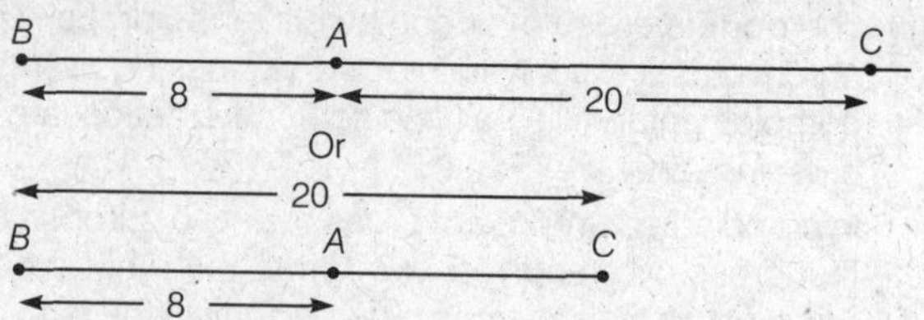

A test cross between *B* and *C* is carried out to find whether the cross over value is 12% or 28%. Thus, by means of three test crosses, the relative positions of three genes are determined.

A complication may arise with genes lying great distances apart because there is a good chance that there may be two or more chiasmata between them, e.g.

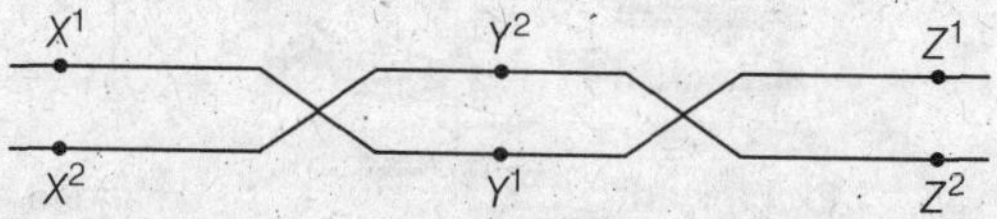

The presence of two chiasmata may apparently cancel out the effects of **recombination**. This problem is best overcome by the mapping of a large number of genes at short distances from each other (since, double cross overs are unlikely between closely linked genes), thus allowing the composite picture to be constructed.

Chapter Exercise

1. **Sutton and Boveri used .... movement to explain Mendel's laws.**

(a) gene
(b) chromosome
(c) chromatid
(d) spindle

2. **According to chromosomal theory of linkage of Morgan and Castle**

(a) strength of linkage between two successive genes is inversely proportional to distance between two genes
(b) genes lie in a linear order in the chromosomes
(c) linked genes are arranged in *cis* or *trans* manner *cis* arrangement $\frac{A\ B}{a\ b}$ brings coupling (parental combinations) whiletrans-arrangement $\frac{A\ b}{a\ B}$ brings recombinations (repulsions)
(d) All of the above

3. **Consider the following statements regarding linkage.**

I. The linked genes are located on the same chromosome.
II. Crossing over between linked genes is rare.
III. Linked genes are always inherited together.
IV. Linkage affects the percentage of homozygosity following hybridisation.

Of these statements are correct.

(a) I and III
(b) I and II
(c) II, III and IV
(d) I, III and IV

4. **Variability may originate during meiosis due to**

(a) crossing over
(b) chromosomal aberrations
(c) polyploidy
(d) mutations

5. **Sutton and Boveri argued that the pairing and separation of a pair of chromosomes would lead to**

(a) the segregation of a pair of factors they carried
(b) the segregation of the characters
(c) recombination of the factors they carried
(d) Both (b) and (c)

6. **The coupling and repulsion theory of Bateson and Punnett later on modified in linkage and crossing over by Morgan. Two completely linked genes shows a dihybrid ratio of**

(a) 3 : 1
(b) 1 : 1
(c) 1 : 1 : 1 : 1
(d) 9 : 3 : 3 : 1

7. Criss-cross inheritance means
(a) X-chromosome from female will pass to female of next generation
(b) X-chromosome from a male will pass to a female of next generation
(c) X-chromosome from male will pass to a male of next generation
(d) None of the above

8. Crossing over occurs at four stranded stage. This was proved by the observation that
(a) usually two gametes resulting from meiosis are recombinants
(b) all the four gametes resulting from meiosis are recombinants
(c) chiasmata are seen only at four stranded stage
(d) chiasmata are seen only at two stranded stage

9. Sex-linked inheritance was discovered in 1910 by
(a) McClung (b) Mendel (c) Landsteiner (d) Morgan

10. The first attempt to show linkage in plants was done in
(a) *Zea mays* (b) *Oenothera lamarckiana*
(c) *Pisum sativum* (d) *Lathyrus odoratus*

11. The genetic maps are used as a starting point in the sequencing of whole genomes as in the case of
(a) human project
(b) human gene sequencing project
(c) human genome sequencing project
(d) All of the above

12. Sutton gave chromosomal theory of inheritance he united the knowledge of chromosomal segregation with
(a) recombination (b) crossing over
(c) Both (a) and (b) (d) Mendelian principles

13. The crossing over between homologous chromosomes never exceeds beyond
(a) 50% (b) 75%
(c) 100% (d) None of these

14. Linkage and cytological maps for the same chromosome
(a) are both based on mutant phenotypes and recombination data
(b) may have different sequences of genes
(c) have both the same sequences of genes and intergenic distances
(d) have the same sequences of genes but different intergenic distances

15. A linkage group is explained as
(a) different groups of genes located on the same chromosomes
(b) that all the linked genes of a chromosome pair
(c) that all the linked genes of a chromosome
(d) None of the above

16. According to the chromosomal theory of inheritance
(a) chromosomes are decondensed in telophase
(b) genetic codes are found on chromosomes
(c) chromosomes carry alleles that are segregated as chromosomes and are pulled apart during the cell division
(d) genes are imitable

17. A pair of genes are linked if their frequency of recombination in a test cross is
(a) 75% (b) 50%
(c) 100% (d) lower than 50%

18. Linkage in *Drosophila* was first discovered by
(a) Bateson and Punnett (b) Bridges
(c) Morgan (d) Strutevant

19. Sutton and Boveri noted that the behaviour of chromosomes was parallel to the behaviour of
(a) genes (b) chromatids (c) nucleus (d) nucleolus

20. Number of linkage in a polynucleotide would be
(a) same as number of nucleotides
(b) twice as number of nucleotides
(c) one less than the number of nucleotides
(d) one

21. Complete linkage is observed in
(a) frog (b) birds (c) lizards (d) *Drosophila*

22. The distance between the genes A, B, C and D in mapping units are A-D = 3.5, B-C = 1, A-B = 6, C-D = 1.5, A-C = 5. Findout the sequence of arrangement of these genes.
(a) ACDB (b) ABCD (c) ADCB (d) ADBC

23. Chromosomal theory of inheritance was proposed by
(a) Gregor Johann Mendel (b) Hugo de Vries
(c) Bridges (d) Sutton and Boveri

24. When a cluster of genes show linkage behaviour they
(a) do not show independent assortment
(b) induce cell division
(c) do not show a chromosome map
(d) show recombination during meiosis

25. Variability may originate during meiosis due to
(a) crossing over (b) chromosomal aberrations
(c) polyploidy (d) mutations

26. The chromosomal theory of heredity means that
(a) chromosomes are composed of genes
(b) chromosomes are made up of DNA and protein
(c) genes are located on the chromosome
(d) All of the above

27. Crossing over brings about
(a) recombination of genes
(b) no significant change
(c) study offspring
(d) cytoplasmic reorganisation

28. Experimental verification of the chromosomal theory of inheritance was given by
(a) Gregor Johann Mendel (b) Hugo de Vries
(c) Langdon Down (d) Thomas Hunt Morgan

29. Unit of distance between genes in a chromosome is known as
(a) cDNA (b) Morgan
(c) centiMorgan (d) Spacer

30. Crossing over is nil between genes located near
(a) telomere (b) centromere
(c) Both (a) and (b) (d) None of these

31. Two linked genes, i.e. a and b shows about 20% of recombination, the individuals of a dihybrid cross between ++/++ × ab/ab shall show gametes
(a) ++80 : ab : 20
(b) ++50 : ab : 50
(c) ++40 : ab40 : +a10 : +b : 10
(d) ++30 : ab30 : + a20 : +b : 20

32. Two dominant non-allelic genes are 50 map units apart. The linkage is
(a) *Cis* type (b) *trans* type
(c) complete (d) absent/incomplete

33. Which one of the following is considered to be the suitable experiment on linkage?
(a) aaBB × aaBB (b) AABB × aabb
(c) AaBb × AaBb (d) AAbb × AaBB

34. Crossing over in diploid organism is responsible for
(a) segregation of alleles
(b) recombination of linked alleles
(c) dominance of genes
(d) linkage between genes

35. If there are three genes a, b, c. And the percentage of crossing over between a and b is 20%, b and c is 28% while. a and c is 8%. Predict the sequence of genes on chromosomes from the following
(a) b, a, c (b) a, b, c
(c) a, c, b (d) None of these

36. Mendel fails to observe linkage due to
(a) mutation (b) crossing over
(c) synapsis (d) independent assortment

37. Crossing over is more frequent in
(a) males (b) females
(c) Both (a) and (b) (d) None of these

38. Complete linkage has been reported in
(a) human female (b) male *Drosophila*
(c) female *Drosophila* (d) maize

39. How many linkage groups of chromosomes will be present in case of maize, if all its genes are mapped?
(a) 5 (b) 10
(c) 50 (d) 100

40. In which plant phenomenon of linkage was observed first?
(a) *Lathyrus odoratus* (b) *Pisum sativum*
(c) *Datura* (d) *Mirabilis jalapa*

41. Coupling and repulsion phenomenon was concerned with

A ——— B
a ——— b
Coupling (*Cis*)

A ——— b
a ——— B
Repelsion (*Trans*)

(a) crossing over (b) mutation
(c) linkage (d) All of these

42. Crossing over in diploid organism is responsible for
(a) dominance of genes
(b) segregation of alleles
(c) recombination of linked genes
(d) linkage between genes

43. Who coined the term 'linkage'?
(a) Correns (b) Mendel (c) Morgan (d) de Vries

44. Linkage prevents
(a) homozygous condition (b) heterozygous condition
(c) segregation of alleles (d) hybrid formation

45. Crossing over occurs when the homologous chromosomes contain
(a) one chromatid (b) two chromatid
(c) four chromatid (d) eight chromatid

46. Genetic map is one that
(a) shows the stages during cell division
(b) establishes various stages in gene evolution
(c) shows the distribution of various species in a region
(d) establishes sites of the genes on a chromosome

47. In a dihybrid cross, two recessive genes shows 10% recombinants. The distance between two genes is
(a) 10 map unit (b) 20 map unit
(c) 5 map unit (d) 100 map unit

48. How many linkage groups will be present in the case of human beings
(a) 23 (b) 46 (c) 44 (d) 22

49. 1 map unit or centiMorgan (cM) is equal to
(a) 1% recombination (b) 10% recombination
(c) 100% recombination (d) 0.1% recombination

50. Linked genes are
(a) located on different chromosomes of same size and shape
(b) located on the same chromosome
(c) rarely inherited together
(d) None of the above

Answers

1. (b)	**2.** (d)	**3.** (a)	**4.** (a)	**5.** (a)	**6.** (b)	**7.** (b)	**8.** (a)	**9.** (d)	**10.** (d)
11. (c)	**12.** (d)	**13.** (a)	**14.** (d)	**15.** (b)	**16.** (a)	**17.** (a)	**18.** (c)	**19.** (a)	**20.** (d)
21. (d)	**22.** (c)	**23.** (d)	**24.** (a)	**25.** (a)	**26.** (c)	**27.** (a)	**28.** (d)	**29.** (c)	**30.** (c)
31. (c)	**32.** (d)	**33.** (b)	**34.** (b)	**35.** (a)	**36.** (d)	**37.** (b)	**38.** (b)	**39.** (b)	**40.** (a)
41. (c)	**42.** (c)	**43.** (c)	**44.** (c)	**45.** (c)	**46.** (d)	**47.** (a)	**48.** (a)	**49.** (a)	**50.** (b)

24

Sex Determination, Mutation and Chromosomal Aberration

Determination of Sex

The first definite evidence of chromosome theory came from sex determination. Biologically, sex is an aggregate of those morphological, physiological and behavioural qualities that differentiate the organisms producing eggs from the organisms that produces sperms. This separation of species or section of animals into two sexes is called as **gonochorism**.

The sex behaves as a Mendelian character and its inheritance follows the **law of segregation**. Many mechanisms are given to explain the determination and differentiation of sex. Most of them are under the genetic control. These mechanisms are:

(i) Chromosomal theory of sex determination or sex chromosome mechanism
(ii) Genic balance mechanism
(iii) Haplodiploidy mechanism
(iv) Single gene control of sex or single gene effect
(v) Cytoplasmic sex determination

Chromosomal Theory of Sex Determination

According to this theory of sex determination, the male (♂) and female (♀) individuals normally differ in their chromosomal constituents. There may be two types of chromosomes present in these individuals. These are:

(i) **Sex chromosomes** or **allosomes** are the chromosomes which are responsible for the determination of sex in an individual, e.g. X and Y-chromosomes.

(ii) **Autosomes** are the chromosomes which have no relation with the sex and contain genes which determine the somatic characters. They are same in both males and females.

The X-chromosome was first observed by the German biologist Henking in 1891, during his studies on spermatogenesis in male bug. He termed it as X-body. The chromosomal theory of sex determination was worked out by EB Wilson and Stevens (1902-1905). They named X and Y-chromosomes as allosomes or sex chromosomes and other chromosomes as autosomes.

The parents can be of following two types:

(i) **Homogametic** with similar gametes.

(ii) **Heterogametic** with different gametes.

In some cases, males are heterogametic, e.g. human, *Drosophila,* bug, etc., while in some other cases, females are heterogametic, e.g. birds.

Thus, we have following two types of systems among diploid organisms, on the basis of chromosomal theory of sex determination:

(a) System having heterogametic males.
(b) System having heterogametic females.

In the first system (i.e. having heterogametic males) the females have two X-chromosomes, while males, have one X-chromosome because of which male during gametogenesis produces two types of gametes (sperms), i.e. 50% that carry X-chromosome and another 50% that are those carry Y-chromosome and also lack X-chromosome.

This chromosomal pattern varies in different organisms in the way that male either contain Y-chromosome (i.e. XX-XY system or lygaeus type) or contain no chromosome along with the X-chromosome, i.e. XX-XO type or protenor type.

XX-XY Pattern

The XX-XY method is the most common method of sex determination. It is commonly found in man, other mammals and certain insects like *Drosophila,* etc.

In these animals, the female possess two homomorphic X-chromosomes (i.e. XX) and being homogametic, i.e. produces eggs of one kind only. These all eggs have one X-chromosome in them.

And the males, they possess one X and one Y-chromosome (hence, called XY) and produces two kinds of sperms, i.e. 50% with X-chromosome and 50% with Y-chromosome. The sex of embryo depends upon the kind of sperm.

Thus, as a result, female individual is produced, if egg fertilises with a sperm having X-chromosome and a male is produced if same is fertilised with Y-chromosome bearing sperm.

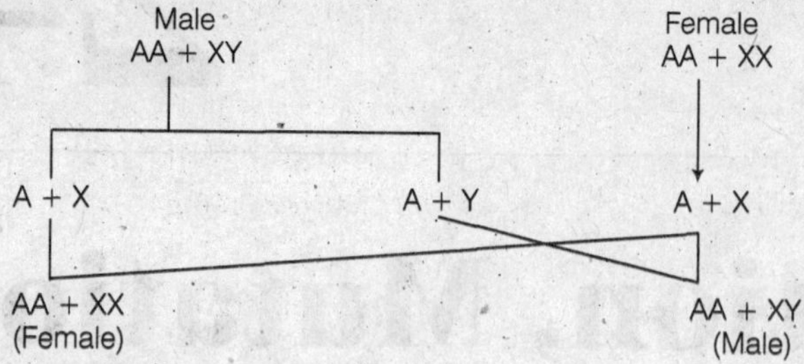

XX-XO Pattern

An other method, i.e. (XX-XO type) is found in certain **insects** especially those of class–**Hemiptera** (true bugs) and order–**Orthoptera**.

In this method, the female has two X-chromosomes it is called XX, while male has only one X-chromosome it is called as XO. The Y-chromosome is completely lacking here thus, the presence of an unpaired X-chromosome determines the **masculine sex**.

The female just like lygaeus method produces only one type of egg and male produces two types of sperms, i.e. 50% with one X-chromosome and 50% without any sex chromosome.

Hence, the sex of an offspring depends upon the type of sperm, which fertilises the egg, e.g. in grasshopper.

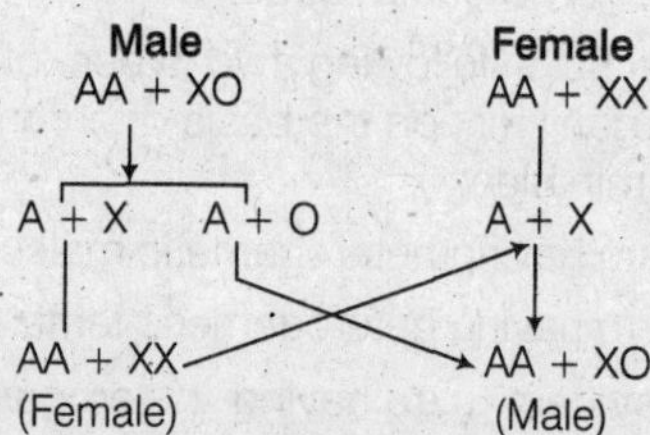

ZZ-ZW and ZZ-ZO Pattern

In the second system, i.e. female is heterogametic. The male possesses two homomorphic X-chromosomes and female possesses a single X-chromosome or X-chromosome supported by the Y-chromosome.

Thus, it produces two types of gametes, i.e. 50% with X-chromosome and rest 50% either with Y-chromosome or without X-chromosome.

To avoid confusion with XX-XY and XX-XO type here, X is designated as Z and Y as W. Hence, the systems here are called ZZ-ZW and ZZ-ZO systems.

The ZW-ZZ method is found in certain insects and vertebrates, e.g. fishes, reptiles, birds, etc., and ZZ-ZO method is found in certain plants as well along with some insects like *Fumea* (a moth).

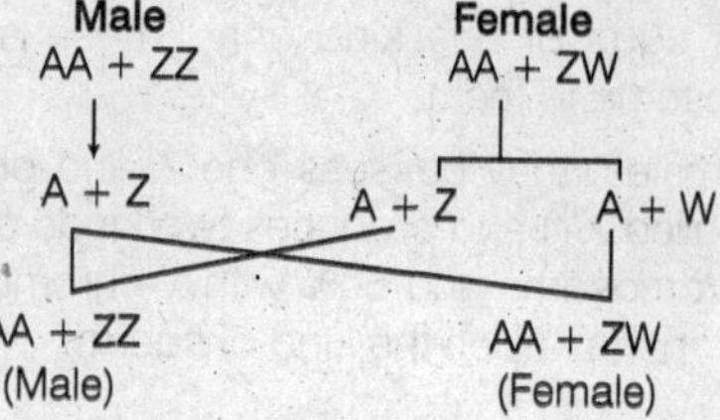

The chromosomal mechanism discussed above is not always correct. It was **Patterson**, who told us that in *Drosophila*, the Y-chromosome is heterochromatic and plays no significant role in sex-determination.

Sex-Determination in Plants

Most flowering plants are monoecious in nature hence are without sex chromosomes. The well known examples of dioecious plants are usually under the direct control of single gene locus as described previously under single gene control of sex. However, chromosomal activity has been reported in many plant species, e.g. *Melandrium album*, *Coccinia indica*, *Vallisneria spiralis*, etc.

In *Coccinia* and *Melandrium*, XX-XY type of mechanism is seen, i.e. here the male plant have XY combination, while the female plant have XX combination.

The sex chromosomal mechanism of sex determination is also observed in certain monoploid bryophytes, (e.g. *Sphaerocarpus*) in which XX female and XY male is seen (the sporophyte containing XY combination produces two types of **meiospores**, i.e. some with X and other with Y-chromosome).

The meiospores with X-chromosome form female gametophyte and those with Y-chromosome produces male gametophyte.

Mutation

Any heritable change in the genetic makeup of an individual belongs to the word mutation. It is a sudden change of a gene or chromosome from one form to another.

Mutagen

A chemical that cause mutation is called **mutagen**. Various types of mutagens are as under:

(i) Physical mutagen–X-rays and UV-rays.

(ii) DNA reactive chemical–Superoxides and nitrous acid.

(iii) Base analogs–5 bromouracil.

(iv) Inter cloting agents–ETbR and proflavin.

(v) Metals–Arsenic and cadmium.

(vi) Biological agents–Virus and bacteria.

Mutation produces an alteration in the character under its control. The term 'mutation' was introduced by **Hugo de Vries**, who used it for large spontaneous inheritable changes which occurs suddenly in naturally reproducing population.

Dobzhansky stated that mutation is a mistake or misprint in a normal cell division. It also refers to the process by which such a change takes place. Such changes under mutation includes changes in chromosomes either in number or in structure (chromosomal mutations) and changes in the structure or in composition of genes (gene mutations). Chromosomal mutations are also called chromosomal aberrations and gene mutations are known as **point mutations**.

Types of Mutation

Mutation can be of following types:

1. **Point Mutations**

The intragenic or point mutation involves the alterations in the structure of gene by altering the structure of DNA.

Thus, we can say that in point mutation, there is a change in the normal base sequencing of a DNA molecule. This change may results into the modification in structural characteristics of an individual.

Point mutations are mainly of two following types:

(i) **Base pair mutation** These are the most common type of mutations in which one base is substituted with the another in a DNA molecule.

These can be further categorised into:

(a) **Transitions** in which a purine base is replaced by another purine and the pyrimidine base is replaced by another pyrimidine.

(b) **Transversion** in which a purine is replaced by a pyrimidine and *vice versa*.

(ii) **Frameshift mutation** It is derived from the fact that there is a shift in reading frame either backward or forward. It occurs by following ways:

(a) **Deletion** is the phenomenon of removal of one or more nitrogenous bases from a nucleotide chain. As the code is commaless, the complete reading of frame after deletion is shifted towards its right as shown below:

Original Sequence →

GAT CAT GAT CAT GAT CAT ......
↑
If deletion of this nitrogenous base occur
↓

Modified Sequence →

GAT (unchanged) ATG ATC ATG ATC AT (Frameshifts towards (right) →)

(b) **Insertion** is the phenomenon of addition of one or more nitrogenous bases to the nucleotide chain. Here, again as the code is commaless, the complete reading frameshifts towards it's left as shown below.

Original Sequence →

CAG TAG CAG TAG CAG TAG
↑
Insertion point of adenine

Modified Sequence →

CAG (Unchange) ATA GCA GTA GCA GTA G (Shifted frame towards left (←))

Note If simultaneous deletion and insertion of nitrogenous bases is seen, then the shift in the reading frame will be visible only between the deletion and insertion points (as shown below).

Original Sequence → CAG TAG CAG TAG CAG TAG

Deletion point removing T ↑ ↑ **Insertion point inserting T**

Modified Sequence → CAG (Unchanged) AGC AGT AGT (Changed Frame) CAG TAG (Unchanged)

2. **Chromosomal Aberration**

Change in number and structure of chromosome is called chromosomal aberration.

(i) **Variation in chromosomal number** The variation in chromosome number is called **heteroploidy** and can be categorised to

(a) **Euploidy** The cases, where the variation of entire set is seen. The euploidy can be either monoploidy (i.e. haploid set of chromosomes) or polyploidy (i.e. more than two sets of chromosomes). The polyploidy can be triploidy ($3n$), tetraploidy ($4n$), pentaploidy ($5n$), hexaploidy ($6n$), etc.

- **Monoploidy** (Haploidy) These organisms contain only one set of chromosomes, i.e. these are hemizygous in nature. The monoploids are normally smaller in size than the corresponding diploids with less vigour. Monoploidy plants have their leaves, flowers, fruits and seeds are also comparatively smaller. These have just univalent chromosomes without any homologous to pair. These univalents are distributed at random during anaphase-I of meiosis producing monosomics.
- **Polyploidy** The polyploids may be classified as primary and secondary. A primary polyploid is always **even** numbered because it arises by doubling of chromosomes of a particular organism.

e.g. $2n \xrightarrow[\text{To}]{\text{Doubled}} 4n$

or $3n \xrightarrow[\text{To}]{\text{Doubled}} 6n$, etc.

A **secondary polyploid** on the other hand arises by the crossing between organisms of different ploidy levels.

e.g. $4n$ (primary) $\times 2n \xrightarrow[\text{Rise to}]{\text{Gives}} 8n$ (secondary)

(b) **Aneuploidy** It can be seen, where variation of individual chromosomes occurs. Aneuploid plants and animals have incomplete genomes. These can be of two types mainly:

- **Hyperploids**, are those organisms which have one or more chromosome in addition to the normal diploid genome.
- **Hypoploids**, are those organisms which have one or more chromosome less than the normal diploid genome.

(ii) **Variation in chromosome structure** The variation in chromosome structure is due to following four main processes

(a) **Deficiency and Deletion** The deficiency is basically the mechanism of absence of a chromosomal portion, whether terminal or interstitial resulting into loss of genes or a gene. Depending upon the length of the lost segment, the number of genes lost may vary from one to many. The process which results into deficiency is called deletion.

(b) **Duplication** The presence of one block of genes more than one in a haploid complement is known as duplication and this additional block is known as **repeat**.

(c) **Translocation** The process of translocation is a kind of rearrangement in which a block of genes from one chromosome are transferred to other non-homologous chromosome. It occurs due to the irregularities during crossing over. Breaks in the chromosomes are followed by union in ways different from the normal one.

(d) **Inversion** It is the process, in which a section of chromosome becomes changed after rotation through **180°**. Thus, the order of genes in the section become reversed.

(iii) **Aneuploidy of chromosomes** This results in humans due to the absence, excess or an abnormal arrangement of one or more chromosomes.

It is classified as

(a) **Down's syndrome** It was described by J Langdon Down in 1866. It is caused due to the additional copy of chromosome number 21 or trisomy of chromosome 21 in humans and also seen in chimpanzees and other mammals.

(b) **Klinefelter's syndrome** It is caused due to the presence of an additional copy of X-chromosome resulting in the karyotype 45+XXY which results into 47 chromosomes.

(c) **Turner's syndrome** It is caused due to the absence of one of the X-chromosome, i.e. 45 with XO.

Chapter Exercise

1. A segment of the DNA has a base sequence AAG GAG GAC CAA CCA, which of the following sequences represents a frameshift mutation?

(a) AGG AGG ACC AAC CA
(b) AAG GCG GAC CCA AC
(c) ACG GAC GAC CAG CCA
(d) AAG GAG GAC CAA CCA

2. Who has introduced X-ray mutations in barley and maize?

(a) Stadler (b) Muller
(c) Morgan (d) All of these

3. If the DNA codons are ATG ATG ATG and a cytosine base is inserted at the beginning, which of the following will result?

(a) CA TGA TGA TG (b) CATG ATG ATG
(c) CAT GAT GAT G (d) A non-sense mutation

4. If a mutation occurs in a gamete, it would influence

(a) only a single individual
(b) sterility in the progeny
(c) all successive generation of the parents
(d) only the particular sex of the progeny, whose gamete had undergone mutation

5. ZO-ZZ type of sex determination is

(a) opposite of XX-XO (b) opposite of XX-XY
(c) opposite of ZZ-ZW (d) None of these

6. Which one of the following is not an example of chromosomal disorder?

(a) Down's syndrome (b) Turner's syndrome
(c) Klinefelter's syndrome (d) None of these

7. Any change during cell division that results in loss or gain of one or more chromosomes is known as

(a) aneuploidy (b) euploidy
(c) monoploidy (d) hypoploidy

8. X-chromosome is

(a) telocentric (b) metacentric
(c) acrocentric (d) acentric

9. A mutant rice is

(a) allopolyploid (b) autopolyploid
(c) autoallopolyploid (d) segmental polyploid

10. The site of the gene, at which the mutations occur with usually high frequency are

(a) recons (b) hot spots
(c) mutons (d) palindromes

11. Type of gene mutation which involves the replacement of purine with pyrimidine or *vice versa* or the substitution of one type of base with another type of base is

(a) transition (b) transduction
(c) translation (d) transversion

12. The gene that controls the rate of mutation of another gene is
(a) mutator gene (b) inducer gene
(c) mutable gene (d) regulator gene

13. Which of the following will cause a more effective mutation?
(a) One codon (b) One base deletion
(c) Base substitution (d) Base deamination

14. A single gene mutation affecting more than one phenotype is called
(a) azotrophic (b) pleiotropic
(c) auxotropic (d) pleiotrophic

15. A mutation that change a codon specifying one amino acid to a termination codon is called
(a) mis sense mutation (b) transition mutation
(c) non-sense mutation (d) frameshift mutation

16. Tendency of genes to get inherited together is called
(a) differential inheritance (b) sex linkage
(c) linkage (d) None of these

17. Exchange of segments between non-homologous chromosome is called
(a) crossing over (b) inversion
(c) duplication (d) translocation

18. Father of *Neurospora* genetics (haploid genetics) is
(a) Lindegren (b) Dodge
(c) Morgan (d) Muller

19. Hereditary variations in plants have been produced by the use of
(a) DDT (b) auxins
(c) X-rays (d) gibberellic acid

20. Which of the following is a type of base pair mutation?
(a) Deletion (b) Insertion
(c) Transitions (d) None of these

21. Recessive mutations are expressed normally in
(a) heterozygous condition
(b) homozygous condition
(c) neither in homozygous nor in a heterozygous condition
(d) Both (a) and (b)

22. Mutations are mainly responsible for
(a) increasing population rate
(b) maintaining genetic continuity
(c) controlling variation in organisms
(d) controlling extinction of organisms

23. A frameshift mutation could result from
(a) a base deletion only
(b) a base insertion only
(c) either an insertion or a deletion of base
(d) deletion of three consecutive bases

24. Inheritable mutations can be studied by
(a) pure line breeding
(b) Punnett square
(c) using chemical mutagens
(d) generating a pedigree of a family

25. Exchange of genetic material from non-sister chromatid is called
(a) transformation
(b) crossing over
(c) translocation
(d) deletion

26. Sex chromosome of a female bird are represented by
(a) XO (b) XX
(c) XY (d) ZW

27. Genes when combined together on a chromosome are known
(a) linkage (b) mutation
(c) translation (d) transcription

28. Frameshift mutation occurs when
(a) base is deleted
(b) base is added
(c) base is added or deleted
(d) None of the above

29. Sex-chromosomes of a male silkworm is
(a) X (b) y (c) XX (d) XY

30. Which one is caused by the mutation in animals?
(a) Change in habit
(b) Change in Golgi body
(c) Change in gene structure
(d) Change in peroxysome

31. A well known autosomal abnormality at birth is
(a) Klinefelter's syndrome
(b) Down's syndrome
(c) Patau's syndrome
(d) Turner's syndrome

32. The genetic defect Adenosine Deaminase (ADA) deficiency can be cured permanently by
(a) the enzyme replacement therapy
(b) periodic infusion of genetically engineered lymphocytes having functional ADA CDNA
(c) introducing bone marrow cells producing ADA into cellsat early embryonic stages
(d) administering adenosine deaminase activators

33. Of a normal couple, half the sons are haemophilic while half the daughters are carriers. The gene is located in
(a) X - chromosome of father
(b) Y-chromosome of father
(c) one X- chromosome of mother
(d) Both X-chromosomes of mother

34. Of both normal parents, the chance of a male child becoming colourblind are

(a) no
(b) possible only when all four grand parents had normal vision
(c) possible only when father's mother was colourblind
(d) possible only when mother's father was colourblind

35. A fruitfly is heterozygous-for sex-linked genes when mated with normal female fruit fly, the males specific chromosome will enter egg cell in the proportion.

(a) 3 : 1 (b) 7 : 1
(c) 1 : 1 (d) 2 : 1

36. Which of the following disease sets an example of pleiotropy?

(a) Haemophilia (b) Thalassemia
(c) Sickle cell anaemia (d) Colour blindness

37. XY sex chromosomes were discovered by

(a) Mendel (b) R Brown
(c) Nettie Stevens (d) MJD White

38. XX male and XY females are very few in humans. They are due to

(a) genes of Y- chromosomes are on X- chromosome
(b) crossing over between X and Y
(c) Non disjunction of chromosomes
(d) All of the above

39. Genes located on Y-chromosome are

(a) autosomal genes (b) sex influential genes
(c) holandric genes (d) Mutant genes

40. One of the genes present exclusively on the X-chromosome in humans is concerned with

(a) baldness (b) ved green colour blindness
(c) night blindness (d) None of these

41. Genic balance theory of sex determination was given by

(a) Morgan (b) Bridges
(c) Darwin (d) Mendel

42. Sex determination was first of all studied in which plant?

(a) *Rumex* (b) *Datura*
(c) *Mirabilis* (d) *Melandrium*

43. The probability that a male inherited his Y-chromosome from his maternal grand father is

(a) 0% (b) 25%
(c) 50% (d) 100%

44. Which one of the following genetic condition is not sex-linked?

(a) Ichtyosis (b) Colour blindness
(c) Sickle cell anaemia (d) Both (a) and (c)

45. A women whose brother has haemophilic is concerned about passing this trait to her offspring. What is the risk that she will have a son with haemophilia?

(a) $\frac{1}{8}$ (b) $\frac{1}{4}$ (c) $\frac{1}{2}$ (d) 1

46. Can a male be a carrier for a sex-linked disease?

(a) Yes, if the trait is recessive
(b) Yes, if the male's father and mother were carriers
(c) No, males have only a single copy of sex-linked genes
(d) No way to predict

47. X-inactivation can be used to identify individuals who are

(a) homozygous unaffected (b) heterozygous
(c) homozygous affected (d) Missing X-linked genes

48. Beard growth in humans is an example of a .... trait.

(a) sex linked (b) sex influenced
(c) sex limited (d) Y- linked

49. What would be the sex of an xxy individual?

(a) Male (b) Female
(c) Hermaphrodite (d) Mosaic

50. A female whose father was colourblind marries a normal male whose father was also colourblind. What is the probability that their son will be colour blind?

(a) 0% (b) 25% (c) 50% (d) 75%

Answers

1. (a)	**2.** (a)	**3.** (c)	**4.** (c)	**5.** (a)	**6.** (d)	**7.** (a)	**8.** (b)	**9.** (c)	**10.** (b)
11. (d)	**12.** (a)	**13.** (b)	**14.** (b)	**15.** (c)	**16.** (c)	**17.** (d)	**18.** (b)	**19.** (a)	**20.** (c)
21. (b)	**22.** (c)	**23.** (c)	**24.** (a)	**25.** (b)	**26.** (d)	**27.** (a)	**28.** (c)	**29.** (c)	**30.** (c)
31. (b)	**32.** (c)	**33.** (c)	**34.** (d)	**35.** (c)	**36.** (c)	**37.** (c)	**38.** (a)	**39.** (c)	**40.** (b)
41. (b)	**42.** (d)	**43.** (a)	**44.** (c)	**45.** (b)	**46.** (c)	**47.** (b)	**48.** (c)	**49.** (a)	**50.** (c)

25

Gene Expression

Central Dogma

Central dogma of molecular biology states that the information always flows in one way or in unidirectional, i.e. from the master copy DNA to the working copy RNA (transcription) and from working copy RNA to the building plan polypeptide (translation)..

$$\text{DNA} \xrightarrow{\text{Transcription}} m\text{RNA} \xrightarrow{\text{Translation}} \text{Polypeptide}$$

Central dogma of molecular biology was proposed by Crick (1958). It is also written as

$$\underset{\text{Replication}}{\text{DNA}} \xrightarrow{\text{Transcription}} m\text{RNA} \xrightarrow{\text{Translation}} \text{Polypeptide}$$

Replication of DNA

Replication of DNA is the basis of life and it takes place during the S-phase (synthetic phase) of an interphase in cell cycle.

The process of replication is semiconservative in nature, i.e. out of two strands present, one is the old strand while the other is the new one. Dispersive and conservative modes of replication can be possible mode of replication but the semiconservative nature of DNA replication was proved by the experiment of Meselson and Stahl (1958).

In a prokaryotic cell, the process of replication is catalysed mainly by the enzyme called DNA polymerase along with certain other enzymes.

Mechanism of Replication

Replication of DNA begins at certain fixed unique point called origin of replication (*ori*) in a cell. The other requirements of replication include:

(i) DNA template (ii) Deoxyribonucleotides

(iii) Enzymes and proteins

At origin, the replication begins with a nick or incision made by an incision enzyme called **endonuclease**.

The two strands of the DNA double helix unwinds with the help of DNA unwinding enzyme called helicase. The unwinding occurs as this enzyme begins its binding with the DNA strands, thus breaking the hydrogen bonds between the complementary nitrogenous bases. This process is also known as **melting**.

The unwinding of DNA imposes tension or strain on the distal end of DNA molecule. This strain is released by superhelix relaxing enzyme (also called DNA gyrase or topoisomerase or swiveling protein). This enzyme has nick and joining activity. It only works on one strand of DNA and produces nicks (cuts) on this strand.

When strain or tension increase on the distal part of DNA due to unwinding of proximal part, this enzyme produces nicks on one strand of DNA which then rotates over the other end to relax the strain or coiling. This strand is then joined by the activity of the enzyme known as DNA ligase.

Initiation of DNA synthesis always requires a smaller segment of RNA called **RNA primer**. This primer is synthesised by the enzyme **primase** (also called specialised RNA polymerase).

The synthesis always occurs from 3′-end → 5′-end (on DNA template) or from 5′-end → 3′-end (in newly synthesised strand). The length of the primer varies from 10-60 nucleotides. Afterwards, from 3'-end of DNA, synthesis begins similarly with the help of enzyme DNA polymerase.

DNA polymerase progressively adds deoxyribonucleotides to the 3′–OH end of this growing polynucleotide chain.

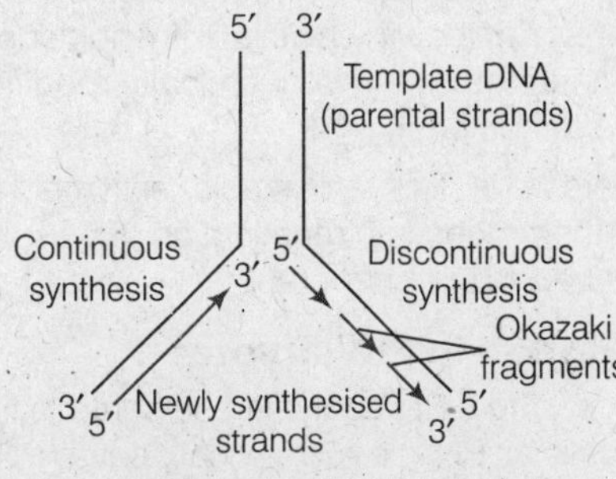

Replicating fork of DNA

The enzyme DNA polymerase-I now degrades the RNA primer and simultaneously catalyses the synthesis of short DNA segments to replace the primer. This segment is then joined to the main DNA strand by a DNA ligase enzyme.

Previoulsy, it was thought that there occurs a simultaneous continuous synthesis of DNA at both the strands of template in 5′⟶ 3′ direction of newly synthesised strand.

But Okazaki et.al. (1968) suggested that it is only one strand which shows such a continuous replication and is called the **leading strand**, while the other strand replicates in a discontinuous manner and synthesises short fragments called **Okazaki fragments**. This discontinuous strand is also called the **lagging strand**.

The Okazaki fragments are further joined by polynucleotide ligase (a joining enzyme) to form a continuous segments.

The survival of the cell depends upon its ability to repair the DNA damage.The DNA damage is of two general types:

(i) **Monoadducts**, alterations in a single nitrogenous base.

(ii) **Diadducts**, alterations in more than one nitrogenous bases. These alterations or damages are removed with the help of repair replication. The enzymes which participates in the repair replication are called correctional nucleases.

Genes

Johannsen has defined gene as an elementary unit of inheritance assigning to a particular trait. A gene can also be defined as a unit of inheritance, composed of a segment of DNA or chromosome situated at a specific locus (gene locus) which carries coded information associated with a specific function and can undergo crossing over as well as mutation.

Transcription of DNA

The central dogma states that the directed assembly of amino acids into a polypeptide chain is a multistep process which begins with the formation of RNA, by using DNA of the gene as **template**.

This process of RNA synthesis is called **transcription**.

The RNA formed here is a messenger RNA (*m*RNA) but the process of protein synthesis requires all the three RNAs, i.e.

(i) The *m*RNA which takes actual message from DNA.

(ii) The *t*RNA which helps in bringing and binding amino acids.

(iii) And the *r*RNA which forms ribosomes after combining with proteins.

The fresh RNA formed from DNA template is often referred to as **primary transcript** or which is still in nucleus as pre RNA, because, it is edited (i.e. it is chemically modified, splice and reassembled in nucleus).

RNA polymerase is the remarkable enzyme responsible for carrying out the process of transcription. It was isolated from *E. coli* by Chamberlin and Berg (1962).

DNA Transcription in Prokaryotes

In prokaryotes, single RNA polymerase controls the synthesis of all types of cellular RNA, e.g. in E. coli, same RNA polymerase catalyses the synthesis of all RNAs, i.e. mRNA, tRNA and rRNA.

This RNA polymerase is very large, consisting of five main units, i.e. two α-subunits (which binds regulatory proteins), two β -subunits, and one ω subunit (smallest subunit which facilitates assembly of RNA P (RNA polymerase) and stabilises assembled RNA P) out of them one binds RNA nucleoside, while another is joined to DNA template and one σ (sigma) factor.

The first four subunits, i.e. α, α', β and β' forms the core enzyme, while the σ factor is not firmly attached to the core enzyme and can be isolated easily.

This σ factor recognises the start signal on the DNA template and directs the binding of RNA polymerase to the promoter region. It thus, increases the specificity of core polymerase binding to DNA template.

The above description is meant for E. coli and anacystis (blue-green alga).The core enzyme in bacterium Bacillus subtilis have seven polypeptide chains, i.e. $\beta, \beta', \alpha, \alpha', \delta, \omega_1$ and ω_2.

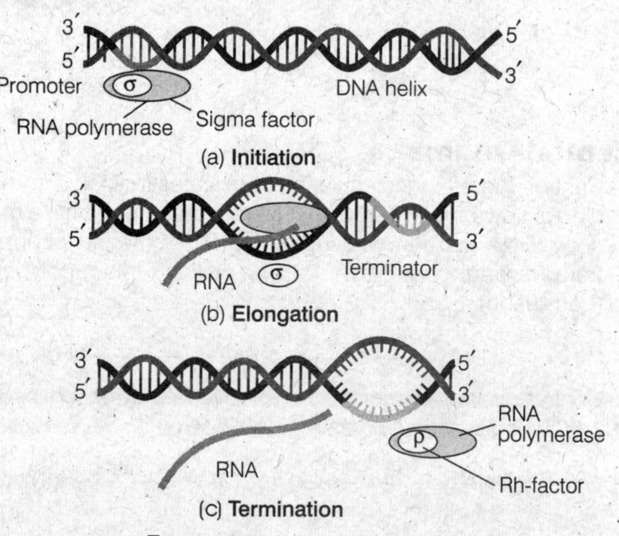

Transcription in prokaryotes

Genetic Code

More than 150 amino acids are found in nature of which about 20 only are involved in protein synthesis. To specify these twenty amino acids, four nitrogenous bases are present in a DNA molecule. Now, the question arises, how it is possible that the names of twenty amino acids are encoded by only four nitrogenous bases. The most accepted theory to solve this question was given by FHC Crick. According to this theory, there exists a genetic code as the smallest unit to code one amino acid.

This genetic code is defined as set of if 3 the nucleotide or nucleotide sequence of three nitrogenous bases which specifies the amino acid sequence in a polypeptide molecule.

Features of a Genetic Code

The genetic code is written in the linear form, using the ribonucleotide bases that compose mRNA molecule as 'letters'. This ribonucleotide sequence is derived from their complementary bases in DNA.

The general features of a genetic code can be summarised as

(i) Each 'word' of a codon consists of three letters, i.e. the codon is **triplet**. Thus, we can say that each amino acid is specified by three nitrogenous bases in DNA/RNA.

The triplet nature of codon can be understood as:

(a) Since, there are only four nitrogenous bases that codes for twenty amino acids, a single nitrogenous base is insufficient to express four combinations.

(b) Similarly, a combination of two nitrogenous bases, i.e. doublet code provides $4 \times 4 = 16$ codons which are still insufficient for twenty amino acids.

(c) It was Gamow (1954) who pointed out the possibility of a three letter code, i.e. triplet codon. This will give $4 \times 4 \times 4 = 64$ codons which are more than enough to code for twenty amino acids.

(d) Several of the triplets have same letters but in different sequences. These codes for different amino acids, which means it is the sequence of nitrogenous bases in a codon, which is important in determining the name of an amino acid.

(ii) The genetic code inside the cell medium is said to be non-ambiguous, this is because a particular codon always codes for the same amino acid. But in certain rare cases, the genetic code is found to be ambiguous, i.e. some codons code for different amino acids under different conditions.

e.g. in streptomycin sensitive strain of *E. coli,* the codon UUU, normally codes for phenylalanine but it may also code for the animal amino acid isoleucine, leucine or serine. This ambiguity is enhanced at high Mg ion concentration, low temperature and in the presence of ethyl alcohol.

(iii) The code is degenerate, i.e. a given amino acid can be specified by more than one codons.

(iv) The degeneracy of codon is not random. Mostly, different codons for a particular amino acid and have the same first two letters. For example the first two letters of all the four codons for valine are GU and that for alanine are GC. However, leucine, serine and arginine are exceptions.

(v) The code contains start and stop signals, i.e. certain triplets are necessary to initiate and to terminate the process of translation.

The initiation and termination process occurs in the following way:

(a) The codon present in the beginning of a functional gene is called initiation codon. It marks the message for a polypeptide chain.

(b) This initiation codon is AUG in most of the cases and codes for methionine (*met*) in eukaryotes and for N-formyl methionine (f-met) in prokaryotes.

(c) Less often, GUG also serves as initiation codon in bacterial protein synthesis. It is the normal codon for valine. This codon also found to initiate protein synthesis when the normal AUG codon is lost by deletion or parallel processes.

(d) Both AUG and GUG shows ambiguity in one sense, i.e. each of them codes for two different amino acids. But when any of these is present at initiation point, it codes for N-formyl methionine.

(e) In interstitial positions, AUG codes for methionine and GUG for valine.

(f) The last codon of a functional gene helps in the termination of the polypeptide chain. This codon is known as termination codon. There are three termination codons, i.e. UAG, UGA and UAA.

Earlier, when the function of these codons was not known these were called non-sense codons. These codons were also called as amber, opal and ochre, respectively. The termination codons do not code for any amino acid, hence, they cause termination of polypeptide chain and release of chain.

(g) The initiation and termination codons are generally known as signals and this phenomenon is known as punctuation.

(vi) The code is said to be commaless, i.e. no internal punctuation (commas) is used in the code. Once the translation begins, the codons are read one after the other without breaks or demarcating signals in between them.

(vii) The code is non-overlapping, i.e. any single ribonucleotide at a specific location within *m*RNA is the part of only one triplet.

(viii) The code has polarity, i.e. a code has fixed starting and termination ends. The available evidences indicates that the message in *m*RNA is reads in the $5' \longrightarrow 3'$ direction and the polypeptide chain is synthesised in $N \longrightarrow C$ direction, i.e. from amino group (NH_2) to the carboxylic (COOH) group.

(ix) The code is almost universal, i.e. with only minor exceptions, a single coding dictionary is used by almost all viruses, prokaryotes, archaea and eukaryotes.

In 1979, investigators began to determine the complete DNA sequences of mitochondrial genome in humans, cattles and mice.

It came as a shock, when these investigators learned that the genetic code used by these mammalian mitochondria was not same as the universal code.

e.g. the non-sense codon UGA was read for amino acid tryptophan, AUA codes for methionine rather than isoleucine, while AGA and AGG were used as non-sense codons rather than coding for arginine. Similarly, minor difference from universal code have also been found in the genomes of chloroplasts and some ciliate protists.

Translation

It is the process through which the genetic information present in *m*RNA directs the order of specific amino acid to form a polypeptide. Before initiating the synthesis of polypeptide chains, activation and transfer of amino acid to *t*RNA are the essential processes.

The main steps in translation as:

Activation of Amino Acid

The activation of amino acids takes place through their carboxyl groups. Each amino acid activation is catalysed by its own activating enzyme called **aminoacyl *t*RNA synthetase**. The complete reaction is

Amino acid + ATP + Enzyme $\longrightarrow$ Enzyme – Amino acid –AMP
+ PP inorganic phosphate

Transfer of Activated Amino Acid to *t*RNA

During the process of transference, a high energy ester bond is formed between the carboxyl group of amino acid and 3'–OH group of terminal adenosine of *t*RNA.

The complete reaction is carried out by the enzyme amino acyl *t*RNA synthetase which has two active sites, i.e. one for recognising *t*RNA and another for recognising specific amino acid molecule.

Enzyme – Amino acid – AMP + *t*RNA → Amino acid *t*RNA + AMP + Enzyme

Initiation of Polypeptide Chain Synthesis

(Initiation of Chain Formation)

The initiation of synthesis of polypeptide chain involves the following steps:

(i) The smaller ribosomal unit (30S or 40S) attaches to the *m*RNA molecule forming *m*RNA-smaller subunit complex.

(ii) Most often the starting amino acid is methionine (met) in eukaryotes and N-formyl methionine (f-met) in prokaryotes. The activated amino acid *t*RNA-complex attaches to the initiation codon (AUG) on *m*RNA to form smaller subunit initiation complex.

(iii) The larger subunit then joins this smaller subunit initiation complex to form a complete initiation complex.

Elongation of Polypeptide Chain

The elongation of polypeptide chain involves the following steps:

(i) The second amino acid tRNA-complex now occupies the A-site of larger ribosomal subunit. It is followed by the formation of the peptide bond between both the amino acids present at P-site as well as A-site. The enzyme which helps in the process of peptide bond formation is **peptidyl transferase**. Bond formation is simultaneous with the transfer of amino acid from P-site to A-site.

(ii) After peptide bond formation translocation occurs which involves the movement of second amino acid *t*RNA-complex from A-site to P-site.

In actual sense, this involves the movement of ribosome relative to *m*RNA in 5′ ⟶ 3′ direction. The initiator *t*RNA molecule is unloaded from the ribosome. As a result of translocation, the P-site becomes vacant and the second amino acid *t*RNA-complex along with met or f-met occupies the P-site.

(iii) The third amino acid *t*RNA-complex occupies the A-site. This is followed by peptide bond formation, transfer of chain and translocation in the similar manner resulting to a chain of three amino acid.

Termination of Chain Synthesis

Chain elongation continues until a termination codon (UAA, UAG or UGA) reaches the ribosome. Afterwards, the chain terminates and released from the ribosome. This process requires certain release factors (i.e. RF-1, RF-2 and RF-3 in prokaryotes and RF in eukaryotes).

The process can be simplified as :

(i) The termination codon provides signals to the ribosome for the attachment of release factors.

(ii) The release factor then interacts with the peptidyl transferase causing hydrolysis of bond between *t*RNA and polypeptide chain and the chain is released from the ribosome.

(iii) Hydrolysis of GTP results in the dissociation of the release factor from the ribosome molecule. The *t*RNA gets also unloaded and ribosomal subunits dissociates releasing the *m*RNA for the breakdown to nucleotides.

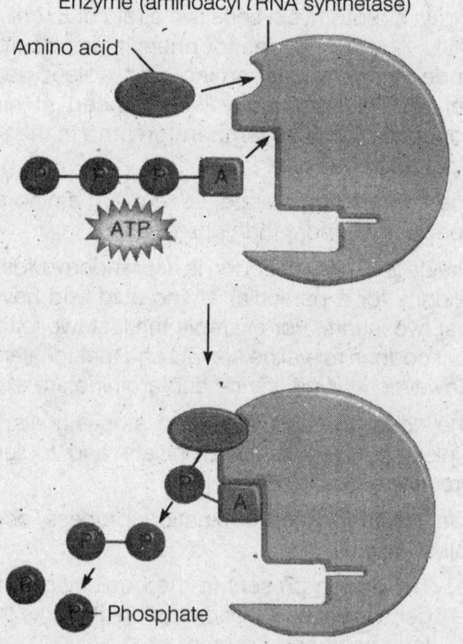

Amino acid activation

Chapter Exercise

1. Sigma factor is a component of
(a) DNA ligase (b) DNA polymerase
(c) RNA polymerase (d) Dissociation factor

2. RNA polymerase
(a) is the protein responsible for the production of ribonucleotides
(b) is the enzyme that creates hydrogen bonds between nucleotides on the DNA template strand and their complementary RNA nucleotides
(c) is the enzyme that transcribes exons but does not transcribe introns
(d) begins transcription at a promoter sequence and moves along the template strand of DNA, elongating the RNA molecule in a $5' \longrightarrow 3'$ direction

3. Transcription involves the transfer of information from
(a) DNA to RNA
(b) RNA to DNA
(c) *m*RNA to an amino acid sequence
(d) the nucleus to the cytoplasm

4. Which of the following is true for RNA processing?
(a) Exons are excised before the *m*RNA is translated
(b) Assemblies of protein and *sn*RNPs, called spliceosomes, may catalyse splicing
(c) The RNA transcript that leaves the nucleus may be much longer than the original transcript
(d) All of the above

5. The strongest evidence that DNA is the genetic material comes from the
(a) fact that the chromosomes are made of DNA
(b) studies on the transformation of bacterial cells
(c) knowledge that the DNA is present in the nucleus
(d) findings that the DNA is not present in the cytoplasm

6. Initiation of polypeptide chain in protein synthesis is induced by
(a) methionine (b) leucine (c) lysine (d) glycine

7. UAA, UAG and UGA are known as non-sense codons because they
(a) code for amino acids, which are not essential amino acids
(b) code for the initiator amino acid (methionine), which is later removed from the polypeptide chain
(c) are never present on *m*RNA
(d) do not code for any amino acid

8. Gene is
(a) a segment of DNA
(b) a segment of DNA and histone
(c) a segment of DNA, RNA and histone
(d) All of the above

9. The first translated codon in any *m*RNA binds to an anticodon is
(a) 3′ UAC 5′ (b) 5′ UAC 3′ (c) 3′ UAU 5′ (d) 5′ CAU 3′

10. The minimum length of cistron in base pair, which synthesises a polypeptide of 50 amino acids is
(a) 50 bp (b) 100 bp (c) 150 bp (d) 200 bp

11. Translocation involves
(a) the movement of the ribosome in the P-site to the A-site
(b) the movement of the *m*RNA strand one triplet length in the A-site
(c) the release of the unattached *t*RNA from the E-site
(d) All of the above

12. If the sequence of bases in DNA is ATTCGATG, then the sequence of bases in its transcript will be
(a) GUAGCUUA (b) UAAGCUAC
(c) CAUCGAAU (d) AUUCGAUG

13. The direction of synthesis of DNA, RNA and protein are always
(a) $3' \rightarrow 5'$ (b) $5' \rightarrow 3'$
(c) Both (a) and (b) (d) None ot these

14. Experiments using N^{15} (heavy nitrogen) to confirm the semiconservative replication of DNA were carried out by
(a) Meselson and Stahl (b) Hershey and Chase
(c) Beadle and Tatum (d) Watson and Crick

15. Regulatory genes are the genes that
(a) code for repressor proteins
(b) are transcribed continuously
(c) are not contained in the operon they control
(d) All of the above

16. Which one of the following pairs is correctly matched?

(a)	Fredrick Griffith	–	Discovered the phenomenon of transformation
(b)	Linus Pauling	–	Isolated DNA for the first time
(c)	Francis Crick	–	Proposed one gene-one polypeptide hypothesis
(d)	George Beadle	–	Proposed the concept of inborn errors

17. Genetic code translated the language of
(a) RNA into that of proteins (b) proteins into that of RNA
(c) amino acids into that of RNA (d) RNA into that of DNA

18. A gene is made up of
(a) DNA (b) RNA
(c) either DNA or RNA (d) amino acids

19. The growing polypeptide chain is released from the ribosomes when
(a) a chain terminating codon is reached
(b) a chain terminating *t*RNA binds to the ribosome
(c) the 7 methyl guanosine cap is reached
(d) the poly A tail is reached

20. The peptidyl transferase reaction occurs
(a) on the large subunit (b) on the small subunit
(c) between the subunit (d) None of these

21. The pathway of a *t*RNA during polypeptide elongation on the ribosome is

(a) A – site → P – site → Exit – site
(b) A – site → P – site → Entry – site
(c) P – site → A – site → E – site
(d) P – site → Entry – site → Exit – site

22. How many energy bonds are expended in the formation of a peptide bond?

(a) 3 (b) 4 (c) 2 (d) 6

23. An antibiotic which inhibits translation in both prokaryotes and eukaryotes is

(a) tetracycline (b) actinomycin -D
(c) chloromycetin (d) puromycin

24. The core enzyme constituting the *E. coli* RNA polymerase consists of following subunits

(a) β, β', a, w (b) β, β', σ, a (c) β, β', σ, w (d) β, σ, a, w

25. Which of the following will be translated into a protein?

(a) UAA AUG CCC TCC CAT TATG
(b) AUG CCC UCU AUA GUA GTC
(c) UAA CCC UCU AUA GUA GUC
(d) AUG CCC UCU AUG GUC UAG

26. Gene controls

(a) protein synthesis and heredity
(b) protein synthesis but not heredity
(c) biochemical reactions of some enzymes
(d) heredity but not protein

27. RNA polymerase is capable of catalysing

(a) initiation (b) elongation (c) termination (d) All of these

28. *m*RNA is also known as

(a) information RNA (b) template RNA
(c) messenger RNA (d) All of these

29. The RNA constituting about 70-80% of the total RNA is

(a) ribosomal RNA (b) messenger RNA
(c) transfer RNA (d) template RNA

30. *t*RNA attaches amino acids at its

(a) 3′-end (b) 5′-end (c) anticodon (d) loop

31. 'Jumping genes' that have the ability to move within and between chromosomes are called

(a) introns (b) oncogenes
(c) transposons (d) retroviruses

32. In transcriptional control in eukaryotic cells

(a) a different combination of DNA binding proteins (transcriptional factors) may regulate the activity of a particular gene
(b) enhancers may be involved in the promotion as well as regulation of gene transcription
(c) may be due to the phosphorylation of transcriptional factors by a kinase
(d) All of the choices are correct

33. Transcription takes place in

(a) matrix (b) cytosol
(c) nucleus (d) cytoplasm

34. DNA polymerase III is thought to add nucleotides

(a) to the 5′-end of the RNA primer
(b) to the 3′-end of the RNA primer
(c) in the place of the primer RNA after it is removed
(d) on single stranded templates without need for an RNA primer

35. The sequence of one strand of DNA is 5′ TCGATC 3′. The sequence of the complementary strand would be

(a) 3′ AGCTAG 5′ (b) 5′ TCGATC 3′
(c) 5′ CTAGCT 3′ (d) 5′ GCTAGC 3′

36. The RNA polymerase that produces the primer necessary for DNA synthesis is called the

(a) origin of replication (b) convertase
(c) primase (d) ligase

37. DNA replication in vivo is discontinuous due to

(a) polymerase slippage
(b) trinucleotide repeats
(c) being restricted to synthesis in the 5′ ⟶ 3′ direction
(d) topoisomerases cutting the DNA in a random fashion

38. Which enzyme that catalyzes the formation of a covalent bond between adjacent 5′-P and 3′-OH termini of separate fragments of DNA?

(a) Origin of replication (b) Convertase
(c) Primase (d) Ligase

39. The process of producing a RNA polymer from a DNA template is called

(a) replication
(b) transcription
(c) translation
(d) duplication

40. Which of the following features is common to both DNA replication and transcription?

(a) Nucleotides are added to the 5′-end of the newly synthesised strand.
(b) A sugar-phosphate bond is formed between the 3′-OH and the 5′-phosphate.
(c) Deoxyribonucleotides are incorporated into the growing sequence
(d) Both RNA and DNA polymerase require oligonucleotide priming
(e) Both RNA and DNA polymerase initiate at promoter sequences

Answers

1. (c)	**2.** (d)	**3.** (a)	**4.** (d)	**5.** (b)	**6.** (a)	**7.** (d)	**8.** (a)	**9.** (b)	**10.** (c)
11. (a)	**12.** (b)	**13.** (b)	**14.** (a)	**15.** (a)	**16.** (a)	**17.** (a)	**18.** (a)	**19.** (a)	**20.** (a)
21. (a)	**22.** (b)	**23.** (d)	**24.** (d)	**25.** (d)	**26.** (a)	**27.** (b)	**28.** (d)	**29.** (a)	**30.** (a)
31. (c)	**32.** (d)	**33.** (c)	**34.** (b)	**35.** (a)	**36.** (c)	**37.** (c)	**38.** (d)	**39.** (b)	**40.** (a)

26

Regulation of Gene Expression

As we have studied that the genes of the organism produce specific proteins for its particular phenotype. Thus, the regulation of gene expression controls the amount and time of formation of the gene product.

The genes are considered as **constitutive** in nature which expresses themselves constantly in a cell as their products are required for the normal cellular activities and **non-constitutive genes,** which does not expresses themselves constantly in a cell, hence called silent genes.

The regulation of gene expression involves two main aspects. These are as follows:

(i) Making a quantitative difference in the amount of protein or enzymes of different genes.

(ii) Measuring the amount of protein or enzyme produced by a gene at different time interval or in different tissues.

Regulation in Prokaryotes

The rate of transcription in prokaryotes is controlled itself at the initiation of the process of transcription by the use of the on and off mechanisms.

A hypothesis for the regulation synthesis of protein in prokaryotes was given by F Jacob and J Monod. This hypothesis is called an **operon model**.

Operons are of the following main two types:

(i) **Inducible operon** In this operon, the repressor is active and is made inactive inducer, e.g. *lac* operon.

(ii) **Repressible operon** In this operon, the repressor is inactive and called as **aporepressor**. It is made active by the end product known as **co-repressor**. This operon is switched off by the end product, e.g. *trp* operon.

Lac Operon

While studying catabolism of lactose in *E. coli,* two microbiologists from Pasteur Institute in Paris, named F Jacob and J Monod (1961) proposed a hypothesis for protein synthesis called **Operon model.** They were awarded with the Nobel Prize for this discovery in 1956.

An operon is a part of DNA which acts as a single regulatory unit having one or more structural gene, operator gene, promoter gene, regulator gene, a repressor, an inducer or co-repressor.

Structure of *Lac* Operon

A *lac* operon consists of following main components serving the particular functions

(i) **Structural genes** These are the fragments of DNA which transcribes *m*RNA for the polypeptide synthesis. Three commonly found structural genes are:

(a) **Z gene**, which codes for the enzyme β-galactosidase (β-gal) that helps in the hydrolysis of disaccharides into monomeric units, i.e. lactose into galactose and glucose.

(b) **Y gene**, which codes for the enzyme permease that increases the permeability of the cell to β-galactosidase.

(c) **A gene**, which codes for the enzyme transacetylase that catalysis the transacetylation of the lactose in its active form.

These three structural genes are located in a row adjacent to each other and are linked together. They contain the information to determine the sequence of amino acid present in the protein.

(ii) **Promoter** It is the sequence of DNA where RNA polymerase binds and initiates the transcription of structural genes.

(iii) **Operator** It is the sequence of DNA lying adjacent to promoter at which specific repressor protein binds.

(iv) **Regulator** It is the gene (*i* gene) which codes for the repressor protein binding to the operator and suppresses the activity, thus switching off the gene.

(v) **Inducer** It is the substrate which prevents the repressor from binding to the operator. It may be in the term the chemical of diverse nature like metabolite, hormone substrate, etc. Thus, binding of inducer causes the gene to switch on.

Thus, the complete *lac* operon looks like

Regulator	Promoter	Operator	Structural		
			Z	Y	A
≈ 1200 bp	≈ 30 bp	35 bp	3063 bp	800 bp	800 bp

Mechanism of Inducible System

Two conditions explain the functioning of *lac* operon model:

(i) **When lactose** (inducer) **is absent**

Gene '*i*' regulates and produces repressor *m*RNA
↓
Repressor protein binds with the operator gene
↓
This binding thus blocks the activity of operator gene and gene is switched off
↓
Due to this, structural genes do not express themselves, hence no enzyme synthesis occurs

(ii) **When lactose** (inducer) **is present**

Lactose or inducer binds to the repressor forming an inactive repressor
↓
Repressor fails to bind with the operator gene
↓
RNA polymerase binds to the operator and transcribes *lac* *m*RNA
↓
Polycistronic *lac* *m*RNA, produces all three enzymes β-galactosidase, permease and transacetylase
↓
Lac operon is switched on

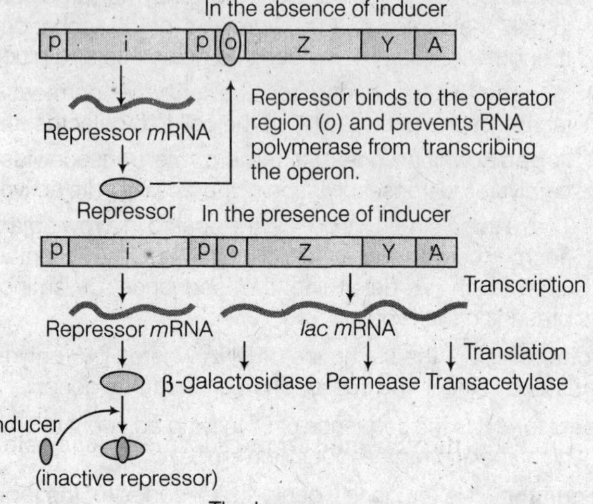

The lac operon

Regulation in Eukaryotes

The control of gene expression is complex in eukaryotic cell as operons are controlled by more than one promoter region. In eukaryotes, gene regulation takes place by different mechanisms. Here, histones play a major role in gene regulation.

Histones are the gene repressor that masks the activity of gene in a non-specific manner. Histones are responsible for the supercoiling of DNA (in eukaryotes), RNA polymerase molecule would not be able to move along the supercoiled DNA, hence no transcription is possible. Thus, the removal of histones is necessary from the genic site to produce a protein.

For the removal of histones, two possibilities are suggested:

(i) **Frenester's model** (1965) **of gene specific de-repressor RNA** According to this model, histones are removed from a particular genic site with the help of **nuclear polyions** that forms complexes with histones. It permits the separation of DNA strand in a random manner. Gene specific de-repressor RNA hybridises with the non-transcribing DNA strand, thus stabilising it in open position. It supports the other DNA strand and permits it to transcribe *m*RNA.

(ii) **Paul** (1971) *et, al*. **and Stein** *et, al*. **model** In this model the non-histones work as repressors. The events are more or less similar as suggested by the Frenester's model with the little difference that in this model, the gene specific acid proteins displace histones. These acid proteins are infact, non-histone chromosomal protein and being acidic, these have affinity for the histones. Both the models written above are almost equally supported today.

According to Britton and Davidson in eukaryotes, the operon contains four basic type of genes:

(i) **Sensor genes** These genes are infact detectors which are sensitive to the state of cell and its environment. These are stimulated by the change in environment or state of the cell. Thus, the sensors are responsible for sensing the need of an enzyme to be produced.

(ii) **Interrogator genes** The sensor inturn after stimulation sends out a specific signal (can be in the form of molecule) to other genes through the structure known as **interrogators**. Thus, the interrogators are the genes associated with sensors. There may be two possibilities, i.e. either one interrogators is associated with one sensor or many interrogators are associated with one sensor. When several interrogators are associated with a single sensor, the latter can initiate a variety of signals.

(iii) **Receptor genes** The signals send out by an interrogators are received by the receptor gene which normally remains associated with the producer.

(iv) **Producer genes** The producer genes are the output controls of the regulation mechanism. These genes get activated when the receptor receives the signals from interrogators. Thus, producers are at last responsible for the synthesis of that protein.

Chapter Exercise

1. **Expression of gene is controlled in eukaryotes**

(a) at one place in the nucleus
(b) at many places in the nucleus
(c) at many places in the cell
(d) cannot be controlled at all

2. **The example of inducible operon is**

(a) *trp* operon (b) *lac* operon
(c) Both (a) and (b) (d) None of these

3. **In *trp* operon, the co-repressor is**

(a) tryptophan (b) lactose
(c) β-galactoside (d) glucose

4. **Which one of the following is not a component of *lac* operon model?**

(a) Promoter gene (b) Structural gene
(c) Primer gene (d) Regulator gene

5. **The sequence of structural gene in *lac* operon is**

(a) *lac* A, *lac* Y, *lac* Z (b) *lac* A, *lac* Z, *lac* Y
(c) *lac* Y, *lac* Z, *lac* A (d) *lac* Z, *lac* Y, *lac* A

6. **Operon system consists of**

(a) operator and structural genes
(b) operator, regulator, repressor
(c) promoter and all of the above
(d) None of the above

7. **Which molecule is not produced by *E. coli* in the lactose medium?**

(a) Lactose dehydrogenase
(b) Thiogalactoside transacetylase
(c) β-galactosidase
(d) Lactose permease

8. **The lactose operon consists of**

(a) four genes–one inhibitor, one operator and two structural genes
(b) five genes–one inhibitor, one operator and three structural genes
(c) six genes–one inhibitor, one operator and four structural genes
(d) three genes–one inhibitor, one operator and structural genes

9. **Aporepressor is inactive repressor in tryptophan repressible operon system. It is generally a**

(a) protein (b) substrate
(c) metabolite (d) None of these

10. **The regulatory genes are located**

(a) in between operator and the structural genes
(b) in the middle of the structural genes
(c) at the end of the structural genes
(d) in front of the structural genes

11. **Addition of lactose in a culture of *E. coli* induces enzymes ……… to breakdown the lactose into glucose and galactose.**

(a) β-galactosidase and transacetylase
(b) transacetylase and permease
(c) β-galactosidase, permease, transacetylase
(d) None of the above

12. **In *lac* operon system, *lac* gene-z codes for**

(a) permease (b) repressor
(c) transacetylase (d) β-galactosidase

13. **Which one of the following statements regarding *lac* operon is correct?**

(a) *Lac* operon is inducible system
(b) *Lac* operon is repressible system
(c) *Lac* operon of *E. coli* contain four structural genes
(d) None of the above

14. **In both inducible and repressible systems, the synthesis of protein is controlled by**

(a) operator gene through regulator gene
(b) regulator gene through operator gene
(c) promoter gene through regulator gene
(d) operator gene through regulator and promoter genes

15. **In eukaryotes, the regulation could be exerted at**

(a) transcriptional and processing level
(b) processing and translational level
(c) transport of *m*RNA from nucleus to the cytoplasm
(d) All of the above

16. **Who were the first to elucidate a transcriptionally regulated system?**

(a) Francois Jacob and Jacque Monod
(b) Alfred Hershey and Martha Chase
(c) Oswald Avery and MacLeod
(d) Sever Ochoa and Nirenberg

17. **Regulation of *lac* operon by repressor is referred as**

(a) positive regulation (b) negative regulation
(c) feed back regulation (d) None of these

18. **Which of the following statements is correct?**

(a) Glucose acts as inducer for *lac* operon
(b) Galactose acts as inducer for *lac* operon
(c) Glucose or galactose act as inducers for *lac* operon
(d) Glucose or galactose cannot act as inducers for *lac* operon

19. **Lactose is the substrate for the enzyme β-galactosidase, it regulates**

(a) switching on of the operon
(b) switching off of the operon
(c) Both (a) and (b)
(d) synthesis of polypeptide chain

20. The operator gene of *lac* operon is 'turned on' when lactose molecules bind to

(a) operator gene (b) repressor protein
(c) *m*RNA (d) promoter site

21. When tryptophan is present in the medium, the transcription of tryptophan producing genes in *E. coli* is stopped by a helix-turn-helix regulator binding to the

(a) *trp* repressor (b) *trp* operator
(c) *trp* operon (d) *trp* promoter

22. In order for a gene to be transcribed, RNA polymerase must have access to the DNA helix and be able to bind to the genes

(a) operator (b) promoter (c) activator (d) regulator

23. In the function of the lac operon in *E. coli*, the lac genes are transcribed in the presence of lactose because

(a) the repressor cannot bind to the operator
(b) RNA polymerase binds to the operator
(c) an isomer of lactose binds to operator
(d) CAP does not binds t the operator

24. The most common form of gene expression regulation in both bacteria and eukaryótes is

(a) translational control
(b) transcriptional control
(b) post-transcriptonal control
(d) post-translational control

25. *E. coli* is able to use foods other than glucose in the absence of available glucose, because falling levels of glucose cause an increase of

(a) cAMP (b) CAP
(c) glu operon (d) lactose

26. Which of the following is true of the *lac* operon in *E. coli*?

(a) The operon is only switched on in the absence of lactose in the growth medium
(b) The *lac* operon messenger RNA is a polycistronic *m*RNA (it carries information for synthesis of several proteins)
(c) The enzyme β-galactosidase is only produced in large quantities when the *lac* repressor is bound to the operator
(d) The promoter is the binding site for the *lac* repressor

27. In the absence of glucose, *E. coli* can import lactose to change into glucose and galactose because CAP binds to the

(a) cAMP (b) DNA
(c) *lac* operon (d) operator

28. In eukaryotic gene regulation, the location of the promoter is always before and the location of the enhancer always after the gene being regulated.

(a) True (b) False
(c) Both (a) and (b) (d) None of these

29. Which of the following is part of an operon?

(a) Structural gene
(b) An operator
(c) A promoter
(d) All of these

30. In the context of prokaryotic gene expression, which of the following is the most appropriate definition of an operator?

(a) A cluster of genes that are regulated by a single promoter
(b) A DNA binding protein that regulates gene expression
(c) A non-coding, regulatory DNA sequence that is bound by RNA polymerase
(d) A non-coding, regulatory DNA sequence that is bound by a repressor protein

Answers

1. (c)	**2.** (b)	**3.** (a)	**4.** (c)	**5.** (d)	**6.** (a)	**7.** (a)	**8.** (b)	**9.** (a)	**10.** (d)
11. (c)	**12.** (d)	**13.** (a)	**14.** (a)	**15.** (d)	**16.** (a)	**17.** (b)	**18.** (d)	**19.** (a)	**20.** (b)
21. (d)	**22.** (b)	**23.** (c)	**24.** (b)	**25.** (a)	**26.** (b)	**27.** (a)	**28.** ()	**29.** (d)	**30.** (d)

27

Evolution

The branch of science dealing with the two aspects of life, i.e. origin and evolution is called **evolutionary biology**.

Origin of Universe

According to the Big Bang theory proposed by George Lemaitre (1931), our universe has arosed as a catalymic eruption of hot energy rich sub-atomic particle.

Origin of Life

Earth has evolved about 4.5 billion years ago from the left over dust and gases of the sun. It was in the form of a solidified, red hot cloud containing free atoms and approximate temperature of 5000-6000°C. The surface of the earth was in Molten state with no atmosphere. After about 750 million years, the earth got cooled and the present solid crust developed.

Some important theories were given by various scientists in context of origin of life. These are:

(i) **Theory of special creation** It was proposed by Father Saurez and was supported by Milton. It was believed that a supernatural power (god) has created all living beings on the earth.

(ii) **Theory of spontaneous generation** It is also known as theory of abiogenesis or autobiogenesis. According to this theory, life had originated from non-living matter automatically.

(iii) **Theory of biogenesis** It states that the life had originated from the pre-existing life.

(iv) **Theory of cosmozoic origin** It was supported by Richter (1865) which states that the life is coeternal with matter without any beginning, thus, both living and non-living objects has originated simultaneously.

(v) **Theory of panspermia/extraterrestrial origin** It was proposed by Arrhenius which states that the life extraterrestrial origin on earth was infected from these advanced planets. This theory was supported by genetic code and role of metal molybdenum supported by Cuvier (1769-1832).

(vi) **Theory of catastrophism** It states that the world has passed through many ages. A catastrophic event at the end of each age killed most of the living beings and at beginning of next age new creation evolved.

Experimental Studies of Origin of Life

Some of the common experiments performed regarding the origin of life were given by Louis Pasteur (1864) through, which he concluded that the life comes from pre-existing life but fails to explain the experimental verification of the origin of life.

Later on, AI Oparin and JBS Haldane (1920) postulated that natural chemicals and physical process have led to the spontaneous development of simple organic compounds that are the components of a primitive atmosphere.

Modern System for Origin of Life

Miller and Urey (1953), gave and experimental explanation for the theory postulated by AI Oparin and JBS Haldane.

According to this theory, life has originated on earth after a long abiogenic evolution of organic and inorganic compounds (chemical evolution) followed by the biological evolution.

Hence, this theory is also known as **modern theory** or **abiogenic** origin or **naturalistic theory** or **Physicochemical evolution**.

Evidences of Evolution

Evidences from various palaentological, morphological, biochemical, biogeographical concepts embrace the concept of evolution.

These are discussed below in detail:

(i) Morphological Evidences

These evidences are concerned with the structural and functional properties of different organisms.

(a) **Homologous organs** These are the organs which have same embryonic structure and origin but, may or may not perform the same function, homology in organs indicates the common ancestry,
For example in animals, the pentadactyl forelimb of vertebrates consist of same bones, i.e. humerus, radius, ulna, carpals, metacarpals and phalanges but differ in their function in different organisms.

(b) **Analogous organs** These organs are the organs which are different in their basic structure and developmental origin but it performs similar function. Analogy is based on convergent evolution due to adaptive convergence.
e.g. in animals, wings of birds, insects and bat, all are used for flying.

Palaeontological Evidences

Palaeontology is the **study of fossils** where fossils are defined as the relics of former living life (plants and animals) embedded in or dug out of superficial deposits in past geological periods. Fossils records also provides a missing link between two group of organism, e.g. *Archaeopteryx* is a connecting link between reptiles and birds.

Biogeographical Evidences

It suggests that species restricted to a region develops a unique features and also species that are present in the widely separated region, shows similarities of ancestry.

Note The process of evolution of different species starting from a point in a geographical area and finally radiating its other area of geography. Darwin's finches provide and excellent example of adaptive radiation supporting the concept of evolution.

Theories of Evolution

Several theories have been put forward in order to explain the phenomenon of evolution or origin of species.

Lamarckism

The first complete explanation of evolution was given by the French biologist, Jean Baptiste de Lamarck (1744-1829) in 1809, the year of Darwin's birth. He made the first convincing case for the idea that the fossils were the remains of extinct animals. Lamarck's evolutionary mechanism, 'the inheritance of acquired characteristics', was engagingly simple.

It states that the organisms by striving to meet the demands of their environments, acquired adaptations and passed them by heredity to their offsprings, e.g. the genetic studies now show that the traits acquired by an organism during its lifetime, such as strengthened muscles are not inherited by the offsprings.

Postulates of Lamarckism

Lamarck's theory regarding evolution was published in *Philosophie Zoologique* (1909). It has the following postulates:

(i) Living organisms and their component parts tends to increase in size as a result of internal forces of life.

(ii) Highly evolved flowering plants are much smaller in size, while many extinct conifers and ferns were gigantic.

(iii) A person constantly busy in reading and writing develops impaired sight.

Evidences in Favour of Law of Inheritance of Acquired Characters

(i) During vegetative propagation of plants and regeneration in animals, somatic cells can produce the germ cells which do not follow the Weismann's theory of continuity of germplasm.

(ii) **Tower** observed the direct effect of environment on germ cells when he treated potato to very high temperature and humidity during the development of their reproductive organs.

(iii) **Harrison** observed the effect of environment on germ cells through the somatic cells, when he found a dark variety of moth in manganese polluted localities whose ancestors were pale in colour.

(iv) Sudden heritable variations or mutations were obtained by Auerbach *et. al*. On the exposure of *Drosophila* to high energy radiations like UV-rays, X-rays, Y-rays, mustard gas, etc.

(v) Radish is a biennial crop in cold countries but completed its growth in a year in tropical areas.

(vi) European peach which is deciduous becomes evergreen in India.

(vii) Hormonal effects do not isolate somatic cells and germ cells.

Examples in Favour of Lamarckism

According to Lamarck, the giraffe evolved its long neck because its ancestors who were like horses with small, neck and for climbs lengthened their necks by stretching to obtain food and then passed the lengthened neck to their offsprings. Over many generations, these changes accumulated to produce the long neck of the modern giraffe.

Neo-Lamarckism

A group of scientists has further studied the Lamarck's theory and some scientists like Spancer, Giard, Lawrence (1831), Mac Dougall, Cope, Rechard, Wells Gadow, Dali, etc., supported the modified form of Lamarckism which is termed as Neo-Lamarckism.

Postulates of Neo-Lamarckism

(i) According to Neo-Lamarckism, there is a causal relationship between the structure, function and environment.

(ii) Some of the variations acquired by an individual can be transmitted to its offsprings.

(iii) The role of internal vital forces in evolution has been discarded.

(iv) Only the variations that are associated with the germ cells are inherited.

Darwins Theory of Evolution

After collecting a vast amount of information, Darwin began to realise that under the intense competition of members in a population, any variation which favoured survival in a particular environment would increase the individual's ability to reproduce and leave fertile offsprings.

Less favourable variations would be at a disadvantage and organism possessing them would therefore have their chances of successful reproduction decreased.

These data provided Darwin with the framework to formulate, a theory of evolution by Natural Selection in 1839. He did not publish his findings at that time. Indeed, Darwin's greatest contribution to science was not so much to show that evolution occurs but how it might occur.

Postulates of Darwinism

The main postulates which formed the basis of Darwin's theory, are as follows:

(i) **Overproduction** The multiplication of individual of a species occurs in a geometric proportion. Some organisms produces very large number of species. If this tendency of multiplication was unchecked, in a very short time earth would be overcrowded.

Despite of rate of reproduction of a species, its number remains about constant under fairly stable environment.

The production of more offsprings by some organisms and fewer by others is termed as **differential reproduction**.

(ii) **Limited resources** The resources like food, space, etc., remain limited inspite of rapid multiplication of the individuals of all the species

(iii) **Struggle for existence** Struggle for existence may be of following three levels:

(a) **Intraspecific struggle** It is the type of struggle found among the individuals of the same species for their common requirements like food, shelter, breeding places, etc. Intraspecific struggle is very severe, e.g. young trees in a forest, cannibalism (to take the individual of its own species as a food) and human wars.

(b) **Interspecific struggle** It is the struggle between the individuals of different species for their similar requirements like food and space.
e.g. a frog feeds on insects and is preyed upon by snakes while kite feeds on both frogs as well as snakes.

(c) **Struggle with environment** (inanimate nature) It is the struggle of living forms against the environmental conditions like extreme heat, cold, drought, earthquakes, storms, disease, volcanic eruption, etc.

(iv) **Variations** These are the difference among the individuals. As all individuals are dissimilar in some characters except the identical twins, these variations can help to adjust with their environment.

Darwin recognised the following two types of variations:

(a) **Continuous variations** These variations show the whole range of variations in a particular character.

(b) **Discontinuous variations** These are the variations that appears suddenly and show on gradation. These variations were termed as sports by Darwin and mutation by Hugo de Vries.
Darwin regarded continuous variations to be more important because the discontinuous variations being most harmful would not be selected again.

Natural Selection

As a result of struggle for existence, only those organisms could survive, which have favourable variations to adapt with the environmental conditions or with so, many variations in populations.

In other words, the struggle for existence results in survival of the fittest (a term coined by Herbert Spencer) which refers to the idea that natural selection tends to favour those organisms that are most fit to survive with the reproductive age in a particular environment and to produce offsprings.

Hence, the survival of fittest is a result of selection and proliferation of only those organisms which were most suitably adapted to the environment and successful in mating, i.e. natural selection.

Formation of New Species

Darwin considered that the origin of new species from old species occurs as a result of struggle for existence, variability (continuous variations) and inheritance. As a result, new species became better adapted to their environment. These beneficial adaptations are preserved and accumulated in the individuals of species and resulted into the origin of new species.

Neo-Darwinism

The theory of evolution was proposed by Darwin and Wallace, has been modified in the light of modern evidence from genetics, molecular biology, palaeontology, ecology and ethology (the study of behaviour) and is known as Neo-Darwinism (*neo* new). This may be defined as the theory of **organic evolution** by the natural selection of inherited characteristics.

Mutation Theory of Hugo de Vries

Hugo de Vries (1848-1935) was a Dutch botanist. He proposed the mutation theory in order to explain the mechanism of evolution. This theory was published in 1901, in his book entitled **Die Mutation Theory**.

Hugo de Vries gave much importance to the discontinuous variations or saltatory variations and proposed that the new species arises not by the accumulation of minor and continuous variations through the natural selection but by suddenly appearing saltatory variations for which he coined the term mutation.

Salient Features of Mutation Theory

(i) Mutations act as a raw material for evolution.
(ii) Mutations are heritable and subjected to natural selection.
(iii) Mutations are large, heritable changes in contrast to small, directional fluctuating variations of Darwin.
(iv) Mutations may occur in any direction and may be useful or harmful. The harmful mutations can be eliminated, while useful mutations are selected by the nature.
(v) Sometimes new species are produced by a single mutation.

Criticism Against Mutation Theory

(i) Natural mutations are not the common phenomenon.
(ii) Most of the mutations are recessive and retrogressive.
(iii) Development of mimicry cannot be explained satisfactorily.
(iv) Theory failed to explain the role of nature.

Modern Synthetic Theory of Evolution

The modern theory of origin of species or evolution is known as modern synthetic theory of evolution. It is the combination of Darwinian selection and genetic theory.

The modern synthetic theory of evolution evolved in 1937, with the publication of Dobzhansky's genetics and the Origin of Species which was supported by Huxley (1942), Mayr (1942), Simpson (1944,1953) and Stebbins (1950).

The fundamentals of modern synthesis theory arrived largely independently by Chetverikov, Fischer, Haldane and Wright between 1926 and 1932.

Main Postulates of Modern Synthetic Theory of Evolution

This theory recognises four basic types of processes

(i) Gene mutation

(ii) Changes in chromosome structure and number

(iii) Genetic recombination

(iv) Natural selection

(v) Besides these, three accessory process affect the working of above four basic processes migration, hybridisation and chance

Hardy-Weinberg Equilibrium

This is the fundamental law of population genetics which states that the genes and genotypic frequencies in Mendelian populations remains constant generations after generations, if there is no selection, mutation, migration or random drift. It is used to compare allele frequencies of a given population over a period of time.

Sum total of all the allelic frequencies is equals to 1, e.g. in a diploid, if p and q represent the frequency of an allele A and allele a. The frequency of AA individuals in a population is p^2 of aaq^2 and of Aa is $2pq$.

Hence, it can be expressed by the following reaction,

$$p^2 + 2pq + q^2 = 1.$$

Human Evolution

The fossil evidence clearly indicates that origin of man occurred in Central Asia, China, Japan and India (Shivalik hills).

(i) The common ancestor of apes and man is a primate *Dryopithecus*. The next stage was *Ramapithecus* in the hominid evolution.

(ii) Both were hairy and walked like gorillas and chimpanzees. *Ramapithecus* was more like man and was the fore-runner of hominid evolution but *Dryopithecus* was more ape-like.

The human evolution took place as follows:

Human ancestors	Time of origin	General features
Dryopithecus	20-25 mya	Ape-like, hairy, ate soft fruits and leaves. Arms and legs of same length, large brain.
Ramapithecus	14-15 mya	More man-like, walked more erect, teeth like modern man.
Australopithecus	3-4 mya	Fossils found in Tanzania and Ethiopia, man-like primates, tall feet, walked upright, ate fruits, hunted with stone, weapons, brain capacity was 400-600 cc.
Homo habilis	2 mya	Fossils found in East Africa grasslands, first human like being, brain capacity 650-800 cc, did not eat meat.
Homo erectus (Java man)	1.5 mya	Fossils found in Java (Java man), ate meat brain capacity 900 cc, probably.
Homo sapiens neanderthalesis (Primitive man)	1,00,000-40,000 years ago	Known as neanderthal man (*Homo sapiens neanderthalesis*). Fossils found in East and Central Asia, brain size 1400 cc, used hides to protect body, burries their dead, became extinct 25000 years ago and must have lived between 1,00,000-40,000 years before.
Homo sapiens (Modern man)	75,000-10,000 years ago	Arose during the ice age between 75,000-10,000 years ago. Developed cave art around 18,000 years before, agriculture started around 10,000 years back. Human settlement and civilisation started.

Chapter Exercise

1. Which of the following statements is correct?

(a) Accumulation of useful mutations through ages has led to the creation of new species of living beings
(b) Genetic variability provides raw material or the operation of natural selection and reproductive isolation
(c) Genetic variability is produced by gene mutations
(d) All of the above

2. The theory of evolution is based on

(a) mutations (b) reproductive isolation
(c) sexual reproduction (d) All of these

3. Concept of inheritance of acquired characters was disapproved by the experiments of

(a) Oparin (b) Weismann
(c) Fox (d) Darwin

4. The theory of natural selection failed to explain the

(a) survival of the fittest
(b) arrival of the fittest
(c) over reproduction
(d) balance in size of population

5. Darwin in his natural selection theory did not believe in the role of which one of the following in organic evolution?

(a) Parasites and predators as natural enemies
(b) Survival of the fittest
(c) Struggle for existence
(d) Inheritance of acquired characters

6. Lamarckism fails to explain which of the following?

(a) The absence of all the four limbs amongst amphibians
(b) Exceptionally long neck of giraffe
(c) An intellectual having a dull son
(d) Loss of vision in cave-dwelling mammals

7. The most important evidence that bears on Darwin's theory has been in the area of

(a) palaeontology
(b) genetics
(c) comparative anatomy
(d) comparative embryology

8. According to Charles Darwin, evolution is

(a) descent with modification
(b) *Natural Selection*
(c) survival of the fittest
(d) irreversible change

9. Who claimed that the *Oenothera lamarckiana* was a hybrid plant which could be obtained by hybridisation of two wild species and it is not a normal plant?

(a) Theodosius (b) Julian
(c) Hugo de Vries (d) TR Malthus

10. Darwin believed that the certain parts of body get larger and become more complex through the generations because they

(a) are used more extensively than other parts
(b) contribute to greater reproductive success
(c) are predetermined to do so
(d) are most similar to God's perfections

11. Darwin believed that the giraffe has the long neck because

(a) the creator designed it that way
(b) the catastrophes eliminated short-necked forms
(c) its ancestors stretched their necks to get food
(d) ancestral giraffes with slightly longer necks than others got more food and left more surviving offsprings

12. An adaptive radiation, which involves major changes as in organisation plane resulting in the formation of new families, orders, classes and phyla

(a) microevolution (b) macroevolution
(c) megaevolution (d) intraspecific evolution

13. Hugo de Vries worked on the plant

(a) *Pisum sativum* (b) *Lathyrus odoratus*
(c) *Oenothera lamarckiana* (d) *Antirrhinum majus*

14. TR Malthus is famous for his book on

(a) population (b) mathematics
(c) geography (d) genetics

15. The use and disuse principle of evolution or theory of 'inheritance of acquired characters' was proposed by

(a) Hugo de Vries (b) Lamarck
(c) Weismann (d) Darwin

16. Key point of Lamarck's view about organic evolution is that every offspring

(a) is similar to its parents
(b) inherits characters acquired by the parental generation
(c) shows struggle for existence
(d) repeats phylogeny in its ontogeny

17. The evolution of a species is based upon sum total of adaptive changes preserved by

(a) natural selection
(b) isolation
(c) speciation
(d) human conservation

18. Retrogressive evolution is shown by

(a) man (b) birds
(c) tunicates (d) fish

19. Who put forward the first theory of evolution?

(a) Darwin (b) Lamarck
(c) Hugo de Vries (d) Cuvier

20. The study of Galapagos finches show
(a) use and disuse of organs
(b) adaptive radiations
(c) convergent evolution
(d) macroevolution

21. Evolutionary history of an organisms is
(a) ontogeny (b) orthogenesis
(c) phylogeny (d) None of these

22. The evidences of evolution based on the knowledge of fossils are
(a) palaeontological (b) anatomical
(c) Both (a) and (b) (d) None of these

23. Genetic drift is a
(a) Interbreeding of genetically different individuals
(b) Random change in gene frequency
(c) Sudden change in the genotype
(d) Orderly change in gene frequency

24. Non-inheritable variations are
(a) variations in the gametes
(b) variations in the germ cell
(c) variations in the zygote
(d) variations in the somatic cell

25. The term 'survival of the fittest' was contributed by
(a) Darwin (b) Wallace
(c) Spencer (d) Lamarck

26. Lamarck believed that certain parts of the body get larger and more complex through the generations because they
(a) contribute to greater reproductive success
(b) are similar to God's perfection
(c) are predetermined to do so
(d) are used more extensively than other parts

27. G Cuvier realised that history of life is recorded in fossils and believed that replacement of one species by another is caused by
(a) mutation
(b) the wrath of God
(c) genetic inbreeding
(d) extinction due to catastrophes such as floods

28. The term 'evolution' was coined by
(a) Haldane (b) Darwin
(c) Herbert Spencer (d) AL Oparin

29. Beak in birds had teeth in
(a) *Ostrich* (b) cassowary
(c) *Archaeopteryx* (d) kiwi

30. A lizard bird, *Archaeopteryx* lithographical is a
(a) connecting link between birds and reptiles
(b) missing link between birds and reptiles
(c) missing link between birds and mammals
(d) connecting link between reptiles and mammals

31. Reversible colloidal complex formed during early stages of origin of life was
(a) eobiont (b) protocell
(c) protobiont (d) coacervate

32. The most accepted theory of origin of life is
(a) theory of special creation (b) theory of panspermia
(c) naturalistic theory (d) theory of catastrophism

33. Birth of human baby with a small tail is an example of
(a) connective link (b) vestigial organ
(c) atavism (d) None of these

34. Big Bang theory was proposed by
(a) Kant (b) Oparin
(c) Lemaitre (d) Miller

35. In man which of the structure is not vestigial
(a) wisdom teeth (b) auricular muscles
(c) nails (d) vermiform appendix

36. What was the most significant trend in the evolution of modern man (*Homo sapiens*, from his ancestrous)?
(a) Increasing cranial capacity
(b) Upright posture
(c) Shortening of jaws
(d) Binocular Vision

37. Sewall wright effect is the another name of
(a) gene pool (b) gene flow
(c) genetic drit (d) isolation

38. Identify the correct statements out of the following
(a) There is no evidence for the existence of gills during embryogenesis of mammal
(b) Stems cells are specialised cells
(c) Ontogeny repeats phylogeny
(d) All plants and animal cells are totipotent

39. When two species of different genealogy come to resemble each other as a result of adaptation the phenomenon is termed as
(a) micro evolution
(b) divergent evolution
(c) convergent evolution
(d) coevolution

40. An important evidence in favour of organic evolution is the occurrence of
(a) homologous organs
(b) homologous and analogous organs
(c) analogous and vestigial organs
(d) homologous and vestigial organs

41. Jurassic period of the Mesozoic era is characterised by
(a) dinosaurs become extinct and angiosperms appear
(b) radiation of reptiles and origin of mammals like reptiles
(c) gymnosperms are dominant plants and first birds appear
(d) flowering plants and first dinosaur appear

42. Which one of the following experiments suggests that simplest living organisms could not have originated spontaneously from non-living matter.
(a) Microbes did not appear in stored meat
(b) Meat was not spoiled, when heated and kept sealed *via* vessel
(c) Lava could appear in decaying organic matter
(d) Microbes appeared from unsterilised organic matter

43. According to de Vries theory, evolution is
(a) jerky (b) continuous and smooth
(c) discontinuous (d) Both (a) and (c)

44. Lamarck theory of organic evolution is usually known as
(a) natural selection
(b) descent with change
(c) inheritance of acquired characters
(d) continuity of dermplasm

45. Sympatric speciation develops reproductive isolation without
(a) geographic barrier (b) barrier to mating
(c) barrier to gene flow (d) genetic change

46. Balancing selection is concerned with successful reproduction of
(a) homozygous recessive
(b) homozygous individuals
(c) heterozygous individuals
(d) All of these

47. Which one is used for knowing whether or not a population is evolving?
(a) Genetic drift
(b) Hardy-Weinberg equation
(c) Degree of evolution
(d) Proportion between acquire variation

48. Illustrate the correct order of the era
(a) paleozoic - auchaeozoic - coenozoic
(b) archaeozoic - palaeozoic - proterozoic
(c) palaeozoic - mesozoic - coenozoic
(d) mesozoic - archaeozoic - proterozoic

49. Quick change in phenotype in a small band of colonisers is called
(a) founder effect (b) genetic drift
(c) genetic bottle neck effect (d) gene flow

50. Mutation may be described as
(a) continuous genetic variation
(b) phenotypic change
(c) discontinuous genetic variation
(d) change due to hybridisation

51. Humming birds and hawk illustrates
(a) convergent evolution (b) homology
(c) adaptive radiation (d) parallel evolution

52. 'Continuity of germplasm.' theory was given by
(a) de Vries (b) Weismann (c) Darwin (d) Lamark

53. Parallelism is known as
(a) adaptive divergence
(b) adaptive divergence of widely separated species
(c) adaptive convergence of widely different species
(d) adaptive convergence of closely related groups

54. Basic principles of embryonic development were pronounced by
(a) Von Baer (b) Weismann (c) Haeckel (d) Morgan

55. According to one of the most widely accepted theory, earth's atmosphere before origin of life consisted of a mixture of
(a) O_2, CH_4, O_3 and NH_3
(b) H_2, CO_2, NH_3 and CH_4
(c) CH_4, NH_3, H_2 and H_2O vapour
(d) CH_4, O_2, H_2O and NH_3

56. Miller and Urey performed an experiment to prove origin of life. They took gases NH_3 and H_2 along with
(a) N_2 and H_2O (b) H_2O and CH_4
(c) CH_4 and N_2 (d) CO_2 and NH_3

57. Who prepared germplasm theory against Lamarck's principle of use and disuse?
(a) de Vries (b) Darwin (c) Spencer (d) Weismann

58. *Pithecantropus erectus* fossil was found in
(a) Java (b) China (c) Japan (d) Texus

59. Which is most important in 'speciation'?
(a) Geographical isolation (b) Ecological isolation
(c) Reproductive isolation (d) Ethological isolation

60. Theory of pangenesis was by
(a) Darwin (b) Hugo de Vries
(c) Lamarck (d) Oparin

Answers

1. (d)	**2.** (d)	**3.** (b)	**4.** (b)	**5.** (d)	**6.** (c)	**7.** (b)	**8.** (a)	**9.** (c)	**10.** (b)
11. (d)	**12.** (c)	**13.** (c)	**14.** (a)	**15.** (b)	**16.** (b)	**17.** (b)	**18.** (c)	**19.** (b)	**20.** (c)
21. (c)	**22.** (c)	**23.** (c)	**24.** (d)	**25.** (c)	**26.** (d)	**27.** (d)	**28.** (b)	**29.** (c)	**30.** (a)
31. (d)	**32.** (c)	**33.** (b)	**34.** (c)	**35.** (b)	**36.** (a)	**37.** (c)	**38.** (c)	**39.** (c)	**40.** (d)
41. (c)	**42.** (b)	**43.** (d)	**44.** (c)	**45.** (a)	**46.** (c)	**47.** (b)	**48.** (c)	**49.** (a)	**50.** (c)
51. (c)	**52.** (b)	**53.** (d)	**54.** (a)	**55.** (c)	**56.** (b)	**57.** (d)	**58.** (a)	**59.** (c)	**60.** (a)

28

Plant-Water Relation and Transpiration

Water acts as the most important and unique molecule in all living organisms. It is a tiny molecule that can easily pass through the biological membranes. It forms the basic cytoplasmic medium of the cell. Thus, all physiological functions of plant are directly or indirectly dependent on water. Hence, to understand how water plays an important role in the universe some of the basic fundamental phenomena are being described.

Diffusion

The process that determines the movement of water into, within and out of the plants is called **diffusion**. In other words, diffusion is the phenomenon of net movement of particles of a substance from a region of greater or higher concentration to a region of lower concentration due to their own intrinsic kinetic energy. The rate of diffusion of gases is faster than liquid or solutes.

Thus, diffusion is a mechanism of passive movement of molecules along their concentration gradient. It does not require expenditure of energy. During diffusion, each molecule diffuses down its own concentration gradient and remains unaffected by the concentration gradient of other molecules.

The tendency of the molecules, atoms or ions of gases, liquids and solids get evenly distributed, throughout the available space is due to the kinetic energy present in them.

The rate of diffusion is being affected by many factors such as temperature, density of diffusing substances, density of the medium, diffusion pressure gradient, etc. It is significantly important phenomenon because it helps in the gaseous exchange in plants and animals.

e.g. the fragrance of flowers also spreads out due to the process of diffusion which attracts insects for pollination.

Diffusion Pressure (DP)

The force or pressure developed by the movement of diffusible particles of a substance as they move from a place of higher concentration to an area of lower concentration is called diffusion pressure. The diffusing particles cause pressure which is proportional to the kinetic movement and number of such particles.

Thus, diffusion always takes place from a region of higher diffusion pressure to a region of lower diffusion pressure, i.e. along a diffusion pressure gradient.

It is directly proportional to the concentration and temperature. Diffusion pressure of pure water is maximum, its value at 20°C is 1336.40 atm.

Permeability

The term 'permeability' refers to the degree to which a membrane permits movement of gases, liquids and dissolved substances across it. A membrane can be highly permeable to one substance, moderately permeable to second, while can be slightly permeable to the third one.

Thus, the permeability of a protoplasmic membrane depends on the kind and variety of its constituents, i.e. its state of aggregation, fluidity, thickness, degree of hydration, etc. Because all plant cells do not have same properties towards a given substances.

In other words, it can be said that permeability of protoplasmic membranes of different plant cells is different.

Plasmolysis

If a plant cell is placed in a hypertonic solution (i.e. a solution of lower water potential than the cell's water potential) **exosmosis** takes place. Thus, the water moves out of the cells and the cell wall is no longer under tension. Further, the loss of water results in the contraction of the protoplasm which eventually assumes a spherical shape.

This phenomenon of shrinkage of protoplasm from the cell wall under the influence of some hypertonic solution is known as **plasmolysis**. Initially, when the protoplasm begins to contract from the cell wall. It is called as **incipient plasmolysis**.

Again, if a plasmolysed cell or tissue is placed in water, process of endosmosis takes place. Due to which the water enters into the cell sap and the cell becomes turgid assuming the protoplasm again to its normal shape. This mechanism is called **deplasmolysis** and is just reversed of plasmolysis.

Osmosis

It is a kind of diffusion that plays an important role in the movement of water into and within the plants. The phenomenon of osmosis was discovered by Nollet in 1748.

In plants, it is defined as the diffusion of water molecules through a differentially permeable membrane or a semi permeable membrane. A differentially permeable membrane is the one that has different permeability properties for different substances. It permits free passage of water molecules but restricts the passage of dissolved solutes.

All protoplasmic membranes are differentially permeable or semi permeable. The net direction and rate of osmosis depends on both the **pressure gradient** and **concentration gradient**.

Depending upon the concentration of solution and the cell sap, solution are of three main types:

(i) **Isotonic solution** When the concentration of the solution is equal to that of the cell sap is called isotonic solution. e.g. no change is observed in a size of raisin when placed in the water.

(ii) **Hypotonic solution** When the concentration of the solution is lower than that of the cell sap is called hypotonic solution, e.g. raisins gets swelled up when placed in water.

(iii) **Hypertonic solution** When the concentration of the solution is higher than that of the cell sap is called hypertonic solution, e.g. raisins gets shrinked when placed in concentrated solution.

Osmotic Pressure (OP)

The term 'osmotic pressure' was proposed by Pfeffer. It is defined as the actual pressure that develops in a solution, when it is separated from pure water by the means of semi permeable membrane.

In other words, it is the pressure needed to prevent the passage of water into the solution through the semi permeable membrane thus, preventing an increase in the volume of solution.

The osmotic pressure depends upon the concentration, ionisation, hydration and temperature of solute particles. Plant cells exhibit a considerable range of variation in osmotic pressure.

In general, the osmotic pressure of plant cell varies from 4-5 atmosphere. The cells of inner cortex of roots possess higher OP than the outer cortex.

Osmotic pressure can be calculated by

$$\boxed{OP = miRT}$$

where, m = molar concentration

i = ionisation constant

R = gas constant

T = temperature

Osmotic pressure is numerically equal to the osmotic potential or solute potential (ψ_s) but has an opposite (negative) value.

The various methods used for the measurement of osmotic pressure are as follows:

(i) Plasmolytic method by Hugo de Vries

(ii) Pfeffer osmometer

(iii) Berkeley/Hartley osmometer

(iv) Cryoscopic osmometer

Osmotic Potential (ψ_s)

It is the degree of decrease in chemical potential of pure water due to the presence of solute particles in it.

Osmotic pressure and osmotic potential values are numerically equal. But the value of osmotic pressure has positive sign and osmotic potential always carries a negative sign (if $OP = 20$ atm, then $\psi_s = -20$ atm).

The value of osmotic potential or solute potential is calculated by the following formula:

$$\boxed{\psi_s = C \times R \times T}$$

Turgor Pressure (TP)

If a living plant cell or tissue is placed in hypotonic solution, water enters into the cell sap by the process of osmosis. Due to the result of the entry of water into the cell sap, a pressure is developed. It presses the protoplasm against the cell wall and the cell becomes turgid. This pressure is called as **turgor** (hydrobates) **pressure** or **pressure potential** (ψ_p).

Consequence of the turgor pressure is the wall pressure which is defined as the force exerted by the cell wall over the protoplasm. Normally, wall pressure is equal and opposite to turgor pressure except when cell becomes flaccid.

At a given time, turgor pressure equals to the wall pressure.

$$TP = WP$$

The value of these two opposing forces continue to rise till the cell becomes fully swollen or turgid. At this time, the value of wall pressure or turgor pressure becomes equal to osmotic potential.

$$\boxed{\psi_p = \psi_s}$$

Diffusion Pressure Deficit (DPD)

The difference between the diffusion pressure of the solution and its solvents at a particular temperature and atmospheric conditions is called diffusion pressure deficit. It determines the direction of net movement of water.

The term DPD was first coined by Meyer (1938). It is the reduction in diffusion pressure of solvent in a system over its pure state. DPD has a positive value, e.g. 5 atm.

DPD is directly proportional to the concentration of the solution, thus, if the solution is more concentrated, then its DPD will increase and gets decreased with the dilution of the solution.

$$DPD \propto \text{Concentration of solution}$$

DPD is also known as suction pressure as it is a measure of the ability of the cells to absorb water. It is related with osmotic pressure, turgor pressure and wall pressure, in the following ways:

$$\boxed{DPD/SP = OP - WP}$$

Imbibition

If certain substances like seeds which are when dropped into water (say about for the few hours) tends to absorb water and swell up considerably. Apart from seeds many other materials also get swelled up in a similar way when get immersed in the water, e.g. starch, cellulose, agar, gelatin, gum, dry wood and new dry seeds, etc. These substances are called **imbibants** and the phenomenon or process of swelling is known as **imbibition**.

Imbibition is basically a diffusion process which may be regarded as a difference in the diffusion pressure between the liquid in the external medium and the liquid in the imbibant (substance). An equilibrium will be reached only, when the diffusion pressure of the water in the two parts of the system has attained the same value. As a result of imbibition, a pressure is developed known as **imbibition pressure**. The magnitude of this pressure is tremendous that a rock can be splitted, if some dry wooden pieces are inserted in a small crack in rock and then soaked with water. The water potential of an imbibant is equal to its matric potential (always – ve) plus turgor pressure or pressure potential.

Transpiration

The process of evaporation of water molecules from the aerial or upper portions of the living plants and also the movement of water vapour into the atmosphere is known as transpiration. Since, transpiration reduces the level of water in the soil, it is considered as a **necessary evil** because it has its necessity in absorbing for water and minerals from the soil and for the ascent of sap.

About 98-99% of water absorbed by the plant, is lost in transpiration. Hardly, 0.2% is used in photosynthesis, while the remaining is retained in the plant during growth.

Types of Transpiration

Water loss by the phenomenon of transpiration may take place from any part of a plant exposed to the external atmosphere. However it occurs principally from the leaves and is known as **foliar transpiration**. While only the relatively small quantities of water is transpired from stem, is called **cauline transpiration**.

Hence, depending upon the plant surface, transpiration is of following four types:

(i) **Stomatal transpiration** About 85-90% of transpiration occurs through the structure called **stomata** which forms a link or channel between the internal tissues of leaf and the surrounding atmosphere. So, these are the passage, through which water vapour escapes from the leaf. Depending upon the area of light absorbance stomata are found on different surfaces of leaves in different plants.
e.g. in monocots (such as grasses) they are equally distributed on all sides. While, in the aquatic plants with floating leaves, they are present on the upper surface only.

(ii) **Cuticular transpiration** It occurs through the cuticle or epidermal cells of the leaves and other exposed parts of the plant. It is also known as **peristomatal transpiration**.
Although cuticle is impervious to water, still some water may be lost through it. In common land plants, cuticular transpiration is only 3-10% of the total transpiration. In herbaceous shade living plants, cuticle is very thin hence, the cuticular transpiration continues throughout day and night.

(iii) **Lenticular transpiration** It is commonly found in the woody branches of trees where lenticels are present. These lenticels helps in transpiration as they always remain open and water vapour escape through the loose mass of cell (i.e. the complementary cells) of each lenticel.

(v) **Bark transpiration** It occurs through the corky covering of the stems of the woody plants. Bark transpiration is very little but its measured rate is often more than the lenticular transpiration due to the larger area. It can also occurs continuously during the day and the night time.

Stomatal Apparatus

The term 'stoma' (Gk. *Stoma* – mouth) refers to the pair of guard cells and a pore between them. Stomata are present not only in the epidermis of leaves but also in certain fruits (such as banana, citrus, cucumber, avocado, etc).

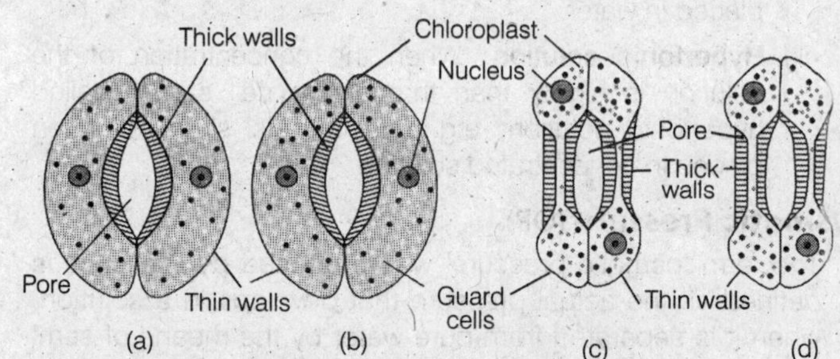

Open and close stomata (a) Dicotyledonous (b) Monocotyledonous (c-d) Dumb-bell shaped

Each stoma is surrounded by two small, specialised green epidermal cells known as **guard cells**, that are small in size, and are rapidly influenced by the turgor changes. They are kidney-shaped in outline and are joined at their ends.

Number of stomata on a typical leaf is about 1,000 to 60,000 per sq. cm. The size of stomatal pore (fully open) measures about 3-12 μm in diameter and 10-40 μm in length. It helps in the exchange of respiratory and photosynthetic gases (i.e. CO_2 and O_2) between the leaves and the external atmosphere. Also the transpiration occurs primarily through the stomata in the epidermis.

Types of Stomata

Stomata are categorised mainly into three main categories:

(i) **Barley or cereal type** These are dumb-bell shaped and are found equally on both the surfaces (amphistomatic leaf) of the leaf. They usually remain open during the day for few hours, e.g. maize, wheat, other cereals, etc.

(ii) **Leucerne or alfa-alfa type** These stomata open mainly during the daytime and gets close during the night under mesophytic conditions, e.g. pea, radish, apple, grapes, turnip, mustard, etc.

(iii) **Potato type** These stomata are found on both the surfaces of leaves (i.e. more on lower surface than on upper surface). Under mesophytic conditions, the stomata remains open throughout the day and night but close down for different periods at various times under the conditions of less water availability, e.g. *Cucurbita*, *Portulaca*, tulip, onion, banana, potato, etc.

Stomatal Movement

The guard cells differ from the other epidermal cells in a way of containing chloroplast and having peculiar thickenings on their adjacent surfaces. A radial micellation of cellulose microfibrils is found in the walls of guard cells.

The cellulosic microfibrils radiates out from the pore around their circumference. This allows the guard cells to elongate lengthwise thus, preventing their elongation crosswise when they become turgid.

Due to the increase in the Osmotic Pressure (OP) and Diffusion Pressure Deficit (DPD) of guard cells, osmotic diffusion of water from the surrounding epidermal and mesophyll cells takes place into the guard cells.

The turgor pressure of guard cell increases and becomes turgid. The guard cells thus swells, increases in length due to which their thickened surface gets stretched forming a pore hence opening of the stomata occurs When OP and DPD of guard cell decreases in comparison to the surrounding epidermal and mesophyll cells, the guard cells become flaccid. Thickened surface of guard cells come close to each other, closing the stomatal pore.

Factors Affecting Transpiration

Several external and internal factors that affects the rate of transpiration are as follows:

(i) **External** (Environmental) **Factors**

These factors includes temperature, light, relative humidity, availability of water and wind.

These are described here under:

(a) **Light** The light specifically, the light intensity is probably the most obvious among the environmental factors affecting transpiration implants. It has a controlling effect on the opening of stoma through which the water primarily escapes in gaseous state. In general, transpiration rate is high during daytime, particularly when light is bright, than during the night time.

(b) **Relative humidity** This environmental factor affects transpiration by regulating stomatal movement and atmospheric demand. At high RH (moist air), the stoma tends to close and thus limit the exit of water vapour from the plant. Further, high RH means that the water-potential gradient (also water vapour concentration and vapour pressure gradient) from plant to the atmosphere will be minimal compared to when RH is low.

(c) **Temperature** The rate of transpiration is fastest when air temperature is between 20-30° C (Moore *et. al.* 2013). At these temperatures the stomatal apertures or opening are generally widest. In general the stomata close at temperatures about 0°C and progressively increase in aperture up to about 30° C temperature as an environmental factor affecting transpiration also relates to water potential and relative humidity.

(d) **Soil water** Where the supply of water from the soil is limiting, the rate of transpiration tends to slow down. This is more pronounced where other conditions, such as bright light and warm temperature, favour escape of water from the plant.

(e) **Wind** This environmental factor affects transpiration by removing that thin moist layer of air, called boundary layer, which lies next to the surface of a leaf. This moist air causes a lesser water potential gradient from the leaf resulting to reduced rate of transpiration. This layer also reduces light penetration into the leaf. But with wind, this boundary layer is replaced with drier air thus increasing water potential gradient and enhancing transpiration.

(ii) **Plant** (internal) **factors**

Several plant factors that affects the rate of transpiration are root-shoot ratio, leaf area, stomatal frequency, structure of leaf, etc.

These are described here under.

Theories Related to Stomatal Movement

Theories related to the stomatal movement are as follows:

Starch-Sugar Interconversion Theory

The main features of this theory was given by **Sayere** (1923) and was modified by **Steward** (1964). It states that the guard cells contains starch when stomata are closed.At the time of opening, starch is found to be converted into glucose. This theory is based on the effect of pH on starch phosphorylase enzyme, which reversibly catalyses the conversion of starch and inorganic phosphate into glucose-1-phosphate.

As a result of photosynthesis, CO_2 concentration in the guard cells decreases leading to the increase in pH value in them during the day light. This favours the hydrolysis of starch (insoluble) into glucose-1-phosphate (soluble), so that the osmotic potential becomes lower in the guard cells. As a result water thus, enters into the guard cells by osmotic diffusion from the surrounding epidermal and mesophyll cells.

Guard cells now becomes turgid and the stomata gets opened up while, during the night (dark), reverse process occurs, i.e. glucose-1-phosphate converts into starch in the guard cells thus, increasing the osmotic potential. The guard cells release water, become flaccid and stomata gets closed.

During light,

$$\text{Starch} + n\text{H}_3\text{PO}_4 \xrightarrow{\text{Phosphorylase}} n\text{Glucose-1-phosphate}$$

$$\text{Glucose-1-phosphate} \overset{\text{Mutase}}{\rightleftharpoons} \text{Glucose-6-phosphate}$$

$$\text{Glucose-6-phosphate} + \text{water} \xrightarrow{\text{Phosphatase}} \text{Glucose} + \text{H}_3\text{PO}_4$$

During dark,

$$\text{Glucose} + \text{ATP} \xrightarrow{\text{Hexokinase}} \text{Glucose-6-phosphate} + \text{ADP}$$

$$\text{Glucose-6-phosphate} \overset{\text{Mutase}}{\rightleftharpoons} \text{Glucose-1-phosphate}$$

$$\text{Glucose-1-phosphate} \longrightarrow \text{Starch} + \text{H}_3\text{PO}_4$$

Objections of Starch-Sugar Interconversion Theory

There are certain objects for starch-sugar interconversion theory, due to which it is not universally applicable and may operate only under certain circumstances.

Some major objections to this theory are:

(i) Glucose is not found in the guard cells at the time of stomatal opening.

(ii) Starch $\rightleftharpoons$ Sugar changes are chemically slow, while opening and closing of stomata are rapid in nature.

(iii) Change in pH cannot be explained on the basis of CO_2 concentration.

Malate or K^+ ion Pump Theory

The main feature of this theory were put forward by Levitt (1974). This is also known as **modern theory**. According to this theory there is an accumulation of K^+ ions in the guard cells during day time.

A rise in the pH during daytime causes the hydrolysis of starch to form organic acid, i.e. phosphoenol pyruvate which gets combines with the available CO_2 to produce malic acid. That further dissociates to produce H^+ and malate. The protons (H^+) are pumped out from the guard cells into the subsidiary cells and in the exchange, K^+ ions are pumped into them, from the adjustment subsidiary cells.

This exchange of H^+ and K^+ ions is followed by the entry of Cl^- anions into the guard cells with the help of cAMP, ATP and cytokinins. K^+ ions immediately combines with malate and increase the osmotic concentration of the guard cells.

As a result, of which the guard cells absorb water from the nearby epidermal (subsidiary) cells, gets swell up leading to the opening of the stomata.

The reverse of this situation prevails during dark when stomata are closed, i.e. there is no accumulation of K^+ ions in guard cells.

Chapter Exercise

1. Process of selective transmission of a liquid through semipermeable membrane is called
(a) diffusion (b) osmosis
(c) plasmolysis (d) transmission

2. When a cell is fully turgid which of the following will be zero?
(a) Wall pressure (b) Osmotic pressure
(c) Turgor pressure (d) Water potential

3. Selective permeability identifies the phenomenon of
(a) imbibition (b) osmosis
(c) diffusion (d) plasmolysis

4. Change in diffusion pressure of a pure solvent, when it is added with a solute, is called
(a) osmosis (b) diffusion
(c) DPD (d) imbibition

5. The process of osmosis involves
(a) movement of solute through a semipermeable membrane
(b) movement of solvent through a semipermeable membrane
(c) movement of solution through a semipermeable membrane
(d) None of the above

6. With the increase in a turgidity of a cell, the wall pressure will
(a) increase (b) decrease
(c) fluctuates (d) remain unchanged

7. If a cell kept in a solution of unknown concentration gets deplasmolysed, then the solution is called
(a) hypotonic (b) hypertonic
(c) isotonic (d) detonic

8. DPD is equal to
(a) OP – WP (b) OP + WP
(c) TP (d) OP

9. Water potential in the leaf tissue is positive (+) during
(a) excessive transpiration (b) low transpiration
(c) excessive absorption (d) guttation

10. In a flaccid cell
(a) TP = OP (b) DPD = TP (c) OP = O (d) DPD = OP

11. If turgor pressure becomes equal to the wall pressure then
(a) water leaves the cell
(b) no exchange of water takes place
(c) water enters the cell
(d) solute goes from the cell into water

12. Osmotic pressure is responsible for the turgidity of plant cells which

(a) causes cell elongation
(b) causes opening of stomata
(c) prevents wilting of leaves
(d) All of the above

13. Root pressure can be measured by means of

(a) porometer (b) potometer
(c) auxanometer (d) manometer

14. DPD stands for

(a) Diffusion Pressure Deficit
(b) Diffusion Pressure Demand
(c) Daily Photosynthetic Depression
(d) Daily Phosphorus Demand

15. A cell becomes turgid, when placed in

(a) isotonic solution (b) hypertonic solution
(c) hypotonic solution (d) None of these

16. Osmosis is

(a) diffusion of solvent (b) diffusion of solute
(c) imbibition of solution (d) None of these

17. Which of the following do not transpire?

(a) Algae (b) Fungi
(c) Submerged hydrophytes (d) All of these

18. The diameter of stomatal pore ranges between

(a) 25-50 μ (b) 15-25 μ
(c) 3-10 μ (d) 1-2 μ

19. The rate of transpiration is affected by

(a) root-shoot ratio (b) leaf area
(c) shape of leaf margin (d) Both (a) and (b)

20. The rate of transpiration increases with the decrease of

(a) temperature (b) moderate wind
(c) light (d) atmospheric pressure

21. Absorption of water by roots is increased when

(a) transpiration rate is less
(b) salt absorption is increased
(c) transpiration rate increases
(d) photosynthetic rate increases

22. When stomata open in night only, they are called

(a) photoactive stomata (b) scotoactive stomata
(c) nyctinastic stomata (d) All of these

23. Such a membrane, which permits only the selected materials to pass through it termed as

(a) permeable (b) impermeable
(c) semipermeable (d) selectively semipermeable

24. Swelling of grapes in water confirms to

(a) exosmosis (b) endosmosis
(c) diffusion (d) imbibition

25. In a fully turgid cell, the values (arbitrary values are shown) of DPD, OP and TP will show the tendency as

(a) DPD 10 atm; OP 15 atm; TP 5 atm
(b) DPD 02 atm; OP 07 atm; TP 5 atm
(c) DPD 05 atm; OP 12 atm; TP 7 atm
(d) DPD 00 atm; OP 15 atm; TP 15 atm

26. The real force responsible for the movement of water form one cell to other cell is mainly

(a) OP (b) TP
(c) DPD (d) Imbibition

27. Under which condition, does DPD becomes more than OP?

(a) OP < TP (b) OP = TP
(c) TP is – ve (d) OP > TP

28. When a plasmolysed cell is placed in hypotonic solution then water enter in cell due to which force

(a) DPD (b) OP
(c) WP (d) None of these

29. When roots are kept in a hypotonic solution, the process of root pressure will

(a) not be affected (b) increase
(c) decrease (d) stop

30. In a plasmolysed cell, the space between cell wall and plasma membrane is occupied by

(a) external solution (b) water from cell sap
(c) Both (a) and (b) (d) None of these

31. Water potential in the leaf tissue is positive (+) during

(a) excessive transpiration (b) low transpiration
(b) excessive absorption (d) guttation

32. Selective permeability identifies the phenomenon of

(a) imbibition (b) osmosis
(c) diffusion (d) plasmolysis

33. When stomata remain open throughout the day and night, they are called

(a) alfa-alfa type (b) potato type
(c) barley type (d) *Equisetum* type

34. Transpiration is regulated by

(a) guard cells (b) subsidiary cells
(c) Both (a) and (b) (d) cuticle and lenticel

35. Mark the correct relationship

(a) $\Psi_w = \Psi_p - (\Psi_\pi + \Psi_m)$ (b) $\Psi_w = \Psi_p + \Psi_\pi + \Psi_m$
(c) $\Psi_w = \Psi_p + \Psi_\pi - \Psi_m$ (d) None of these

36. Swelling of grapes in water confirms to

(a) exosmosis (b) endosmosis
(c) imbibition (d) diffusion

37. In a flaccid cell

(a) *DPD = PO* (b) *DPD = TP*
(c) *TP = OP* (d) *OP = O*

38. A cell kept in hypertonic solution then cell will be
(a) turgid (b) flaccid
(c) plasmolysed (d) impermeable

39. By which greek letter the osmotic pressure is indicated
(a) Ψ (b) π (c) δ (d) λ

40. If the cell wall is elastic instead of being rigid and if the cell is put in a medium of sugar solution of higher concentration that in the cell then
(a) the cytoplasm will be pulled away from the wall as water moves out
(b) cell shape and size will not change
(c) the wall will shrink along with the cytoplasm
(d) the cell wall will rupture as the cytoplasm shrinks

41. Initial absorption of water by root hair is through
(a) osmosis (b) endosmosis
(c) absorption (d) imbibition

42. The apparatus used to measure osmotic pressure of a solution is known as
(a) osmometer (b) photometer
(c) calorimeter (d) manometer

43. The value of osmotic pressure becomes equals to turgor pressure, when a cell is placed in a
(a) hypertonic solution (b) isotonic solution
(c) hypotonic solution (d) None of these

44. Bacteria cannot survive in highly salted pickle because
(a) salt inhibits reproduction
(b) bacteria do not get enough light for photosynthesis
(c) they become plasmolysed and consequently killed
(d) the pickle does not contain nutreints necessary for bacteria to live

45. Soil particles cling to mucigel and increase the roots contact with the soil. These properties of mucigel help in
(a) maintaining the continuity between the roots and soil water
(b) movement of cell contents out of the cell
(c) movement of particles continues
(d) plasmolysis in root cells

46. A 10% solution of which of the following substances shall have maximum OP?
(a) NaCl (b) Sucrose
(c) Glucose (d) Fructose

47. Plant cells do not burst in distilled water because their
(a) cell wall is elastic, rigid and gets stretched
(b) cell wall is living
(c) cell wall is the outermost layer in plant cell
(d) cell walls is permeable

48. In rainy season, the doors and windows swell up due to
(a) imbibition of water
(b) endosmosis of water
(c) separation of xylem vessels
(d) All of the above

49. Potato slices are immersed in a series of solutions of different osmotic concentrations, no change in its volume or weight is observed with slices in a 0.4 M solution. The osmotic concentration of the vacuolar sap is therefore,
(a) 0.4 M.
(b) greater than 0.4 M
(c) less than 0.4 M
(d) not related at all to the outside solution

50. The water available for root absorption in soil is
(a) hygroscopic water (b) combined water
(c) gravitational water (d) capillary water

51. During osmosis, water moves through a semipermeable membrane.

	From	To
(a)	Low water potential	High water potential
(b)	High solute concentration	Low solute concentration
(c)	High osmotic potential	Low osmotic potential
(d)	A hypotonic solution	A hypertonic solution

52. When a cell is fully turgid, which of the following will be zero
(a) wall pressure (b) SP/DPD/WP
(c) turgor pressure (d) osmotic pressure

53. Most abundant solute in the phloem sap is
(a) potassium ions (b) glucose
(c) starch (d) sucrose

54. Swelling of gums on being placed in moistened environment confirms
(a) diffusion (b) imbibition
(c) endosmosis (d) exosmosis

55. Soil water potential is measured with the help of the instrument called
(a) porometer (b) tensiometer
(c) paedometer (d) vacuum gauje

56. The cell A has an osmotic potential of – 20 bars and a pressure potential of + 6 bars. What will be its water potential?
(a) – 14 bars (b) + 14 bars
(c) – 20 bars (d) + 20 bars

57. The OP and TP of two pairs of cell A-B and X-Y are as under,

Cell A	Cell B	Cell X	Cell Y
OP = 10 atm	OP = 10 atm	OP = 10 atm	OP = 8 atm
TP = 4 atm	TP = 6 atm	TP = 4 atm	TP = 4 atm

The net movement of water shall be from
(a) A to B and X to Y
(b) A to B and Y to X
(c) B to A and X to Y
(d) B to A and Y to X

58. Addition of a solute to pure water causes
(a) negative water potential
(b) more negative water potential
(c) positive water potential
(d) more positive water potential

59. The term water potential was coined by
(a) Sayre and Dixon
(b) Llyod and Bose
(c) Von Mohl
(d) Slayter and Taylor

60. Dry wooden stables, if driven in to a small crash in a rock and then soaked, can develop enough pressure to split the rock. Such pressure is built up through the phenomenon of
(a) imbibition (b) turger pressure
(c) deplasmolysis (d) osmatic pressure

61. Stomata open by the
(a) influx of Ca^{++} ions (b) influx of K^+ ions
(c) outflux of K^+ (d) influx of mg^+ ions

62. In the process of osmosis, the volume of solvent
(a) remain the same (b) increases
(c) decreases (d) None of these

63. Passive absorption of mineral salt is not dependent
(a) diffusion (b) osmosis
(c) donnans equilibrium (d) ionic exchange

64. If two solutions have the some osmoregularity they are said are said to be
(a) hypertonic (b) hypotonic
(c) isotonic (d) None of these

65. Transpiration facilitate
(a) absorption of water by roots
(b) excretion of minerals
(c) electrolyte balance
(d) opening of stomata

66. Of all the environmental factors which is the most influential in determing the rate of transpiration.
(a) Light
(b) Water
(c) Relative rumidity of atmosphere
(d) Superative

67. Oozing out of water drops from injured edges or tips is called
(a) oozation (b) bleeding
(c) gutation (d) transpiration

68. When stomata closes which of the following events does not occur
(a) guard cell become flaccid
(b) sygan→starch
(c) OP of guard cell decreases
(d) accumulation of O_2

69. Which of the following plant material is an efficient water imbibant?
(a) Lignin (b) Pectin
(c) Agar (d) Cellulose

70. Which of the following theory gives the latest explanation for closure of stomata?
(a) ABA theory (b) Starch → Glucose theory
(c) Munch theory (d) Active K^+ transport theory

Answers

1. (b)	**2.** (d)	**3.** (b)	**4.** (c)	**5.** (b)	**6.** (a)	**7.** (a)	**8.** (a)	**9.** (c)	**10.** (d)
11. (b)	**12.** (d)	**13.** (d)	**14.** (a)	**15.** (c)	**16.** (a)	**17.** (d)	**18.** (b)	**19.** (d)	**20.** (d)
21. (c)	**22.** (b)	**23.** (d)	**24.** (b)	**25.** (d)	**26.** (c)	**27.** (c)	**28.** (a)	**29.** (d)	**30.** (b)
31. (b)	**32.** (b)	**33.** (d)	**34.** (a)	**35.** (d)	**36.** (b)	**37.** (a)	**38.** (c)	**39.** (a)	**40.** (c)
41. (d)	**42.** (a)	**43.** (b)	**44.** (c)	**45.** (a)	**46.** (a)	**47.** (a)	**48.** (a)	**49.** (a)	**50.** (d)
51. (d)	**52.** (b)	**53.** (d)	**54.** (b)	**55.** (b)	**56.** (a)	**57.** (d)	**58.** (a)	**59.** (d)	**60.** (a)
61. (b)	**62.** (c)	**63.** (b)	**64.** (c)	**65.** (a)	**66.** (c)	**67.** (b)	**68.** (b)	**69.** (c)	**70.** (d)

29

Mineral Nutrition and Transport of Water and Organic Solute

Living organisms including plants requires a number of macromolecules or nutrients for their growth and development. Nutrients are of two main types, i.e. inorganic and organic. The product of photosynthesis are considered as **organic nutrients,** while those absorbed by the roots *via* various methods are known as **inorganic nutrients**.

Culture Experiments

Till 18th century, the role of micronutrients to plants was unclear to the scientists. Hence, to study the role of micronutrients, a series of experiments were initiated by a eminent German botanist Julius Von Sachs and Knops (1865) for the very first time.

These culture experiments are useful in determining

(i) Deficiency symptoms and toxicity caused by an element.

(ii) Role of essential element in the maintenance of plant physiology, anatomy and morphology.

(iii) Essentiality of mineral ion/element.

Hydroponics

It is one of the most widely used method in culture experiments. It is known to be the technique of soil less production of plants. This technique is eventually proved to be very useful in raising different flowers and vegetable varieties. In some countries like USA, Abudhabi, it is often used commercially. Hence, out of season vegetables (such as lettuce, seedless cucumber and tomato) and fruits can be obtained easily through this technique.

The technique of hydroponics can be of further three main types:

(i) Tank system (ii) Film system

(iii) Aeroponics

Classification of Elements

In the middle 19th century, the mineral elements were grouped under the two broad categories, i.e. essential and non-essential elements. The term 'essential element' was proposed by Arnon and Stout (1939). They concluded three criteria for an element to be considered as essential (vegetative or reproductive stage).

(a) The element which must be so essential for the normal growth and reproduction that the plant fails to proceed its life without this element.

(b) The requirement of an element must be specific and cannot be replaced by the any other element.

(c) The requirement of the element must be direct that is not the result of any indirect effect, e.g. for relieving toxicity caused by some other substance.

Chemical analysis has shown that the plant contains about 40 different elements. Some of them are indispensible, i.e. necessary for the normal growth and development of the plants and without which the plants are unable to complete their life cycle. These are called as **essential elements**. Rest of the elements are called as **non-essential elements**.

At present 16 chemical elements are known to be essential for the growth of all higher plants, these are C, H, O, N, P, K, S, Mg, Ca, Fe, B, Mn, Cu, Zn, Mo, Cl. Calcium, potassium and magnesium are present in soil as cations (Ca^{2+}, K^+ and Mg^{2+}). Recently, (Ni) nickel has also been added in this list, making the number of essential elements 17. Similarly nitrogen, phosphorus and sulphur are present in soil as anions (NO_3^-, $H_3PO_4^-$, SO_4^-). Besides these 16 essential elements, Al, Si, Na, Co and Ga may be essential for some plants.

Essential elements are present in different proportions in the plants. They are differentiated into two main categories as

(i) **Macronutrients or Major elements** are those elements that are generally found in the plant tissues in large amounts (in excess of around 10m/mol Cg^{-1} or 10 mg per gm of dry matter) and are involved in the synthesis of organic molecules and development of an osmotic potential. Main groups, i.e. primary and secondary nutrients out of which the primary nutrients are nitrogen types phosphorus and potassium (i.e. N,P,K). The secondary nutrients are calcuim, magnesium and sulphur. Macronutrients which become' commonly deficient in the soils are called **Critical elentents** such as (N,P,K). Most of the fertilisers contains these critical elements and are hence, called **Complete fertilisers**. The deficiency caused by these critical elements is known as **primary deficiency**.

(ii) **Micronutrients or minor elements** are those elements that are found in the plant tissue in traces or very small amounts, (i.e. less than about 10 m/mol Kg^{-1} or less than 10 mg per gm of dry matter) and are mostly involved in the functioning of enzymes as cofactors or activators of metals.

Essential Elements, their Functions and Deficiency Symptoms

Elements (Macroelements)	Functions	Deficiency symptoms
Nitrogen (N)	Components of proteins, nucleic acids, porphyrins and coenzyme.	Chlorosis appears first in older leaves, dormancy of lateral buds, delaying of flowering, Interfere in protein synthesis.
Phosphorus (P)	Components of nucleic acids, proteins, phospholipids, ATP and NADP, Involved in energy transfer.	Reduces growth, promotes leaf fall and delay flowering. Vascular tissues are poorly developed with increased pith tissue.
Sulphur (S)	Components of sulphur containing vitamins (biotin, thiamine) and coenzyme-A. It is also, the component of allyl sulphide (onion, garlic) and sinigrin (mustard).	Chlorosis in young leaves and reduced growth, extensive root growth, formation of hard and woody stem. Reduction in juice content of *Citrus* fruit. **Diseases** Yellow disease of tea, marsh spot (peas), etc.
Calcium (Ca)	Component of middle lamella helps to stabilise the structure of the chromosomes, required for the development of root and stem apex, conversion of toxic oxalic acid to calcium oxalate.	Death of root and tips, chlorosis along the margins of young leaves, premature fall of flowers in legumes. **Diseases** Wither tip, blossoms end rot (tomato), etc.
Magnesium (Mg)	Component of chlorophyll, enzyme activator of hexokinase and phosphorylase, required for ribosomes.	Interveinal chlorosis occurs first in older leaves, appearance of necrotic spots, tips and margins of leaves turn upward.
Potassium (K)	Required as a cofactor for about 40 enzymes. Essential for permeability, stomatal movement and for translocation of food.	Inhibits protein synthesis and causes scorching of older leaves, rosette or bushy habit due to reduced apical bud growth, decreased resistance to diseases.
Iron (Fe)	Essential for chlorophyll synthesis. Component of flavoproteins, cytochromes and ferredoxin helps in energy generation.	Interveinal chlorosis occurs first in young leaves, inhibition of chloroplast formation, weaker stems.
Manganese (Mn)	Photolysis of water, activates carboxylases, dehydrogenases and reductases.	Chlorotic and necrotic spots first in young leaves. **Diseases** Grey speck (oats), specked yellow (beets), heart rot of sugarbeet, water core in turnip.
Molybdenum (Mo)	Involved in nitrogen metabolism including nitrogen fixation. Constituent of enzyme nitrate reductase.	Mottling and necrosis of leaves. **Diseases** Yellow spot of *Citrus*, scald of beans.
Boron (B)	Carbohydrate metabolism, sugar translocation, nodulation in legumes, pollen germination and cell differentiation.	Reduces meristematic growth, causes premature leaf fall. Disintegration of internal tissues leading to brown heart disease of beets and top sickness of tobacco. **Diseases** Corky core (apples), heart rot of beets, brown heart (turnip), top sickness (tobacco), khaira, little leaf.
Zinc (Zn)	Constituents of carbonic anhydrase and auxin.	Mottled leaves decreases auxin content and causes little leaf. **Diseases** Little leaf (apple, peaches), white bud (maize), die back (*Citrus*).
Copper (Cu)	Component of plastocyanin, cytochrome oxidase, activator of some enzymes.	Causes die back disease in *Citrus* and leaf tip necrosis (reclamation in legumes). **Diseases** Exanthema in fruit trees, reclamation of cereals, legumes, whiptail of crucifers.
Chlorine (Cl)	Photolysis of water, solute concentration and ionic balance.	Bronze coloured leaves, wilting, chlorosis and necrosis.

Toxicity of Micronutrient

Plant can tolerate a specific amount of micronutrient. A slight lesser amount of its can cause pa 807, (exp vol I) deficiency symptom and a slight higher amount of it can cause toxicity. The mineral ion concentration which reduces the dry weight of tissue by 10% is called toxic concentration.

This concentration is different for different micronutriens as well as for different plant, e.g. Mn^{2+} is toxic beyond 600 μgg^{-1} for soybean and beyond 5300 μgg^{-1} for sunflower. It is very difficult to identify the toxicity symptoms of mineral ion. It is because excess uptake of one element can reduce the uptake of other element at times.

For example Manganese (Mn^{2+}) becomes toxic when absorbed by plants in higher amounts. The toxicity is expressed in form of draw spots surrounded by chloretic vein. It is due to the following reasons:

(i) Reduction in uptake of Fe^{3+} and Mg^{2+}.

(ii) Inhibition of binding of Mg^{2+} to specific enzymes.

(iii) Inhibition of Ca^{2+} translocation in shoot apex.

Thus, excess of Mn^{2+} causes deficiency of iron, magnesium and calcium.

Absorption of Water

Water may be absorbed through almost any surface of plant but most of the absorption of water and mineral salts occurs in terminal portion of the roots. Because of ex⁺ nsive branching of roots, there are numerous root tips that helps in obtaining water and minerals from the soil. The root-hair zone which bears slender hair-like outgrowths of the epidermal cells are known as **root hairs**. There are the area of young roots where most of the absorption of water takes place.

The walls of root hairs are permeable and consist of pectic substance and cellulose which are strongly hydrophilic (water loving) in nature. Root hairs contains vacuoles filled with cell sap. These are delicate structures that do not lasts for more than a few days or a week.

Pathways of Water Movement

The epidermis to the endodermis of the roots, there are mainly three pathways through which the water can flow, i.e. apoplast, transmembrane and symplast pathway.

(i) **Apoplast pathway** It is the pathway of adjacent cell walls (i.e. interconnecting cell walls, intercellular spaces, cellwall of endodermis) that occurs continous throughout the plant in which the water moves exclusively through the cell wall without crossing any membranes. Apoplast pathway is particularly important for water uptake by young corn roots.

The apoplast pathway does not provide any barrier to the movement of water, which occurs through the mass flow due to adhesion and cohesion of water molecules. It is to be noted that in the apoplastic pathway there is no involvement of osmosis.

(ii) **Transmembrane pathway** It is the route followed by water that sequentially enters a cell on one side, exits the cell on the other side and enters the next in the series and so on. During this pathway, the water crosses atleast two membranes for each cell in its path (the plasma membrane on entering and on exiting). Transport across the tonoplast may also be involved.

(iii) **Symplast pathway** It is the pathway in which the water travels from one cell to the next via plasmodesmata. This pathway includes the living part of the plant cells made up of interconnected protoplasts of neighbouring cells connected through cytoplasmic strands extending through plasmodesmata. The water enters into the cell sap of root hair as a result of active absorption and moves into the underlying cortex cells bounded by a selectively permeable membrane. This movement is down the potential gradient.

As the apoplast pathway is blocked by the bond of a suberised matrix called **casparian strip** in the inner boundary of the cortex, the endodermis impermeable to water. This cytoplasmic streaming help on the quicker movement across individual cells.

Hence, water in this symplast pathway moves from the cell wall and enter the cell cytoplasm travelling from cytoplasm of one cell to the another cells cytoplasm. It thus, finally reach the xylem elements.

Ascent of Sap

The phenomenon of upward movement of water molecules from roots to the aerial parts of a plant is known as ascent of sap.

Various experiments demonstrates that the actual pathway of ascent of sap is through xylem tissue in terrestrial plants. Xylem contains vessels, tracheids and xylem parenchyma.

Xylem vessels are made up of thick-walled cells joined end to end. The partition wall between the cells gets dissolved, cell contents disappears forming a hollow tubular structure that resembles the water pipes.

The thories explaining to the ascent of sap. Number of theories have been put forward to explain the mechanism of ascent of sap. These are as follows :

1. Root Pressure Theory

This theory was put forward by Priestley in 1916. Although root pressure developed in the xylem of the roots can raise water to a certain height but it does not seems to be an effective force in ascent of sap due to certain reasons like low magnitude (about 2 atms) of root pressure, etc. It is because even in the absence of root pressure, the ascent of sap continuous. In gymnosperms root pressure has rarely been observed.

2. Cohesion Tension Theory

This theory was originally proposed by Dixon and Jolly (1894) and was greatly supported and elaborated by Dixon (1914-1924).

This theory is based on the following main features:

(i) **Continuous water column** From root hairs to the tip of the plant, water forms a continuous column. The water column is present in the tracheary elements.

(ii) **Cohesive and adhesive properties** The water molecules remain joined to each other due to a force of attraction called **cohesion force**. Due to this force, the water column can bear a tension or pull upto 100 atm. Therefore, it is also called **tensile strength.** This attraction is due to the presence of hydrogen bonds (H-bonds) between them.

The magnitude of this force is very high, therefore, a continuous water column in the xylem cannot be broken easily due to the force of gravity or other obstructions offered by internal tissues in the upward movement of water.

(iii) **Transpiration pull** It is known that when transpiration takes place, water evaporates from the intercellular spaces of the leaves to the outer atmosphere through stomata. Due to this water loss, osmotic pressure of leaf cells increases. As a result of which, the cells develops the low water potential and water from the leaf veins (xylem) moves into the leaf cells.

The xylem vessels in turn draws water from the xylem of the main stem. A pressure (pull) is thus, exerted by all the leaves on the stem. This combined pressure is called **transpirational pull** which is strong enough to pull up the column of water to a greater height.

The transpiration pull overcomes the following criterias :

(a) Gravitation pull.

(b) Resistance of narrow xylem channels and their end walls.

(c) Resistance offered by water coming out of the narrow capillary pores to the soil.

Translocation

The transport of food from the production center (leaves) to the consumption centres (apices, roots, fruits, tubers) through the phloem tissue is called translocation of organic solutes. The phloem sap is mainly composed of water and sucrose, but other sugars, hormones and amino acids are also transported or translocated through it.

Mechanism of Translocation

Several theories have been put forward to explain the mechanism of solute movement in the phloem. The important one is given below as :

Mass or Pressure Flow Theory

This hypothesis of solute transport in phloem tissue is also called as pressure flow hypothesis or Munch hypothesis. It was first proposed by Ernst Munch (1930)

According to this theory, the organic solutes are translocated through the sieve tubes from the supplying end or source (leaves) to the consumption end or sink (root, fruits and tubers).

Generally, the source is understood to be that part of the plant which synthesises the food, i.e. the leaf and the sink is the part that needs or stores the food. But the source and sink may be reversed depending on the season or the plant needs.

Since, the source-sink relationship varies, the direction of movement in the phloem can be upwards or downwards, i.e. bidirectional. This is different than the direction of movement in the xylem in which the movement is always unidirectional.

Water enters the mesophyll cells from the xylem as a result, turgor pressure of the mesophyll cells increases. Sugars dissolved in water move from mesophyll cells into the symplast system of sieve tubes. Solutes are carried through the symplast to finally reach the consumption centres. At the consumption end, the food materials (solutes) are either used up (as in roots) or are stored in insoluble forms (as in fruits).

Hence, the osmotic concentration and consequently the turgor pressure in the these cells will be low.

Thus, a continuous pressure gradient establishes between the source and sink. Water returns to the source through the apoplast system.

The sucrose is transported from the mesophyll cells of the leaf to the sink (such as growing cells at a root tip) through three main processes. These are

(i) **Phloem loading** is an active transport mechanism. It is carried out by a specific carrier protein molecules in the cell surface membrane of the companion cells that uses the energy of ATP from the photosynthetic mesophyll cells to the sieve tubes in the veins of a leaf. Phloem loading involves active transport against a concentration gradient. The sugar moves from the photosynthetic cells into a chain of mesophyll cells, transfer cells and to the phloem of a vascular bundle.

(ii) **Long distance transport** of sucrose occur in the stem and root phloem.

(iii) **Phloem unloading** is a passive transport mechanism from the sieve tubes to the cells at the root tip. It takes place passively down a concentration gradient of sucrose. The transfer cells are often present at unloading sites. Phloem unloading also requires metabolic energy that is used by sink organs for respiration and biosynthetic reactions.

Chapter Exercise

1. Chlorosis is caused due to the deficiency of
   (a) calcium (b) boron
   (c) iron (d) copper

2. Active uptake of mineral requires
   (a) respiration (b) expenditure of energy
   (c) root metabolism (d) photophosphorylation

3. The most important use of potassium is that it
   (a) provides red colour to fruit
   (b) aids photosynthesis
   (c) influences enzymatic activity, which regulates many plant processes
   (d) helps in the formation of the cambium

4. Active transport of molecules from outside to inside across a membrane requires
   (a) cyclic AMP
   (b) acetylcholine
   (c) ATP
   (d) stimulation exit of like molecules

5. The concentration of Ca^{2+} in phloem sap in comparison to the xylem sap is
   (a) more (b) less
   (c) same (d) variable

6. 'Die back' *Citrus* and 'reclamation' of legumes and cereals disease occurs due to the deficiency of
   (a) copper (b) zinc
   (c) sodium (d) molybdenum

7. The accumulation of phosphorus is more in
   (a) older leaves (b) older roots
   (c) younger roots (d) younger leaves

8. What is needed for the active transport?
   (a) Evolution of heat (b) Availability of light and heat
   (c) The presence of light (d) Availability of energy

9. Element present in protoplasmic proteins is
   (a) K (b) N
   (c) O (d) Mg

10. Boron in green plants assists in
    (a) activation of enzymes (b) acting as enzyme cofactor
    (c) photosynthesis (d) sugar transport

11. Critical elements are
    (a) N, P and S (b) N, S and K
    (c) N, P and K (d) P, K and S

12. Minerals are absorbed by the plants through
    (a) active transport (b) diffusion
    (c) osmosis (d) pressure flow

13. The most common ion found freely in the cell is
    (a) phosphorus (b) potassium
    (c) iron (d) sulphur

14. Closure of stomata is due to
    (a) efflux of K^+ (b) influx of K^+
    (c) influx of Na^+ (d) efflux of Na^+

15. The brown colour of leaves of cabbage is due to the deficiency of
    (a) phosphorus (b) nitrogen
    (c) sulphur (d) calcium

16. Enzyme catalysed reaction can be stopped by
    (a) Zn^{2+} (b) Cu^{2+}
    (c) Hg^{2+} (d) Mg^{2+}

17. Reduced availability of nitrogen during late growth causes
    (a) increase in the carbohydrate content of fruits and storage organs
    (b) decrease in the carbohydrate content of fruits and storage organs
    (c) increase in the ethylene content of fruits and storage organs
    (d) decrease in the ethylene content of fruits and storage organs

18. A fertiliser has a formula of three figures 15-9-9, it stands for the percentage of
    (a) N, Ca and Mg (b) Mg, P and K
    (c) Ca, N and Fe (d) N, P and K

19. A number of minerals like Ca, Mg and K are held over the surface of clay particles because the later are
    (a) negatively charged
    (b) positively charged
    (c) neutral
    (d) having both positive and negative residual valencies

20. Which of the following is not a micronutrient?
    (a) Mo (b) B
    (c) Zn (d) Mg

21. Molybdenun is involved in plant metabolism in
    (a) translocation of solutes (b) nitrate reduction
    (c) tryptophan synthesis (d) ascorbic acid synthesis

22. Cohesion transpiration pull theory operates only in
    (a) passive water absorption
    (b) active water absorption
    (c) conditions favouring transpiration
    (d) conditions restricting transpiration

23. The term 'root pressure' was coined by
    (a) Priestley (1729) (b) Stephen Hales (1727)
    (c) Sachs (1809) (d) JC Bose (1923)

24. According to the transpiration cohesion theory, the upward pull of water is transmitted to other water molecules by cohesion, which is caused by
(a) hydrogen bonds (b) hydrophilic cell walls
(c) turgor pressure (d) osmosis

25. The force of tension-cohesion exceeds root pressure on
(a) rainy day (b) fog morning
(c) sunny day (d) full moon night

26. Who was the first to suggest that the upward movement of water takes place by imbibition?
(a) Sachs (1874) (b) Boehm (1809)
(c) Weatherley (1963) (d) Unger (1868)

27. Difference between the term 'adhesion' and 'cohesion' is
(a) adhesion is union of similar parts
(b) cohesion is the adhesion of dissimilar parts
(c) both are wrong
(d) both are right

28. Supply ends in transport of solutes are
(a) green leaves (b) roots and stem
(c) xylem and phloem (d) hormones and enzymes

29. The main mechanism(s) determining the direction of short distance transport within a potato tuber is (are)
(a) determined by the structure and function of the tonoplast of tuber cells
(b) diffusion due to concentration differences and bulk flow due to pressure differences
(c) not affected by temperature and pressure
(d) pressure flow through the phloem

30. Ascent of sap takes place through
(a) cambium (b) phloem (c) xylem (d) epidermis

31. The direction of the conduction of food through phloem is
(a) from bottom to top
(b) from top to bottom
(c) Both of the above
(d) phloem never conducts food

32. Root pressure does not takes an active part in the ascent of sap because it is
(a) not found in all plants and water can rise in the absence of root
(b) seen during most favourable periods like spring or rainy season
(c) never found to exist in the plants
(d) Both (a) and (b)

33. The element required by an enzyme urease to hydrolyse urea by living organisms is
(a) boron (b) calcium
(c) phosphorus (d) nickel

34. Enzyme catalysed reaction can be stopped by
(a) Hg^{2+} (b) Cu^{2+}
(c) Mg^{2+} (d) None of these

35. Brown heart disease in turnip is due to the deficiency of
(a) K (b) Mo (c) B (d) Fe

36. Identify the well known trace element
(a) Fe (b) Mg
(c) Cu (d) S

37. Sulphur is required by the plants for
(a) protein synthesis (b) glucose synthesis
(c) ATP synthesis (d) vitamin synthesis

38. Bacteria involved in denitrification is
(a) *Nitrococcus* (b) *Nitrosomonas*
(c) *Pseudomonas* (d) *Azotobacter*

39. Zinc is an activater of
(a) amino acid oxidase
(b) PEPCO
(c) tryptophan synthetase
(d) succinic acid dehydrogenase

40. Element directly absorbed by plants from soil is
(a) C (b) H_2 (c) N (d) B

41. What special mode of transport *via* phloem is applicable in angiospherms?
(a) Action theory (b) Flow theory
(c) Pressure flow theory (d) Pressure theory

42. ATP and proteins is supplied to the companion cells through
(a) sieve tubes (b) cell wall
(c) porous wall (d) vascular tissues

43. Aeroponics is called as
(a) soil less cultivation of plants
(b) vivipary
(c) parthenocarpy
(d) phytotorn

44. Which of the following is a correct list of contexts of organic material in soil ?
(a) Humus, roots, and decomposed material
(b) Roots and small animals
(c) Minerals, small animals and roots
(d) Small animals, humus and roots

45. In plants, water rises beyond the point supported by the atmospheric pressure mostly because of
(a) capillarity (b) gravity
(c) evaporation (d) active transport

46. Scientists take advantage of .... in studying translocation by phloem
(a) ants (b) aphids
(c) bees (d) butterflies

47. In the absence of transpiration water into and up the xylem because of
(a) root pressure (b) guttation
(c) turgor pressure (d) mass flow

48. Air is transported in xylem in the form of bubbles by

(a) proton pump (b) active transport
(c) bulk flow (d) None of these

49. When a plant is flooded it often increases its

(a) ethylene (b) cytokines
(c) gibberellins (d) All of these

50. The plant hormone .... plays a role in closing of stomata.

(a) ethylene (b) abscise acid
(c) cytochrome (d) auxin

51. The reason behind the fact that a column of water in a tall tree does not break because of

(a) due to the formation of hydrogen bonds with the plant vessels
(b) tensile strength of column of water
(c) bubbles formed are too large to be transported
(d) All of the above

52. Carbohydrates moves through the sieve as a result of the movement of water due to

(a) cohesion (b) adhesion
(c) diffusion (d) osmosis

53. Micronutrient .... is needed for chlorophyll synthesis.

(a) boron (b) iron
(c) sulphur (d) Iodine

54. Organic is called as fertiliser with

(a) macro nutrients (b) micro nutrients
(c) potassium (d) humus

55. Which of the follwoing are not matched correctly?

(a) Chlorine-micronutrient and is a major component of organic molecules
(b) Potassium-macronutrient and involved in opening of stomata
(c) Nitrogen-macronutrient and component of amino acid
(d) Iron-micronutrient and involved in Chlorophyll synthesis

56. The uptake of water in plants involves

(a) active transport (b) passive transport
(c) Both of these (d) None of these

57. The diffusion of ions along with water also takes place by mass flow system along the .... path.

(a) toroplast (b) apoplast
(c) symplast (d) vacuolar

58. In plants, the neighbouring cells are connected with one another by

(a) plasmodesmata (b) cellwalls.
(c) vacuoles (d) Both (a) and (b)

59. Exceptionally, the positive hydrostatic pressure generated by root pressure is about

(a) 600 kpa (b) 800 kpa
(c) 1000 kpa (d) 900 kpa

60. Pressure flow theory was proposed by

(a) M-van Monlul (b) Dixon
(c) Van der Waal (d) Ear not Munch

61. Gray spot of oat are caused by deficiency of

(a) Cu (b) Zn
(c) Mn (d) Fe

62. N_2 fixation in root nodules of *Alnus* is brought about by

(a) *Frankia* (b) *Azorhizobium*
(c) *Brodyrhizobium* (d) *Clostridium*

63. Zn, Mo, Fe, Cu are

(a) trace elements (b) non-essential elements
(c) mono-nutrients (d) None of these

64. Which is essential for root hair growth?

(a) Zn (b) Ca
(c) Mo (d) S

65. Plant ash has maximum content of

(a) Mg (b) Fe
(c) K (d) B

Answers

1. (c)	**2.** (b)	**3.** (c)	**4.** (c)	**5.** (b)	**6.** (a)	**7.** (d)	**8.** (d)	**9.** (b)	**10.** (d)
11. (c)	**12.** (a)	**13.** (b)	**14.** (a)	**15.** (a)	**16.** (c)	**17.** (a)	**18.** (d)	**19.** (a)	**20.** (d)
21. (b)	**22.** (a)	**23.** (a)	**24.** (a)	**25.** (c)	**26.** (d)	**27.** (c)	**28.** (c)	**29.** (b)	**30.** (c)
31. (c)	**32.** (a)	**33.** (d)	**34.** (a)	**35.** (c)	**36.** (c)	**37.** (d)	**38.** (c)	**39.** (d)	**40.** (d)
41. (c)	**42.** (a)	**43.** (a)	**44.** (a)	**45.** (c)	**46.** (b)	**47.** (a)	**48.** (d)	**49.** (a)	**50.** (b)
51. (b)	**52.** (d)	**53.** (b)	**54.** (d)	**55.** (a)	**56.** (b)	**57.** (b)	**58.** (a)	**59.** (b)	**60.** (d)
61. (c)	**62.** (a)	**63.** (a)	**64.** (b)	**65.** (c)					

30

Photosynthesis and Respiration

Photosynthesis

All animals including human being depends upon plants and their prepared food for their living and better survival. These green plants carry out photosynthesis, a physiochemical process by which they uses the light energy in order to drive the synthesis of organic compounds. Thus, photosynthesis is an anabolic process by which green plants, some bacteria and protistans prepares or manufactures their own food in the form of sugar in the presence of sunlight. The sugar thus, produced in this process is further used through **cellular respiration** in order to produce ATP (a fuel used by all living organisms on earth).

Requirements of Photosynthesis

On the basis of outline knowledge of light, it is to be noted that besides the requirement of light for photosynthesis, green plants also requires CO_2, water and a green pigment called chlorophyll for the complete process to take place.

(i) **Necessity of chlorophyll** When we test for the presence of starch in the experiment, it is predicted that photosynthesis occurs only in the green parts of leaves in the presence of chlorophyll.

(ii) **Necessity of carbondioxide** When we test for the presence of starch in the experiment, it is predicted that the leaf area exposed to light tested positive for starch while, the area of leaf being enclosed in the test tube containing cotton seated in KOH tested negative for the starch. This indicate requirement of CO_2 for photosynthesis.

(iii) **Necessity of water** Through radiolabelling of O_2 in water molecule it is confirmed that the O_2 released during photosynthesis comes from H_2O not from CO_2.

Site and Pigments of Photosynthesis

Chloroplast is the cell organelle which acts as a site for photosynthesis. Internally, the chloroplast contains thylakoids in groups called **grana** and a proteinaceous matrix or fluid called **stroma**. The membranes of thylakoids contains chlorophyll molecules which are basic molecule to carry out photosynthesis. The pigments involved in the process of photosynthesis are called **photosynthetic pigments** which provides different shades of green colour to the plants.

On the basis of their significance following types of pigments take part in photosynthesis :

(i) **Main pigments** Which forms the main molecule of photosystem such as chlorophyll-*a*, *b* and bacterio-chlorophyll-*a*, *b*.

(ii) **Accessory pigments** Which support the function of main pigment such as carotenoids (β-carotene and lutein) and phycobilins.

These pigment molecules absorb the light energy at different wavelengths in the following way:

Pigment Molecule	Wavelength of Light Absorbed (nm)
Chl-*a*	430, 660
Chl-*b*	463, 643
B chl-*a*	364, 770
B chl-*b*	373, 795

The pigments discussed above are not involved independently in a process but works together in a group. This active group of pigments is called **photosynthetic unit**. The accessory pigments absorbs light and transfer the energy to a centre point, where chlorophyll-*a* is available to receive this energy. Naturally, the photosynthetic pigments are bounded with the protein in the thylakoid membrane where they form a photosystem or pigments system, which absorbs the light at fixed wavelength and then perform the process of photosynthesis.

Photosystems

These are the light harvesting complexes having thousands of pigment molecules bounded by different protein molecules.

Common photosystems found are:

(i) **Photosystem-I** that works at the wavelength of about 700 nm. It is found exclusively on the membrane of stroma lamellae (stroma thylakoids) and at the edges of grana lamella (grana thyllakoids).

(ii) **Photosystem-II** that works at the wavelength of about 680 nm. It is mostly found at the grana thylakoids.

Process of Photosynthesis

Photosynthesis is a two stage process, i.e. it involves two phases in which the synthesis of food takes place. These includes light reaction and dark reaction.

(i) **Light reaction** (Photochemical phase) It is the process of photosynthesis that takes place in the thylakoid of the chloroplast and is mainly involved in the process of phosphorylation and formation of NADPH and ATP.

The process of photophosphorylation in light reaction occurs by the two main methods :

(a) **Non-cyclic photophosphorylation** In which both the photosystem (i.e. PS-I and PS-II) cooperate in the light driven synthesis of ATP, also the electrons that are released during this cycle from PS-II, does not returned back to it. Production of NADPH also takes place in this.

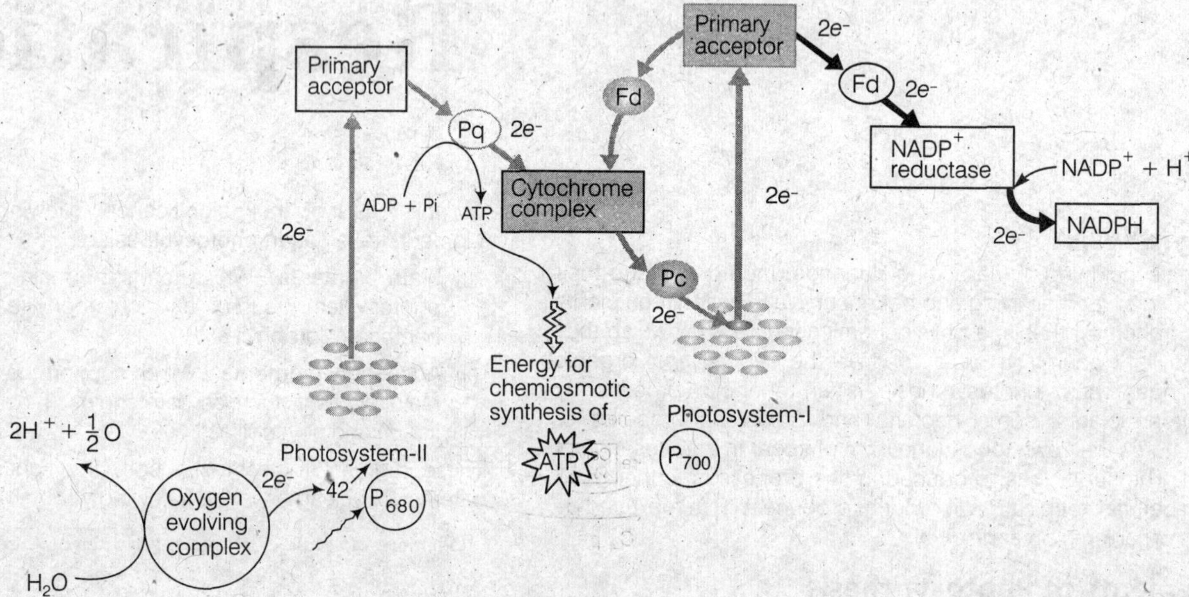

(b) **Cyclic photophosphorylation** In which only PS-I takes part and also the electron that is released during this cycle from PS-I is returned back to it after passing through a series of carrier molecules.

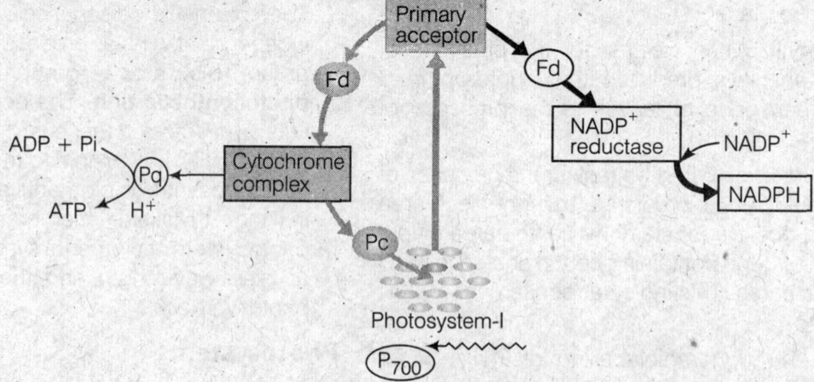

(ii) **Dark reaction** (Biosynthetic phase) This phase of photosynthesis is independent of light, i.e. it does not require direct sunlight but depend on the products of light reaction, (i.e. ATP and NADPH), traps CO_2 and convert it to carbohydrates. This process is also known as carbon-fixation or photosynthetic carbon reduction cycle. The assimilation of CO_2 during photosynthesis occurs by two main pathways, i.e. C_3 pathway in which C_3 acid is produced as the first product and C_4 pathway in which C_4 acid is produced as the first product.

(a) **C_3 pathway** (Calvin cycle) It is a cyclical biochemical pathway of reduction of CO_2 which runs in all photosynthetic plants. It occurs mainly in the stroma of the chloroplast. It is completed mainly in three major steps, i.e. carboxylation of RuBP, i.e. fixation of CO_2 into a stable organic intermediate. Reduction of PGA which leads to the formation of glucose through a series of reactions and regeneration of RuBP. Thus, in order to produce one molecule of glucose through the Calvin pathway, 18 ATP's and 12 NADPH's are required.

(b) **C_4 pathway** It is the pathway which takes place in C_4 plants, i.e. plants having a special type of leaf anatomy called **kranz anatomy** that can tolerate high temperature and shows responses in high light intensities.

These leaves have two types of cells, i.e. outer mesophyll and inner bundle sheath cells, unlike C_3 leaves which have only mesophyll cells.

This pathway is also known as **Hatch** and **Slack Cycle** which occurs in the following main steps :

- Initial fixation of CO_2 to form phosphoenol pyruvate.
- Formation of C_4 organic acid, i.e. Oxalo Acetic Acid (OAA).
- Transportation of OAA to bundle sheath cells to form a 3-carbon molecule releasing CO_2.
- Transportation of C_3 molecule back again to the mesophyll cells to regenerate PEP.

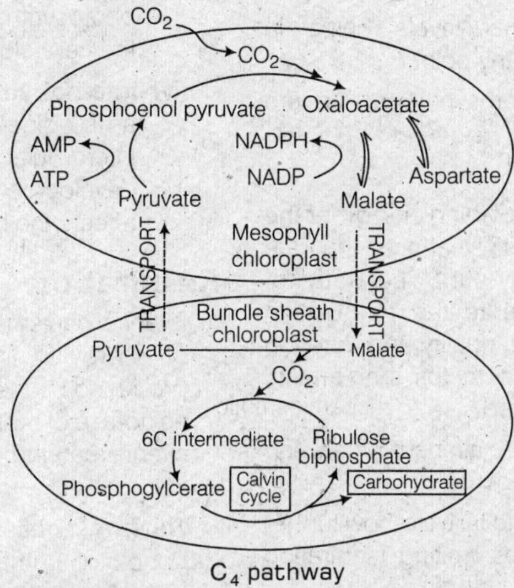

C_4 pathway

Photorespiration

Inhibition of photosynthesis in C_3-plants by O_2 is called photorespiration. It is also known as the process of oxygenation of RuBP and release of CO_2 by the photosynthetic organs of a plant. It occurs inside the chloroplast, mitochondria and peroxisomes are also involved in this process. At low concentration of CO_2, RuBisCO performs oxygenase activity and leads to the production of 3-PGA and 2-PGA. It is considered as a wasteful process because:

(i) There is no synthesis of sugar and ATP or NADPH

(ii) CO_2 is released with the utilisation of ATP.

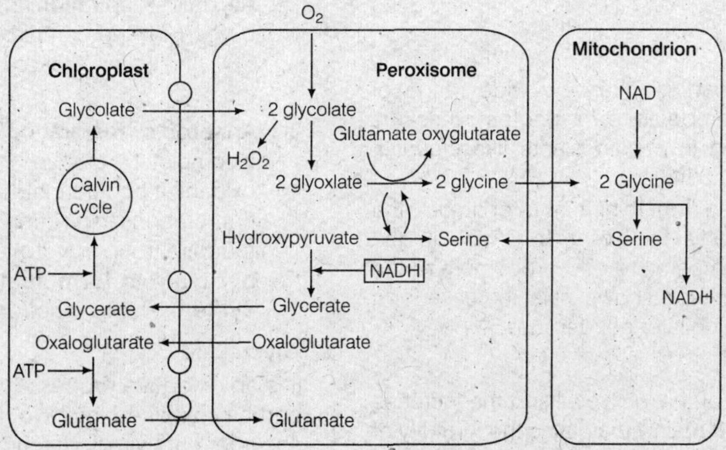

Factors Affecting Photosynthesis

The rate of photosynthesis is essential in determining the plants yield including the crop plants. It is generally influenced by the number of factors including both external, i.e. environmental and internal (plant) factors.

These are as follows:

1. External Factors

These factors generally includes an environmental factors like availability of sunlight, temperature, concentration of CO_2 and water.

Hence, factors available at sub-optimal levels helps in determining the rate of photosynthesis at any point.

Some of the factors that affect the photosynthetic rate are discussed below in detail :

(i) Light

The light it self provides the major role in providing energy for the process of photosynthesis. Hence, it mainly affects the rate of photosynthesis in or the other way-out. The rate of photosynthesis is generally affected due to the three main characteristics of the light, i.e. the intensity, quality and duration of light. Thus, discussing the light as a factor that affects photosynthesis, two conditions are observed.

(a) At **low light intensities**, the linear relationship between incident light and rate of CO_2 fixation occurs.

(b) While at **higher light intensities**, the rate fails to show further increase as the other factors becomes limiting (saturation point).

(ii) Carbon Dioxide

Initially increase in the concentration of CO_2 increases the rate of photosynthesis. However, higher concentration of this gas has an inhibitory effect on photosynthesis. Thus, CO_2 acts as the major limiting factor of photosynthesis. The C_4-plants show saturation at 360 ppm while C_3 show saturation at about 450 ppm of CO_2 in atmosphere.

Note It is important to not that plants like tomatoes and bell pepper when grown it increases level of CO_2.

(iii) Temperature

Process of photosynthesis occurs in a very wide range of temperature. The optimum temperature for photosynthesis is maximum at 25-30°C, though in certain plants like *Opuntia*, photosynthesis also occurs at 55°C.

When temperature is increased from minimum to optimum, the rate of photosynthesis doubles for every 10°C rise in the temperature. But when the temperature above optimum range reaches, the rate of photosynthesis shows initial increase for a very short duration of time which later get declines.

(iv) Water

An increase in water content of the leaf results in the increase in the rate of photosynthesis. It mainly maintains the turgidity of the assimilatory cells and the proper hydration of their protoplasm to wilt thus, reducing their surface and also their metabolic activity.

2. Internal Factors

The internal factors affecting rate of photosynthesis :

(i) **Chlorophyll content of leaves** The chlorophyll only acts as a biocatalyst and hence, a small quantity is enough to maintain the large bulk of the reacting substances.

(ii) **Protoplasmic factor** The plants which are transferred from dark to light do not start photosynthesis immediately. It is due to some protoplasmic factors, the exact nature of which is not known.

(iii) **Accumulation of by products** The final product in the photosynthetic reaction is sugar and its accumulation in the cells slows down the process of photosynthesis.

(iv) **Internal structure of leaf** The thickness of the cuticle and epidermis of the leaf, the size, distribution of intercellular spaces and distribution of stomata, the development of chlorenchyma and other tissues also affects the rate of photosynthesis.

Respiration

Every organisms for its survival requires a continuous supply of energy in the form of O_2. About 50% of the energy produced by the cell is utilised by the cellular activities (such as uptake of materials, absorption, growth, development and even breathing) and rest of it is changed into heat and get lost.

Thus, respiration is the process of breaking of the C—C bonds of complex compounds within the cells through oxidation and release of the considerable amount of energy. It occurs inside the cytoplasm and mitochondria of the cell. The major respiratory substrates involved in the respiration is carbohydrate which gets oxidise and releases high amount of energy.

Types of Respiration

On the basis of the dependence of cells on O_2, respiration is divided into two main types :

(i) **Aerobic respiration** In which living organism utilises oxygen for the complete oxidation of organic food into CO_2 and water. It takes places inside the mitochondria of the cell.

(ii) **Anaerobic respiration** In which an organic food is incompletely oxidised without utilising energy as an oxidant. It occurs inside the cytoplasm of the cell. The first step in respiration is the **glycolysis** (oxidation of glucose to pyruvate acid), after that **oxidation of pyruvate to form acetyl Co-A** and lastly the **Krebs' cycle** (aerobic respiration).

1. Glycolysis

It is an oxidative process in which one mole of glucose is partially oxidised into two moles of pyruvate in a series of enzyme catalysed reactions. Glycolysis occurs in the cytosol of all cells. In plants, this glucose is derived from sucrose which is the end product of photosynthesis.

Steps involved in Glycolysis

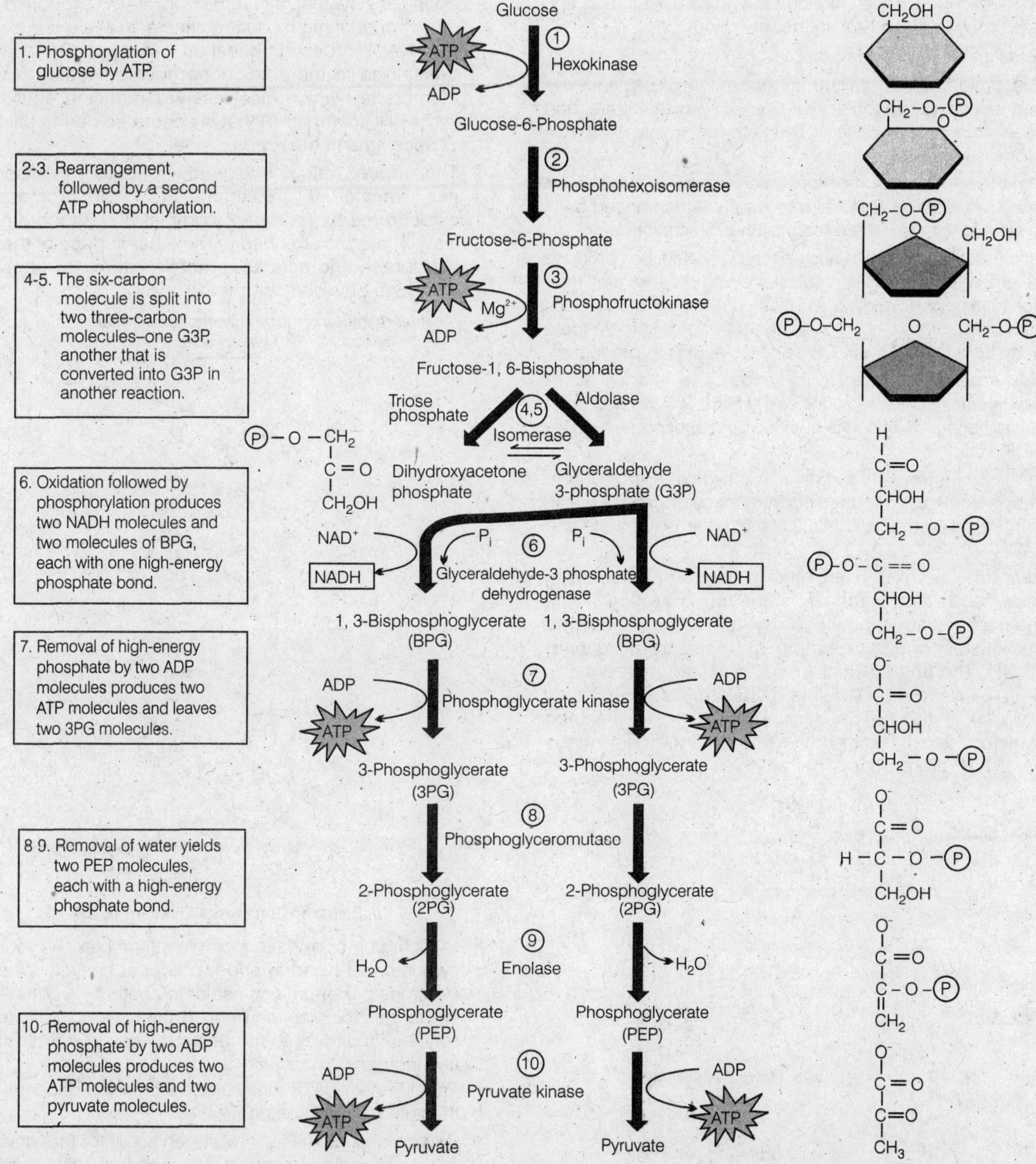

The energy yield from one molecule of glucose in glycolysis is 2 ATP and 2 NADH. In the presence of oxygen, one NADH reoxidise to yield 2.5 ATP. Hence, in aerobic glycolysis 7 ATP are produced. While, in anaerobic glycolysis, only 2 ATP are produced.

2. **Pyruvate Oxidation**

It is the process of conversion of pyruvate into acetyl Co-A and is catalysed by the pyruvate dehydrogenase complex.

During this process, it is made up of three enzymes, i.e. pyruvate dehydrogenase, dehydrolipoyl transacetylase and dehydrolipoyl dehydrogenase. During this process, the carboxyl group of pyruvate is released as CO_2 forming an acetyl Co-A, which is a connecting link between the glycolysis and citric acid cycle.

3. **Krebs' Cycle** (TCA)

It is also known as Tricarboxylic Acid Cycle (TCA). It was first elucidated by sir Hans Kreb in the year 1940.

It occurs into following main steps:

- The cycle begins with the condensation of oxaloacetate and acetyl group of acetyl Co-A in which citrate and coenzyme-A are formed. The enzyme involved in this process is citrate synthase.
- Then isomerisation reaction catalysed by *cis*-aconitase occurs in which H_2O is first removed and then added back to yield isocitrate. Fluoroacetate inhibits aconitase.
- isocitrate is oxidised and decarboxylated to produce α-ketoglutarate. This is catalysed by an enzyme isocitrate dehydrogenase. Another oxidative decarboxylation results in the formation of succinyl Co-A (enzyme-α- ketoglutarate dehydrogenase, NADH, CO_2 and Co-A-SH are produced).
- Next reaction is catalysed by an enzyme succinyl Co-A synthetase to form succinate which is coupled with phosphorylation of purine nucleoside diphosphate, usually GDP.
- In the next step FAD removes two hydrogen atoms from succinate by the enzyme succinate dehydrogenase which is inhibited by malonate. This reaction forms fumarate and $FADH_2$.
- After this, the molecule of H_2O is added to fumarate to form malate and in the last of the four oxidation steps, the carbon carrying the hydroxyl group is converted to carbonyl group, regenerating oxaloacetate producing NADH. The energy yield from 2 pyruvate molecules is then oxidised to $6CO_2$ *via* pyruvate dehydrogenase complex and the TCA cycle. The electrons are then transferred to O_2 *via* oxidative phosphorylation as many as 25 ATP are produced.

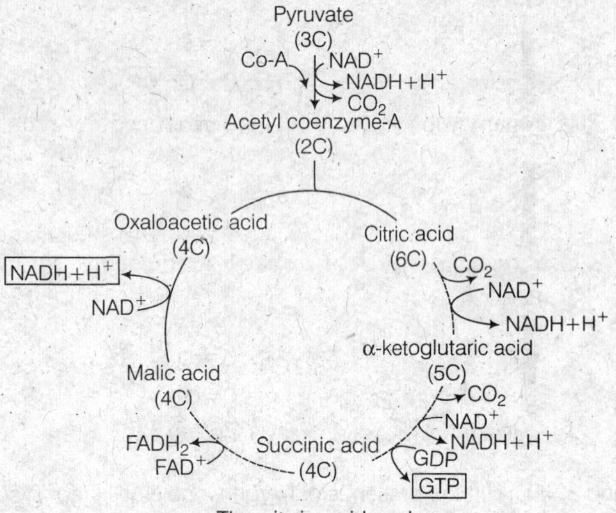

The citric acid cycle

Electron Transport Chain (ETC)

The electron transport chain is a process in which the NADH and $FADH_2$ produced during glycolysis, TCA cycle, etc., are oxidised thus, releasing the energy in the form of ATP. It occurs in the inner mitochondrial membrane of the cell.

It occurs in the following steps :

- During this system, electrons from NADH produced in the mitochondrial matrix during citric acid cycle are oxidised by NADH dehydrogenation (complex I) and are transferred to the ubiquinone (located within the inner membrane), which also receives reducing equivalents *via* $FADH_2$ (complex II) that is generated during oxidation of succinate in the Krebs' cycle.
- The reduced ubiquinone (ubiquinol) is then oxidised with the transfer of electrons to cytochrome-*c* *via* cytochrome-*b*,c_1 complex (complex III). Cytochrome-*c* is a small protein attached to the outer surface of the inner membrane and acts as a mobile carrier for transfer of electrons between complex III and IV.

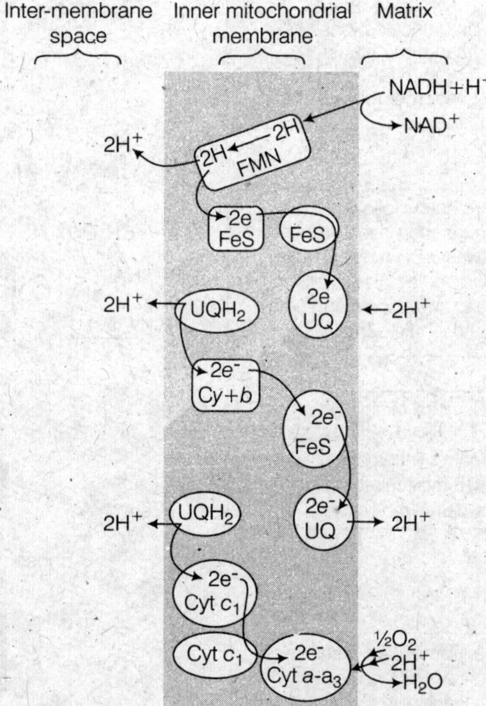

Electron Transport System (ETS)

- Cytochrome-c oxidase complex (complex IV) contains cytochromes-*a* and a_3 and two copper centres. When the electrons pass from one carrier to another via complex I-IV in the electron transport chain, they are coupled to ATP synthase (complex V) for the production of ATP from ADP and inorganic phosphate.
- The number of ATP molecules synthesised depends upon the nature of electron donor.
- Oxygen acts as the final hydrogen acceptor that drives the whole process by removing hydrogen from the system. The energy released during the electron transport system is, thus utilised in synthesising ATP with the help of ATP synthase.

Note ATP synthase is the enzyme used in ETC. It has two main components, i.e. F_1 and F_0. The passage of protons through the channel is coupled to the catalytic site of the F_1 component for the production of ATP. For each ATP produced, $2H^+$ passes through F_0 from the intermembrane space to the matrix down the electrochemical proton gradient.

Chapter Exercise

1. For each glucose molecule that is broken down in glycolysis there is a net grain of
(a) 1 ATP molecule (b) 2 ATP molecules
(c) 3 ATP molecules (d) 4 ATP molecules

2. Which of the following enzymes catalyses a non-reversible reaction in glycolysis?
(a) Phosphofructokinase
(b) Enolase
(c) Triose phosphate isomerase
(d) Phosphohexose isomerase

3. Glucose-6-phosphate inhibits which of the following enzymes?
(a) Hexokinase
(b) Glucokinase
(c) Phosphorylasekinase
(d) Fructose-1, 6-bisphosphatase

4. The end product of fermentation of molasses by yeast is
(a) pyruvate (b) ethyl alcohol
(c) lactate (d) methyl alcohol

5. The oxidation of 1 mol of glucose by anaerobic glycolysis yield a net of
(a) 2 mol of acetyl Co-A and 2 mole of ATP
(b) 2 mol of lactate and 2 mol of ATP
(c) 2 mol of lactate, 2 mol of NAD^+ and 6 mole of ATP
(d) 2 mole of lactate, 2 mol of NADH and 2 mol of ATP

6. Glucokinase is characterised as
(a) it has a much higher Michaelis constant for glucose than hexokinase
(b) it is an allosteric enzyme
(c) it is found in most cells
(d) it phosphorylates most hexoses

7. Fluoride inhibits which glycolytic enzyme
(a) enolase (b) glucokinase
(c) phosphofructokinase (d) lactate dehydrogenase

8. Iodoacetate inhibits which glycolytic enzyme
(a) hexokinase
(b) enolase
(c) glyceraldehyde-3-phosphate dehydrogenase
(d) phosphoglycerate kinase

9. Glycolysis occurs
(a) mitochondria (b) cytosol
(c) Both (a) and (b) (d) nucleus

10. Which enzyme is not present in muscle?
(a) Hexokinase (b) Glucose-6-phosphatase
(c) Glycogen synthase (d) Phosphorylase-*b*

11. Which glycolytic enzyme is used in gluconeogenesis?
(a) Pyruvatekinase (b) Glucokinase
(c) Aldolase (d) Phosphoglycerate kinase

12. Which enzyme cleaves C—C bond in glycolysis?
(a) Aldolase (b) Enolase
(c) Phosphoglyceratekinase (d) Phosphoglyceratemutase

13. Which enzyme is inhibited by arsenate?
(a) Glyceraldehyde-3 phosphate dehydrogenase
(b) Lactate dehydrogenase
(c) Hexokinase
(d) Phosphofructokinase

14. An embryo with complete deficiency of pyruvatekinase, how many net moles of ATP would be generated in conversion of one mole of glucose to one mole of pyruvate?
(a) 0 (b) 1
(c) 2 (d) 3

15. Malonate is an inhibitor of
(a) citrate synthase
(b) aconitase
(c) succinate dehydrogenase
(d) malate dehydrogenase

16. Which enzyme in TCA cycle is involved in formation of GTP from GDP?
(a) Succinate dehydrogenase (b) Malate dehydrogenase
(c) Citrate synthase (d) Succinate thiokinase

17. Which vitamin is required for the synthesis of co-factor, required in conversion of succinate to fumarate?
(a) Riboflavin (b) Thiamine
(c) Lipoic acid (d) Niacin

18. Which of the following is utilised as precursor for cholesterol biosynthesis?
(a) Succinyl Co-A (b) Acetyl Co-A
(c) Succinate (d) Malate

19. How many molecules of CO_2 are produced per mole of acetyl Co-A in TCA cycle?
(a) 2 (b) 1 (c) 3 (d) 4

20. Which of the following is a component of succinate dehydrogenase in an electron transport chain?
(a) Niacin (b) FMN
(c) FAD (d) Co-enzyme-Q

21. Which of the following ETC components accepts only one electron?
(a) Cytochrome-*b* (b) Oxygen
(c) FAD (d) FMN

22. Which of the following enzymes catalyses the conversion of hydrogen peroxide to water?
(a) Hydratase (b) Catalase
(c) Dioxygenase (d) Dismutase

23. Which process does not involve cytochrome-*c*?
(a) Apoptosis (b) TCA cycle
(c) Electron transport (d) Oxidative phosphorylation

24. Light reaction is dependent on light and produce
(a) ATP (b) NADPH
(c) Both (a) and (b) (d) None of these

25. Which of the following pigments is found in the reaction centre?
(a) Phycobilins (b) Chlorophyll
(c) Carotenoids (d) All of these

26. The wavelength absorbed by chlorophyll-*a* is
(a) 430 (b) 660
(c) Both (a) and (b) (d) None of these

27. Non-cyclic photophosphorylation utilise which of the following?
(a) Only PS-I (b) Only PS-II
(c) Both PS-I and PS-II (d) None of these

28. Which of the following ion is involved in ATP synthesis?
(a) Ca^{2+} (b) Zn^{2+} (c) Cl^- (d) H^+

29. During photosynthesis, which process releases electrons that return chlorophyll molecules to their reduced state?
(a) Nutrition of photosystem-I
(b) Oxidation of reduced NADP
(c) Phosphorylation of ADP
(d) Photolysis of water

30. Which of the following describes conditions in a photosynthesising cell, exposed to high light intensity and low carbon dioxide concentration?

	Concentration acceptor	Concentration of ATP	Concentration of GP (PGA)
(a)	High	High	Low
(b)	High	Low	Low
(c)	Low	High	High
(d)	Low	Low	High

31. Which statement is correct about compensation point?
(a) No, growth occurs
(b) Rate of photosynthesis is equal to respiration
(c) Neither O_2 nor CO_2 is given off
(d) Occurs, when light intensity falls to 20% of normal

32. PS-II is not concerned with
(a) photolysis of water
(b) flowering
(c) reduction of CO_2
(d) release of energy

33. Agranal chloroplasts are characteristics of
(a) mesophyll of pea leaves
(b) bundle sheath of mango leaves
(c) mesophyll of maize leaves
(d) bundle sheath of sugarcane leaves

34. If intensity of light increases 20 times, the rate of photosynthesis will
(a) increase
(b) constant
(c) decrease
(d) increase till feed back inhibition

35. Which of the following is essential for the process of photosynthesis?
(a) Carbohydrates and oxygen
(b) Water, minerals and CO_2
(c) Light, chlorophyll and optimum temperature
(d) Both (b) and (c)

36. Chloroplast contains the maximum concentration of which of the following?
(a) Starch (b) Fructose
(c) 3-PGA (d) RuBP carboxylase

37. All photosynthetic green plants contain
(a) chlorophyll-*a* (b) chlorophyll-*b*
(c) chlorophyll-*c* (d) chlorophyll-*e*

38. Photo centres in higher plants are
(a) P_{700} (b) P_{680}
(c) Both (a) and (b) (d) Chlorophyll-*a*

39. Which of the following is most effective in photosynthesis?
(a) Blue light (b) Green light
(c) Red light (d) Yellow light

40. Chlorophyll contains
(a) Fe (b) Mg (c) K (d) Mn

41. The number of oxygen molecules produced per quantum of light absorbed is
(a) oxygen yield (b) photosynthesis yield
(c) quantum yield (d) organic yield

42. The pigment chlorophyll-*a* is absent in
(a) gymnosperms (b) bacteria
(c) algae (d) bryophyta

43. PS-II is concerned with
(a) photolysis of water (b) flowering
(c) reduction of CO_2 (d) release of energy

44. Photosynthesis is a process in which
(a) light energy is converted into chemical energy
(b) chemical energy is converted into light energy
(c) chemical energy is converted into chemical energy
(d) organic chemicals are broken down

45. A cell that lacks chloroplast does not
(a) evolve CO_2 (b) liberate O_2
(c) require water (d) utilise carbohydrates

46. Cyclic photophosphorylation is confined to
(a) PS-I (b) PS-II
(c) Both (a) and (b) (d) None of these

47. Non-cyclic photophosphorylation involves
(a) PS-I (b) PS-II
(c) Both (a) and (b) (d) None of these

48. In blue-green algae, photosynthesis occurs in
(a) chromatophores (b) chloroplast
(c) photosynthetic lamellae (d) chromoplast

49. The electrons liberated from PS-II go to
(a) carotenoids (b) plastocyanin
(c) ferredoxin (d) plastoquinone

50. The organic acid concentration in CAM plants
(a) decreases during day (b) increases at night
(c) Both (a) and (b) (d) remains same always

51. The substrate for photorespiration is
(a) malic acid (b) oxaloacetic acid
(c) glycolic acid (d) PGA

52. How many ATP and $NADPH_2$ molecules are used for fixing one molecule of CO_2?
(a) Three and two respectively (b) Three each
(c) Two each (d) Two and three respectively

53. Rate of photosynthesis is independent of
(a) duration of light (b) intensity of light
(c) temperature (d) respiration

54. During photosynthesis
(a) water gets reduced and carbon dioxide gets oxidized
(b) CO_2 gets reduced and water gets oxidised
(c) Both (a) and (b)
(d) both carbon dioxide and water get reduced

55. RuBP carboxylase enzyme catalyses the carboxylation reaction between
(a) CO_2 and ribulose 1, 5-diphosphate
(b) acetyl Co-A and oxaloacetic acid
(c) PGA and dihydroxy acetone phosphate
(d) ribulose diphosphate and phosphoglyceraldehyde

56. The first enzyme of Calvin cycle is
(a) cytochrome oxidase
(b) photo-pentokinase
(c) RuBP carboxylase
(d) triose phosphate dehydrogenase

57. NADP is converted into $NADPH_2$ in
(a) photosystem-II
(b) photosystem-I
(c) Calvin cycle
(d) non-cyclic photophosphorylation

58. In blue-green algae, photosystem-II contains important pigment concerned with photolysis of water. It is a
(a) phycocyanin (b) cytochrome-*c*
(c) chlorophyll-*b* (d) β-carotene

59. Rate of photosynthesis is
(a) equal to that of respiration
(b) less than respiration
(c) depends upon chlorophyll content
(d) faster than that of respiration

60. Photosynthesis is a
(a) catabolic and reductive reaction
(b) catabolic and oxidative reaction
(c) anabolic and reductive reaction
(d) catabolic and exothermic reaction

61. The C_4-plants differ from C_3-plants in
(a) type of pigments involved in photosynthesis
(b) number of ATP's evolved in formation of sugar
(c) the substance that accepts CO_2 during assimilation
(d) type of end products of photosynthesis

62. During photochemical reaction of photosynthesis
(a) release of O_2 and formation of ATP+$NADPH_2$ takes place
(b) formation of ATP and NADP takes place
(c) assimilation of CO_2 takes place
(d) liberation of oxygen takes place

63. Tropical plants like maize show high efficiency of CO_2 fixation because of
(a) TCA cycle (b) EMP pathway
(c) Hatch-Slack cycle (d) Calvin cycle

64. During dark reaction of photosynthesis
(a) CO_2 reduced to organic compounds
(b) water splitting
(c) chlorophyll is activated
(d) 6-carbon sugar is broken down to 3-carbon sugar

65. Chlorophyll-*b* is
(a) $C_{54}H_{70}O_6N_4Mg$ (b) $C_{55}H_{70}O_6N_4Mg$
(c) $C_{55}H_{72}O_5N_4Mg$ (d) $C_{45}H_{72}O_5N_4Mg$

66. A photosynthesising plant is releasing ^{18}O more than the normal. The plant must have been supplied with
(a) O_3 (b) H_2O with ^{18}O
(c) CO_2 with ^{18}O (d) $C_6H_{12}O_6$ with ^{18}O

67. $6CO_2 + 12H_2O + 12NADPH + A \longrightarrow C_6H_{12}O_6 + 12NADP + 6H^+ + B + 18Pi$ in this reaction, A and B are
(a) 6 ADP and 6 ATP
(b) 18 ADP and 18 ATP
(c) 18 ATP and 18 ADP
(d) 18 ADP and 18 GTP

68. Which of the following types of plant fixes carbon dioxide by the way of CAM pathway?
(a) Oak tree (b) Grass
(c) Red algae (d) Cactus (succulents)

69. The two enzymes which are used for primary carboxylation in C_3 and C_4 respectively are
(a) RuBP carboxylase and PEP carboxylase
(b) RuBP carboxylase and RuBP carboxylase
(c) PEP carboxylase and pyruvate carboxylase
(d) PEP carboxylase and RuBP carboxylase

70. The production of energy rich phosphate bonds as in ATP (adenosine triphosphate) is known as
(a) photophosphorylation
(b) photorespiration
(c) photooxidation
(d) oxidative photophosphorylation

71. Which one of the following is the most appropriate reason for storing green-coloured apples at low temperature?

(a) The rate of photosynthesis is reduced
(b) The rate of respiration is reduced
(c) The rates of photosynthesis and respiration are reduced
(d) Respiration and photosynthesis are completely inhibited

72. Which of the following product of glycolysis is consumed in alcoholic fermentation?

(a) $NADH_2$ (b) CO_2
(c) ATP (d) Both (a) and (c)

73. Lactic acid fermentation does not produce

(a) CO_2 (b) NADH
(c) Both (a) and (b) (d) ATP

74. Which of the following connects the glycolysis and Krebs' cycle?

(a) Pyruvic acid (b) Glucose
(c) Acetyl Co-A (d) ATP

75. Respiration is different from the combustion in that

(a) energy from carbohydrates is released in one step
(b) energy from carbohydrates is released in different steps
(c) efficiency is very low
(d) All of the above

76. Krebs' cycle starts with the formation of a six carbon compound by reaction between

(a) fumaric acid and pyruvic acid
(b) OAA and acetyl Co-A
(c) malic acid and acetyl Co-A
(d) succinic acid and pyruvic acid

77. Site of ATP synthesis occurs on the

(a) outer membrane of mitochondrion
(b) matrix
(c) inner membrane of mitochondrion
(d) None of the above

78. Respiration is accompanied by

(a) decrease in weight (b) increase in weight
(c) weight remains constant (d) decrease in size

79. Pyruvate dehydrogenase is used in converting

(a) glucose to pyruvate
(b) pyruvate to glucose
(c) pyruvate to acetyl Co-A
(d) lactic acid to pyruvic acid

80. End product of fermentation is

(a) pyruvic acid and H_2O (b) H_2O and CO_2
(c) C_2H_5OH and CO_2 (d) CO_2

81. The maximum energy in the cell is liberated when

(a) pyruvic acid is converted to acetyl Co-A
(b) pyruvic acid is converted into CO_2 and H_2O
(c) sugar is converted to pyruvic acid
(d) glucose is converted to pyruvic acid

82. Removal of hydrogen and CO_2 from a substrate is called

(a) decarboxylation
(b) oxidation
(c) oxidative decarboxylation
(d) reductive decarboxylation

83. When a molecule of pyruvic acid is subjected to anaerobic oxidation and produced lactic acid, there is a net

(a) loss of three ATP molecules
(b) loss of six ATP molecules
(c) gain of two ATP molecules
(d) gain of six ATP molecules

84. ATP is a

(a) component of nucleic acid
(b) molecule which contains high energy phosphate bonds
(c) Both (a) and (b)
(d) protein

85. Out of 36 ATP molecules, produced per glucose molecule during respiration

(a) two are produced outside in glycolysis and 34 during respiratory chain
(b) two are produced outside mitochondria and 34 inside mitochondria
(c) two during glycolysis and 34 during Krebs' cycle
(d) all are formed inside mitochondria

Answers

1. (b)	**2.** (a)	**3.** (a)	**4.** (b)	**5.** (b)	**6.** (a)	**7.** (a)	**8.** (c)	**9.** (b)	**10.** (b)
11. (d)	**12.** (a)	**13.** (a)	**14.** (a)	**15.** (c)	**16.** (d)	**17.** (a)	**18.** (b)	**19.** (a)	**20.** (c)
21. (a)	**22.** (b)	**23.** (b)	**24.** (c)	**25.** (b)	**26.** (c)	**27.** (a)	**28.** (d)	**29.** (d)	**30.** (a)
31. (b)	**32.** (b)	**33.** (d)	**34.** (d)	**35.** (d)	**36.** (d)	**37.** (a)	**38.** (c)	**39.** (c)	**40.** (b)
41. (b)	**42.** (b)	**43.** (a)	**44.** (a)	**45.** (b)	**46.** (a)	**47.** (c)	**48.** (a)	**49.** (d)	**50.** (c)
51. (d)	**52.** (b)	**53.** (d)	**54.** (b)	**55.** (a)	**56.** (c)	**57.** (d)	**58.** (a)	**59.** (c)	**60.** (c)
61. (c)	**62.** (a)	**63.** (c)	**64.** (a)	**65.** (b)	**66.** (b)	**67.** (c)	**68.** (d)	**69.** (a)	**70.** (a)
71. (c)	**72.** (a)	**73.** (b)	**74.** (c)	**75.** (d)	**76.** (b)	**77.** (c)	**78.** (a)	**79.** (c)	**80.** (d)
81. (b)	**82.** (c)	**83.** (b)	**84.** (c)	**85.** (b)					

31

Plant Growth and Development

Plant Growth

It can be defined as an irreversible permanent increase in size of an organ or its parts or even an individual cell.

It is accomplished by metabolic processes that utilise the energy obtained by the process of nutrition. All cells of a plant are descendents of the zygote (fertilised egg). The development is actually the sum total of two main processes, i.e. growth and differentiation.

Pattern of Plant Growth

The development of a mature plant from a zygote follows a precise and highly ordered succession of events. During this process, a complex body organisation is formed that produces roots, leaves, branches, flowers, fruits, seeds and dies after completing its life span.

Regions of Plant Growth

In plants, growth is confined only to meristems, i.e. **root apex** and **shoot apex**. Plant growth is unique because plants retain the capacity for unlimited growth throughout their life. This property of plants is due to the presence of **meristem tissue**.

The meristem cells have the capacity to divide and self-perpetuate. These cells later form plant body when they loose the capacity of dividing.

Determinate and Indeterminate Structures

Some plant structures are determinate, others are indeterminate. A determinate structure grows to a certain size and then stops eventually undergoing senescence and death, (e.g. leaves, flowers and fruits).

On the other hand, the vegetative stem and roots are indeterminate structures which grows by meristems that continuously replenish themselves.

Growth Curve

Many investigat. s have plotted size or weight of an organism against time, producing a growth curve. Often the curve can be fitted with a simple mathematical function such as a straight line or simple S-shaped curve (sigmoid growth curve).

Phases of Plant Growth

The grand period of plant growth or sigmoid curve shows three phases:

(i) **Lag phase or slow phase** In this phase, the size of the cell increases exponentially with time. It means that the growth rate is slow at first but then continuously increases, e.g. the constantly dividing cells, both at the root apex and the shoot apex, represent the meristematic phase or lag phase of the growth.

(ii) **Log phase or exponential growth phase** In the linear phase, increase in size continues at a constant rate, usually maximum for sometime and eventually slows down. The cells which are just next away from the tip to the meristematic zone represents the phase of elongation. Increase vacuolation, cell enlargement and new cell deposition are the characteristics of the cells in this phase.

(iii) **Senescence phase or steady phase** It is characterised by a decreasing growth rate as the plant reaches maturity and begin to senescence. This is more proximal to the phase of elongation and lies along the portion of axis, undergoing the phase of maturation.

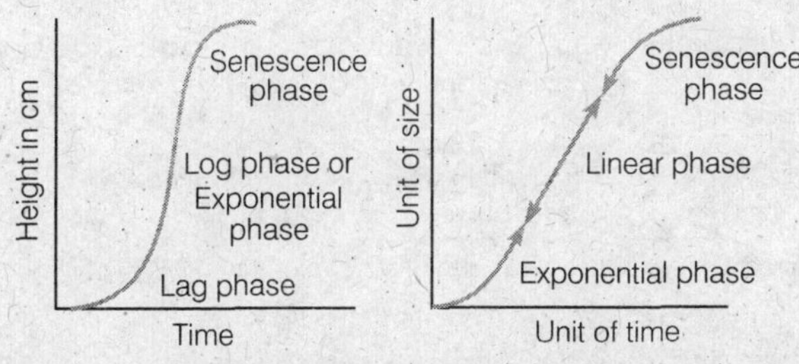

Sigmoid growth curve in plants

Growth of a plant is influenced by several factors such as water, oxygen, CO_2 concentration, temperature, nutrients, etc.

Plant Development

All changes that a plant undergoes during its life cycle initiating from the germination of seed till the senescence (ageing) constitutes its development or phase of development.

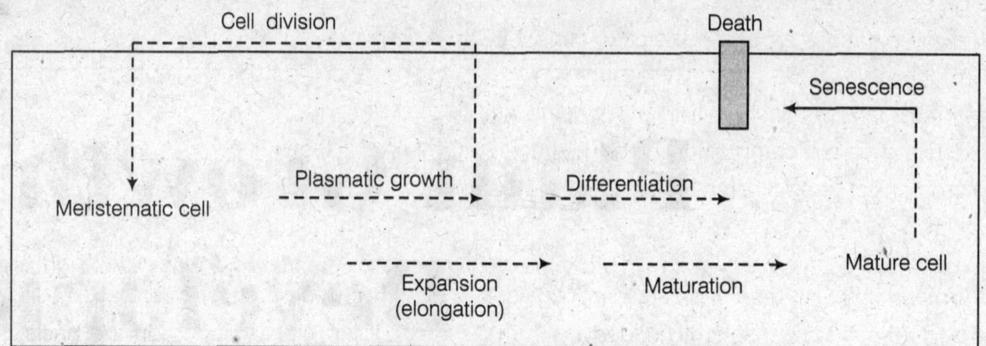

Development process in plants

Plasticity and Heterophylla

Plants follow different pathways in response to environment or phases of life to form different kinds of structure. This ability is called **plasticity**, i.e. heterophylla in cotton, coriander and larkspur. In such plants, the leaves of the juvenile plant are different in shape from those in mature plants.

On the other hand difference in shapes of leaves produced in air and those produced in water in buttercup also represent the heterophyllous development due to the environment. This phenomenon of **heterophylla** is an example of plasticity. Thus, growth, differentiation and development are very closely related events in the life of a plant.

Development can be considered as the sum of growth and differentiation. The development in plants is under the control of intrinsic and extrinsic factors.

The intrinsic factors include both intracellular (genetic) or intercellular factors/chemicals such as plants growth regulators, while the extrinsic factors include light, temperature, water, oxygen, nutrition, etc.

Plant Hormones

Growth and most of the physiological activities in plants are regulated by the action and interaction of some chemical substances present in them, i.e. **plant hormones** or **phytohormones**. A plant hormone is an organic compound which is synthesised in one part of a plant and is translocated to the another part, where they in very low concentrations, causes physiological response.

These phytohormones are grouped generally into five main categories such as auxins, cytokinins, gibberellins, ethylene and abscisic acid.

Plant Hormones, their Functions and Location

Hormone	Regions found in Plant	Major Functions
Auxins (IAA)	Apical meristems, other immature parts of plants.	Promotion of stem elongation and growth, formation of adventitious roots, inhibition of leaf abscission, promotion of cell division (with cytokinins), inducement of ethylene production, promotion of lateral bud dormancy.
Cytokinins	Root apical meristems, immature fruits.	Stimulation of cell division but only in the presence of auxin, promotion of chloroplast development delay of leaf ageing, promotion of bud formation.
Gibberellins	Roots and shoot tips, young leaves, seeds.	Promotion of stem elongation, stimulation of enzyme production in germinating seeds.
Ethylene	Roots, shoot apical meristems, leaf nodes, ageing flower, ripening fruits.	Control of leaf, flower and fruit abscission, promotion of fruit ripening.
Abscisic acid	Leaves, fruits, root caps, seeds.	Inhibition of bud growth, control of stomatal closure, some control of seed dormancy, inhibition of effects of other hormones.
Brassinosteroids*	Pollen, immature seeds, shoot, leaves.	Overlapping function with auxins and gibberellins.
Oligosaccharides*	Cell walls.	Pathogen defence, possibly reproductive development.

*Two latest additions have been done in the list.

Seed Dormancy

A dormant seed is said to be the seed that is unable of germinating into normal or a viable seed even after providing the most favourable environmental condition that are normally suitable for the germination of the non-dormant seed. The phenomenon shown by such seed, it is called seed dormancy.

Dormancy of seed may be caused by one or the other factors such as

(i) Impermeability of the seed coat.

(ii) Immature embryo of some seeds at the time of shedding.

(iii) Production of certain chemical compounds inhibiting the process of germination.

A seed can be ruptured or get broken down either by the **natural** or by **artificial means**.

Photoperiodism

It is the mechanism of requiring certain day length in order to induce flowering. The period of a day when light is available to the plant is called **photoperiod** and this response of plants to periods of day/night is called photoperiodism.

Critical photoperiod is the duration of light needed by the plant in order to get flowered. On the basis of the photoperiodic responses, flowering plants have been classified as **long day plants** which begin to flower when day length exceeds a critical length and **short day plants** which requires the shorter photoperiod to flower.

Vernalisation

The temperature also acts as a vital phenomena of flowering beside light. Thus, chilling requirement of the plants for flowering is called vernalisation. It is commonly seen in biennial plants which usually flower in one season and die in another season.

Uses of Vernalisation

(i) Vernalisation shortens the vegetative period of plant.

(ii) It increases yield, resistance to cold and diseases.

(iii) Kernal wrinkles of *Triticale* can be removed by vernalisation.

Plant Movements

Apart from growing in length and width, plants also shows movements, i.e. they have the capacity changing their positions in response to the external and internal stimuli.

Major types of movements shown by plants are given in the following diagram below:

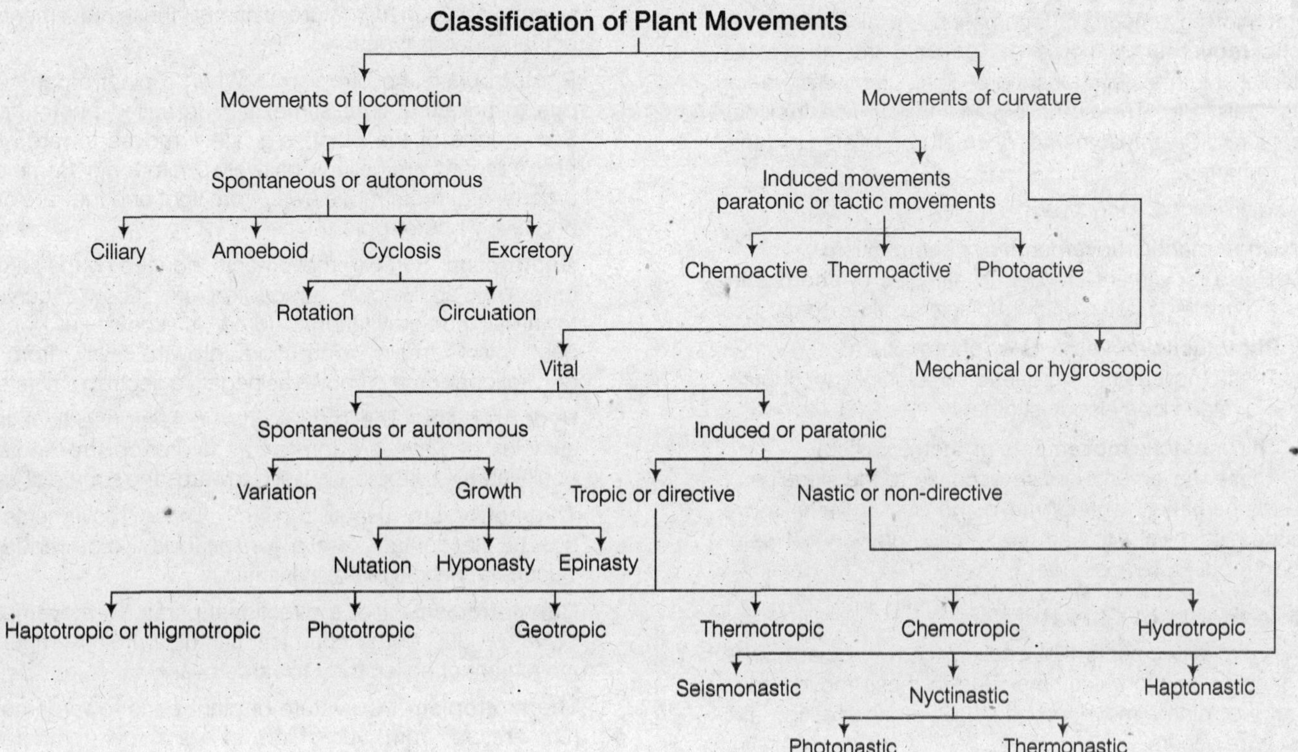

1. Movements of Locomotion

Physical movement of whole plant body or an organ from one place to another is called movement of locomotion. These movements include movements of protoplasm and different cell organelles, unicellular plants like *Chlamydomonas,* multicellular plants like *Volvox* or the movements of their free organs, e.g. gametes or zoospores.

Movements of locomotion are of two types

(i) Autonomous or Spontaneous Movements

This movement of plant is responsible itself by the internal factors (stimulus).

They are of following types:

(a) Ciliary movements, e.g. *Chlamydomonas, Euglena* zoospores.

(b) Amoeboid movements, e.g. gametes of ***Spirogyra.***

(c) Cyclosis is the movement of cytoplasm within a cell (also called protoplasmic streaming).

These are of two main types:

- **Rotation** When the whole protoplasm moves around a single central vacuole, it may only in one direction, e.g. leaf cells of *Hydrilla*, *Vallisneria*, etc.
- **Circulation** When the whole protoplasm moves around the different vacuoles in different directions, it may be either clockwise or anti-clockwise, e.g. staminal hair shoot hairs of gourds.

(d) Excretory, e.g. apical part of *Oscillatoria.*

(ii) Induced or Paratonic Movements

Paratonic movements of locomotion are also known as **tactic movements**. These movements are due to the external factors (stimulus) like light, temperature or chemicals, etc. These movements take place in small plants, e.g. *Chlamydomonas* or small free ciliated organs, e.g. gametes.

These are of following types:

(a) **Chemotactic movements or chemotaxism**
These movement are due to stimulus of **chemicals**, e.g. male gametes of lower plants cantherozoids.

(b) **Phototactic movements or phototaxism**
These movements are due to stimulus of light source, e.g. algae by eyespot (stigma).

(c) **Thermotactic movements or thermotaxism**
These movements are in response to the stimulus of temperature, e.g. *Chlamydomonas*, *Euglena* move from cold water to warm water and from very hot water to medium temperature.

2. Movements of Curvature

These are movements that occur only in the plant parts (organs) in relation to others. Due to these movements, organs of plants move in one or the other direction, i.e. curvature occurs.

These movements are of the following two types:

(i) Mechanical Movements or Hygroscopic Movements

Such movements are found in non-living organs of the plant and depend upon the presence or absence of water. e.g. peristomial teeth of moss, the spores of the *Equisetum* which coils and uncoils in the presence or absence of water respectively.

(ii) Vital Movements

These movements are found in living organs of plants and are of two types:

(a) **Autonomous movements** These movements are of further two types:

- **Movements of variation** In Indian Telegraph plant, a member of papilionatae the leaf has a terinal leaflet and two lateral leaflets. During day time, the lateral leaflets move up and down due to variation in turgidity.
- **Movements of growth** These movements are due to the unequal growth in different parts of an organ.

These are of the following types:

Nutation, e.g. tendrils become spiral due to more growth on the outer side.

Hyponasty These movements are usually found in leaves, petals, etc.

Epinasty When a leaf or petal shows more growth on the adaxial or morphologically upper surface, the organ curves downwards.

(b) **Paratonic or induced movements**

These movements are of the following types:

- **Tropic movements** (tropism) The direction of this movement is always related to the direction of stimulus. Depending upon the nature of stimuli, these movements are of the following types:

Phototropism (heliotropism) When a plant organ curves due to unilateral light stimulus, it is called phototropism. Some parts of the plant, e.g. stem moves towards light. These organs are called positively phototropic. Some other organs, e.g. roots move away from light and they are called negatively phototropic.

Geotropism Growth movements induced by the stimulus of gravity are known as geotropism. Growth curvature towards the gravitational force is known as positive geotropism (main root) and growth away from the gravitational force is known as negative geotropism, (stem).

Hydrotropism The movements in response to external stimulus of water are termed as hydrotropism. Roots are positively hydrotropic, i.e. bend towards the source of water.

Thigmotropism (Haptotropism) These movements are due to the contact with a foreign body. e.g. Tendrils of *Cucurbita*, petiole of *Clematis,* etc.

Chemotropism It is a directional paratonic movement of curvature due to the chemical stimulus. For example, the movement of pollen tube towards ovary.

Thermotropism Curvature of plant parts towards normal temperature from very high or very low temperature. e.g. peduncles of *Tulip*.

- **Nastic movements** These movements are due to stimulus of light, temperature or contact.

  Seismonastic movements These movements occur due to touch, shock, rain, electric current, etc. e.g. leaves of sensitive plant *Mimosa pudica*.

  Nyctinastic (sleeping) **movements** The diurnal (changes in day and night) movements of leaves and flowers of some species, which take up sleeping position at night are called nyctinastic movements. Depending upon the stimulus, they may be photonastic (light stimulus) or thermonastic (temperature stimulus).

  Haptonastic movements When marginal glandular hair of *Drosera* come in contact with some foreign body, e.g. body of insect, they show haptonastic movements.

Chapter Exercise

1. Apical dominance means

(a) suppression of growth of apical buds by the axillary buds
(b) suppression of growth of axillary buds by the presence of apical buds
(c) stimulation of a growth of apical buds by the removal of axillary buds
(d) inhibition of growth of axillary buds by the removal of axillary buds

2. Which of the following made pineapple to flower in off season?

(a) Zeatin (b) Ethylene
(c) Temperature (d) Short-days

3. Which one of the following hormone is involved in phototropism?

(a) IAA (b) GA_3 (c) Kinetin (d) 2, 4-D

4. The synthetic hormone used as a weedicide is

(a) indole 3-acetic acid
(b) gibberellic acid
(c) 2,4-dichlorophenoxyacetic acid
(d) malic hydrazide

5. Induction of cell division and delay in senescence is done by

(a) cytokinin (b) kinetin (c) gibberellin (d) auxin

6. Which one of the following is an agent orange?

(a) Weedicide with dioxin
(b) Chemical used in luminous paint
(c) Biodegradable insecticide
(d) Colour used in fluorescent lamp

7. A hormone used for inducing morphogenesis in plant tissue culture is

(a) abscisic acid (b) gibberellins
(c) cytokinins (d) ethylene

8. Highest concentration of auxin exists

(a) at the base of various plant organs
(b) in growing tips of plants
(c) in leaves
(d) in xylem and phloem cells only

9. Which of the following is the commercial use of auxin?

(a) Apical dominance
(b) Enhancement of RNA
(c) Root formation
(d) Protein synthesis

10. Which of the following is an auxin?

(a) Malic acid (b) Abscisic acid
(c) Pyruvic acid (d) Indole acetic acid

11. The plant substances which become inhibitory on accumulation are

(a) abscisic acid (b) phenolic inhibitors
(c) auxins (d) cytokinins

12. Which one is the test for gibberellin?

(a) Bolting in cabbage
(b) Morphogenesis in tobacco callus
(c) Rapid divisions in carrot cells
(d) Elongation of oat coleoptile

13. The phenomenon of growth is immediately followed by

(a) senescence (b) differentiation
(c) development (d) maturation

14. Vernalisation is applicable to plants like

(a) *Chrysanthemum* (b) *Eucalyptus*
(c) *Narcissus* (d) *Gladiolus*

15. Dwarfness can be controlled by treating plants with

(a) cytokinin (b) gibberellic acid
(c) auxin (d) antigibberellin

16. Which one of the following growth regulator is used for rooting the leaves of *Bryophyllum*?

(a) IAA (b) Gibberellic acid
(c) Carotene (d) Lycopene

17. Indole-3-acetic acid, called auxin, was first isolated from

(a) human urine (b) corn germ oil
(c) *Fusarium* (d) *Rhizopus*

18. Sprouting of potato in storage can be prevented by

(a) malic hydrazide (b) 2,4-D
(c) indole acetic acid (d) naphthalene acetic acid

19. **Which of the following physiological effect is caused by gibberellins in plants?**

(a) Lengthening of internodes (b) Maleness in plants
(c) Femaleness in plants (d) Chlorophyll preservation

20. **Growth inhibitor hormone is**

(a) auxin (b) abscisic acid
(c) gibberellin (d) naphthalene acetic acid

21. **A plant hormone is**

(a) an ion that alters turgor pressure
(b) a pigment that responds to environmental changes
(c) a chemical messenger that coordinates body cells
(d) a secondary metabolic compound

22. **On the basis of excision experiments, it is concluded that the**

(a) shoot gibberellins are derived from roots
(b) root gibberellins are derived from fruits
(c) root gibberellins are derived from shoot
(d) root and shoot gibberellins are derived from fruits

23. **Abscisic acid is primarily synthesised in**

(a) lysosomes (b) Golgi complex
(c) chloroplast (d) ribosomes

24. **In tissue culture differentiation of root and shoot can be controlled by**

(a) modifying auxin and cytokinin ratios in the medium
(b) using tissue of right size
(c) giving temperature shocks
(d) change in light intensity

25. **Who first suggested the presence of growth regulatory chemicals in plants?**

(a) Went (b) Sachs
(c) Darwin (b) Paal

26. **Seed dormancy occurs due to the presence of**

(a) ethylene (b) abscisic acid
(c) IAA (d) starch

27. **Stimulation of cell division is concerned with**

(a) cytokinin (b) gibberellin
(c) auxin (d) florigen

28. **2, 4-D stands for**

(a) 2, 4-dimethyl phenoxy acetic acid
(b) 2-methyl-4-chlorophenoxy acetic acid
(c) 2-chloro-4 methyl phenoxy acetic acid
(d) 2,4-dichlorophenoxy acetic acid

29. **Gibberellins causes**

(a) curvature of coleoptile (b) elongation of internodes
(c) initiation of lateral roots (d) cell division

30. **Increased availability of abscisic acid stimulates**

(a) lipid formation
(b) protein formation in cell
(c) starch synthesis in guard cells
(d) cellulose synthesis

31. **Negative geotropism in horizontal stems is caused by**

(a) accumulation of auxins on the lower side
(b) accumulation of auxins on the upper side
(c) cell shrinkage on the lower side
(d) cell enlargement on the upper side

32. **Seedless fruits can be made by treating the unpollinated ovaries with**

(a) auxins (b) colchicine
(c) sucrose solution (d) pure lanolin

33. **Cut or excised leaves remain green for long if induced to root or dipped in**

(a) gibberellins (b) cytokinins
(c) auxins (e) None of these

34. **Which of the following phytohormone does not occur naturally in plants?**

(a) 2, 4-D (b) Gibberellic acid
(c) 6-furfuryl amino purine (d) IAA

35. **Growth rhythms are set in a plant when it is exposed to**

(a) CO_2 environment (b) light
(c) low temperature (d) UV radiation

36. **Etiolation of plants is caused when they**

(a) are grown in dark (b) have mineral deficiency
(c) are grown in intense light (d) are grown in blue light

37. **In short-day plants, flowering is inhibited by**

(a) interruption of dark by white or red light
(b) dark interruption by far-red light
(c) dark interruption red light followed by far-red light
(d) not possible

38. **Bakanae disease is caused by**

(a) fungus (b) algae (c) bacterium (d) virus

39. **Growth is a**

(a) irreversible increase in size
(b) reversible increase in size
(c) reversible increase in shape
(d) increase in volume

40. **The seed which possess the ability to germinate in suitable favourable condition is called**

(a) quiescent (b) viable
(c) dry (d) dormant

41. **Dormancy of seed is**

(a) phase of active growth (b) phase of Inactive growth
(c) phase stimulated growth (d) dead condition

42. **The growth in dormant organs is**

(a) completely stopped (b) continuous
(c) temporarily arrested (d) permanently arrested

43. **Dormancy in seeds of *Xanthium* is due to**

(a) seed coat imperable to water
(b) seed coat imperable to gases
(c) need for after ripening period
(d) mechanically resistant seed coat

44. Softening or rupturing of hard seed coat for breaking dormancy is called

(a) villing (b) vernalization
(c) stratification (d) scarification

45. The chemical substances like phenolics, caumarins, ferulic acid are

(a) growth hormones (b) growth inhibitors
(c) germination inhibitors (d) germination promoters

46. Which of the following growth regulator is not used to break dormancy?

(a) Ethylene (b) Gibberellins
(c) Cytokinins (d) Abscisic acid

47. Which of the following is a growth inhibitor?

(a) Auxin (b) Cytokinin
(c) Gibberellins (d) ABA

48. The phase of growth which is the first phase and represents lag phase of growth curve is

(a) formative phase (b) cell enlargement phase
(c) maturation phase (d) stationary phase

49. The lateral meristem in plant is responsible for

(a) primary growth (b) secondary growth
(c) exponential growth (d) growth in elongation

50. The rate of growth is called as

(a) growing index (b) increasing index
(c) effective index (d) efficiency index

51. Auxins were first isolated from the plants by

(a) Darwin (b) FW Went
(c) Boysen-Jensen (d) Sachs

52. Apical dominance is due to

(a) abscisic acid (b) gibberellic acid
(c) auxin (d) cytokinin

53. Hormone related with cell divisions is

(a) NAA (b) IAA
(c) Cytokinin (d) GA_3

54. Which property is specifically attributed to gibberellins?

(a) Apical bud growth inhibition
(b) Root initiation in stem cuttings
(c) Elongation of genetically dwarf plants
(d) Act as selective weedicide

55. Which of the following can stop flowering in short day plants?

(a) Longer but discontinuous dark period
(b) Photoperiod longer than critical duration
(c) Very short period
(d) Any of these

56. When night temperature is low during winter season, the size of potato tuber will

(a) increase (b) decrease
(c) Both (a) and (b) d) no effect

57. If the seedlings are grown in darkness they are

(a) similar to those grown in light
(b) taller than those grown in light
(c) of the same size as those grown in light
(d) much healthier than those grown in light

58. For producing flowers, the plant must pass the following sequence of phases

(a) photophase $\rightarrow$ thermophase $\rightarrow$ flowering
(b) thermophase $\rightarrow$ photophase $\rightarrow$ thermophase $\rightarrow$ flowering
(c) photophase $\rightarrow$ thermophase $\rightarrow$ photophase $\rightarrow$ flowering
(d) thermophase $\rightarrow$ photophase $\rightarrow$ flowering

59. Many transplanted seedlings may not survive because

(a) they do not like the new soil
(b) most of the root hair are lost during transplantation
(c) they do not get the required mineral salts
(d) the leaves get damaged

60. Bananas can be prevented from over ripening by

(a) maintaining them at room temperature
(b) refrigeration
(c) dipping in ascorbic acid solution
(d) storing in a freezer

61. With the increase in intensity of light, the NAR

(a) increases (b) decreases
(c) remains the same (d) becomes zero

62. Match the following columns.

Column I	Column II
A. Girdling experiment	1. Translocation in phloem
B. Cobalt-chloride paper method	2. Unequal transpiration on leaf surfaces
C. Crescograph	3. Growth
D. Bell-jar experiment	4. Transpiration

Codes

	A	B	C	D		A	B	C	D
(a)	1	2	3	4	(b)	4	3	1	2
(c)	4	1	2	3	(d)	2	1	4	3

63. In decapitated plants, axillary buds are activated because

(a) auxin is made available to them
(b) more food material is available to them
(d) amount of cytokinin increase in them
(d) they get more light

64. Seedless fruits can be made by treating the unpollinated ovaries with

(a) auxins (b) colchicine
(c) sucrose solution (d) pure lanolin

65. Which hormone plays an important role in the formation or roots?

(a) IAA (b) ABA
(c) NAA (d) 2, 4-D

66. Negative geotropism in horizontal stems in caused by
(a) accumulation of auxins on the lower side
(b) accumulation of auxins on the upper side
(c) cell shrinkage on the lower side
(d) cell enlargement on the upper side

67. TIBA is
(a) a gibberellin (b) an antiauxin
(c) an auxin (d) a kinetin

68. High concentration of synthetic auxins would
(a) prevent lateral buds to grow
(b) kill plants
(c) control cell division
(d) cause phototropism

69. The fungus causing foolish seedling desease, was discovered by
(a) a Japanese farmer (b) an Indian botanist
(c) a Japanese pathologist (d) an American botanist

70. The impermeability to oxygen by the seed coat can be found in
(a) Leguminous seeds (b) *Brassica* seeds
(c) *Xanthium* seeds (d) *Pisum* seeds

71. If a growing plant is decapitated
(a) axillary buds are inactivated
(b) axillary buds are activated
(c) leaves become yellow and have a tendency to fall off
(d) growth stops

72. The movement of cytokinins is
(a) non-polar and acropetal
(b) polar and acropetal
(c) polar and basipetal
(d) non-polar and basipetal

73. Most of the plants are seasonal due to
(a) photoperiodism (b) phototropism
(c) photosynthesis (d) photolysis

74. Best flowering occurs in
(a) red light (b) day light
(c) green light (d) blue light

75. Vernalisation treatment can be given to partially germinated seeds in one of the following plants
(a) wheat (b) rye (c) rice (d) cotton

76. Vernalisation is
(a) growth curve related to light
(b) effect of photoperiods of plant growth
(c) speeding up ability to flower by low temperature treatment
(d) diurnal photoperiodicity

77. Hormone responsible for vernalisation is
(a) florigen (b) caulocaline
(c) abscissin (d) vernalin

78. The term 'vernalin' was coined by
(a) Lysenko (b) Chourad (c) Purvis (d) Melchers

79. The technique of vernalization has found its use in
(a) India (b) Australia (c) Mexico (d) Russia

80. In annuals, the site of vernalisation is the
(a) root (b) seed
(c) hypocotyl (d) flower bud

81. For producing flowers, vernalisation overcomes
(a) thermophase (b) photophase
(c) developmental phase (d) growth phase

82. Devernalisation can be done by
(a) ary air (b) heat treatment
(c) red light exposure (d) chilling treatment

83. The banyan (*Ficus benghalensis*) tree shows
(a) whole tree senescence
(b) shoot senescence
(c) synchronous leaf senescence
(d) sequential leaf senescene

84. The first visible sign of senescence is
(a) yellowing of leaves (b) curling of leaves
(c) bunchy top (d) contraction of leaves

85. In the lag phase, the size of the cell......... exponentially with time.
(a) increases (b) decreases
(c) both 'a' and 'b' (d) None of these

Answers

1. (b)	**2.** (d)	**3.** (a)	**4.** (c)	**5.** (a)	**6.** (a)	**7.** (c)	**8.** (b)	**9.** (c)	**10.** (d)
11. (b)	**12.** (d)	**13.** (d)	**14.** (a)	**15.** (b)	**16.** (d)	**17.** (a)	**18.** (d)	**19.** (b)	**20.** (b)
21. (c)	**22.** (a)	**23.** (c)	**24.** (a)	**25.** (c)	**26.** (b)	**27.** (a)	**28.** (d)	**29.** (b)	**30.** (c)
31. (a)	**32.** (a)	**33.** (b)	**34.** (a)	**35.** (b)	**36.** (a)	**37.** (a)	**38.** (a)	**39.** (a)	**40.** (b)
41. (b)	**42.** (c)	**43.** (b)	**44.** (d)	**45.** (c)	**46.** (d)	**47.** (d)	**48.** (a)	**49.** (b)	**50.** (d)
51. (b)	**52.** (c)	**53.** (c)	**54.** (c)	**55.** (d)	**56.** (b)	**57.** (b)	**58.** (c)	**59.** (c)	**60.** (c)
61. (a)	**62.** (a)	**63.** (a)	**64.** (a)	**65.** (a)	**66.** (a)	**67.** (d)	**68.** (a)	**69.** (c)	**70.** (c)
71. (b)	**72.** (c)	**73.** (a)	**74.** (b)	**75.** (a)	**76.** (c)	**77.** (d)	**78.** (d)	**79.** (d)	**80.** (b)
81. (a)	**82.** (b)	**83.** (d)	**84.** (a)	**85.** (a)					

32

Role of Microbes in Human Welfare (Part I)

Microorganisms are the well known microscopic, single-celled, tiny creatures that are individually too small to be seen with the naked eyes. These constitute, the major groups of biological world and are found everywhere including soil, water, plants, animals, humans, etc. While many microbes are the source of most infectious diseases, most of them are also used by the humans in their daily life for their welfare in either one or the another way out.

Thus, these days microorganisms have become a part of many useful articles, tools, antibiotics, organic compounds, etc. These microorganisms ferment sugar, acids, gases or alcohol. Fermentation was first described by Louis Pasteur. Some common microbes and their useful products used in everyday life are discussed below:

Microbes in Household Products

For thousands of years, the microorganisms have been used to make products such as curd, cheese, butter, vinegar, alcohol, etc.

Some important products produced by microorganisms are given below:

- **Curd microorganisms** such as *Lactobacillus* and others commonly called **Lactic Acid Bacteria** (LAB) grow in milk and convert it to curd. LAB produces acids that coagulate and partially digest the milk proteins.

  It also involves nutritional quality by increasing vitamin-B_{12}. In our stomach too, the LAB plays a very beneficial role in checking disease causing microbes.
- **Dosa and idli** are also fermented by bacteria. The dough which is used for making bread, is fermented by using baker's yeast (*Saccharomyces cerevisiae*). A number of traditional drinks and foods are also made by fermentation by the microbes.
- **Toddy,** a traditional drink found in some parts of southern India is made by fermenting sap from palm's tree.
- Microbes are also used to ferment fish, soyabean and bamboo shoots to prepare different food items.
- **Cheese** is one of the oldest food items in which microbes are used. The large holes in 'Swiss cheese' are due to the production of a large amount of CO_2 by a bacterium named as *Propionibacterium sharmanii*.
- The 'Roquefort cheese' is ripened by growing a specific fungus on them which gives them a particular flavour. In order to prepare special cheeses, such as blue Roquefort and soft camembert, the blue mould *Penicillium* is commonly used.
- *Saccharomyces cerevisiae*, the common yeast is used in baking as a leavening agent where it converts the sugars present in dough into carbon dioxide.
- **Bread** is prepared when yeast is mixed with flour, salt and warm water or milk. The dough is kneaded until it is smooth, then left to doubled in size for sometime.
- The microorganisms such as *Streptococcus lactis* and *Leuconostoc citrivorumare* are used to prepare **butter** by churning of sweet or sour cream. The characteristic butter aroma develops due to a volatile substance, i.e. diacetyl. It is produced by the action of *Streptococcus* on pasteurised milk.

  Apart from all the food stuffs, bacteria are also used in the separation of fibres of flax, hemp and jute. For this purpose, the stems of plants are submerged in water, where the bacterial activity results in the rotting of softer parts. The tough bast fibres become loosened and easily separated from each other.

Industrial Uses of Microorganisms

Microorganisms such as bacteria, fungi, yeasts, etc are now used in a wide range of industrial processes. The study of microorganisms in industrial products production is known as **industrial microbiology**.

The microorganisms are usually cultured in large fermentation chambers known as **fermentors** under controlled conditions.

The important products manufactured by microorganisms are antibiotics, alcohol, various organic acids, enzymes and other bioactive molecules.

Antibiotics

These are the chemical substances being produced by some microbes that can kill or can retard the growth of other disease causing microbes. The term antibiotic was first coined by Waksman (1942). Alexander Fleming discovered the first antibiotic called penicillin, while working on the bacteria *Staphylococci*. The antibiotics are mainly obtained from eubacteria, Actinomycetes, fungi, etc..

Some Common Antibiotics Synthesised by Microbes and their Uses

Antibiotics	Obtained from	Used against
Neomycin	*Streptomyces fradiae*	Gram positive, Gram negative and TB bacteria.
Amphomycin	*S. carus*	Gram positive bacteria.
Amphotericin	*S. nodosus*	Yeast and fungi.
Leucomycin	*S. bitasoensis*	Gram positive bacteria.
Trichomycin	*S. hachijoensis*	Yeast and fungi.
Viomycin	*S. floridae*	Gram positive, Gram negative and TB bacteria.
Bacitracin	*Bacillus subtilis*	Gram positive bacteria.
Gramicidin	*B. brevis*	Grape positive bacteria.
Tyrothricin	*B. brevis*	Gram positive and Gram negative bacteria.
Polymyxin B	*Aerobacillus polymyxa*	Gram negative bacteria.

Alcoholic Beverages

These are the fermented liquor say wine, beer or distilled spirit containing ethanol or ethyl alcohol as an intoxicating agent, i.e. It is described one of the most exotic oxygen containing organic chemical due to its unique combination of properties as solvent, a grumicide, a beverage an antifreeze at well and a versatile chemical intermediate for other organic chemical. Alcoholic beverages are fermented from the sugars in fruits, berries, grain and such other ingredients as plant saps, tubers, honey, milk and may be distilled to reduce the original water liquid to a liquid of much greater alcoholic strength, under the action of microorganisms mainly yeasts species such as *Saccharomyces cerevisiae* which ferments the sugar common from different sources.

Some Common Alcoholic Beverages with their Raw Materials

Name	Raw material	Percentage of alcohol
Undistilled alcohols (simple fermentation products)		
	Fruit juice and grains	03-24 %
Cider	Apple juice and sugar	03-06%
Beer	Barley (malt)	04-05%
Claret (red wine)	Grape juice	07-13%
Champagne (white wine)	Grape juice	08-10%
Port (heavy, sweet, fortified, wine, dark, red or white)	Grape juice	15-24%
Sherry (fortified wine, plate gold to dark brown)	Grape juice	18-24%
Toddy	Palm sap	–
Arrack	Coco, palm and rice	–
Distilled alcohols (Distilled fermented lines to raise strength)		
Hollands	Rye	40%
Brandy	Apple, peach and cherry juice	40-50%
Whisky	Barley, rye, maize and potatoes	40-50%
Rum	Molasses	45-55%
Gin	Barley	40-55%
Vodka (used in Russia)	Rye	40-55%
Sake (used in Japan)	Rice	40-55%

Organic Acids

These are the organic compounds having acidic properties. They are produced industrially in microbial fermentations by the action of different microbes such as yeast, bacteria, fungi, etc.

Organic Acids Synthesised by Various Microbes

Organic acids	Microorganisms
Lactic acid	*Lactobacillus delbreukie, L. bulgaricus, Streptococcus lactis* and *Rhizopus* sp.
Acetic acid (vinegar)	*Acetobacter aceti*
Citric acid	*Aspergillus niger, Penicillium* sp. and *Mucor* sp.
Gluconic acid	*Acetobacter aceti, Aspergillus niger, Penicillium* and *Chrysogenum*
Propionic acid	*Bacterium propioni*
Butyric acid	*Clostridium acetobutyricum*
Oxalic acid	*Aspergillus* sp.
Gallic acid	*Aspergillus niger*
Same amino acids	*Escherichia coli*

Enzymes

Fermentation is also used as a method of generating enzymes for the various industrial purposes. Many industries now-a-days are dependent on enzymes for the production of their goods and articles.

Enzymes Synthesised by Various Microbes

Enzymes	Microorganisms
Amylase	*Bacillus subtilis, B. mecarans* and *B. diastoticus* (Bacteria). *Aspergillus oryzae, A. niger* and *Rhizopus oryzae* (Fungi).
Protease	*Bacillus subtilis* and *B. licheniformis* (Bacteria). *Aspergillus oryzae, Penicillium roqueforti* and *Mortieralla reinspora* (Fungi).
Pectinase	**Fungi—*Aspergillus niger* and *Byssochlamys fulvo.***
Cellulase	*Myrothecium verrucasia, Trichoderma reesi* and *Aspergillus niger* (Fungi).
Lactase	*Saccharomyces fragilis* and *Torula cremoris* (Fungi).
Lipase	*Candida lipolytica, Geotrichum candidum, Aspergillus niger* and *Mucor javanicus* (Fungi).
Invertase (sucrase)	*Saccharomyces cerevisiae* (Fungi).
Streptokinase	Haemolytic bacterial species of *Streptococcus.*
Renin/Rennet	*Mucor pusilus* and *Endothia parasitica* (Fungi).
Glucose oxidase	*Aspergillus niger* and *Penicillium notatum* (Fungi).

Other Important Products

Apart from all the compounds and products mentioned previously, some other important products are also produced industrially by the mechanism of fermentation.

Some Other Important Products of Microorganisms

Products	Microorganisms
Cyclosporin-A 11 membered cyclic oligopeptide, an immunosuppressive that inhibits the activation of T-cell response to transplanted organs.	*Trichoderma polysporum* and *Tolypocladium inflatum* (Fungus).
Statins Inhibitor of enzyme HMG Co-A reductase of liver which also lowers LDL cholesterol level.	*Monascus perpureus* (Yeast).
Vitamin-B_2 (riboflavin) synthesised by intestinal bacteria.	*Ashbya gossypii* (Mould). *Eremothecium ashbye* (Yeast), *Clostridium butyricum, C. acetobutylicum* (Bacteria).
Vitamin-B_{12} (cobalamin/cyanocobalamin).	Bacteria (*Propionibacterium frendeureichii, Pseudomonas denitrificans, Bacillus megathelium, Streptomyces oliaecus*).
Vitamin-C (L-ascorbic acid).	*Acetobacter.*

Microbes in Waste Treatment

Microorganisms are also used to treat the vast quantities of wastes generated by modern and rural societies. Bio-treatment, the processing of wastes using living organisms, is an environmental friendly, relatively simple and cost effective alternative to physico-chemical clean up options. Bioreactors are the equipments or vessels employed in bio-treatment processes.

Chapter Exercise

1. Sterilisation is an autoclaving that
   (a) destroys the bacterial activity
   (b) denatures the enzymes
   (c) Both (a) and (b)
   (d) None of the above

2. An antifungal drug is obtained from
   (a) *Penicillium griseofulvum* (b) *P. camemberti*
   (c) *P. roqueforti* (d) *P. notatum*

3. First hormone obtained from genetically engineered bacteria is
   (a) adrenaline (b) thyroxine
   (c) insulin (d) testosterone

4. *Saccharomyces cerevisiae* is used in
   (a) tanning (b) brewing
   (c) baking (d) Both (b) and (c)

5. Father of fermentation is
   (a) Louis Pasteur (b) Nageli
   (c) Porter (d) None of these

6. Curd, cheese and butter are produced by
   (a) Yeast (b) *Penicillium*
   (c) *Streptococcus* (d) None of these

7. Distribution of clean and quality milk throughout the world has been made possible through the work of
   (a) Leeuwenhoek (b) Koch
   (c) Pasteur (d) Blackman

8. During alcoholic fermentation, conversion of sugar into alcohol is due to direct action of
   (a) temperature
   (b) microorganisms
   (c) concentration of sugar solution
   (d) zymase enzyme

9. Vitamin-B_2 is produced by
   (a) *Propionibacterium* (b) *Ashbya gossypii*
   (c) *Saccharomyces* (d) *Rhizopus*

10. Vinegar is produced by
    (a) two step processes, i.e. first fermentation of sugar by yeast and second fermentation of ethyl alcohol by acetic acid bacteria
    (b) fermentation of sugar by *Lactobacillus*
    (c) fermentation of sugar by *Aspergillus*
    (d) fermentation of sugar by *Saccharomyces*

11. Father of microbiology is
    (a) Robert Koch (b) Lister
    (c) Pasteur (d) Robert Brown

12. Distillation of wine is necessary as it
    (a) increases quality
    (b) prevents further fermentation and spoilage
    (c) decreases toxicity
    (d) None of the above

13. Tetracycline is extracted from
    (a) *Streptomyces ramosus*
    (b) *Streptomyces aurofaciens*
    (c) *Streptomyces scolens*
    (d) *Bacillus lichenformis*

14. Cyclosporin is obtained from
    (a) *Bacillus subtilis*
    (b) *Trichoderma polysporum*
    (c) *Tolypocladium inflatum*
    (d) Both (b) and (c)

15. An amylase enzyme is produced by fungus
    (a) *Aspergillus oryzae* (b) *Aspergillus flavus*
    (c) *Aspergillus niger* (d) *Trichoderma polysporum*

16. Which vitamin is synthesised by bacteria in human gut?
    (a) Vitamin-A (b) Vitamin-C
    (c) Vitamin-D (d) Vitamin-K

17. A broad spectrum antibiotic is
    (a) penicillin (b) streptomycin
    (c) erythromycin (d) All of these

18. A bioreactor is
    (a) a fermentation tank
    (b) culture containing radioactive isotopes
    (c) culture for synthesis of new chemicals
    (d) hybridoma

19. Streptomycin is used for the treatment of
    (a) Gram + ve
    (b) Gram – ve
    (c) Both (a) and (b)
    (d) Any type of bacterial infection

20. Penicillin was discovered by Alexander Fleming in the year
    (a) 1901-02 (b) 1919-20
    (c) 1928-29 (d) 1899-1900

21. Fermentation to produce alcohol by yeast (*Saccharomyces*) is due to
    (a) amylase (b) invertase
    (c) zymase (d) All of these

22. *Streptococcus* is employed in the manufacture of
    (a) wine (b) bread
    (c) cheese (d) All of these

23. Which is not correct about antibiotics?
(a) Fleming discovered the first commercial antibiotic
(b) Waksman coined the term 'antibiotic' in 1942
(c) Allergy may develop against an antibiotic
(d) An antibiotic is effective against one specific pathogen

24. *Clostridium butyricum* has been used in the synthesis of
(a) vitamin-B (b) vitamin-A
(c) vitamin-C (d) vitamin-D

25. *Mucor javonicus* synthesises
(a) lipase (b) amylase
(c) protease (d) streptokinase

26. Dosa and idli are prepared by the action of
(a) *Lacto bacillus*
(b) *Saccharomyces cerivisae*
(c) *Bacillus subtilis*
(d) *Rhizopus oryzae*

27. The microorganisms grown on molasses and sold as a food flavouring substance is
(a) *Saccharomyces* (b) *Rhizopus*
(c) *Acetobacter* (d) *Lactobacillus*

28. Tetracycline is extracted from
(a) *Streptomyces ramosus* (b) *S. aurofaciens*
(c) *S. scolens* (d) *Bacillus lichenformis*

29. Anaerobic respiration of yeast produces
(a) alcohol
(b) carbon dioxide
(c) alcohol, carbon dioxide and other beverages
(d) None of the above

30. Antibiotics are mostly obtained from
(a) fungi (b) Actinomycetes/bacteria
(c) Both (a) and (b) (d) cyanobacteria

31. An important drug is obtained from the bark of
(a) *Withania* (b) *Cinchona*
(c) *Papaver* (d) *Momordica*

32. Morphine, which is used as an analgesic is obtained from
(a) *Taxus brevifolia* (b) *Papaver somniferum*
(b) *Cinchona officinalis* (d) *Berberis nilghiriensis*

33. The term antibiotic was coined by
(a) Howard Florey (b) John Tyndall
(c) Selman Waksman (d) Gerhard Domagk

34. The yeast that is used in alcohol production and bread making is
(a) *Escherichia coli* (b) *Saccharomyces cerevisiae*
(c) *Bacillus subtilis* (d) *Pseudomonas putida*

35. Which of the following bacterium produces butyric acid?
(a) *Acetobacter aceti* (b) *Clostridium butylicum*
(c) *Lactobacillus* (d) None of the above

36. Which one of the following is a wrong matching of a microbe and its industrial product, while the remaining three are correct?
(a) *Clostridium butylicum* – Lactic acid
(b) *Aspergillus niger* – Citric acid
(c) Yeast – Statins
(d) *Acetobacter aceti* – Acetic acid

37. Ethanol is commercially produced through a particular species of
(a) *Aspergillus* (b) *Saccharomyces*
(c) *Clostridium* (d) *Trichoderma*

38. Bread becomes porous due to release of CO_2 by the action of
(a) yeast (b) bacteria
(c) virus (d) protozoans

39. Yeast cannot ferment starch and complex carbohydrates because they
(a) lack diastase (b) lack lipase
(c) lack zymase (d) lack invertase

40. Yeast grows abundantly on
(a) cowdung and cow urine
(b) organic matter rich in sugar
(c) organic matter rich in fats
(d) nectar

41. Gluconic acid is produced by fungi
(a) *Aspergillus niger* (b) *A. purpurogenum*
(c) Both (a) and (b) (d) *A. fumigatus*

42. Which one is not correctly matched?
(a) *Streptomyces* — Antibiotic
(b) *Serratia* — Drug addition
(c) *Rhizobium* — Biofertiliser
(d) *Spirulina* — Single cell protein

43. Match the following columns.

Column I	Column II
A. *Aspergillus niger*	1. Lactic acid
B. *Acetobacter aceti*	2. Butyric acid
C. *Clostridium butylicum*	3. Acetic acid
D. *Lactobacillus*	4. Citric acid

Codes

	A	B	C	D
(a)	2	3	4	1
(b)	2	4	3	1
(c)	4	3	2	1
(d)	4	1	3	2

44. The bacterium which helps in the degradation of man-made and natural organic pollutants is
(a) *Pseudomonas* (b) *Clostridium*
(c) *Leuconostoc* (d) None of these

45. Match the following columns.

Column I	Column II
A. *B. propioni*	1. Gallic acid
B. *Aspergillus*	2. Propionic acid
C. *E. coli*	3. Butyric acid
D. *C. acetobutyricum*	4. Oxalic acid

Codes

	A	B	C	D		A	B	C	D
(a)	1	2	4	4	(b)	4	1	2	3
(c)	2	4	1	3	(d)	2	1	4	3

46. An antifungal drug is obtained from

(a) *P. griseofulvum*
(b) *R. camemberti*
(c) *R. roqueforti*
(d) *P. notatum*

47. Which of the following is not correct?

(a) Cyanobacteria are photosynthetic autotrophs
(b) Both photosynthesis and nitrogen fixation take place in cyanobacteria
(c) Heterocysts and hormogone are found in archaebacteria
(d) Mesosomes are found in eubacteria

48. During alcoholic fermentation, conversion of sugar into alcohol is due to direct action of

(a) temperature
(b) pH
(c) concentrate of sugar solution
(d) zymase enzyme

49. Sausages are fermented

(a) vegetables (b) meats
(c) milk (d) sauce

50. Which group is responsible for the formation and flavour of yoghurt?

(a) *Bacillus megathermus* and *Xanthomonas* sp.
(b) *Bacillus subtilis* and *E. coli*
(c) *Rhizobium meliloti* and *Azotobacter*
(d) *Lactobacillus casei* and *Streptococcus thermophilus*

51. An enzyme produced commercially from *Saccharomyces cerevisiae* is

(a) lactase (b) invertase
(c) amylase (d) maltase

52. During fermentation by yeast, alcohol is produced from

(a) sugars (b) proteins
(c) mucopolysaccharides (d) fats

53. Cheese and yogurt are the products of

(a) *S.greisas* (b) *Lactobacillus*
(c) *Streptococcus* (d) *R. oryzae*

54. Which is used up in the production of alcohol?

(a) *Streptomyces cerevisiae* (b) *Torulopsis utilis*
(c) *Clostridium botulinium* (d) *Leuconostoc citrovorum*

55. Match the following columns.

Column I	Column II
A. Neomycin	1. *Streptomyces fradiae*
B. Terramycin	2. *Penicillium notatum*
C. Viridin	3. *Streptomyces ramosus*
D. Penicillin	4. *Gliocladium virens*

Codes

	A	B	C	D		A	B	C	D
(a)	3	1	4	2	(b)	3	3	1	2
(c)	1	3	4	2	(d)	4	1	3	3

Answers

1. (c)	**2.** (a)	**3.** (c)	**4.** (d)	**5.** (a)	**6.** (c)	**7.** (c)	**8.** (d)	**9.** (b)	**10.** (a)
11. (c)	**12.** (a)	**13.** (b)	**14.** (d)	**15.** (a)	**16.** (d)	**17.** (d)	**18.** (a)	**19.** (c)	**20.** (c)
21. (c)	**22.** (c)	**23.** (d)	**24.** (a)	**25.** (a)	**26.** (c)	**27.** (a)	**28.** (b)	**29.** (c)	**30.** (b)
31. (b)	**32.** (b)	**33.** (c)	**34.** (b)	**35.** (b)	**36.** (a)	**37.** (b)	**38.** (a)	**39.** (a)	**40.** (b)
41. (a)	**42.** (b)	**43.** (c)	**44.** (a)	**45.** (c)	**46.** (d)	**47.** (c)	**48.** (d)	**49.** (b)	**50.** (d)
51. (b)	**52.** (a)	**53.** (c)	**54.** (a)	**55.** (c)					

33

Microbes in Healthcare and Medicine

'Microbes' are also used to produce different products of healthcare and in the treatment of various diseases in the field of medicines such as insulin, growth hormones and antibodies. They are also helpful in the treatment of diseases such as cancer, etc.

Some common examples of different microorganisms producing important products are given below:

- Research shows that clostridia can selectively target cancer cells. Various strains of non-pathogenic clostridia have been shown to infiltrate and replicate within solid tumours. Clostridia, therefore, have the potential to deliver therapeutic proteins to tumours.
- *Lactobacillus* species and other lactic acid bacteria possess numerous potential for therapeutic properties including anti-inflammatory and anti-cancer activities.
- Many bacteria are now used in the preparation of serum and vaccines. These substances induces immunity to various diseases in man.
- Serums are effective against certain diseases like diphtheria, pneumonia, etc.
- Vaccines are effective against diseases like typhoid, smallpox, cholera, etc.

Ergot, a most useful drug is obtained from sclerotia of *Claviceps purpurea*, a fungus which is parasitic in grasses especially rye.

The disease is called **Ergot of rye**. The ergot contains an alkaloid, called **ergotinine** and many other substances like ergostetrine, ergotonic acid, cornutine, space lotoxin, etc. This drug stimulates muscles of uterus and used to assist childbirth. It is also used to increase blood pressure, controls uterine haemorrhage and for many other uterine disturbances.

Microorganisms as Biofertilisers

Biofertilisers are the nutrient materials been obtained from the living organisms or their remains and are used for enhancing the fertility and quality of the soil.

Microorganisms used as biofertilisers bring about soil nutrient enrichment, maximise the ecological benefits and minimise environmental hazards.

For this purpose, the specific microorganisms (bacteria, cyanobacteria, fungi, etc) are cultivated in scientific laboratories and then supplied to the farmers so that they can be utilised in the fields.

Microbes Biofertilisers

Plants sometimes shows the deficiency of an element called **phosphorus**, even if they grow in phosphorus rich soils. Because phosphorus in soil occurs either in immobile or in the bounded forms. Soil phosphates become available to the plants by soil microorganisms through the secretion of organic acids.

Microbes biofertilisers includes such microorganisms which solubilises the bound phosphate of rocks and soil to increase the availability of soluble phosphate to the plants. The main among such microorganisms are bacteria, e.g. *Pseudomonas striata*, *Bacillus polymyxa* and fungi, e.g. *Aspergillus*, etc.

The phosphate dissolving bacteria reduces the pH of the substrate by the secretion of number of organic acids such as formic, acetic, propionic, lactic, glycolic, fumaric and succinic acids. Thus, causing the release of soluble inorganic phosphate (H_2PO_4) into soil through the decomposition of phosphate rich organic compounds.

Advantages of using Biofertilisers in Agriculture

- *Azotobacter* and other microbes, biofertilisers increases about 15-35% additional yield in most of the vegetative crops.
- Beside fixing atmospheric nitrogen, cyanobacteria synthesises and excrete several growth hormones (auxin and ascorbic acid) and vitamin (vitamin-B_{12}), which enhance the seed germination and growth of crop plants.

- Some of the biofertilisers enhance crop yield even under ill irrigated conditions, where chemical fertilisers are of not much advantage.
- Biofertilisers are cheaper and more economical than the chemical ones and thus can be used even by poor farmers.
- Some biofertiliser excrete antibiotics and thus act as biopesticides.
- They improves the physical and chemical properties of the soil (water holding capacity, buffer capacity, etc).

Sources of Biofertilisers

Biofertilisers contains some organisms which can bring about nutrient availability to the crop plants. The main sources of biofertilisers are

(i) Nitrogen-fixing bacteria (free living and symbiotic)
(ii) Nitrogen-fixing cyanobacteria (free living and symbiotic)
(iii) Mycorrhizal-fungi

1. Nitrogen-Fixing Bacteria

These can be either free living or symbiotic in nature.

Free Living Nitrogen-Fixing Bacteria

Non-symbiotic (free living) nitrogen-fixing bacteria are found particularly in high concentrations in rhizosphere. These includes the members of *Klebsiella, Azotobacter, Clostridium, Rhodospirillum, Azospirillum*, etc.

Free living nitrogen-fixing bacteria may be of following main types:

(i) **Aerobic nitrogen-fixing bacteria** These help to maintain the reduced oxygen conditions (micro-aerobic conditions), through their high levels of respirations, e.g. *Azotobacter*.

(ii) **Anaerobic nitrogen-fixing bacteria** They do not require oxygen. These bacteria can survive only in anaerobic conditions, since oxygen is not present in their habitat.

(iii) **Facultative organisms** They are able to grow under both aerobic and anaerobic conditions. These organisms generally fix nitrogen only under anaerobic conditions.

(iv) **Anaerobic free living nitrogen-fixing bacteria** They can be either photosynthetic (e.g. *Rhodospirillum*) or non-photosynthetic (e.g. *Clostridium*). Free living nitrogen-fixing bacteria add about 10-25 kg of nitrogen per hectare per annum.

Symbiotic Nitrogen-Fixing Bacteria

Many plants form a symbiotic relationship with the nitrogen-fixing bacteria that contain an enzyme complex called nitrogenase which has the ability to reduce free nitrogen into ammonia, e.g. *Rhizobium* group of bacteria, which forms nodules on the roots of leguminous plants.

The nodules so formed contains an oxygen binding haem protein called **leghaemoglobin** (present in the cytoplasm of infected nodule cells at high concentration and helps in providing the nodules a pink colouration).

2. Nitrogen-Fixing Cyanobacteria

These can also be either free living or symbiotic in nature.

(i) **Free Living Nitrogen-Fixing Cyanobacteria**

Anabaena, Nostoc, Aulosira, Cylindrospermum, Trichodesmium, etc., are some free living nitrogen-fixing blue-green algae or cyanobacteria. They fix 20-30 kg of nitrogen per hectare of soil and water. *Aulosira fertilissima* is the most active nitrogen fixer in rice field, whereas *Cylindrospermum* is active in sugarcane and maize fields.

(ii) **Symbiotic Nitrogen-Fixing Cyanobacteria**

Anabaena and *Nostoc* species of cyanobacteria are common symbionts in lichens, *Anthoceros, Azolla* and cycad's roots. The water fern *Azolla* has *Anabaena azollae* as the symbiont, which fixes nitrogen within the fronds.

3. Mycorrhiza

Mycorrhizae (sing. mycorrhiza) are known to be the mutualistic association of a fungus and the roots of higher plants. They are found in most groups of vascular plants. They can be either **ectomycorrhiza** or **endomycorrhiza**.

Microorganisms as Biocontrol Agents

Microorganisms such as bacteria, fungi, viruses, protozoan, etc., and their products are used for pest control and are known as **biopesticides**.

These are the most important, effective and commercially viable agents as these are inexpensive, cause no pollution, pose no risk to human health and have a long term capability to control pests.

Bacteria as Biological Control

Bacillus thuringiensis is the most widely applied species of bacteria used for biological control with atleast three sub-species used to control Lepidopteran (moth and butterfly), Coleopteran (beetle) and Dipteran (true flies) insect pests.

Some strains of *Bacillus thuringiensis* produce proteins that kill certain insects. *B. thuringiensis* forms protein crystals during a particular phase of their growth. These crystals contain a toxic insecticidal protein.

The *Bt* toxin protein exist as inactive protoxins, but once an insect ingest the inactive toxin, it is converted into an active form of toxin due to alkaline pH of the gut which solubilise the crystals. The activated toxin binds to the surface of midgut epithelial cells and create pores, cause cell swelling, lysis and eventually death of the insect.

Doom is another bacterial pesticide, which is known as the a mixture of *Bacillus papillae* and *Bacillus lentimorbus*, and has been commercially used for controlling Japanese beetle Popilliae. *Bacillus sphaericus* is toxic to larvae of *Anopheles* mosquito.

Fungi as Biocontrol Agents

The first **bioherbicide** was a mycoherbicide which was based on a fungus *Phytophthora palmivora*.

First herbicide developed in 1981, is used to control the growth of milk weed in citrus orchards. Since, then fungi has been extensively used as biocontrol agent.

Fungi that causes diseases in insects are known as **entomopathogenic fungi**, including atleast fourteen species of entomophthoraceous fungi which attack aphids.

Species of the genus *Trichoderma* are used to manage some soil borne plant pathogens. *Beauveria bassiana* is used to manage different types of pests such as white flies, thrips, aphids and weevils.

Trichoderma viridae fungus has been used against Dutch Elm disease and to treat the spread of fungal and bacterial growth on tree wounds. It may also have potential as a means of combating silver leaf disease.

Nematodes as Biocontrol Agent

Recently, a chemical named Nemaslug is obtained from a microscopic nematode *Phasmarhabditis hermaphrodita* has been developed for the control of slugs. This nematode reproduce inside slugs and kill them.

The nematode is applied by watering into the moist soil thus giving protection for about six weeks in optimum conditions. Nemaslug nematodes are mainly effective with small and young slugs under the soil surface.

Microbes in Sewage Treatment

A large quantity of wastewater is generated everyday in cities and towns. The major component of this wastewater is human excreta. This municipal wastewater is also known as **sewage**, containing large amounts of organic matter and microbes.

In sewage, most of the microorganisms are pathogenic. This huge quantity of sewage or urban wastewater is discharged into natural water bodies like rivers and streams after treating in Sewage Treatment Plants (STPs) so, as to make it less polluting.

Treatment of wastewater is done by the heterotrophic microbes naturally present in the sewage.

The treatment of sewage is carried out in following two main stages:

1. Primary Treatment

This is the basic step of treatment that mainly involves the physical removal of particles, i.e. large and small, from the sewage through filtration and sedimentation. These are involved in stages.

Initially, floating debris is removed by sequential filtration then the grit (soil and small pebbles) is removed by sedimentation. All solids that settle, forms the primary sludge and the supernatant forms the effluent. The effluent from the primary settling tank is taken for the secondary treatment.

2. Secondary Treatment or Biological Treatment

The primary effluent is passed into large aeration tanks, where it is constantly agitated mechanically and air is pumped into it. This allows vigorous growth of useful aerobic microbes into flocs (masses of bacteria associated with fungal filaments to form mesh-like structures). While growing these microbes consume the major part of the organic matter in the effluent. This significantly reduces the BOD (Biochemical Oxygen Demand) of the effluent.

Once, the BOD of sewage or wastewater is reduced significantly. The effluent is then passed into a settling tank, where the bacterial 'flocs' are allowed to sediment. This sediment is now called the **activated sludge**.

A small part of the activated sludge is pumped back into the aeration tank to serve as the inoculum. The remaining major part of the sludge is pumped into large tanks called **anaerobic sludge digesters**. Here, other kinds of bacteria, which grow anaerobically, digest the bacteria and fungi in the sludge.

Microbes in Biofuels

Biofuels are known as the fuels of biological origin. These are infact, used for the production of heat and other forms of energy. The energy derived from the biofuels is called bioenergy. Fuel wood and biogas are the sources of bioenergy. Biologically generated hydrogen, methane, ethanol, butanol and diesel are referred as biohydrogen, biomethane, bioethanol, biobutanol and biodiesel, respectively. The advantages of biofuels are:

1. These are renewable energy resources.
2. They release relatively low greenhouse gases including carbon dioxide emission than fossil fuels.
3. The raw materials used in biofuel production are often wastes, including municipal waste. Therefore, it helps in pollution control.

Biogas or Gobar Gas

Biogas is known to be the methane rich fuel gas being produced by the anaerobic breakdown or digestion of biomass with the help of methanogenic bacteria. It is mainly produced from the animal wastes.

In fact animal and plant waste contains a large amount of glucose in the form of cellulose which is utilised by the microbes to produce methane gas.

$$\underset{\text{Glucose}}{C_6H_{12}O_6} \longrightarrow \underset{\text{Methane}}{3CH_4} + \underset{\text{Carbon dioxide}}{3CO_2}$$

Energy value $16 kJg^{-1}$ $56 kJg^{-1}$

Various microorganisms are involved in production of biogas, their properties listed below:

Advantages of Biogas

(a) It is cheaper than other fuels.
(b) It is ideal for small scale application.
(c) It helps to clean environment because waste materials can be used as substrate.
(d) Aseptic conditions are not required.
(e) Recovery of methane is spontaneous.
(f) Renewable source of energy.

Disadvantages of Biogas

(a) It has low heat value.
(b) It has low yield of biogas.
(c) Large scale production is not economical.
(d) It contains some gases as impurities.

Ganga Yamuna Action Plan

The Ministry of Environment and Forests has initiated Ganga Action Plan (GAP) and Yamuna Action Plan (YAP) to save the major rivers of our country from pollution. Under these plants, it is proposed to build a large number of sewage treatment plants so that only treated sewage may be discharged in the rivers.

Chapter Exercise

1. Primary sludge is used for
(a) preparation of compost (b) preparation of manure
(c) biogas production (d) All of the above

2. Some blue-green algae are used as biofertilisers because they
(a) are photosynthetic (b) have mucilage
(c) grow everywhere (d) fix nitrogen

3. The bacterium which is used in the prepartion of vitamin riboflavin is
(a) *Leuconostoc* (b) *Clostridium butylicum*
(c) *Lactobacillus* (d) None of these

4. Passage of effluents into oxidation tank is for
(a) primary treatment (b) secondary treatment
(c) tertiary treatment (d) Both (a) and (b)

5. Cyclospdrin-A is used for
(a) dissolving blood clots
(b) lowering cholesterol level
(c) immunosuppression
(d) enhancing tenderness of meat

6. Lactic acid is used in
(a) preservation (b) tanning
(c) plastic making (d) All of these

7. Biological control component is controlled to advanced agricultural production, which of the following is used as a third generation pesticide?
(a) Insect hormone analogues (b) Insect repellents
(c) Pheromones (d) Pathogens

8. A common biocontrol agent for the control of plant disease is
(a) *Baculovirus* (b) *Bacillus thuringiensis*
(c) *Glomus* (d) *Trichoderma*

9. Which of the following is not used as biopesticide?
(a) *Bacillus thuringiensis* (b) *Trichoderma harzianum*
(c) *Nuclear polyhedrosis* virus (d) *Xanthomonas compestris*

10. Sporeine is a bioinsecticide, based on *Bacillus thuringiensis (Bt)*. It was developed in Germany and kills insects by inhabiting ion transport in
(a) foregut (b) midgut
(c) hindgut (d) digestive system

11. A non-photosynthetic aerobic nitrogen fixing soil bacteria
(a) *Rhizobium* (b) *Clostridium* (c) *Azotobacter* (d) *Klebsiella*

12. A fungus used for the biocontrol of plant pathogens is
(a) *Trichoderma viridae* (b) *Mucor mucedo*
(c) *Penicillium notatum* (d) *Phytophthora parasitica*

13. Lepidopterous are used as
(a) biofertilisers (b) healthcare agents
(c) biological control (d) None of these

14. Which of the following is used as biological control?
(a) *Albugo* (b) *Puccinia*
(c) *Phytophthora* (d) None of these

15. *Azotobacter* is used as
(a) biofertilisers (b) biological control
(c) Both (a) and (b) (d) None of these

16. Primary treatment of sewage is
(a) physical process (b) biological process
(c) chemical process (d) biochemical process

17. Farmyard manure and compost are prepared by bacterial action from
(a) refuse (b) garbage
(c) animal excreta (d) All of these

18. Which one is a biofertiliser?
(a) NPK mixture
(b) Rhizobia in legume roots
(c) Rhizobia in farmyard manure
(d) Green manure

19. Passage of effluents into oxidation tank is used for
(a) primary treatment (b) secondary treatment
(c) Both (a) and (b) (d) tertiary treatment

20. Which of the following can be controlled by using biopesticides?
(a) Insects (b) Diseases
(c) Weeds (d) All of these

21. Most common nitrogen-fixing cyanobacterium of paddy fields is
(a) *Cylindrospermum* (b) *Aulosira*
(c) *Oscillatoria* (d) *Nostoc*

22. Cheese and yogurt are the products of
(a) *S. greisas* (b) *Lactobbacillus*
(c) *Streptococcus* (d) *R. orzyae*

23. 'Palmivoralt' is a
(a) herbicide (b) fungicide
(c) nematocide (d) rodenticide

24. Some blue-green algae are used as biofertiliser as they
(a) fix nitrogen (b) are mucilaginous
(c) can grow everywhere (d) are deficient in nutrient

25. Soil microorganisms are mainly comprised of
(a) fungi
(b) fungi and bacteria
(c) fungi, bacteria and Actinomycetes
(d) None of the above

26. Which one of the following groups of bacteria supplies nitrogen to our crop plants irrespective of crop rotation or manuring?
(a) *Nitrosomonas* (b) *Nitrobacter*
(c) *Rhizobium* (d) *Azotobacter*

27. An organism used as a biofertiliser for raising soybean crop production is
(a) *Azospirillum* (b) *Rhizobium*
(c) *Nostoc* (d) *Azotobacter*

28. Biofertilisers give better yield when added along with
(a) phosphate (b) calcium
(c) nitrogen (d) urea

29. BGA is chiefly used as biofertiliser in the crops of
(a) wheat (b) gram (c) paddy (d) mustard

30. Symbiotic nitrogen fixing bacteria belongs to
(a) *Xanthomonas* (b) *Acetobacter*
(c) *Pseudomonas* (d) *Rhizobium*

31. Which one of the following is used as biological insecticide?
(a) Silkmoth (b) Caterpillar (b) Tiger bettle (c) Mazra bok

32. During anerobic digestion of organic waste, such as in producing biogas, which one of the following is left undergraded?
(a) Hemicellulose (b) Lignin
(c) Lipids (d) Cellulose

33. *Bacillus thuringiensis* (*Bt*) strains have been used for designing novel
(a) biofertilisers
(b) bio-mineralisation processes
(c) bio-insecticidal plants
(d) bio-metallurgical techniques

34. Leghaemoglobin is a
(a) respiratory pigment in humans
(b) pigment present in leguminous plant
(c) Both (a) and (b)
(d) None of these

35. The common nitrogen-fixer in paddy fields in
(a) *Frankia* (b) *Rhizobium*
(c) *Azospirillum* (d) *Oscillatoria*

36. Which of the following is mainly produced by the activity of anaerobic bacteria on sewage?
(a) Marsh gas (b) Laughing gas
(c) Propane (d) Mustard gas

37. Agriculture by using only biofertilisers is called
(a) manuring (b) composting
(c) inorganic farming (d) organic farming

38. Cyanobacteria serves as important biofertilisers in the fields of
(a) rice (b) maize (c) wheat (d) sugarcane

39. Physical removal of large and small from the sewage through filtration and sedimentation is called
(a) primary treatment (b) secondary treatment
(c) tertiary treatment (d) quaternary treatment

40. Secondary sewage treatment is mainly a
(a) biological process (b) physical process
(c) mechanical process (d) chemical process

41. The ergot of rye is caused by chemical
(a) ergotinine (b) heparin (b) histamine (d) arginine

42. Nitrogen-fixation is performed by
(a) green alga and fungi
(b) mycorrhiza
(c) mycoplasma
(d) blue-green algae and bacteria

43. Nitrifying bacteria are
(a) chemosynthetic (b) photoautotrophic
(c) parasitic (d) saprotrophic

44. *Nitrosomonas* is a
(a) photoautotroph (b) chemoautotroph
(c) chemoheterotroph (d) photoheterotroph

45. Symbiotic nitrogen-fixing bacteria belong to
(a) *Xanthomonas* (b) *Acetobacter*
(c) *Pseudomonas* (d) *Rhizobium*

46. One of the following is a biofertiliser rich in free living nitrogen-fixing bacteria
(a) *Rhizobium* (b) *Micrococcus*
(c) *Pyrethrin* (d) *Azotobacter*

47. Biogas is
(a) cheaper fuel (b) ecofriendly fuel
(c) renewable source of energy (d) All of the above

48. Gobar gas is produced by
(a) nitrifying bacteria (b) methanogenic bacteria
(c) lactogenic bacteria (d) oxidising bacteria

49. Thuricide is a proteinaceous toxin, which is obtained from
(a) biofertiliser (b) bacterial origin
(c) green manure (d) farmyard manure

50. Farmers have reported 50% higher yields of rice by using the biofertiliser
(a) mycorrhiza
(b) *Azolla pinnata*
(c) cyanobacteria
(d) legume-*Rhizobium* symbiosis

Answers

1. (d)	**2.** (d)	**3.** (b)	**4.** (b)	**5.** (c)	**6.** (d)	**7.** (a)	**8.** (d)	**9.** (d)	**10.** (b)
11. (d)	**12.** (a)	**13.** (c)	**14.** (c)	**15.** (a)	**16.** (a)	**17.** (d)	**18.** (b)	**19.** (b)	**20.** (d)
21. (b)	**22.** (c)	**23.** (a)	**24.** (a)	**25.** (c)	**26.** (b)	**27.** (d)	**28.** (a)	**29.** (c)	**30.** (d)
31. (c)	**32.** (b)	**33.** (c)	**34.** (d)	**35.** (c)	**36.** (a)	**37.** (d)	**38.** (a)	**39.** (a)	**40.** (a)
41. (a)	**42.** (d)	**43.** (a)	**44.** (b)	**45.** (d)	**46.** (d)	**47.** (d)	**48.** (b)	**49.** (b)	**50.** (b)

34

Control Methods of Plant Disease

As we have already studied that plants are affected by various pathogens. Thus, a good management system is required for controlling the pathogens and protecting the plants. The major aspect of plant disease is to eliminate or eradicate the pathogen from plant/crop, pre or post-harvest to control the disease and loss of crop yield by the pathogen.

Thus, plant disease control is an eminent process in which crop yielding can be done without complete or partial eradication of pathogen.

Methods of Controlling Diseases in Plants

Following common methods are used for managing/controlling the plant diseases:

(i) **Avoidance** Prevent disease by selecting a time of the year or site where there is no inoculum or where the environment is not favourable for infection. It may be drought, stress, salt avoidance, etc.

(ii) **Pathogen exclusion** This basically prevents the transportation and introduction of plant pathogens. It is done by following measures:

(a) Checking of transport of pathogens.

(b) Sale and purchase of certified seeds.

(c) Treatments of seeds (fungicides on seed coating) and disinfestations.

(d) Modify actions to prevent spread into uncontaminated areas.

(iii) **Pathogen eradication and reduction of inoculum** It can be done by various practices such as:

(a) **Sanitation** Also the infected plant debris is removed.

(b) **Roguing and Pruning of diseased plants** It is done in the infected plant parts and chemical pesticides/fungicides are applied on the cut part to prevent the secondary infection.

(c) **Crop Rotation** Planting a non-host crop can reduce (but not eliminate) the density of the prevailing pathogen. It is to be noted that plant crops of the same family in the same position should not be planted in the garden every year.

(d) Use of techniques that disfavour movement of disease spreading vectors. The techniques such as reflective mulches for aphids, sticky cards for other insect vectors, trench barriers for root diseases are commonly used.

(e) **Soil sterilisation** It is done by the use of 1-2 mm thick plastic for 4-6 weeks can reduce inoculum density by heating the top 6" of the soil to lethal temperature.

(f) **Biocontrol** It is done mainly with the use of biocontrol agents.

(iv) **Plant Protection** It can be done by following practices:

(a) Use of pesticides, fungicides, bactericidal agents, etc.

(b) Use of resistant varieties and integrated pest management.

(c) Use of updated information about concerned crop.

(d) Reduction/Elimination of initial infection.

(e) Use of latest pesticides, agents, etc.

(v) **Cross Protection** It is a phenomenon in which the tolerance or resistance of a plant to one virus strain is often allowed to induced by systemic infection with the second one. Hence, for many years serological and cross protection tests are used as routine methods to determine the strain inter relationships in different plant viruses.

Methods used in Plant Disease Management

Management of various diseases in plants can be done by the following methods such as:

- Cultural methods, involving agricultural cropping, harvesting and storage, tillage, crop rotation, soil management, growing of resistant varieties, planning of land use and other related practices.
- Biological control methods.
- Breeding methods for disease resistance.
- Chemical methods.
- Plant disease assessment methods.
- Post-harvest disease control.

Some method used in plant disease mangement are described below:

Biological Control Methods

The term 'biological control' clearly implies to the control of diseases by the means of a biological agency, i.e. living micro or macroorganisms.

According to **Gaullett** (1965) biological control of plant disease can be defined as the practice or condition in which the activity or survival of a pathogen is reduced by the action of microorganism.

Some common agents used to control plant disease are given below:

- **Mycorrhizal fungi** These agents have received a considerable attention in the recent years. They are symbiotic fungi that predominates in the roots and soils of agricultural crop plants.
- **Soil amendments** Many root diseases can be controlled by ploughing the organic material into the soil which increases the activity of saprophytic organisms resulting into the liberation of CO_2 in the abundant amount.
- **Hyperparasitism** In hyperparasitism, the pathogen is directly attacked by a specific BCA that kills it or its propagules. In general, there are four major classes of hyperparasites obligate bacterial pathogens, Hypovirus, facultative parasites and predators. ***Pasteuria penetrans*** is an obligate bacterial pathogen of **root-knot nematodes** that has been used as a BCA.

  Hypoviruses are also hyperparasites. A classical example is the virus that infects *Cryphonectria parasitica*, a fungus causing **chestnut blight** which causes hypovirulence (a reduction in disease producing capacity of the pathogen).
- **Bacteriophages** A number of phages has been developed for many phytopathogenic bacteria such as *Agrobacterium tumefaciens*, etc., but no plant disease has been cured yet by the treatment with phage after the diseases has developed.

Chemical Control Methods

Apart from biologial agents, many chemicals are also used to control plant diseases. They are known as fungicides, weedicides, etc. Inspite of their efficiency these are not used much these days due to the harmful action of chemical spray, dusts, etc.

Some the chemicals used as fungicides are

- Inorganic copper fungicides
- Organic sulphur compounds
- Inorganic mercuric compounds
- Inorganic sulphur compounds
- Quinone and phenolic compounds
- Benzene compounds
- Organophosphate fungicides
- Synthetic fungicides, etc.

Principles of Plants Disease Management

Integrated management is the practice of using all pertinent information regarding crops, pathogens, etc. It also includes the knowledge about the history of disease, varietal resistance, different environmental conditions, level labour and the cost of any particular treatment.

Goals of IM

(i) To eliminate or to reduce inoculum.

(ii) To reduce the effectiveness of initial inoculum.

(iii) To increase the resistance with in the host.

(iv) To delay the onset disease.

(v) To slowdown the secondary cycles.

Chapter Exercise

1. Fungicides are used for checking the infection of
(a) bacteria (b) fungi
(c) virus (d) Both (b) and (c)

2. Which of the following is not a method of plant disease control?
(a) Pathogen exclusion (b) Pathogen eradication
(c) Inoculation of pathogen (d) Plant protection

3. Fungicide is a
(a) amino acid (b) hormone
(c) pesticide (d) mineral

4. Sterilisation of soil is a part of
(a) pathogen exclusion (b) pathogen eradication
(c) plant protection (d) None of these

5. Sanitation is a part of which of the following method of plant disease control?
(a) Plant exclusion
(b) Plant eradication
(c) Plant protection
(d) None of the above

6. Use of various techniques of plant disease control under one roof is called
(a) IM (b) CPM
(c) TPM (d) None of these

7. Use of latest pesticide is the part of
(a) pathogen exclusion (b) pathogen eradication
(c) plant breeding (d) integrated management

8. Roguing and pruning of affected plants is necessary for
(a) hybridisation (b) pathogen eradication
(c) plant breeding (d) tissue culture

9. The bordeaux mixture, an inorganic copper fungicide used for the treatment of plant diseases was developed by
(a) Paul Berg (b) Louis Pasteur
(c) Joseph Lister (d) JH Millardet

10. In plant disease control, protection of the host from any pathogen is
(a) curative measure (b) roughing
(c) prophylactic method (d) therapeutic method

11. Which of these is a fungicide?
(a) BHC (b) DDT
(c) Bordeaux mixture (d) 2, 4-D

12. A biological control practice is
(a) tillage (b) crop rotation
(c) roughing (d) hyperparasitism

13. Exclusion of plant disease by legislation is known as
(a) disease resistance
(b) plant quarantine
(c) biological control of plant
(d) cultural control

14. The chemical present in flit is
(a) malathion (b) DDT
(c) BHC (d) Aldicrab

15. The common mode of action of herbicide is
(a) blocking of phloem (b) blocking of xylem
(c) blocking of photosystem-I (d) blocking of photosystem-II

16. Clchineal insect has been used, in checking the growth of
(a) *Opuntia* (b) *Eichhornia*
(c) aphids (d) screw worm

17. Which of the following is weedicide?
(a) Mala thione (b) Carbamates
(c) Bordeaux mixture (d) 2, 4D

18. Which of the following is not matched?
(a) Organophosphate — Malathion
(b) Carbamate — Carbofuran
(c) Organochlorine — DDT
(d) Pyrethroid — Atrazine

19. Natural insecticide obtained from plant
(a) azadirachtin (b) rotenone
(c) pyrethrum and cinerin (d) All of these

20. The pesticides having very low biodegradation and strong affinity for fatty tissue are
(a) Organochlorine
(b) Triazenes
(c) Organophosphate
(d) Pyrethroids

Answers

1. (b)	**2.** (c)	**3.** (c)	**4.** (b)	**5.** (b)	**6.** (a)	**7.** (d)	**8.** (b)	**9.** (d)	**10.** (c)
11. (c)	**12.** (d)	**13.** (b)	**14.** (a)	**15.** (d)	**16.** (a)	**17.** (d)	**18.** (d)	**19.** (d)	**20.** (a)

35

Introduction to Ecology and Environment

'Ecology' was first described in 1869 by Ernst Haeckel, a German biologist. It is described as the branch of biology dealing with 'the study of an organism and its environment'.

Branches of Ecology

Ecology has been divided into several branches such as:

- **Autecology** It was coined by Schroeter and Kirchner (1896). The term is basically concerned with the study of an individual organism or an individual species.
- **Synecology** It was also coined by Schroeter and Kirchner (1896). The term deals with the study of a group (groups) of organisms, associated together as a unit. It is further sub-divided into population ecology, community ecology and the ecosystem ecology.
- **Demecology** It refers to the ecology of population.
- **Applied ecology** It is the study of specialised field of ecology, been concerned with the conservation and economic exploitation of organisms.
- **System ecology** It is the branch of ecology dealing with the interpretation of ecological concepts and processes in terms of mathematical models and formulae.
- **Genecology** It is the study of genetic composition and changes in relation to the origin of ecads, ecotypes, new species, etc.
- **Ethology** It is the study of relations of organisms within the family and society in an aggregate and community.
- **Hexicology** It is the study of relations, that exists between the organisms and their environment.
- **Conservation ecology** It deals with the methods of proper management of natural resources such as land, water, forests, sea, mines, etc., for the benefits of human beings.
- **Production ecology** It is the modern sub-division of ecology, dealing with the gross and net production of different ecosystems such as freshwater, sea water, etc.
- **Radiation ecology** It deals with the study of gross effects of radiations and radioactive substances over the environment and living organisms.
- **Taxonomic ecology** It is concerned with the ecology of different taxonomic groups such as plant ecology, insect ecology, human ecology, etc.

Environment

The word 'environment' means everything that is around us. It may be defined as the whole complex of climatic, edaphic, physiographic and biotic factor that act upon an organism or an ecological community and ultimately determine its form and survival.

The total environment is made up of some abiotic or non-living components as well as some biotic or living components. The non-living environment includes the non-living factor, both physical and chemical that influences the life of organisms. This Includes temperature, light, energy, water, atmosphere including atmospheric gases and air current, fire, gravity, topography, soil, pH, nutrients, etc.

While the living environment constitutes biotic factors such as animals and plants.

Classification of Environment

The surface of the earth consists of three basic elements, i.e. land, sea and air. On the basis of these elements, it is divided into the lithosphere consists of rock formations; the pedosphere, composed of disintegrating compounds of rock and stone forming soil; hydrosphere, a layer of water penetrating lithosphere and the pedosphere and lastly the atmosphere, a layer of air embracing all three above. These components of environment are interdependent and interact with one another in one or the another way.

Atmosphere

The atmosphere is defined as an insulating blanket of air around the earth which serves as the reservoir of oxygen (needed by animals) and of carbon dioxide (essential for the plants). It serves as a distribution system for moisture. The atmosphere shields the earth from the lethal concentration of ultraviolet radiations.

It selectively filters the sun's rays so that only two small segments of the electromagnetic spectrum penetrates the earth's surface in appreciable amounts.

Composition of Clean, Dry Air or Near sea Level

Components	Percentage by volume	Components	Percentage by volume
Nitrogen	78.04	Hydrogen	0.00005
Oxygen	20.9476	Nitrous oxide	0.00005
Argon	0.934	Xenon	0.0000087
Carbon dioxide	0.0314	Ozone	Summer 0 0.000007 Winter 0 0.000002
Neon	0.001818	Nitrogen dioxide	0.000002
Helium	0.000523	Iodine	0-0.000001
Methane	0.0002	Ammonia	0 to trace
Krypton	0.000114	Carbon monoxide	0.1-0.2 ppm
Sulphur dioxide	0 to 0.0001		

The atmosphere is divided into four different types zones as:

(i) **Troposphere** It is the zone that found nearest to the earth. One property that keeps changing in the troposphere is the temperature. It drops to – 56°C by the time an altitude of about 16-18 km is reached. The rate of temperature drop is about 6.4°C per 1000 meters. This is called the **normal environmental lapse rate**.

The region, through which this lapse rate occurs, defines the troposphere. At the equator, the troposphere is about 18 km thick but at the poles it is only about 8 km thick. The upper boundary of troposphere is termed as **tropopause**.

(ii) **Stratosphere** This zone begins at the top of the tropopause. Extending from 8-16 to 30-50 km.

Temperature shows a gradual increase with the increase in altitude. It includes much of the **ozone layer.** The warning trend in the stratosphere is caused by a cycle of chemical changes in the ozone layer which converts all the incoming high energy UV-rays to heat.

(iii) **Mesosphere** This zone exists above the stratosphere and extends roughly to 80 km. Its temperature drops from – 2°C to between –80 and –94°C. The upper boundary of mesosphere records the lowest temperature at our atmosphere.

(iv) **Thermosphere** This zone exists above 80 km altitude. The temperature becomes very high but the air has such a low density that it can hold very little heat. Any denser object in the thermosphere will be extremely hot in sunlight but very cold at night.

Climate

Climate forms the important feature of the environment. It also helps to determine food, clothing, housing and general mode of our life. The conditions of the atmosphere at any place at a specific time or for a short time with respect to various elements (temperature, sunshine, wind, clouds, fog, etc) is known as **weather**. It varies from time to time at the same place and place to place at the same time.

Climate is studied in terms of the various climatic elements like temperature including radiations, atmospheric pressure, wind, humidity (water vapour, clouds, precipitation, evaporation). Four climatic and vegetation zones have been recognised on the basis of mean temperature along the latitude.

The organisms are classified into four temperature groups on the basis of their occurrence in different climatic zones:

(i) **Megatherms** These are the organisms that are adapted to high temperature throughout the year as found in tropical zone.

(ii) **Mesotherms** These are the organisms that are adapted to mild winters and high summer temperature. These are the organism that live in sub-tropical zone.

(iii) **Microtherms** These are the organisms that they live in temperate areas where the winter temperature is low, but the summer temperature is moderate.

(iv) **Hekistotherms** These are the organism that are adapted to brief summer period of below 10°C and long snowy winter period. This condition occurs in Arctic or alpine zone.

Habitat

It is the natural abode of an organism including total environment. It is an area where all the needs of a population are fulfilled, including the members of that population.

Although, the habitat is the physical environment of the species, the species might not be present in all parts of its habitat at all times. The physical factors like, heat, moisture, nutrients and light are available in the habitat but population in different habitats are affected by different physical factors or by combination of factors.

Earth has four major habitats that is marine, freshwater, terrestrial and estuarine. Each habitat is characterised by the physical conditions or limiting factors such as salinity or temperature that influences the presence and survival of organisms in that habitat and their distribution within it. The organism is adapted to the physical conditions of the habitat.

Microhabitat is a small part of the habitat with its own characteristic environmental features, e.g. forest floors, trees canopies, edges of a pond.

Niche

There are about 2 million species of living organisms on the earth of which 1.5 million are animals and half a million are plants. Each species performs unique functions and occupy specific habitats. The combination of functions and habitat is called its ecological niche. The term 'niche' (from Latin *nidus* meaning nest) was first coined by **J Grinnell** (1917) to describe the place of an organism in the environment.

Charles Elton (1927) has used the term niche in the sense of the 'functional status of an organism in its community'. He defined niche as an animal's place in the biotic environment in relation to its food and enemies.

Ecosystem

An ecosystem is a functional unit of nature where living organisms interacts among themselves and with their surrounding physical environment. The term 'ecosystem' was coined by AG Tansley (1935). It can also defined as the interactive system, where all biotic and abiotic components interacts with each other through the process of exchange of energy and flow of nutrients. It can be **terrestrial** and **aquatic** in nature.

Components of Ecosystem

The ecosystem is divided into two main components.

(i) **Structural Components**

It includes both biotic and abiotic components that influences the properties of each other.

These are as follows :

(a) The **abiotic factors** of structural component includes atmospheric air, light, temperature, water, soil, fire, wind, etc.

(b) The **biotic factors** of structural components includes producers (autotrophs, that can synthesise their own food), consumers (heterotrophs, that feed on others for food). Constituting primary, secondary, tertiary and quaternary consumers and decomposers (saprotrophs), microorganisms feeding on dead and decaying bodies of other organisms.

(ii) **Functional Components**

It includes the following:

(a) **Food chain** It is a sequence of organisms through which the food and its contained energy passes to each member in a sequence and finally become the food of the later member.

The position of each organism in the food chain constitutes the **trophic level**.

Producers occupy the first trophic level, primary consumers (herbivore) the second, and the third trophic level is occupied by the secondary consumers (carnivores).

The food chains are of two main types:

- **Grazing food chain** which extends from producers $\rightarrow$ herbivores $\rightarrow$ carnivores.
- **Detritus food chain** which extends from dead organic matter $\rightarrow$ detritus feeder $\rightarrow$ to the one feeding on dead detritus feeders.

Hence, food chain is important in an ecosystem, because it helps to maintain the ecological balance.

(b) **Food web** Any interconnecting network of different food chains at various trophic levels in an ecosystem, termed as food web. The concept of food web is more real ecologically than the food chain because it increases the adaptability and competitiveness of the organisms.

(c) **Ecological pyramids** The graphical representation of different trophic levels in terms of ecology, biomass and number of organisms is called ecological pyramid. Any given organism may occupy more than one tropic level simultaneously. In this, the base of each pyramid is occupied by the producers, while the apex is been occupied by the consumers. Each trophic level in any pyramid tends to represent functional level.

Types of Ecological Pyramids

Ecological pyramids can be of following three main types :

(i) **Pyramid of number** It represent the total number of organism at each tropic level. It may be upward or may be inverted a tree ecosystem in which it is known to be inverted.

(ii) **Pyramid of biomass** It mainly represents the total weight of the organism in its particular trophic level. Its inverted in case of sea because the biomass of fishes exceeds for much than that of the phytoplankton.

(iii) **Pyramid of energy** It represents the total amount of energy of the organisms in each trophic level. It is down to be always up right, i.e. it can never be inverted due to the fact that when any amount of energy is transferred from a particular trophic level to the next trophic level, some amount of energy is lost in the term of heat at each and every step.

Note Each bar in the pyramid of enrgy indicates the amount of energy at each trophic level in a given time or annually per unit area.

Productivity

Primary production is the rate of biomass production produced per unit area over a time period. It is expressed in terms of weight (g^{-2}) or energy (kcal m^{-2}). The rate of biomass production is called **productivity**. Which is usually expressed in terms of g^{-2} yr^{-1} (kcal m^{-2}) yr^{-1}. Productivity of an ecosystem can be categorised on the basis of its primary as well as the secondary productivity.

(i) **Primary Productivity**

It is the amount of biomass or organic matter produced per unit area over a time period by plants during photosynthesis.

The primary productivity has two aspects given below:

(a) **Gross Primary Productivity** (GPP) rate of production of organic matter during photosynthesis.

(b) **Net Primary Productivity** (NPP) is the amount of energy left in the producers after utilisation of some energy for respiration. In other words, it is weight of organic matter stored by producers in a unit area of volume per unit time. It is actually the available mass for consumption by heterotrophs (herbivores and carnivores).

$$\text{GPP} - R = \text{NPP} \quad \text{where, } R = \text{Respiration losses}$$

Factors Affecting Primary Productivity

Different factors affecting primary productivity are given below:

- The plant species inhabiting in a particular area.
- **Environmental factors** The various environmental factor also contribute in affecting primary productivity like light, temperature, water and photosynthetic capacity/ efficiency of producers.

(ii) **Secondary Productivity**

It is the rate of assimilation and formation of new organic matter by consumers. It is less as compared to primary productivity and tends to decrease with the increase in the trophic level.

Decomposition

It is known to be the phenomenon of breaking down of complex organic matter into inorganic substances like water, carbon dioxide and nutrients.

Detritus is the known as the raw material for decomposition. It includes dead remains of plants (leaves, bark and flowers) and of animals including faecal matter. It is largely an aerobic process (i.e. requires oxygen for its processing).

Factor Affecting Rate of Decomposition

Decomposition is largely an oxygen requiring process. The rate of decomposition is controlled by chemical composition of detritus material and climatic factors.

(i) **Chemical composition of detritus** The rate of decomposition is slower, if detritus is rich in lignin and chitin, while it is quicker if it is composed of nitrogen and water-soluble substances like sugar.

(ii) **Climatic Factors** Temperature and soil moisture are the most important climatic factors that controls decomposition. Warm and moist environment favours decomposition, while low temperature and the rate of anaerobic conditions inhibits the rate of decomposition.

Chapter Exercise

1. Ecology is connected with the study of

(a) environmental factors
(b) effect of plants on environment
(c) plant adaptations
(d) All of the above

2. The term 'ecology' was actually coined by

(a) Aristotle (b) Reiter
(c) Linnaeus (d) Odum

3. Climate zone between 20-40°C is

(a) sub-tropical (b) temperate
(c) Both (a) and (b) (d) temperate and sub-Arctic

4. Winters are absent in climatic zone

(a) sub-tropical (b) tropical
(c) alpine (d) Arctic

5. Aquatic plants generally have a

(a) well-developed vascular system
(b) reduced vascular system
(c) well-developed root system
(d) well-developed stomatal system

6. Succession of communities occurs because

(a) each existing community changes the environment
(b) climatic changes lead to reduced water availability
(c) most population have a limited lifespan and die making room for others
(d) resources in an area are limited

7. A lake with rich nutrients is known as

(a) euphotic (b) eutrophic
(c) aphotic (d) oligotrophic

8. Study of inter-relationships between organisms and their environment is

(a) ecology (b) ecosystem
(c) phytogeography (d) ethology

9. Study of inter-relationship between an entire community and its environment is

(a) autecology (b) resource ecology
(c) species ecology (d) synecology

10. Tropical zone extends from

(a) 0-30°C (b) 0-20°C (c) 10-23°C (d) 5-30°C

11. A plant devoid of roots is

(a) *Wolffia/Ceratophyllum* (b) *Vallisneria*
(c) *Hydrilla* (d) *Azolla*

12. Mean annual temperature of tropical zone is

(a) 10°C (b) 16°C
(c) 24°C (d) 30°C

13. The floating leaves of an aquatic plant has

(a) stomata on the upper surface
(b) stomata on the lower surface
(c) stomata on both the surfaces
(d) no stomata

14. In sub-merged hydrophytes, the stomata occurs

(a) on lower surface (b) on upper surface
(c) no where (d) Both (a) and (b)

15. Thick cuticle occurs on the leaves of plants belonging to

(a) cool habitats (b) warm habitats
(c) dry habitats (d) wet habitats

16. Weather represents

(a) short term properties of atmosphere
(b) seasonal changes in atmosphere
(c) average variations of atmospheric conditions
(d) All of the above

17. Sub-merged hydrophytes have a well developed

(a) vascular system (b) aerenchyma
(c) root system (d) stomatal system

18. Stem of sub-merged hydrophytes is soft and weak because of
(a) the absence of phloem
(b) the absence of stomata
(c) the absence of xylem
(d) feebly developed supporting tissue and xylem

19. Which one is the famous plant ecologist of India?
(a) Charles Darwin (b) Ramdev Mishra
(c) Birbal Sahani (d) Jagdish Chandra Bose

20. The role of an organism in ecological system is known as
(a) habitat (b) herbivory
(c) niche (d) interaction

21. The primary consumers in a pond ecosystem is
(a) phytoplankton (b) zooplankton
(c) fishes (d) bacteria

22. The organic substances which decompose slowly are
(a) chitin (b) lignin
(c) cellulose (d) All of these

23. The food chain which start from microorganism is
(a) terrestrial food chain (b) grazing food chain
(c) detritus food chain (d) predator food chain

24. Decomposers are always
(a) autotrophs (b) primary consumers
(c) saprotrophs (d) heterotrophs

25. The type of food chain in which decomposed organic matter is converted into energy rich compound is called
(a) detritus food chain (b) grazing food chain
(c) cybernatics (d) None of these

26. Consumers are also known as
(a) reducers (b) transducers
(c) consumers (d) producers

27. Biotic community which shows the number of feeding connections is categorised as
(a) food chain (b) pyramid of number
(c) food web (d) trophic levels

28. Organism having wide range of degree of tolerance are
(a) stenotypic (b) eurotypic
(c) ecotypics (d) ecophenes

29. The efficiency of any ecosystem is best depicted by the pyramid of
(a) energy (b) number
(c) biomass (d) volume

30. Energy and nutrients enter a community by the way of
(a) producer (b) consumer
(c) scavenger (d) detritivores

31. Which one is the factor of abiotic environment?
(a) Producer (b) Consumer
(c) Decomposer (d) Air

32. In the pyramid of number, herbivores occupy
(a) 1st position (b) 2nd position
(c) 3rd position (d) nutrient cycle

33. A sequence of species through, which the organic molecule in a community pass is called
(a) pyramid of energy
(b) food chain
(c) food web
(d) nutrient cycle

34. The term 'eco' refers to
(a) biosphere (b) environment
(c) organisms (d) plants

35. Which one of the following soil is rich in nutrients, having good water holding capacity and is adequately aerated?
(a) Silt (b) Clay
(c) Loamy (d) Sandy

36. In which atmospheric zone, an ozone layer exists?
(a) Ionosphere
(b) Thermosphere
(c) Troposphere
(d) Stratosphere

37. The ultimate aim of ecological adaptation is to
(a) adjust to surrounding
(b) seek food and space for the survival of the organism
(c) get space for interbreeding
(d) None of the above

38. The basic unit of ecology is
(a) cell (b) gene
(c) organism (d) community

39. Which of the following does not directly use sunlight as its source of energy?
(a) Cactus (b) Water plant
(c) Fungus (d) Moss

40. Which one of the following components of ecosystem comes from outside?
(a) Oxygen (b) Temperature
(c) Insects (d) Energy

41. A broad-based pyramid-shaped age structure is the characteristic of a population that is
(a) growing rapidly
(b) at carrying capacity
(c) stable
(d) limited by density-dependent factors

42. A lake community suddenly suffers from algal blooms. Using the strategy of biomanipulation, an ecologist may propose
(a) removing zooplankton
(b) adding mineral nutrients to the water
(c) adding fish that eats zooplankton
(d) removing fish that eats zooplankton

43. Which one of the following best describes the base of a pyramid of net production?
(a) Its size depends on the energy available from detritivores
(b) It contains the energy left after the producers have died
(c) It represents the energy available to the secondary consumers
(d) It contains the energy captured by photosynthesis

44. Earth's biosphere is not completely self-contained (closed) because
(a) humans pollute the atmosphere and bodies of water
(b) plants, algae and photosynthetic bacteria obtain energy from sunlight and heat escapes from the biosphere into space
(c) the precipitation that falls on the anterior of Western North America is derived from the Pacific ocean
(d) bacteria that live on snowy mountains receive nutrients blown to them by winds

45. The ecological pyramid was first described by
(a) Tansley (b) Smith
(c) Charles Elton (d) Webster

46. What is the source of energy that flows through the living world?
(a) Photosynthesis (b) Chemical bond
(c) Green plant (d) The sun

47. A group of interconnected food chain is called a
(a) pyramid of energy (b) complex food chain
(c) food web (d) food cycle

48. How much of the solar energy that falls on the leaves of a plant is converted to the chemical energy?
(a) 1% (b) 10% (c) 30% (d) 50%

49. The efficiency of any ecosystem is best depicted by the pyramid of
(a) energy (b) number
(c) biomass (d) volume

50. Biotic components include
(a) producers
(b) consumers
(c) producers and consumers
(d) producers, consumers and decomposers

51. Energy flow in an ecosystem is
(a) unidirectional (b) bidirectional
(c) multidirectional (d) None of these

52. Largest ecosystem in the world is contributed by
(a) forests (b) rivers
(c) oceans (d) grasslands

53. Pyramid of number for forest ecosystem is
(a) upright (b) inverted
(c) Both (a) and (b) (d) None of these

54. The primary consumers in a pond ecosystem is
(a) phytoplankton (b) zooplankton
(c) fishes (d) bacteria

55. In an ecosystem, the population of
(a) primary producer is larger than primary consumers
(b) secondary consumer is largest
(c) primary consumers are more than primary producers
(d) primary consumers are least dependent upon primary producers

56. Which of them can be put in the category of primary consumers?
(a) Eagle and tigers (b) Fishes and whales
(c) Snakes and frogs (d) Insects and cattles

57. Best arrangement of an energy system consists of hawks, snakes, mice and grasses is
(a) grasses $\rightarrow$ mice $\rightarrow$ snakes $\rightarrow$ hawks
(b) grasses $\rightarrow$ snakes $\rightarrow$ mice $\rightarrow$ hawks
(c) grass $\rightarrow$ mice $\rightarrow$ hawks $\rightarrow$ snakes
(d) hawks $\rightarrow$ snakes $\rightarrow$ mice $\rightarrow$ grasses

58. Organisms, which can convert ammonium or nitrates into amino acids are
(a) scavenger (b) producers
(c) primary consumers (d) primary detritivores

59. In the food chain, the saprophytic organisms are
(a) producers (b) consumers
(c) predators (d) decomposers

60. Match the following columns.

Column I	Column II
A. Mobius	1. Biocoenosis
B. Forbes	2. Microcosm
C. Sukachey	3. Biogeocoenosis

Codes

	A	B	C		A	B	C
(a)	3	2	1	(b)	3	2	1
(c)	1	2	3	(d)	1	3	2

61. A pond is a
(a) biome
(b) a community of plants and animals
(c) a natural ecosystem
(d) None of the above

62. Total organic matter present in an ecosystem is called
(a) biomass (b) biome
(c) litter (d) food

63. A food chain consists of
(a) producers (b) consumers
(c) decomposers (d) All of these

64. If phytoplanktons are destroyed in the sea
(a) it will affect food chain
(b) no effect
(c) algae will get more space
(d) primary consumer will grow more

65. In parasitic food chain, the pyramid of number is
(a) linear (b) upright
(c) inverted (d) inverted upright

66. Most of the earth's water is
(a) locked up as salt water or ice
(b) stored as carbon dioxide
(c) radioactive
(d) being lost to atmospheric dehydration

67. Where is the most productive area of the ocean?
(a) Near the shore (b) At the equator
(c) At low latitude (d) In open ocean

68. Most animals that live in deep oceanic water are
(a) producers (b) primary consumers
(c) secondary consumers (d) detritivores

69. An ecological pyramid of biomass is a representation of the ecosystem's
(a) energy flow through each trophic level
(b) tissue at each trophic level
(c) populations in each food web
(d) biological material in relation to abiotic material

70. Which of the following aquatic ecosystems has very little primary productivity?
(a) River (b) Estuary
(c) Lake (d) Marsh

71. Most primary productivity in the surface waters of lakes and oceans is by
(a) floating plants (b) red and brown algae
(c) zooplankton (d) phytoplankton

72. Trophic forests, food plants, coral reefs, are the examples of
(a) average productive ecosystem
(b) high productive ecosystem
(c) less productive ecosystem
(d) low productive ecosystem

73. Decomposers are also known as
(a) reducers (b) transducers
(c) consumers (d) mineralisers

74. Maximum human population can be supported if all the human are
(a) carnivores
(b) pure vegetarian
(c) vegetarian along with use of milk
(d) vegetarian along with use milk and eggs

75. Which one occurs at the same trophic level?
(a) Crow and cow (b) Snake and earthworm
(c) Tiger and wild boar (d) Deer and bees

76. The organisms which physically and chemically break the complex organic remains are known as
(a) scavengers (b) decomposers
(c) parasites (d) saprophytes

77. The products of decomposition process is/are
(a) humus (b) inorganic nutrients
(c) Both (a) and (b) (d) organic nutrients

78. The factors influencing the rate of decomposition are
(a) temperature (b) moisture
(c) composition (d) All of these

79. Productivity is the rate of production of biomass expressed in terms of
I. $(kcal^{-1}m^{-3})yr^{-1}$ II. $g^{-2}yr^{-1}$
III. $g^{-1}yr^{-1}$ IV. $(kcal\ m^{-2})yr^{-1}$

Choose the correct option
(a) II and I (b) Only III
(c) II and IV (d) I and III

80. An inverted pyramid of biomass can be found in which ecosystem?
(a) Forest (b) Marine
(c) Grassland (d) Tundra

Answers

1. (d)	**2.** (b)	**3.** (a)	**4.** (b)	**5.** (b)	**6.** (a)	**7.** (b)	**8.** (a)	**9.** (d)	**10.** (b)
11. (a)	**12.** (c)	**13.** (a)	**14.** (c)	**15.** (c)	**16.** (a)	**17.** (b)	**18.** (d)	**19.** (b)	**20.** (c)
21. (b)	**22.** (d)	**23.** (c)	**24.** (c)	**25.** (a)	**26.** (a)	**27.** (c)	**28.** (b)	**29.** (a)	**30.** (a)
31. (d)	**32.** (b)	**33.** (b)	**34.** (b)	**35.** (c)	**36.** (d)	**37.** (a)	**38.** (c)	**39.** (c)	**40.** (b)
41. (a)	**42.** (d)	**43.** (c)	**44.** (b)	**45.** (b)	**46.** (d)	**47.** (c)	**48.** (a)	**49.** (a)	**50.** (d)
51. (a)	**52.** (c)	**53.** (c)	**54.** (b)	**55.** (a)	**56.** (d)	**57.** (a)	**58.** (b)	**59.** (d)	**60.** (a)
61. (c)	**62.** (a)	**63.** (d)	**64.** (a)	**65.** (c)	**66.** (a)	**67.** (a)	**68.** (d)	**69.** (b)	**70.** (a)
71. (d)	**72.** (b)	**73.** (d)	**74.** (d)	**75.** (d)	**76.** (b)	**77.** (b)	**78.** (d)	**79.** (c)	**80.** (b)

36

Biogeochemical Cycles

Nutrient Cycle

The term **nutrient cycle** or **biogeochemical cycle** is used for the exchange/circulation of biogenetic nutrients between the living and non-living components of the biosphere.

These biogenetic nutrients or biogeochemicals are the essential elements required by the organisms for body building and metabolism. These are provided by earth and are returned to earth again after their death and decay.

Simple organic compounds like carbohydrates are built up of C, H and O obtained from CO_2 and H_2O. Synthesis of other complex organic substances require additional nutrients or essential elements, e.g. P and N in nucleotides, S and N in protein.

The biogeochemical cycles are classified into two main types :

(i) **Gaseous cycles** in which the main reservoirs of chemical are atmosphere and ocean, e.g. carbon cycle, nitrogen cycle, oxygen cycle, etc.

(ii) **Sedimentary cycles** in which the main reservoirs are soil and rocks, e.g. sulphur cycle, phosphorus cycle, etc.

Carbon Cycle

The atmospheric carbon dioxide is virtually the only source of carbon. This gas is used by all the plants in photosynthesis. The end products (organic substance) of this complex process are used in two ways:

(i) One product is used in the construction of living matter. The carbon and oxygen so, supplied by CO_2 remain in the living matter till death. The decay thus subsequently returns CO_2 to the atmosphere and this completes one possible carbon cycle.

(ii) Another product of the organic substance is used as fuel in respiration by both plants and animals. This process releases CO_2 as a byproduct which may be used in photosynthesis again or may return to the environment, completing the second possible carbon cycle.

The CO_2 content of the atmosphere is replenished not only through biological oxidation (respiration), but also through non-living combustion, i.e. forest fires and burning of industrial fuels that releases CO_2 into the air. Occasionally, volcanic eruptions also add CO_2 in the atmosphere.

Oxygen Cycle

The atmospheric O_2 enters the living organism as a gas and during respiration process, CO_2 and H_2O are formed.

$$C_6H_{12}O_6 + 6O_2 \longrightarrow 6CO_2 + 6H_2O + \text{Energy}$$

The metabolic H_2O thus formed is added to all other water present in living organisms and may undergo three possible fates. It may be excreted and added to the water content of the environment. Another part may be used as a building material for the formation of more living matter. A third possible fate of water within the organism is its consumption as a fundamental raw material along with the CO_2 in photosynthesis.

In this process, the oxygen is liberated as shown in following equation:

$$6CO_2 + 6H_2O + \text{Radiant energy} \longrightarrow C_6H_{12}O_6 + 6O_2$$

This free oxygen may now again used in respiration or may returned to the environment as molecular atmospheric oxygen, completing the cycle. Thus, the oxygen enters living organisms only through respiration and leaves through photosynthesis.

In intervening steps, the oxygen is incorporated in water and it can interlink with the water cycle or indirectly with the carbon cycle. The oxygen in the atmosphere is the source of ozone (O_3). The ozone layer thus protects organisms by preventing most of the UV and X-rays from reaching the earth's surface. The most important factor affecting the oxygen cycle of the biosphere and the oxygen budget of the earth is the man himself. In addition to inhaling O_2 and exhaling CO_2, man decreases the oxygen level and increases the CO_2 level by burning of fossil fuels.

Nitrogen Cycle

The atmosphere has about 78% of nitrogen which is the ultimate source of it. It is required by organisms to synthesise proteins, nitric acids and other nitrogenous compounds. It is the most abundant but inert chemical thus, cannot be used as such by the majority of organisms. Hence, it must first be fixed by specialised organisms.

The nitrogen cycle depends upon different kinds of bacteria such as **nitrifiers**, **denitrifiers** and **nitrogen fixers** which regulates the circulation of nitrogen through air, soil, plants and animals.

The most common usable nitrogen source is **nitrate ion** (NO_3^-), which may be used by the plants as a mineral metabolite from the environment and may be converted by them into amino groups and other nitrogen containing compounds of living matter. The nitrates are added to the environment through fertilisers, dissolution of rocks by water, combination of atmospheric nitrogen with O_2 by lighting and through the process of nitrification.

The nitrogen cycle involves several steps such as:

(i) **Ammonification** The organic nitrogen of plants and animals is ultimately broken down by the action of saprophytic fungi and bacteria to ammonia (NH_3) by a process called ammonification.

(ii) **Nitrification** Some of the NO_3 is oxidised to form nitrites (NO_2^-) and nitrates (NO_3^-), through the action of autotrophic (chemotrophic) bacteria, such as *Nitrosomonas* and *Nitrobacter*. This process is called nitrification and the energy released supports the existence of nitrifying bacteria.

(iii) **Denitrification** Other types of anaerobic soil bacteria, (e.g. *Pseudomonas denitrificans*) act on nitrates in the process of denitrification, by which N_2 is liberated into the atmosphere. This type involves the reduction of the nitrate ions (NO_3^-) to nitrogen dioxide (NO_2), dinitrogen oxide (N_2O), nitrogen monoxide (NO) or nitrogen (N_2).

(iv) **Nitrogen-fixation** In this step, the nitrogen is trapped from the air by the nitrogen-fixing bacteria which live either freely in the soil (saprophytic bacteria like *Azotobacter* and *Clostridium*) or as symbionts (nodule bacteria like *Rhizobium leguminosarum*) in the root nodules of legumes. Some blue-green algae, fungi and yeasts also fix nitrogen and converts aerial nitrogen into usable nitrogen. The main steps in nitrogen fixation are as follows:

$$N_2 \longrightarrow 2\,N \qquad \text{(activation of N)}$$
$$2N + 3H_2 \longrightarrow 2NH_3 \qquad \text{(actual fixation)}$$

(a) Before the nitrogen can be fixed, it must be 'activated', i.e. the molecular nitrogen must split into two atoms of free nitrogen. This step requires an investment of energy (about 160 kcal/each molecules of nitrogen).

The actual fixation step, in which two atoms of nitrogen combines with six molecules of hydrogen to form two molecules of ammonia (NH_3), releases about 13 kcal energy. Thus, the two steps together require a net input of atleast 147 kcal. The enzyme that catalyses this step, i.e. nitrogen-fixation is called **nitrogenase**.

(b) Once the ammonium ion (NH_4^+) has appeared in the soil, it can be absorbed by the roots of plants. And nitrogen can be incorporated into amino acids and then into proteins. If the plant is subsequently eaten by an animal, the nitrogen may be incorporated into a new protein. In either cases, the protein ultimately returns to the soil, where it is decomposed by bacteria into its component amino acids.

(c) Nitrogen is also fixed in the industries at a very high temperature (450°C) and pressure (250-100 atm). But, the microorganisms fix atmospheric nitrogen at normal temperature and pressure.

Phosphorus Cycle

Phosphorus cycle lacks an atmospheric component. The basic source and the great reservoir of phosphorus are **rocks** and **other deposits** which have been formedin the past geological ages. These are gradually eroding, releasing phosphates to the ecosystems. But, much phosphate escapes into the sea where part of it is deposited into the shallow sediments and the remaining part of it islost to the deep sediments. However, the means of returning phosphorus to the cycle are inadequate to compensate the loss.

Plants absorb water dissolved inorganic phosphates from the soil for their nutrition and convert them into inorganic phosphates. The latter is transferred to **animal consumers** and **decomposers**. Phosphates absorbed from the soil are returned to it as dead plant and animal organic residues which are then converted to humus by the action of soil microorganisms. The floating algae or phytoplanktons rapidly absorbs soluble inorganic phosphates and convert them into **organophosphates**.

Algae provide food for zooplanktons which inturn are consumed by the other animals. All plants and animals eventually die and in due time, their organic remains or debris decay through the action of microorganisms and the phosphates are released into the water for recycling.

Sulphur Cycle

Plants and animals also requires a continuous supply of sulphur and its compounds in order to synthesise some amino acids and proteins. Only few organisms gain their sulphur requirements in such organic forms as amino acids. Inorganic sulphate (SO_4) is the major source of biologically significant sulphur. Most of the biologically incorporated sulphur is mineralised by the bacteria and fungi in ordinary decomposition by the species of *Aspergillus* and *Neurospora*.

Sulphate is also reduced under anaerobic conditions to elemental sulphur or to sulphides, including H_2S, by some heterotrophic bacteria as species of *Desulfaribrio,* etc. Colourless sulphur bacteria such as species of *Beggiatoa* oxidises H_2S to elemental sulphur and species of *Thiobacillus* oxidise it to sulphate. Other species of *Thiobacillus,* oxidise sulphide to sulphur and still others oxidise sulphur to sulphate as shown in following reactions :

$$H_2S \longrightarrow S \longrightarrow SO_4,\ H_2S \longrightarrow SO_4$$

The sedimentary aspect of the sulphur cycle involves the precipitate of sulphur in the presence of iron under anaerobic conditions. Ferrous sulphide is insoluble in neutral or alkaline water and consequently sulphur has the potential for being bounded up under these conditions to the limits of the amount of iron present.

Because of the thermodynamics of this ferrous sulphide system, other nutrients, important to the biological systems, can also get trapped for varying periods of time. Among these are Cu, Cd, Zn and Co.

Chapter Exercise

1. The phosphorus cycle is unusual in that it is entirely
   (a) within the aquatic ecosystem
   (b) within the terrestrial ecosystem
   (c) sedimentary
   (d) gaseous

2. The diagram shows parts of the carbon cycle.

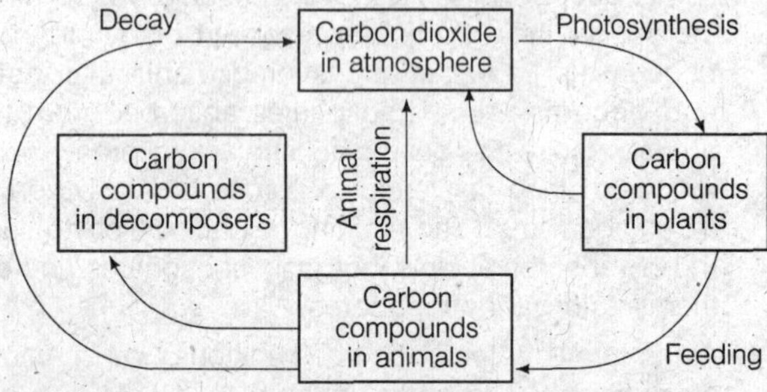

   Which process converts most carbon from one form to the another?
   (a) Animal respiration (b) Decay
   (c) Feeding (d) Photosynthesis

3. Which of the following atoms typically cycles within the most localised area?
   (a) Carbon (b) Water
   (c) Nitrogen (d) Phosphorus

4. The main nitrogen reservoir in the biosphere is
   (a) atmosphere (b) ocean
   (c) organism (d) rock

5. Which of the following atoms most often limits the primary productivity of an ecosystem?
   (a) Carbon (b) Nitrogen
   (c) Sulphur (d) Phosphorus

6. The largest reservoir of phosphorus in the biosphere is the
   (a) atmosphere (b) ocean
   (c) organisms (d) rock

7. Most of the earth's water is
   (a) locked up as salt water or ice
   (b) stored as carbon dioxide
   (c) radioactive
   (d) being lost to atmospheric dehydration

8. In the phosphorus cycle, phosphate becomes available by weathering of rock first to
   (a) consumers (b) producers
   (c) decomposers (d) None of these

9. The global hydrologic cycle supports a net flow of atmospheric water vapour from
   (a) the oceans to land (b) land to the oceans
   (c) polar to tropical regions (d) tropical to polar regions

10. The diagram represents the flow of materials within a balanced ecosystem. The boxes given below are various trophic levels.

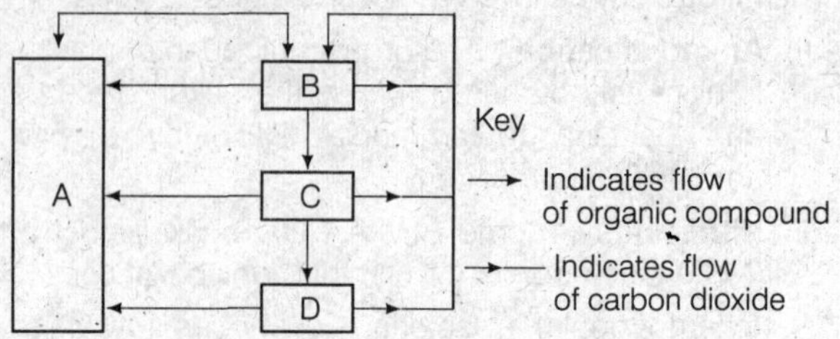

    Which box represents herbivores?
    (a) A (b) B (c) C (d) D

11. Energy and nutrients enter a community by the way of
    (a) producer (b) consumer
    (c) scavenger (d) detritivores

12. Why does after the nutrient enrichment from sewage contamination, a lake often becomes inhospitable to fishes?
    (a) Nutrient input, to a lake causes the explosive growth of algal and cyanobacterial populations. This reduces the penetration of light into the lake, the water temperature falls, and eventually the fish population dies
    (b) Nutrient input to a lake poisons the fish
    (c) Nutrient input to a lake causes the explosive growth of algal and cyanobacterial population. Decomposition of dead algae and cyanobacteria by bacteria results in the depletion of oxygen in the water, which leads to the death of fish
    (d) Nutrient input to a lake poisons the organisms that fish eats

13. The phosphorus cycle lacks .... component.
    (a) an atmospheric (b) an organic
    (c) a mineral (d) an aquatic

14. Phosphorus cycles absorbs phosphates in the form of
    (a) HPO_3^- (b) P_2 (a gas) (c) PO_4^{3-} (d) $AlPO_4$

15. Nitrification is a part of which of the following cycle?
    (a) Oxygen cycle (b) Nitrogen cycle
    (c) Phosphorus cycle (d) None of these

16. Which of the following atoms typically cycles within the most localised area?
    (a) Carbon (b) Water (c) Nitrogen (d) Phosphorus

17. What scientific term is used for the circulation of the essential nutrient?
    (a) Biogeographic cycle (b) Cycling of material
    (c) Biological cycle (d) Biogeochemical cycle

18. Which of the following atoms most often limits the primary productivity of an ecosystem?
    (a) Carbon (b) Nitrogen
    (c) Sulphur (d) Phosphorus

19. Among the following biogeochemical cycles which one does not have losses due to respiration?
(a) Sulphur (b) Phosphorus
(c) Nitrogen (d) All of these

20. The sedimentary cycle is
(a) oxygen (b) nitrogen
(c) phosphorus (d) carbon

21. Sedimentary cycle having a small gaseous component is found in
(a) phosphorus (b) nitrogen
(c) carbon (d) sulphur

22. Biogeochemical cycles are also known as
(a) sedimentary cycling (b) gaseous cycle
(c) material cycling (d) cycles of water

23. Which of the following organisms can fix nitrogen?
(a) Plants (b) Fish (c) Fungi (d) Bacteria

24. Historically, how did nitrogen enter plants?
(a) *Via* rocks (b) As liquid
(c) By bacteria (d) By snake dung

25. The largest reservoir of phosphours in the biosphere is the
(a) atmosphere (b) ocean
(c) organisms (d) rock

26. Phosphorus cycles in the form of
(a) HPO_3^- (b) P_2 (a gas) (c) PO_4^{2-} (d) $AlPO_4$

27. The phosphorus cycle is unusual in that it is entirely
(a) within the aquatic ecosystem
(b) within the terrestrial ecosystem
(c) sedimentary
(d) gaseous

28. Aquatic ecosystems cover approximately how much of the earth's surface?
(a) 20% (b) 40% (c) 75% (d) 90%

29. Which of the following atoms typically cycles within the most localised area?
(a) Carbon (b) Water (c) Nitrogen (d) Phosphorus

30. Nitrogen is a critical elements of the ecosystem because it is
(a) essential element (b) abundant in atmosphere
(c) labile (d) fixed by microbes

31. Which one is sedimentary cycle?
(a) Oxygen cycle (b) Nitrogen cycle
(c) Hydrogen cycle (d) None of these

32. Concentration of nitrogen remains constant by
(a) nitrogen cycle (b) thundering
(c) lightening (d) enzymes

33. Among the following biogeochemical cycles which one does not have losses due to respiration?
(a) Phosphorus (b) Nitrogen
(c) Sulphur (d) All of these

34. Earth is an open system for
(a) minerals (b) energy (c) gases (d) water

35. The supporting services of ecosystem include
(a) soil formation and retention
(b) nutrient cycling
(c) water cycling
(d) All of the above

36. The main nitrogen reservoir in the biosphere is the
(a) atmosphere (b) ocean
(c) organism (d) rocks

37. Fixation occur with the help of
(a) symbiotic bacteria
(b) leguminous nodules
(c) with the help of nitrogenase enzyme
(d) All of the above

38. Which is the most common available form of S to plants?
(a) S (b) SO_2 (c) H_2S SO_4^{2+}

39. Nitrification occur with the help of .... bacteria.
(a) *Nitrosomonas* (b) *Nitrobacter*
(c) Both (a) and (b) (d) None of these

40. Nitrogen fixation is done by free living bacteria
(a) *Azotobacter* (b) *Clostridium*
(c) Both (a) and (b) (d) None of these

41. Most common useable source of nitrogen is
(a) N_2 (b) NO_2^- (c) NO_3^- (d) All of these

42. Which of the following is not gaseous type cycle?
(a) Corbon cycle (b) Oxygen cycle
(c) Phosphorus cycle (d) Nitrogen cycle

43. Main reservoir for sulphur and phosphorus is
(a) air (b) water (c) soil (d) All of these

44. Increase in CO_2 level, leading many problems due to
(a) burning of fossil fule (b) respiration
(c) seveage (d) All of these

45. Ozone layer protect us from
(a) ultra-violet rays (b) infra-red wave
(c) X-rays (d) All of these

Answers

1. (c)	**2.** (d)	**3.** (d)	**4.** (a)	**5.** (d)	**6.** (d)	**7.** (a)	**8.** (c)	**9.** (a)	**10.** (c)
11. (a)	**12.** (c)	**13.** (a)	**14.** (c)	**15.** (b)	**16.** (d)	**17.** (c)	**18.** (a)	**19.** (d)	**20.** (c)
21. (d)	**22.** (c)	**23.** (d)	**24.** (c)	**25.** (d)	**26.** (c)	**27.** (c)	**28.** (c)	**29.** (d)	**30.** (a)
31. (d)	**32.** (a)	**33.** (d)	**34.** (b)	**35.** (d)	**36.** (a)	**37.** (d)	**38.** (d)	**39.** (c)	**40.** (c)
41. (c)	**42.** (c)	**43.** (c)	**44.** (a)	**45.** (a)					

37

Organism, Population and their Community

Organism

In biology, an organism is known as any contiguous living system, such as animal, plant and or bacterium. Hence, it can be said that all organisms are biotic in nature constituting the living part of the environment. Each and every organism on earth have five basic needs without which their survival is not possible such as air, water, nutrients (food), energy and a place to live (Shelter).

The abiotic components of any ecosystem determines, which species of organisms will be able to survive in the given environment. These abiotic factors includes temperature, water, light, soil, etc.

Responses to Abiotic Components

All these above mentioned abiotic factors are known to be highly variable in nature. Thus, in order to exist in such a condition, many species have eventually been evolved a relatively constant internal environment to enhance the overall fitness of the species.

Population

The term population can be defined as a group or assemblage of organisms of the same species that live at a given time in a particular area.

For example, the lion population of the Gir forest or the peacock population of India, etc. The population may be sub-divided into **demes** or **local populations** which are a group of interbreeding organisms. It is the smallest collective unit of a plant or animal population.

The study of population dynamics (changes in population sizes) is called **demography**. The main aim of any demographic study is to quantify the changes in a population by finding out the number of births, deaths, immigrants and emigrants.

Difference between Community and Population

Community	Population
It is a group of individuals of different species inhabiting in a particular area.	It is a grouping of individuals of a single species in an area.
Interbreeding is absent amongst different members of a community.	Individuals interbreed freely.
Different members of a community are morphologically and behaviourly dissimilar.	Morphologically and behaviourly similar species are found in a population.
It is a larger unit of organisation.	It is a smaller unit of organisation.
In a biotic community, there is often a relationship of eating and being eaten.	There is no relationship of eating and being eaten in population.

Characteristics of Population

A population has its own characteristics such as spacing, population density, birth rate, death rate, biotic potential, age, distribution, population pressure, size, growth, cycle, etc. A population must adjust to their environment for their steady and constant survival.

Some important characteristics of population are discussed below:

(i) **Dispersion** There are three possible ways in which the individuals in a population may be spaced relative to one another. It may be uniform, random or clumped. Invertebrates form populations containing uniformly spaced individuals. Random spacing occurs particularly among the plants with dispersal mechanisms.

However, some form of close aggregation of groups of individuals or clumping is the most common pattern of dispersion of individuals in a population.

Even mating pairs and their newborn offsprings may regarded as clumped. Besides such clumping for reproductive purposes, animals aggregate in swarms, herbs, flocks or colonies for greater facility of feeding and protection.

According to **Allee's principle**, each species has an optimum clumping value and over-crowding, beyond this stage, it tends to act as a limiting factor in population regulation.

(ii) **Population density** Density is the number of individuals per unit area or volume. The unit of area used in measuring density varies. The complete count of individuals in an area is called **census.**

The density of a population can be studied in several ways. The abundance of a population within its niche is called its **economic** or **niche density**.

If it is decreasing then obviously the number of individuals in the population who are dying per unit time, exceeds the birth rate and on the other hand, if the population density is increasing, this means that the birth rate is higher than the death rate and lastly, when both rates are equal, the population density appears to be constant for a given time.

(iii) **Natality** It is the rate of production (birth rate) of new individuals to the unit of population per unit time. The maximum natality or birth rate achieved under ideal conditions is called **biotic potential**. In human beings, natality is calculated per thousand per year.

(iv) **Mortality** It is the rate of loss of individuals (death rate) per unit time due to the death of an individuals. Mortality can be either specific or realised death rate that occurs under ideal conditions due to the natural process of senescence.

Realised mortality is the actual rate of death occurring in a unit population due to the different environmental changes, competition, disease, predation, etc. It is seen that maximum mortality occurs in egg/larva/seedling and in old age.

Population Dispersal

It is the inward or outward movement of the individuals from the population for preventing overcrowding, obtaining food, avoiding predators and adverse conditions. It mainly affects the size of population. It can be of following three main types:

(i) **Migration** It is a two way movement of entire population to avoid unfavourable conditions.

(ii) **Emigration** It is the permanent outward movement of individuals from a given population. It actually decreases the size of population.

(iii) **Immigration** It is the permanent inward/coming of individuals from outside into a population. It increases the size of population.

Age Distribution

The distribution of individuals of different age groups in a population is known as age distribution. There are three major ecological ages (age groups) in any population such as pre-reproductive, reproductive and post-reproductive.

The relative duration of these age groups in proportion to the life span varies greatly with different organisms.

In man, the three 'ages' are relatively equal in length, about a third of his life falling in each class. These, three ages were proposed by as Bodenheimer (1985).

Population Growth

It is the number of individuals added per unit population per unit time due to the higher birth rate and immigration over the rate of death and immigration ($\Delta N / \Delta t$). The percentage ratio of natality to mortality is called **vital index** ($N/M \times 100$).

Suppose, N_0 be the initial population and N_t be the population after a time interval 't', B equals to total births, I equals to immigration, D equals to total deaths and E equals to immigration. Then, the population after time 't' will be

$$N_t = N_0 + B + I - D - E$$

Community Ecology

Community is a group of populations of different species. The community ecology deals with study of the interactions between species of different population. The brief detail of different types of interactions of species of different populations is given below:

Population Interaction

There are mainly three types of interactions that are commonly seen among the biotic community. These are of following three types:

(i) **Neutral interaction** in which neither of the species is harmed nor benefitted.

(ii) **Negative** (antagonistic) **interaction** in which one species is harmed, while the other is benefitted, harmed or remains unaffected.

(iii) **Positive** (beneficial) **interaction** in which one species helps the other, either one way or other way.

Antagonistic Relationship

It is a harmful co-action between two different species, in which atleast one of the species is harmed.

The antagonistic relationship between organisms is mainly shown by competition, predation, parasitism and antibiosis.

(i) **Competition** It is a rivalry between two or more organisms for obtaining the same resources such as food, shelter, etc. It can lead to death for some competitors and occurs due to the fighting or being deprived of food or space rather than being killed for food as in predation or by disease as in extreme parasitism. The role of competition in regulating population size is directly effective by causing mortalities through fighting, nest destruction and loss of food supplies.

- **Interspecific competition** is the most critical factor confining a species to a particular niche. This becomes evident with the expansion of the species beyond the usual limits of its niche when this competition is over. The interspecific competition also results in **speciation**, process of evolutionary differentiation between populations, in which one older species may split into two or more new species.

(ii) **Predation** The word 'predation' is used in general sense to describe the eating of one species by another, i.e. the predators consume the other living animals.

A typical predation occurs when the carnivorous prey on herbivorous or on other carnivores. However, predatory organisms are not only the animals, but can be plants also. Typical examples of predatory plants are insectivorous plants like sundew (*Drosera*), aquatic bladderwort (*Utricularia*) and pitcher plant (*Nepenthes*).

(iii) **Parasitism** It is an interaction between two individuals, where in the parasite gets the benefit at the expense of the host. In other words, it is relationship between two living organisms of different species, in which one organism called **parasite**, obtains its food directly from the another living organism called **host**.

The parasites are divided into different categories:

(a) **Ectoparasites** are those parasites that lives on the outside of host, e.g. human body louse (*Pediculus humanus*), human flea, (*Pulex irritans*), rat flea, (*Xenopsylla cheopsis*), etc.

(b) **Endoparasites** are those parasites, which live in the digestive tract, body cavities, various organs blood or other tissues of the host, e.g. *Plasmodium malariae, Trypanosoma gambiense*, *Taenia solium* and *Fasciola hepatica*.

(c) **Facultative parasites** are those parasites, which lives entirely as saprophytes, but become parasites under certain conditions, e.g. oyster, prawn and pea crab.

(d) **Obligate parasites** are those parasites, which can live only parasitically. These obligatory parasites proved to be too detrimental and ultimately caused extinction of their host and thus, must have eventually eliminated themselves by destroying their source of food.

(e) **Antibiosis** It is an interaction between two living individuals of different species in which an organism does not allow other organisms to grow or live near it and inhibition is achieved through the secretion of chemicals called allochemicals.

Symbiotic Relationship

The term 'symbiosis' refers to 'living together'. A symbiotic relationship is a beneficial co-action between two or more different species, in which one or both species are benefitted and neither of them is harmed.

Symbiotic relationship is widespread in nature and is shown through the following connections or associations or relationship.

It can be of following three types:

(i) **Commensalism** It is an association or relationship between two different organisms in which one is always benefitted, while the other is neither benefitted nor harmed. A commensal may live permanently attached to its host or may live within its body. All commensalistic unions are facultative.

(ii) **Protocooperation** It is a relationship between two species, which is favourable but not obligatory to both. An interesting example is shown by the association of **hermit crab** and a **coelenterate** (sea anemone). In which, the sea anemone (*Adamsia palliata*) grows on the back of the hermit crab (*Eupagurus prideauxi*) and protecting the crab with the help of its nematocysts.

(iii) **Mutualism** It is a relationship between two or more individuals of different species in which all are benefitted by one another. Mutualism (Latin *mutu*–reciprocal) may be **facultative**, when the individuals involved in the association are capable of existing independently and **obligatory**, when the relationship is imperative to the existence of one or both species.

Chapter Exercise

1. **The relationship between species and another within a community that has evolved through interaction is based upon**

(a) requirement and mode of obtaining food only
(b) requirement and mode of obtaining shelter only
(c) on the habits of the species only
(d) All of the above

2. **Communities that exist in the aphotic zone ultimately depend on food manufactured by the chemoautotrophic bacteria or**

(a) algae and cyanobacteria that also live in the aphotic zone
(b) algae and cyanobacteria that live in the photic zone
(c) decomposers
(d) minerals found on the ocean bottom

3. **An association of different inter-related populations belonging to different species in common environment, which can survive in nature is called**

(a) herd
(b) family
(c) population
(d) community

4. **Why is a new island more hospitable to the colonisers than an older island?**

(a) Competition is more intense on the newer island
(b) Predation is loss of a factor on older islands
(c) The extinction rate is higher on the newer island
(d) The intensity of both competition and predation is less on the newer island

5. Which of the following is an omnivore?
(a) Man (b) Crow
(c) Both (a) and (b) (d) Rat

6. Humans who have pets are healthier than humans, who do not have pets, which of these terms applies to the relationship between a human and a pet?
(a) Competition (b) Parasitism
(c) Commensalism (d) Mutualism

7. When the carrying capacity is equal to the population size, the net rate of increase in the population is equal to
(a) the carrying capacity
(b) less than zero
(c) the intrinsic rate of growth
(d) zero

8. The parasite, which come in contact with host for only a part of their life or occasionally at the time of feeding are called intermittent parasites. Which of the following is not an intermittent parasite?
(a) Leech (b) Bed bug
(c) Female mosquito (d) *Ascaris*

9. The animal, which captures and feeds on other animals is called
(a) predator (b) prey
(c) parasite (d) host

10. In world populations, individuals most often show a ...... pattern of dispersion.
(a) random (b) density-dependent
(c) equilibrium (d) clumped

11. Which of the following plants is not a predator plant?
(a) *Hydrilla* (b) *Drosera*
(c) *Utricularia* (d) *Dionaea*

12. The sum total of the populations of the same kind of organisms constitute
(a) colony (b) genus
(c) community (d) species

13. Intracellular endoparasite is
(a) *Trichinella* (b) *Plasmodium*
(c) *Trypanosoma* (d) *Ascaris*

14. A synthetic character of biotic community is
(a) fidelity (b) dominance
(c) physiogamy (d) All of these

15. Which one is a qualitative character?
(a) Abundance (b) Frequency
(c) Biomass (d) Phenology

16. A climax community is recognised by
(a) grasses, herbs and shrubs
(b) trees
(c) high biomass
(d) uniform composition

17. The phenomenon of having higher number of species in ecotone is called
(a) edge effect (b) abundance
(c) frequency (d) dominance effect

18. *Wuchereria* is a
(a) ectoparasite of man (b) endoparasite of man
(c) predator of animals (d) parasite of dog

19. What relationship persist between birds and cattle?
(a) Commensal (b) Predator
(c) Protocooperation (d) Mutualism

20. Female *Anopheles* visit the host for sucking blood at intervals. The type of parasitism is
(a) hemiparasite (b) endoparasite
(c) holoparasite (d) intermittent parasite

21. The type of relationship between sea anemone and hermit crab is
(a) commensalism (b) mutualism
(c) symbiosis (d) parasitism

22. Territory is defended against
(a) predator (b) any intruder
(c) intruder of another species (d) intruder of sam species

23. Mycorrhiza, an association of fungus and aniospermic root is referred as
(a) ammensalism (b) mutualism
(c) parasitism (d) commensaim

24. The non-obligatory mutually beneficial elationship is represented by
(a) symbiosis (b) protocooeration
(c) commensalism (d) mimicry

25. Species interaction with negative innence on both is called
(a) predation (b) paratism
(c) ammensalism (d) copetition

26. An association of two species in wich both species are benefitted is called
(a) protocooperation (b) utualism
(c) ammensalism (d) All of these

27. Lichen is a symbiotic associatin of algae with
(a) protozoan () fungi
(c) roots of plants d) bacteria

28. In aquatic mammals, a bluber is present
(a) in the skin
(b) around the different orgas
(c) in the muscles
(d) in the connective tissu

29. The interdependent evlution of flowering plants and pollinating insects togther is known as
(a) mutualism (b) cooperation
(c) commensalism (d) co-evolution

30. **Exotic species sometimes become serious pest because of**
(a) better environmental conditions in new locality
(b) better adaptability to new environment
(c) increase in reproductive potential
(d) the absence of natural competition

31. **Cryptic appearance/camouflage is found in**
(a) praying mantis
(b) leaf insect
(c) stick insect/dead leaf butterfly
(d) All of the above

32. **Relation between viceroy butterfly and monarch butterfly is that of**
(a) camouflage (b) Mullerian mimicry
(c) Batesian mimicry (d) warning colouration

33. **Scavenging is**
(a) an interaction between two living individuals of different species, in which one organism does not allow other organism to grow or live near it
(b) a relationship between two living individuals of different species, in which one is benefitted, while the other is neither harmed not benefitted except to a negligible extent
(c) a rivalry between two or more organisms for obtaining the same resources
(d) feeding on flesh of an animal, which had died naturally or has been killed by another animal

34. **Animals which are parasites on plant are**
(a) nematodes (b) lac insect
(c) aphids (d) All of these

35. **Symbiosis may be defined as a relationship, in which**
(a) both the partners derive benefit from each other
(b) both the partners are harmed
(c) only one partner derives the benefit
(d) none of the partner derives the benefit

36. **The organisms which spend their life on the body of a living host are called**
(a) ectoparasites (b) endoparasites
(c) semi-parasites (d) hyper-parasites

37. **Which of the following shows mimicry?**
(a) *Carausius morosus* (b) *Kalima parolecta*
(c) *Mantis religiosa* (d) All of these

38. **The age structure of population is its**
(a) relative number of individuals at each age
(b) number of newborns each year
(c) number of young reaching puberty each year
(d) relative number of deaths at each age

39. **The periodic departure and return is known as**
(a) migration (b) immigration
(c) emigration (d) mutation

40. **Competition among the plants in a community is for**
(a) water (b) nutrients
(c) sunlight (d) All of these

41. **Protective device in animals is**
(a) competition (b) commensalism
(c) mimicry (d) symbiosis

42. **Population natality is**
(a) realised natality (b) biotic potential
(c) vital index (d) carrying capacity

43. **In ecology, N/S represents**
(a) population density (b) rate of growth
(c) death rate (d) birth rate

44. **dN/dt represents**
(a) birth rate
(b) death rate
(c) change in population size
(d) carrying capacity

45. $\frac{(K-N)}{K}$ **represents**
(a) carrying capacity (b) environmental resistance
(c) vital index (d) crash phase

46. **The non-obligatory mutually beneficial relationship is represented by**
(a) symbiosis (b) protocooperation
(c) commensalism (d) mimicry

47. **Species of a biological community that has higher abundance, cover and biomass is**
(a) keystone species (b) link species
(c) dominant species (d) network species

48. **The occurrence at vertical layers in a forest is**
(a) pattern (b) stratification
(c) zonation (d) ecotone

49. **Species found exclusively in ecotone are called**
(a) morphospecies (b) edges species
(c) allopatric species (d) sympatric species

50. **The number of species per unit number of individuals of a biotic community is called**
(a) abundance (b) frequency
(c) diversity (d) density

51. **Exponential growth in human population is done in**
(a) lag phase (b) log phase
(c) plateau stage (d) first stage

52. **Mutualism and protocooperation are**
(a) positive interactions
(b) negative interactions
(c) Both (a) and (b)
(d) None of these

53. Which one of the following is a qualitative character?
(a) Dispersion (b) Life forms
(c) Growth forms (d) Diversity

54. Insectivorous plants are adapted to soils
(a) rich in water
(b) deficient in water
(c) deficient in nitrogenous compounds
(d) deficient in trace elements

55. Two species cannot live in the same niche due to
(a) Allen's law
(b) Gause's hypothesis
(c) Doll's rule
(d) Weismann's theory

56. At asymptome stage, the population is
(a) stabilised (b) increasing
(c) decreasing (d) changing

57. Which one of the following biomes is dominated by gymnosperm or conifer trees (pines, firs, spruces)?
(a) Taiga (b) Tundra
(c) Desert (d) Broad-leaved forest

58. The age of pyramid with broad base indicates
(a) high percentage of young individuals
(b) low percentage of young individuals
(c) high percentage of old individuals
(d) low percentage of old individual

59. There are two optional ways of exploitation. One way is parasitism. Which is the other one?
(a) Antibiosis
(b) Competition
(c) Predation
(d) Commensalism

60. Match the following columns.

Column I (Population Interaction)	Column II (Examples)
A. Mutualism	1. Ticks on dogs
B. Commensalism	2. *Balanus* and *Chathamalus*
C. Parasitism	3. Sparrow and any seed
D. Competition	4. Epiphyte on a mango branch
E. Predation	5. Orchid, *Ophrys* and bee

Codes

	A	B	C	D	E		A	B	C	D	E
(a)	1	5	4	3	2	(b)	2	1	5	4	3
(c)	3	2	1	5	4	(d)	4	3	2	1	5

61. A high density of elephant population in an area can result in
(a) mutualism (b) interspecific competition
(c) intraspecific competition (d) predation on one another

62. Exponential growth in a population is shown by
(a) J-shaped curve (b) S-shaped curve
(c) Hyperbolic curve (d) None of these

63. Most important parameter for determining population growth is
(a) natality (b) vital index
(c) population size (d) population growth curves

64. The population growth is promoted by
(a) mortality (b) natality
(c) emigration (d) None of these

65. Study of trends in human population growth and prediction of future development is called
(a) demography (b) biography
(c) calligraphy (d) psychology

Answers

1. (d)	**2.** (b)	**3.** (d)	**4.** (d)	**5.** (c)	**6.** (d)	**7.** (d)	**8.** (d)	**9.** (a)	**10.** (d)
11. (a)	**12.** (d)	**13.** (b)	**14.** (d)	**15.** (d)	**16.** (b)	**17.** (a)	**18.** (b)	**19.** (c)	**20.** (d)
21. (b)	**22.** (b)	**23.** (b)	**24.** (b)	**25.** (d)	**26.** (b)	**27.** (b)	**28.** (d)	**29.** (d)	**30.** (d)
31. (d)	**32.** (c)	**33.** (d)	**34.** (d)	**35.** (a)	**36.** (a)	**37.** (d)	**38.** (a)	**39.** (a)	**40.** (d)
41. (c)	**42.** (b)	**43.** (a)	**44.** (c)	**45.** (b)	**46.** (b)	**47.** (c)	**48.** (b)	**49.** (b)	**50.** (b)
51. (b)	**52.** (a)	**53.** (d)	**54.** (c)	**55.** (b)	**56.** (a)	**57.** (a)	**58.** (a)	**59.** (c)	**60.** (d)
61. (c)	**62.** (a)	**63.** (c)	**64.** (b)	**65.** (a)					

38

Biodiversity

'Biodiversity' is a concise term used for 'biological diversity'. Biodiversity means the variability among living organisms from all sources, such as terrestrial, marine, other ecosystems and the ecological complexes of which they are the part. It includes diversity within species, between the species and of the ecosystems.

The term 'biodiversity' was popularised by socio-biologist **Edward Wilson**. It describes all aspects of diversity but especially the richness of species within a specified region or the world. Complexity of ecosystems and diversity differs from place to place as each habitat has its own distinct biota.

Levels of Biodiversity

Biodiversity of any habitat is comprised of three basic levels:

1. **Genetic Diversity**

   Genetic diversity within the species creates different sub-species, variety, breed, forms, etc. It also ensures that greater the genetic diversity among organisms of a species, more sustenance it has against the environment disturbances.

2. **Species Diversity**

   It is known as the measure of the variety of species and their relative abundance present within a region. Species diversity is predicted under two important measures:

   (a) Species richness – No. of species per unit area

   (b) Species evenness – Relative abundance of species in a given area.

3. **Ecological Diversity**

   It refers to the diversity of species. It is also related to the genetic diversity at ecological level. India has greater ecological diversity than scandiravian countries.

Biodiversity on the Earth

According to the IUCN (2004), the total number of plants and animal species described so far is slightly more than 1.5 million but we have no clear idea of how many species are yet to be discovered and described.

Some extreme estimates ranges from 20-50 million but a more conservative and scientifically sound estimate made by Robert May, places the global species diversity at about 7 million. It is estimated that more than 70% of all the species recorded are animals and plants (including algae, fungi, bryophytes, gymnosperms and angiosperms) comprise about 22% of the total.

Among animals, insects are the species most having rich taxonomic group, making up more than 70% of the total. The number of fungi species in the world is more than the combined total of the species of fishes, amphibians, reptiles and mammals.

Approximate Number of Species Present in India

Number of species of plants and bacteria		Number of species of animals	
Group	Number of species	Group	Number of species
Angiosperms	17,500	Mammals	390
Gymnosperms	64	Aves	1,232
Pteridophytes	1,100	Reptiles	456
Bryophytes	2,850	Amphibians	209
Lichens	2,000	Pisces	2,546
Fungi	14,500	Protochordates	119
Algae	6,500	Invertebrates	8,329
Bacteria	850	Arthropodes	68,389
		Molluscus	5,070
		Protozoans	2,577

Approximate Number of Species Identified all Over the World

Group	Number of species
Higher plants	2,70,000
Algae	40,000
Fungi	72,000
Bacteria (including cyanobacteria)	4,000
Viruses	1,550
Mammals	4,650
Birds	9,700
Reptiles	7,150
Fish	26,959
Amphibians	4,780
Insects	10,25,000
Crustaceans	43,000
Molluscs	70,000
Nematodes and worms	25,000
Protozoans	40,000
Others	1,10,000

Biodiversity in India

India with 2.4% of the world's land area, share about 8.1% of the global species diversity. It is among 12 mega diversity countries of the world. Nearly 45,000 species of plants and twice as many of animals have been recorded from India. It is estimated that probably more than 1,00,000 plant species and more than 3,00,000 animal species are yet to be discovered and described.

Extinction (Loss) of Species (Biodiversity)

The total elimination or dying out of a particular species from the earth leads to the extinction of that species. A species becomes prone to extinction due to the categories of attributes, drastic environmental changes and population characteristics, i.e. we can also say that the biological wealth of our country is declining rapidly due to the human activities mainly.

Population traits which make species susceptible to extinction are

- Large body size, e.g. elephant, rhinoceros, lion.
- Small population size.
- Alien (exotic) species invasions.
- Low reproductive potential, e.g. blue whale.
- Higher states of trophic level, e.g. Bengal tiger.
- Fixed migratory route and habitat, e.g. blue whale.
- Narrow range distribution or small geographical range, e.g. woodland.
- Co-extinctions.
- Island species.
- Lack of genetic variability.
- Over exploitation.
- Inability to switch over to alternate foods.

Effect of Loss of Biodiversity

(i) Decline in the plant production.

(ii) Lower resistance and resilience to environment.

(iii) Pertrubutions such as draught and increased variability in ecosystem processes such as plant productivity, water use, diseases, etc.

Conservation of Biological Diversity

Conservation of biodiversity is protection, uplift and scientific management of biodiversity so as to maintain it at its optimum level and derive sustainable benefits for the present as well as future strategies.

Biodiversity is being conserved under two mains approaches, i.e. ***In situ* Conservation** (Conservation and protection of whole ecosystem and its biodiversity at all levels in order to protect the threatened species) and *ex situ* conservation (Conservation by placing the threatened plants and animals in special care units for their protection). The *in situ* conservation includes biodiversity hot spots, protected areas, etc. While, the *Ex situ* conservation includes offsite collection, gene banks (including seed bank, tissue culture, etc).

Main Strategies of Conservation

Some of the well known major strategies to conserve biodiversity are given below:

- All the threatened species should be protected. Priority should be given to those species belonging to the monotypic genera, endangered, overvulnerable, vulnerable and rare over other species.
- All the possible varieties (old or new) of food, forage and timber plants, liverstock, aquaculture animals, microbes should be conserved.
- Wild relatives of economically important organisms should be identified and conserved in the protected areas.
- Critical habitats for feeding/breeding/resting/nursing of each species should be identified and safe guarded.
- Resting/feeding places of migratory/wide ranging animals should be protected, pollution controlled and exploitation regulated.
- **National Wildlife Protection** Lawmust be enacted (in India 1972), wildlife protection strategies formulated (1983) and protection programmes integrated with the international programmes.
- Priority based ecosystems.
- The reproductive capacity of the exploited species and productivity of the ecosystem should be determined.
- International trade in wildlife should be highly regulated.
- Development of reserves or protected areas.
- Controlling introduction of new species.
- Pollution reduction and public awareness.

Chapter Exercise

1. Maintenance which biome requires grazing and periodic fire. It is
(a) scrub forest
(b) grassland
(c) tropical/temperate deciduous forest
(d) taiga

2. In India, per capita forest cover is
(a) 1.06 hac (b) 0.06 hac
(c) 0.6 hac (d) 1.6 hac

3. Deforestation generally decreases
(a) rainfall (b) drought
(c) global warming (d) soil erosion

4. Renewable source of energy is
(a) petroleum (b) coal
(c) trees (d) nuclear fuel

5. Most biodiversity zones are
(a) Western Ghats and North-East region
(b) Deccan Peninsula and Sunderbans
(c) Semi-arid Zone and North-East region
(d) Andaman and Nicobar Islands

6. The key criteria for determining a hot spot are
(a) biological augmentation
(b) disruption of interaction networks
(c) number of endemic species and degree of threat
(d) habitat destruction

7. Silent valley of Kerala is preserved because
(a) it contains very rare species of plants and animals
(b) the soil is rich in minerals
(c) the areas of land were used extensively for agricultural purposes
(d) it has pine trees

8. Estuaries are
(a) freshwater wetlands
(b) salt water wetland
(c) least productive ecosystems
(d) None of the above

9. Forest area in India is about
(a) 9% (b) 19%
(c) 29% (d) 50%

10. The integration of several sciences such as ecology and genetics to sustain biological diversity at all levels is known as
(a) conservation biology (b) restoration ecology
(c) ecology (d) bioremediation

11. Swamps differ from marshes in having
(a) grasses (b) floating leaves plants
(c) trees and shrubs (d) None of these

12. What percentage of national forest policy envisages forest cover for the hills?
(a) 33% (b) 45%
(c) 67% (d) 51%

13. Which of the following factors does not play role in increasing biodiversity?
(a) Complex population interactions
(b) Introduced species
(c) A higher degree of habitat heterogeneity
(d) Narrower niches

14. The three levels of biodiversity are
(a) genetic, species and biome
(b) molecular, species and biome
(c) genetic, population and ecosystem
(d) genetic, species and ecosystem

15. Wetland of world occupies land about
(a) 6% (b) 10%
(c) 12% (d) 14%

16. National forest policy was enunciated in
(a) 1972 (b) 1978 (c) 1982 (d) 1988

17. Mild grazing in grassland
(a) stimulates growth of grasses
(b) arrests growth of grasses
(c) retards its growth
(d) destroys vegetation

18. Dodo of Mauritius became extinct due to the
(a) over-exploitation (b) habitat destruction
(c) global warming (d) predation

19. Important function of wetland is
(a) recycling of groundwater (b) controlling floods
(c) purify run-off water (d) All of these

20. Soil erosion can be prevented by
(a) deforestation (b) afforestation
(c) overgrazing (d) vegetation removal

21. Total land available for grazing in India is
(a) 23% (b) 29% (c) 31% (d) 37%

22. The state which have all the three species of crocodile and started saving crocodile project is
(a) Uttar Pradesh (b) Karnataka
(c) Kerala (d) Maharashtra

23. The convention on biological diversity consists of
(a) 45 articles (b) 42 articles
(c) 52 articles (d) 75 articles

24. The number of ecological hot spots in world is
(a) 14 (b) 16
(c) 20 (d) 34

25. Recently extinct bird is
(a) Carolina parakeet (b) peacock
(c) Kirtland's warbler (d) reed bunting

26. Buffer zone of biosphere reserve is where
(a) no human activity is allowed
(b) human activity is limited
(c) free human activity occurs
(d) wild animals are absent

27. Earth Summit of Rio de Janerio (1992) resulted in
(a) compilation of red list
(b) establishment of biosphere reserves
(c) conventions on biodiversity
(d) IUCN

28. The term hot spots for high diversity ecological region was coined by
(a) IUCN (b) Myer
(c) Odum (d) Kormandy

29. Which of the following is classified as β-diversity?
(a) Habitat diversity (b) Ecosystem diversity
(c) Resource diversity (d) Species diversity

30. Rare endangered and endemic taxa can be found intact and flourishing in
(a) sacred groves (b) buffer zones
(c) tropical forests (d) temperate forests

31. Which example below is not a threat to biodiversity?
(a) Habitat destruction (b) Bioremediation
(c) Introduced species (d) Over-exploitation

32. The integration of several sciences such as ecology and genetics to sustain biological diversity at all levels is known as
(a) conservation biology (b) restoration ecology
(c) ecology (d) bioremediation

33. The key criteria for determining a hot spot are
(a) biological augmentation
(b) disruption of interaction networks
(c) number of endemic species and degree of threat
(d) habitat destruction

34. First National Park of India was
(a) Panna National Park
(b) Jim Corbett National Park
(c) Kaziranga National Park
(d) Satpura National Park

35. Gir sanctuary is mainly for
(a) rhino (b) tiger
(c) lion (d) elephant

36. WCU is
(a) World Conservation Union
(b) Wide Conservation Union
(c) World Conserve Union
(d) None of these

37. Convention of biodiversity came into force on
(a) 29 December, 1993 (b) 6 June, 1992
(c) 3 October, 1994 (d) 4 October, 1993

38. 34 hot spots of the world occupy an area of about
(a) 12% (b) 8% (c) 10% (d) 15.7

39. Arboretum is
(a) a forest patch
(b) a piece of social forestry
(c) botanical garden with trees and shrubs
(d) botanical garden with cacti and herbs

40. A sacred lake is
(a) Dal lake
(b) Khecheopalri lake of Sikkim
(c) Suraj Khand lake
(d) Chilka lake

41. Susceptability to extinction is due to
(a) large body size (b) small population
(c) high trophic level (d) All of these

42. Number of categories recognised by red list is
(a) 4 (b) 5 (c) 6 (d) 10

43. Rate of replacement of species along the gradient of habitats/communities is called
(a) α-diversity (b) β-diversity
(c) γ-diversity (d) ω-diversity

44. Discovery of new species has recently picked up due to project
(a) species 2000
(b) global diversity and information facility
(c) Both (a) and (b)
(d) agenda 21

45. An endangered bird is
(a) passenger pigeon (b) pink-headed duck
(c) great Indian bustard (d) vulture

46. Rhinoceros are found till now-a-days in
(a) Asia (b) Africa
(c) Both (a) and (b) (d) America

47. The only ape found in India is
(a) gorilla (b) chimpanzee
(c) hoolock gibbon (d) leaf monkey

48. MAB stands for
(a) Man And Biosphere
(b) Man, Antibodies and Bacteria
(c) Man And Biotic community
(d) Mayer, Anderson, Bisby

49. Rare animal Hispid hare (*Caprologus hispidus*) is found in
(a) Dachigam National Park
(b) Kaziranga National Park
(c) Manas National Park
(d) Kanha National Park

50. The national park which has maximum number at rare animals is

(a) Manas National Park (b) Kaziranga National Park
(c) Dudhwa National Park (d) Corbett National Park

51. A survey was carried out to investigate the relationship between the human population density and elephant population density in Zimbabwe. Twenty five sites were sampled, all within a 15000 km^2 area in the North of the country.

Elephant populations were estimated by flying over the area and counting elephants seen from the air. The results are shown in the figure. Each point represents the results from one of the sites sampled

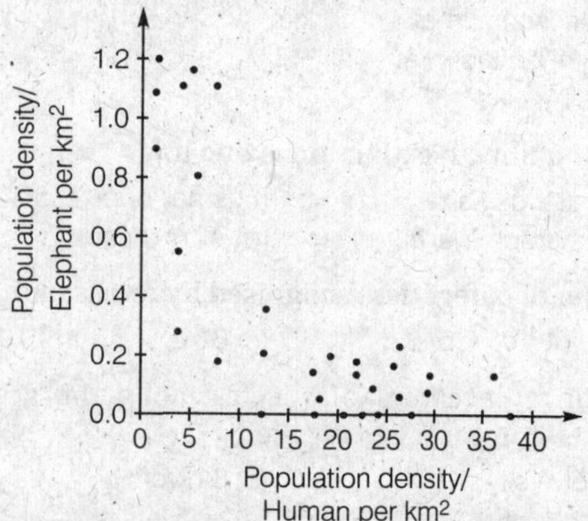

Why elephant populations are very low in areas where human population density is high?

(a) Elephants occupy more grassland areas and this reduces the habitats available to humans
(b) High human population causes an incline in elephant populations
(c) More humans are threat to elephants as they compete for land needed for the agriculture and settlement
(d) All of the above

52. The national animal of USA is

(a) tiger (b) bold eagle
(c) phoenix (d) giant panda

53. Where among the following will you find pitcher plant?

(a) Rainforest of North-East India
(b) Sundarbans
(c) Thar Desert
(d) Western Ghats

54. The term 'alpha diversity' refers to

(a) genetic diversity
(b) community and ecosystem diversity
(c) species diversity
(d) diversity among the plants

55. Silent valley is tropical evergreen forest located in

(a) Kerala
(b) Karnataka
(c) Maharashtra
(d) Odisha

56. Kanha National Park is located in .... and is famous for....... .

(a) Madhya Pradesh; elephant
(b) Madhya Pradesh; tiger
(c) Odisha; tiger
(d) Asom; elephant

57. Identity the odd combination of the habitat and the particular animal concerned.

(a) Dachigam National Park — Snow Leopard
(b) Sundarbans — Bengal Tiger
(c) Periyar — Elephant
(d) Rann of Kutch — Wild Ass

58. The endangered largest living lemur *Idri idri* is inhabitant of

(a) Madagascar (b) Mauritius
(c) Sri Lanka (d) India

59. National 'Wild life Protection Act' was passed in

(a) 1962 (b) 1972 (c) 1982 (d) 1987

60. RDB is released by

(a) IUCN (b) WWF
(c) WCU (d) NEERI

Answers

1. (b)	**2.** (b)	**3.** (a)	**4.** (c)	**5.** (a)	**6.** (c)	**7.** (a)	**8.** (b)	**9.** (b)	**10.** (a)
11. (c)	**12.** (c)	**13.** (b)	**14.** (d)	**15.** (a)	**16.** (d)	**17.** (a)	**18.** (a)	**19.** (d)	**20.** (b)
21. (a)	**22.** (a)	**23.** (b)	**24.** (d)	**25.** (a)	**26.** (b)	**27.** (c)	**28.** (b)	**29.** (b)	**30.** (a)
31. (b)	**32.** (a)	**33.** (c)	**34.** (b)	**35.** (c)	**36.** (b)	**37.** (a)	**38.** (d)	**39.** (c)	**40.** (b)
41. (d)	**42.** (d)	**43.** (b)	**44.** (c)	**45.** (c)	**46.** (c)	**47.** (c)	**48.** (a)	**49.** (c)	**50.** (d)
51. (d)	**52.** (b)	**53.** (a)	**54.** (c)	**55.** (a)	**56.** (b)	**57.** (a)	**58.** (a)	**59.** (b)	**60.** (a)

39

Environmental Pollution and Global Warming

Pollution

The pollution is an undesirable change in physical, chemical or biological characteristics of air, land, water or soil. Agents that bring about such undesirable changes are called **pollutants**. In order to control environmental pollution, the Government of India has passed the Environment Protection Act 1986, to protect and improve the quality of our environment (air, water and soil).

Types of Pollution

Pollution can be divided into following types as given below:

(i) On the basis of part of environment affected
   (a) Air pollution (b) Water pollution
   (c) Soil pollution

(ii) On the basis of origin of the pollutant
   (a) **Natural**, e.g. volcanic eruptions, release of methane (CH_4) by paddy fields and cattle, release of carbon monoxide (CO) by plants and animals, emission of natural gases, ozone, nitrogen oxides, cosmic rays, UV-rays, etc.
   (b) **Anthropogenic**, (man-made) e.g. burning of fossil fuels, deforestation, mining, sewage, industrial effluent, pesticides, fertilisers, etc.

(iii) On the basis of physical nature of pollutants
   (a) Gaseous pollution
   (b) Dust pollution
   (c) Thermal pollution
   (d) Noise pollution
   (e) Radioactive pollution

(iv) On the basis of emission of pollutants, it can be
   (a) **Point source pollution** Pollution from a single source, e.g. chimney, municipal sewer.
   (b) **Line source pollution** Pollution that is passed along a narrow belt, e.g. roads due to the automobile exhaust.
   (c) **Area source pollution**, e.g. mining area, industrial estate.
   (d) **Diffuse source pollution** It extends over a large area, e.g. sprayed pesticides or fertilisers through run off.
   (e) **Fixed source pollution** In this pollution the pollutants are passed from fixed spots as of large factories, small scale industries, mineral smelters, electrical power plants.
   (f) **Mobile source pollution** Pollutants come out from a moving structure like transport vehicles.

Air Pollution

Air pollution is the major environmental problem which has increased with the gradual rise in the number of motor vehicles and factories. Air pollution in big cities and other industrial places has now reached an alarming stage. It is largely confined to the lower atmosphere but its implications are many.

Sources

The major sources of air pollutants are as follows:

- Transportation (42%)
- Fuel combustion in stationary sources (21%)
- Industrial processes (14%)
- Solid waste disposal (5%)
- Forest fires (8%)
- Miscellaneous, including radio-active fallout (10%)

Air Pollutants

The six pollutants that account for most of the air pollution worldwide, are main sources of air pollution. These are carbon monoxide (CO), sulphur dioxide (SO_2), nitrogen oxides (NO_x), ozone (O_3), particulates (PM_{10}) and lead (P_6). These air pollutants cause smog, acid rain and other health hazards. Beside these, there are a large number of hazardous air pollutants including organic and inorganic chemicals.

The above mentioned air pollutants are divided into following types :

Summary of Symptoms and Injury Thresholds for Air Pollution Damage to Plants

Pollutants	Symptoms	ppm	Sustained exposure time
Ozone (O_3)	Fleck, bleaching, bleached spotting, growth suppression. Tips of conifer needles become brown and necrotic.	0.03	4 h
Sulphur dioxide (SO_2)	Bleached spots, bleached areas between veins, chlorosis, growth suppression, reduction in yield.	0.03	8 h
Peroxyacetyl Nitrate (PAN)	Glazing silvering or bronzing on the lower surface of leaves.	0.01	6 h
Hydrogen Fluoride (HF)	Chlorosis, dwarfing leaf abscission, lower yield.	0.0001	5 weeks
Chloride (Cl_2)	Bleaching between veins, tips and leaf abscission.	0.01	2 h
Ethylene (C_2H_4)	Withering, leaf abnormalities, flower dropping and failure of flower to open.	0.05	6 h

Water Pollution

The various impurities present in water are expressed through water quality variables which are broadly classified into three main types of variables:

(i) The **physical variables** which include general appearance, temperature, turgidity, colour, taste, odour, etc.

(ii) The **chemical variables** besides other chemical parameters, which include all possible inorganic and organic substances, such as pH, acidity, alkalinity, hardness, conductivity, chlorides, sulphates, nitrates, nitrites, boron, heavy metals, pesticides, phenol, cyanide, oil, organics, Biochemical Oxygen Demand (BOD), Dissolved Oxygen (DO), etc.

(iii) The **biological variables** which include bacteria, virus, Protozoa, algae, etc.

Some Common Water Polluting Elements and their Toxicity

Elements	Toxicity
Aluminium	Interferes with phosphate metabolism, inhibit absorption of fluorides, Ca and iron compounds.
Arsenic	Loss of appetite, copious secretion of mucus in respiratory tract.
Cadmium	Kidney damage.
Fluorine	About 5-12 ppm is toxic, enamel becomes brittle bones lose their elasticity and are prone to fractures, impairs glycolysis.
Lead	Anaemia and mental retardation due to degenerative changes in motor nerves.
Mercury	Main site of injury is CNS leading to tumours, inability to coordinate, impairment of vision and loss of hearing. Two major episodes of mercury poisoning having occurred in Japan, in Minamata bay and Naigata. In which, mercury was absorbed, bio-accumulated and bio-magnified to high levels. Fish collected from this bay had 10-12 mg of Hg per kg of their flesh and bones. The largest mercury epidemic had occurred in 1971-72 in Iraq, when 6000 people were affected and 500 died.

Soil Pollution

It is the condition of alteration of natural soil environment caused due to the industrial, agricultural and chemical activity.

(i) **Negative soil pollution** It is the process of reduction in soil productivity due to the erosion and overuse.

(ii) **Positive soil pollution** It is the process of reduction in soil productivity due to the addition of undesirable organic substances.

Sources of Soil Pollution

(i) **Biological agents** The various sources of soil pollution are categorised as excreta of human, animals and birds is the major source of soil pollution under this category. Digested sewage sludge. Pathogenic organisms are commonly found in

(a) Excreted matter by man, bacteria and parasitic worms.

(b) Excreted matter by animals cow, pig, sheep, etc., bacteria and other worms.

(c) It can also found naturally in soil due to some edaphic cause they develop.

Diseases Dysentery, cholera, typhoid, etc are the major diseases caused by this type of pollution.

(ii) **Industrial and urban Waste** Various sources of this category includes:

(a) Coal and mineral mines, metal processing industries and engineering industries.

(b) Domestic and community wastes, i.e. sludge.

(c) Garbage, rubbish materials such as paper, residues from home, fuels, street sweepings, glasses, rubber and abandoned vehicles, etc.

Diseases Dumping of solid wastes not only create, aesthetic problems but also public health problems. Such as dysentery, diarrhoea, etc.

(iii) **Radiological agents** Release of radioactive substances from nuclear explosion and radioactive wastes (such as nuclear testing and laboratories). The radioactive elements includes ruthenium 106, rhodium 106, iodine 131, barium 140, lanthanum 140, cerium 144, promethium 144, carbon14, cesium 137. They may create several serious health hazards such as cancer, etc.

(iv) **Agricultural practices** Fertilisers, pesticides, soil condition fumigant and other chemical agents from agricultural fields causes soil pollution such as phosphates, nitrates, DDT, BHC, endrin, aldrin, dieldrin, organosulphurous compounds, organic compounds with Pb, Hg, Ar are also toxic to plants.

Noise Pollution

It is the physical form of pollution that affects the receiver directly. It is mainly caused by loud disturbing sounds which is dumped into the ambient atmosphere without caring for the adverse effect it may have.

Sound level is the logarithm of ratio of ambient intensity to reference intensity. Unit of sound level is **decibel** (dB). Pollutant or a noise sound has a value of 80 dB and above.

Intensity of sound is the rate at which energy from the wave is transferred on to the surface per unit area of W/m^2.

Infrasonic vibrations (below 50 Hz) can be felt because some body parts resonate at this frequency. Ultrasonic vibrations (above 15000 Hz) are used in imaging, cleaning, drilling, etc.

Thermal Pollution

It can be defined as a 'warming up of an aquatic ecosystem to the point, where desirable organisms are adversely affected' (Owen; 1985).

A large number of industrial plants uses cold water from the rivers and discharge it hot in the water bodies. The electric power, steel and chemical industries are the most important users of cooling water.

A single 1000 MW power plant may use one-half million gallons of cooled water per minute. In many countries, atomic energy plants have become a major source of thermal pollution, which is harmful to the fishes and aquatic invertebrates and other aquatic forms.

Sources and Effects of Thermal Pollution

Harmful effects of thermal pollution on aquatic ecosystem are as follows:

- Reduction in dissolved oxygen.
- Interference with reproduction.
- Increased vulnerability to diseases.
- Direct mortality.
- Invasion of destructive organisms.
- Undesirable changes in algal population.
- Destruction of organisms in cooling water.

Radioactive Pollution

Biological effects of radioactive wastes released into the rivers are well known. Absorption are of major importance in the uptake of radioisotopes by the plants.

Fishes take up radioactive materials by the ingestion and assimilation with food and by absorption.

Bioaccumulation of radioisotopes takes place through the aquatic food chains. As a result, river insects have much more radioactivity than the water they inhabit. The most abundant nuclides found in the river are $P^{32}, Cu^{64}, Cr^{51}, Zn^{65}$ and Na^{24}, with the potential for increased build up of radioactive isotopes in fishes and other edible aquatic organisms, it is very important to prevent their contamination, particularly because they may be a source of contamination for human beings.

Global Warming

The atmosphere of earth is comprised of a mixture of gases including nitrogen, oxygen and water vapour.

Small quantities of carbon dioxide, methane from the bacterial decomposition, particularly in the guts of animals, in sediments and in flooded rice paddies and nitrous oxide from fertilisers and animal wastes. The global warming could raise the mean temperatures up to 2-3°C.

The carbon dioxide in our atmosphere acts like a light filter, allowing certain wavelengths (those of visible light) to pass through them by but absorbing others (especially infra-red light).

As the visible light passes from the sun through the atmosphere and strikes the earth's surface, some of it is reflected immediately back into the space and a small amount of it get absorbed by certain biological pigments, such as chlorophyll in leaves or rhodopsin in eyes, where it powers photosynthesis or vision.

But, most of it is absorbed by other molecules and has no effect other than to warm the molecules, causing them to radiate the extra energy away as long wavelength infra-red light.

Many of these low energy infra-red quanta, can pass directly through the atmosphere without hitting a carbon dioxide molecule, because the concentration of carbon dioxide is very low (0.03% of the air).

Greenhouse Effect

The simultaneous warming of one part of the world and the cooling of another is known as the differential greenhouse effect. Many quanta of light do strike atmospheric carbon dioxide molecules and are absorbed, causing them to become warmer rather than being lost to the space. This energy is trapped in the earth and atmospheric system and warms our world. This is called the greenhouse effect, because the glass in greenhouses works the same way, as does the glass in a parked car.

An important balance exists between the atmospheric concentration of carbon dioxide and life. Because of less carbon dioxide, more heat would be lost and earth would be frozen like Mars. With more carbon dioxide, more heat would be trapped and our world would be as hot as Venus, at 800°C, with lakes of molten lead.

We have been adding carbon dioxide to the atmosphere by burning oil, gas and coal and also by destroying the forest trees that can remove the carbon dioxide by photosynthesis. Due to this the concentration of carbon dioxide is increased in the atmosphere and the average temperature also gets increased.

Greenhouse Gases and their Sources

Various greenhouse gases are responsible for causing global warming. These are:

(i) **Carbon dioxide** (CO_2) Its present level in atmosphere is about 368 ppm (parts per million) and the atmospheric life time is 5-200 yr. The amount of CO_2 is increasing due to the burning fossil fuels, deforestation and change in the land use.

High concentration of CO_2 may cause fertilisation effect, i.e. increase in rate of photosynthesis and growth of plants, decrease in stomatal conductance and transpiration rate.

(ii) **Methane** (CH_4) The present level in atmosphere is 1750 ppb (parts per billion).

The methanogen bacteria increases greenhouse effect by producing methane. In nature the major sources are freshwater wetlands, enteric fermentation in cattle. Flooded rice fields along with biomass burning, etc.

(iii) **Nitrous oxide** (N_2O) Present atmospheric concentration is 316 ppb (parts per billion).

Major sources are agriculture, biomass burning, nylon industries, nitrogen rich fertilisers and fuels.

(iv) **Chlorofluorocarbons** (CFCs) The present atmospheric concentration is 282 ppt (part per trillion) and the atmospheric life time is 45-260 yr.

Major sources of CFCs are leaking air conditioners, refrigeration units, evaporation of industrial solvents, production of plastic foams and propellants in aerosol, spraycans.

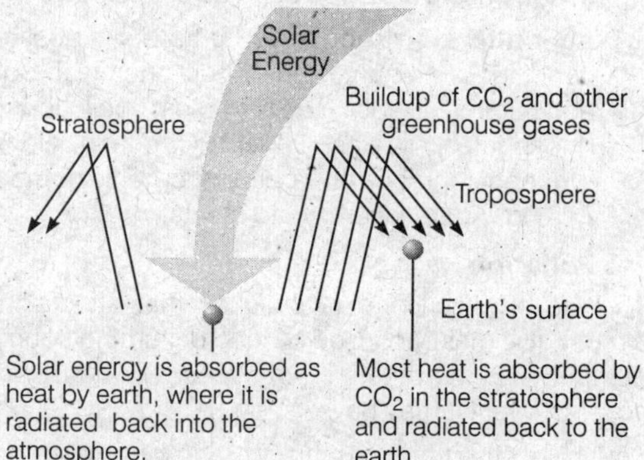

Greenhouse effect

Chapter Exercise

1. Major aerosol pollutant present in the jet plane emission is
(a) SO_2 (b) Fluorocarbon
(c) CCl_4 (d) CO

2. Environment planning organisation is
(a) CSIR (b) NEERI
(c) ICAR (d) CEPHERI

3. Acid rain occurs in areas where
(a) where large plantation occurs
(b) where small plantation occurs
(c) there are big industries and the atmosphere is polluted with SO_2
(d) there are no industries

4. Fly ash is an environmental pollutant produced by
(a) thermal power plant (b) oil refinery
(c) fertilisation plant (d) strip mining

5. Greenhouse effect is enhanced in the environment by the gas
(a) CO_2 (b) CO
(c) Fluorocarbons (d) NO_2

6. Which air pollutant could cause rains to be acidic?
(a) Dust particles from cement factories
(b) Insecticides from crop sprays
(c) Smoke from wood fires
(d) Sulphur dioxide from coal-fired powerstations

7. Acid precipitate kills fish by causing release of
(a) aluminum ions (b) CO
(c) anticoagulants (d) mercury

8. CO is a pollutant because it
(a) inactivates nerves (b) combines with oxygen
(c) combines with Hb (d) inhibits growth

9. Pollutant causing leaf curling is
   (a) SO_2 (b) CO (c) H_2S (d) O_3

0. Which kind of powerstation produces sulphur dioxide, causing air pollution and acid rain?
   (a) Coal-fired (b) Nuclear
   (c) Hydroelectric (d) Wind-powered

1. Most hazardous metal pollutant of automobile exhaust is
   (a) Cd (b) Hg
   (c) Cu (d) Pb

2. Taj Mahal is threatened to the effect of
   (a) Cl (b) SO_2 (c) O_2 (d) H_2

3. The chemical that contributes to the destruction of ozone layer above the earth's surface is
   (a) Hg (b) CO
   (c) SO_2 (d) CFCs

4. Photochemical smog always contains
   (a) Al ions (b) CH_4 (c) O_3 (d) P

5. Why does the oxygen concentration in the water decrease when sewage is discharged into a river?
   (a) Decrease in the number of plants
   (b) Increase in the number of fish
   (c) Less oxygen absorbed from the air
   (d) Rapid growth of bacteria

16. The most harmful types of environment pollutants are
   (a) human organic wastes
   (b) wastes from faecal
   (c) non-biodegradable chemicals
   (d) natural nutrients present in excess

17. Ozone hole is the
   (a) absence of O_3 in troposphere
   (b) absence of O_3 in stratosphere
   (c) deficiency of O_3 in stratosphere
   (d) deficiency of O_3 in troposphere

18. Ozone day is observed on
   (a) January 30 (b) April 29
   (c) September 16 (d) December 25

19. The main indicators of SO_2 pollution are
   (a) lichens (b) ferns
   (c) algae (d) bryophytes

20. Air pollution effects are usually found on the
   (a) leaves (b) flowers
   (c) stems (d) roots

21. If water pollution continue at its present rate, it will eventually
   (a) stop the water cycle
   (b) make O_2 molecule unavailable to water plants
   (c) make nitrate molecules unavailable to water plants
   (d) prevent precipitate

22. Domestic waste is
   (a) effluent
   (b) biodegradable
   (c) non-biodegradable
   (d) air pollutant

23. Noise was recognised as an air pollutant through the amendment of
   (a) Environment Protection Act
   (b) Air Prevention and Control of Pollution Act
   (c) Noise Control and Pollution Act
   (d) All of the above

24. In addition to SO_2, metals and textiles are damaged by
   (a) nitrogen oxides (b) fluorides
   (c) carbon monoxide (d) aldehydes

25. Lead pollution in Indian cities had been mainly due to
   (a) lead pencils (b) batteries
   (c) automobiles (d) electronic gadgets

26. Classical smog was first observed in
   (a) Tokyo (b) New York
   (c) Sydney (d) London

27. Kyoto conference is concerned with
   (a) limiting production of SO_2
   (b) developing alternatives to ODS (Ozone Depleting Substance)
   (c) reduced emission of greenhouse gases
   (d) reduction in use of energy

28. A pollutant is any substance, chemical or other factor that changes
   (a) natural geochemical cycle
   (b) the natural balance of our environment
   (c) the natural flora of our environment
   (d) the natural wildlife of our region

29. Which one of the following diseases not occurs due to contamination of water?
   (a) Hepatitis-B (b) Jaundice
   (c) Cholera (d) Typhoid

30. World's most problematic aquatic weed is
   (a) *Azolla* (b) *Wolffia*
   (c) *Eichhornia* (d) *Trapa*

31. Greenhouse gases
   (a) absorb large wave radiations from the earth
   (b) are transparent to both large and short
   (c) absorb solar radiations for warming the atmosphere of earth
   (d) are transparent to emissions from the earth for passage into outer space

32. The material generally used for sound proofing of rooms like a recording studio or auditorium, etc is
   (a) cotton (b) coir
   (c) wood (d) styro foam

33. Ozone is

(a) harmful in the upper atmosphere
(b) beneficial in the upper atmosphere
(c) beneficial in the lower atmosphere
(d) harmful in the upper atmosphere and beneficial in the lower atmosphere

34. A lake has been polluted by sewage. Compared with unpolluted water, the water in this lake will have

(a) fewer bacteria
(b) more fish
(c) more nitrogen compounds
(d) more oxygen

35. The most outstanding danger at present for the survival of living beings on earth is

(a) glaciation (b) deforestation
(c) radiation hazards (d) desertification

36. If BOD of a river water is found to be very high. This means that the water is

(a) clean (b) highly polluted
(c) having algae (d) contain dissolved minerals

37. BOD stands for

(a) Biotic Oxygen Demand
(b) Biological Organic Decay
(c) Biological Organism Death
(d) Biochemical Oxygen Demand

38. Chemical causing bone cancer and degener: ion of tissues is

(a) I-131 (b) Ca-40 (c) I-127 (d) Sr-90

39. Contamination of radioactive materials are dangerous because it causes

(a) biological magnification (b) gene mutation
(c) photochemical smog (d) ozone destruction

40. The first biocide used on a large scale was

(a) DDT (b) 2, 4-D
(c) mercury (d) melathion

41. Ozone depletion is occurring due to

(a) PCB (b) CO
(c) PAN (d) None of these

42. How have human activities increased the supply of fixed nitrogen available to the primary producers?

(a) Use of nitrogen fertilisers
(b) Increased cultivation of legumes
(c) Deliberate burning of fields
(d) All of the above

43. The destruction of the ozone layer is most directly linked to the

(a) destruction of tropical rain forests
(b) release of chloroflourocarbons (CFCs) into the atmosphere
(c) decrease in the production of ozone precursors by chemical production plants
(d) global warming

44. Eutrophication in lakes is the direct result of

(a) a diminished supply of nitrates and phosphates
(b) industrial poisons
(c) nutrient enrichment such as nitrate and phosphate runoffs from the land
(d) an increase in predators

45. Chances of skin cancer are enhanced due to

(a) smoking
(b) global warming
(c) Pb
(d) stratospheric depletion of O_3 layer

46. Some possible properties of modern insecticides are listed below. When these insecticides are used, which property helps to keep environmental pollution at the lowest level?

(a) Accumulates in the bodies of predators
(b) Broken down by soil bacteria
(c) Easily washed into lakes and rivers
(d) Taken up by the plant roots

47. A lake has been polluted by sewage. Compared with unpolluted water, the water in this lake will have

(a) fewer bacteria
(b) more fish
(c) more nitrogen compounds
(d) more oxygen

48. Which process does not result in the return of carbon dioxide to the atmosphere?

(a) Combustion of fossil fuels
(b) Decomposition of humus
(c) Respiration by bacteria
(d) Photosynthesis by green plants

49. The most harmful types of environment pollutants are

(a) human organic wastes
(b) wastes from faecal
(c) non-biodegradable chemicals
(d) natural nutrients present in excess

50. Which of the following causes biomagnification?

(a) SO_2 (b) Mercury
(c) DDT (d) Both (b) and (c)

51. Ozone layer is present in

(a) mesosphere (b) otratosphere
(c) ionosphere (d) None of these

52. Which type of pollution is caused by scooters?

(a) Air pollution
(b) Soil pollution
(c) Both (a) and (b)
(d) Water pollution

53. If global warming continues, the organism which may face problems more likely

(a) cow (b) lion
(c) snow leopard (d) dolphin

54. CO is a major pollutant of
(a) water (b) air
(c) soil (d) All of these

55. Catalytic converter is used in automobiles
(a) to remove lead (b) for converting NO_x
(c) to remove H_2O (d) All of these

56. Compressed Natural Gas (CNG) is
(a) propane (b) methane
(c) ethane (d) butane

57. The green scum seen in the freshwater bodies is
(a) blue-green algae (b) red algae
(c) green algae (d) Both (a) and (c)

58. Among the following which one causes more indoor chemical pollution?
(a) Burning coal
(b) Burning cooking gas
(c) Burning mosquito coil
(d) Room spray

59. Sewage water can be purified for recycling with the action of
(a) microorganisms (b) penicillin
(c) fishes (d) aquatic plants

60. Choose the incorrect statement.
(a) The Montreal protocol is associated with the control of emission of ozone depleting substances
(b) Methane and carbon dioxide are greenhouse gases
(c) Dobson units are used to measure oxygen content
(d) Use of incinerators is crucial to disposal of hospital wastes

61. In polluted lake, index of pollution is
(a) frog (b) artemia
(c) *Daphnia* (d) None of these

62. If BOD of a river water is found very high. This means water is
(a) clean
(b) highly polluted
(c) having algae
(d) contain dissolved minerals

63. The major drawback of DDT as a pesticide is that
(a) it is not degraded in nature
(b) its cost of production is high
(c) it is significantly less effective than other pesticides
(d) the organisms at once develop resistance to it

64. Thermal pollution of water bodies is due to discharge of
(a) wastes from mining
(b) agricultural run-off
(c) chemicals from the industries
(d) heat from the power plants

65. Minamata disease is produced due to the pollution of water with
(a) silver nitrate (b) methyl isocyanate
(c) crude oil (d) mercuric chloride

66. The major source of noise pollution world wide is due to
(a) office equipment
(b) transport system
(c) sugar, textile and paper industries
(d) oil refineries and thermal power plants

67. The term 'Nuclear war' is associated with
(a) soil pollution (b) water pollution
(c) radioactive pollution (d) All of these

68. Which radioactive isotope is used in the detection of thyroid cancer?
(a) I-131 (b) U-238
(c) C-14 (d) P-32

69. A pollutant is any substance, chemical or other factor that changes
(a) natural geochemical cycle
(b) the natural balance of our environment
(c) the natural flora of our environment
(d) the natural wildlife of our region

70. Contamination of radioactive materials are dangerous because it causes
(a) biological magnification
(b) gene mutation
(c) photochemical smog
(d) ozone destruction

Answers

1. (b)	2. (b)	3. (c)	4. (a)	5. (a)	6. (d)	7. (a)	8. (c)	9. (a)	10. (a)
11. (b)	12. (b)	13. (d)	14. (c)	15. (d)	16. (c)	17. (c)	18. (c)	19. (a)	20. (a)
21. (b)	22. (b)	23. (b)	24. (a)	25. (c)	26. (b)	27. (c)	28. (b)	29. (a)	30. (c)
31. (a)	32. (d)	33. (b)	34. (c)	35. (c)	36. (b)	37. (d)	38. (d)	39. (b)	40. (a)
41. (d)	42. (d)	43. (b)	44. (c)	45. (d)	46. (b)	47. (c)	48. (d)	49. (c)	50. (d)
51. (b)	52. (a)	53. (c)	54. (b)	55. (b)	56. (b)	57. (d)	58. (c)	59. (a)	60. (c)
61. (c)	62. (b)	63. (a)	64. (d)	65. (d)	66. (b)	67. (c)	68. (a)	69. (b)	70. (b)

40

Biotechnology

Biotechnology by Different Organisations

Different organisations defined biotechnology in a different ways some of them are

- **Organisation for Economic Cooperation and Development** (OECD), **1981** 'Biotechnology is the application of scientific and engineering principles to the processing of materials by biological agents to provide goods and services.'
- **European Federation of Biotechnology** (EFB) O' Sullivan, 1981 'Biotechnology is the integrated use of biochemistry, microbiology and engineering sciences in order to achieve technological applications of the capabilities of microorganisms, cultured tissue cells and their parts.'
- **US National Science Federation** (USNSF)' Biotechnology is the controlled use of biological agents such as microorganisms or cellular components for beneficial use.'

Origin of Biotechnology

The use of complete animals and plants is not generally included in biotechnology. Because exploitation of whole plants and animals is the subject of already well-established disciplines of agriculture, horticulture and animal husbandry. If we trace the origin of biotechnology, it is as old as human civilisation. Development of biotechnology can be studied considering its growth that occurred in two phases, i.e. the traditional (old) biotechnology and the modern (new) biotechnology.

Genetic Engineering

It is process by which scientists modify the geome of an organism. Creation of genetically modified organisms requires recombinant DNA. It is accomplished by the use of several tools. These are as follows:

(i) Restriction enzymes
(ii) DNA ligase
(iii) Vectors
(iv) Host organism

Restriction Enzymes

The most important tool in advance genetics are restriction enzymes that belongs to a larger class of enzymes called **nucleases**. These are of two main types, i.e. exonuclease and endonuclease.

Exonucleases remove nucleotides from the ends of the DNA, whereas **endonucleases** make cuts at specific positions within the DNA molecule.

Restriction Endonucleases

Among nucleases, the well known are restrictic endonucleases, the enzymes that bacteria uses to destr the DNA of invading foreign viruses. These endonuclease have the ability to recognise certain nucleotide sequence (i.e. restriction sites or restriction enzyme recognitic sequence) found on the foreign DNA (usually 4-8 base pa long) cleaves DNA at or near those sites.

The bacterium modifies its own restriction sites by th phenomenon of methylation which is mainly used to prote its own DNA from the action of restriction endonucleases.

Phage that survived in one strain could not survived other strain containing different restriction enzyme Restriction enzymes are also referred to as **molecula scissors**.

W Arber, H Smith and D Nathans (1978), awarded Nob Prize in Medicine and Physiology for their pioneering work the study of restriction endonucleases.

The restriction enzymes can be categorised into three ma categories these are

(i) **Type-I** These enzymes interacts with an unmodifie recognition sequence in a double-stranded DN molecule. They cleave single strand of DNA at a apparently random site and have the capability to creat a gap of about 75 nucleotides in length and requir Mg^{2+} ions, **ATP** and **S-adenosyl methionine** cofactor for their functioning.

(ii) **Type-II** These enzymes recognises specific DN nucleotide sequences and cleaves both strands within c near to the **palindrome sequence**, nucleic aci sequence (DNA or RNA), that is same when read in th either direction, i.e. either read 3′ to 5′ of the one stran and 3′ to 5′ of the complementary strand with which it forr a double helix. These enzymes produces DNA fragment of defined length and sequence and require only Mg^{2} for their functioning. The type-II enzymes are mostly use in gene manipulations.

(iii) **Type-III** These enzymes are more or less similar to th type-II enzymes with the slight difference, that thes requires ATP and **S-adenosyl methionine** also alon with the Mg^{2+} for their functioning. Hence, acts as a intermediate between Type-I and Type-II.

DNA Ligase

This enzyme is used to repair the broken DNA by joining two nucleotides in a DNA strand. It is commonly used in genetic engineering to do the reverse of a restriction enzyme, i.e. to join together complementary restriction fragments. Thus, known as **molecular glue** or **molecular binder** in genetic engineering. The enzyme mediates the formation of phosphodiester bonds between adjacent 3′-OH and 5′-phosphate fermini, i.e. polynucleotide strand contain 3′ OH group at one end (3′ end) and phosphate group at 5′ end) thereby, **joining the nicks** in the double-stranded DNA.

The sticky ends allow two complementary restriction fragments to anneal, but only by weak hydrogen bonds which can quite easily be broken, say by gentle heating. As, the backbone is still incomplete, DNA ligase completes the DNA backbone by forming covalent bonds. Restriction enzymes and DNA ligase can, therefore used together to cut and join the lengths of DNA from different sources.

The other enzymes used in various processes of genetic engineering are

(i) **Lyases** used to open the cells or dissolve the cell wall, e.g. lysozymes are used to dissolve the bacterial cell wall.

(ii) **Synthetases** helps in either *in vitro* synthesis of DNA or in the synthesis of complementary DNA, e.g. DNA polymerases and reverse transcriptases.

(iii) **Ligases** which seals or joins the ends of the foreign DNA.

Vectors

Vectors (also known as vehicle DNA) are the DNA molecule capable of self-replicating. These are used as the **carrier of DNA segment** to be cloned (gene). In actual sense, these required for transferring and cloning the gene in a suitable host or bacterium.The gene is inserted in the vector forming a recombinant DNA or a recombinant vector also known as the **hybrid vector** or **chimeric vector** (chimera = a mythological monster with lion's head, goat's body and serpent's tail).

The vectors can be either **cloning vectors** which isolate, characterise and clone genes or **expression vectors,** which does not only allow the multiplication, i.e. cloning but may be manipulated in such a way that the inserted gene may express in the host. A good vector must have:

(i) It has an origin of replication (ori), i.e. ability to replicate autonomously, so as to generate its multiple copies within the host itself. Its representation in many copies in the cell provides more replicates of inserted DNA sequences.

(ii) Small size or low molecular weight for enhanced stability.

(iii) It have single sites for a large number of restriction enzymes.

(iv) It must isolate and purify easily.

(v) Transformation to the host cells should be easy.

(vi) It must have the ability to integrate either itself or through DNA segment which carries it into the genome of the host cell.

(vii) It have the **marker** or markers to permit the selection of the transformed host cells. These markers actually enables the cell carrying vector to get easily distinguished from those who lacks vectors.

Usually the genes encoding resistance to antibiotics such as ampicillin, chloramphenicol, tetracycline or kanamycin, etc are useful selectable markers for E. coli. Because the normal E. coli cells do not carry resistance against any of these antibiotics.

(viii) Must represent itself in many copies in the cell in order to provide more replicates of inserted DNA sequences.

Types of Vectors

Mostly, all these requirements are not found in naturally occurring vectors thus, the vectors are engineered *in vitro* in order to obtain the appropriate features to finally facilitate the gene cloning. The vectors are grouped into bacterial plasmids, bacteriophages, cosmids, phagemids, etc.

(i) **Plasmids** These are the most widely used vectors. Plasmids (like pBR 324, pC194, pBR 322, etc) are extrachromosomal, self- replicating and double-stranded closed, circular DNA segments seen in the bacterial cells. The size of plasmids ranges from 1×10^6 to 200×10^6 dalton.

(ii) **Bacteriophage DNA** Plasmid vectors described above are normally used to clone DNA segments of small size, i.e. up to 10 kilo bases. However, while preparing a genomic library for a eukaryotic cell with a quite large DNA fragments or even whole genomes, the requirements of cloning are fulfilled by the derivatives of the lambda phage which are used for transferring the genetic material from one bacterium to the another.

From these, cloning of 20-25 kilo bases is possible. These consists of linear double-stranded DNA molecules which have been engineered in the way that their **lytic cycle** is possible but lysogenic cycle is not possible.

The **lambda phage** genome is of about 50 kbp circular DNA. It follows either a lytic path or a lysogenic path. Lytic one may be switched towards the lysogeny and *vice versa*.

An important feature of lambda genome is that only about 60% genome is required for the lytic growth of phage and the central 40% can be replaced by the foreign DNA.

The percentage of foreign DNA can also be increased but maximum upto the range of about 75-100% of the wild genome. This is made possible by removing some of the non-essential parts of phage genome.

The lambda cloning vectors are of two main types:

(a) The **insertion vector** which accepts the inserts of only 12 kbp long at a single multiple cloning site, e.g. λ ***gt* 10** and λ ***gt*11** vectors.

(b) The **replacement vector** which accepts the inserts of about 9-23 kbp long with the involvement of replacement of a non-essential part (stuffer) of genome, e.g. **EMBL3** and **EMBL4** vectors.

(iii) **Cosmids** These are the vectors, which can accommodate DNA segments upto the size of 45 kbp. These are actually plasmid particles to which the specific DNA sequences, namely those for the ***cos* sites** are inserted. Since, the *cos* sites enable the DNA to be packed in the phage (λ) particle. So, we can say that the cosmids allow the packaging of DNA in phage *in vitro* thus, permitting their purification.

Like plasmids, these cosmids also perpetuate in bacteria. The advantage of cosmids for cloning is that its efficiency is high enough to produce a complete genome library of 10^6-10^7 clones from a mere 1μg of insert DNA.The only disadvantage of cosmids is that they does not possess the ability to accept more than 40-50 kbp of DNA.

(iv) **Phagemids** These are the plasmids with a fragment of filamentous phage DNA. Thus, combining the desirable features of both. Segments of foreign DNA cloned in them are multiplied as plasmid vectors, but when these plasmid vectors are infected with a filamentous phage called **helper phage**, the phagemid genome being already present behaves like a phage itself.

As a result, the phagemid generates multiple copies of one strand of it and the associated DNA inserted in it.

The single-strand DNA can be purified from these phage particles and can be used for various purposes such as **sequencing**, **site directed mutagenesis** and **synthesis of strand specific probes**

Phagemids have been developed by inserting intergenic regions of a variety of filamentous phages into plasmids. The phagemids have several advantages over other vectors such as

(a) Gives high yield of foreign DNA just like plasmid vectors.

(b) Provides single-stranded DNA alongwith the double- stranded foreign DNA fragments.

(c) Due to their small sizes single-stranded DNA up to 10 kbp can be easily obtained from them, e.g. pUC118, pUC119, etc.

(v) **Ti Plasmids** For plant scientists, the attraction of Ti plasmid provides them a vehicle for the introduction of the genetically engineered DNA into the plants.

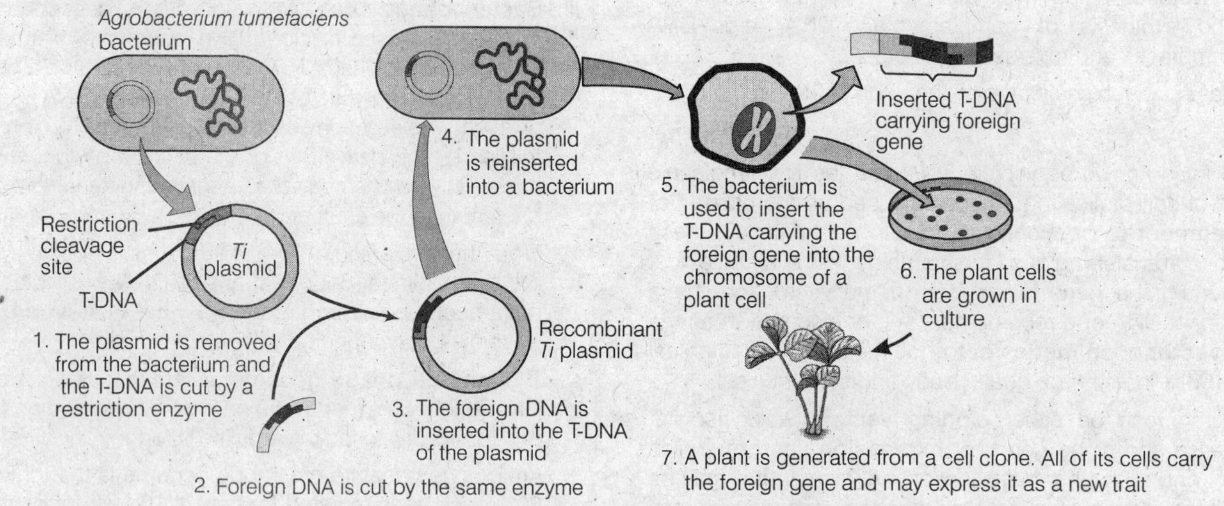

Introduction of DNA segment to plant cell through *Ti* plasmid

Molecular Markers

These are the DNA sequences used for the **mapping** of chromosomes and the location of **specific site** in a chromosomes. The most common type of markers used today are RFLP and RAPD.

(i) **RFLP**, i.e. **Restriction Fragment Length Polymorphism** is a molecular marker based on the differential hybridisation of cloned DNA to DNA fragments in a sample of restriction enzyme digested DNAs. The marker is specific to a single clone/restriction enzyme combination.

(ii) **RAPD**, i.e. **Randomly Amplified Polymorphic DNA** is a molecular marker based on the differential PCR amplification of a sample of DNAs from short oligonucleotide sequences.

Polymerase Chain Reaction (PCR)

PCR is the technique which can easily clone (amplify) DNA samples as small as a single molecule. This technique was developed by **Kary Mullis** (1983) for which he won the Nobel Prize in (1993).

The polymerase chain reaction is simply the method of DNA replication in a test tube. In this, a length of DNA is mixed with four nucleotide molecules (A, T, C and G) along with the enzyme DNA polymerase in a test tube which allows the new DNA to replicate many times. Thus, it relies its importance in today world of biotechnology in copying DNA molecule.

Recombinant DNA Technology

Detailed studies of structure and function of the gene at the molecular level requires large quantities of individual gene in its pure form. A variety of techniques, often referred to as recombinant DNA technology are used in **DNA cloning**, which permits researchers to prepare large numbers of similar DNA molecules.

Recombinant DNA is defined as simply any DNA, which is composed of sequences derived from different sources. The key to cloning DNA molecule, is to link it with a vector DNA molecule.

The major steps involved in this technique are:

(i) The cells having desired DNA molecule are separated and cultured to increase their number. The nuclei of these cells is fragmented after its separation from the cells.

From this fragmented lot, the desired DNA segment is identified and picked up through **electrophoresis** or other techniques like **Southern blotting**. The DNA thus, isolated is called **passenger** or **donor DNA**.

(ii) A suitable vector and donor DNA both are treated with similar restriction endonucleases. Self-ligation is prevented with alkaline phosphatase.

(iii) The recombinant DNA formed is to be introduced into the host cell where it can replicate and form clone. The host cell can be a bacterium or a yeast cell. Three methods used for introduction of recombinant DNA into the host cell, includes, i.e. **transformation, transduction and vectorless gene transfer**.

Gene Transfer

Apart from the transformations with the help of vectors, the gene of interest can also be introduced in a host cell by the direct gene transformation methods. Various direct gene transfer methods are given below:

(i) **Electroporation** A most significant method of gene transfer in which the cells are subjected to **high-voltage pulse** that disrupts the membrane of the cell and allowing the vector to enter the cell.

(ii) **Gene gun** (Biolistics) An extraordinary technique designed to overcome the problem of the strong cell wall in plant tissue.

(iii) **Microinjection** A method in which a cell is held on a glass pipette under a microscope and the foreign DNA is injected directly into the nucleus. This method is used when there are only very few cells available, such as fertilised animal egg cells, etc.

(iv) **Liposomes** It is small membrane vesicles which fuses with the cell membrane to deliver the DNA into the cell. These liposomes works for many types of cells, but is particularly useful for delivering genes to cells *in vivo* (such as in gene therapy), etc.

(v) The bacterial cells with recombinant DNA are selected and isolated from bacterial population either by immunological method or by other methods like nucleic acid hybridisation method, blue white screening, etc.

Gene Libraries

The collections of cloned DNA fragments represent the total genome of a species. These are of following two main types:

(i) **Genomic library** In genomic library total DNA is into small fragments and each one of them is cloned suitable vector. This *r*DNA is amplified in *E. coli* cell.

(ii) ***c*DNA library** *c*DNA (complementary DNA) is synthesised against ***m*RNA** template by using an enzyme **reverse transcriptase**. The library contains only those gene, which is expressed in the cells. The *c*DNA library do not contain introns.

Transgenic Crops (GM Crops)

A transgenic crop is one that contains and expresses a **transgene**. A popular term for the transgenic crop is **genetically modified crop** or GM crops.

The technique of genetic modification in crops has two unique advantages

(i) Any gene (from any organism or a chemically synthesised gene, i.e. synthetic gene) can be used for the transfer.

(ii) Change in genotype can be precisely controlled because only the transgene is added into the crop genome.

When a transgene is introduced into the genome of an organism, it can achieve the following:

(i) **Produces a Protein of Interest** It produces a protein that is the product in which we are interested, e.g. **hirudin** is a protein that prevents blood clotting. The gene which encodes hirudin was chemically synthesised. This gene was then transferred into ***Brassica napus,*** where hirudin accumulates in seeds.

This hirudin is purified and used as medicine. In such cases, the transgene product itself is the product of interest.

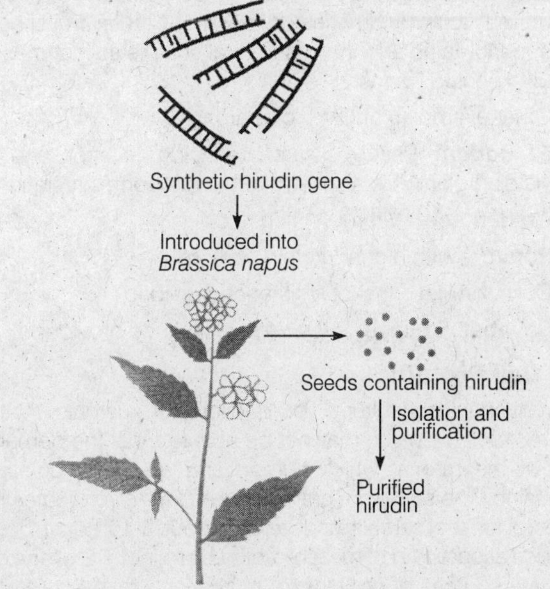

A simplified representation of the production of hirudin from transgenic *Brassica napus* seeds

(ii) **Produces a Desired Phenotype** It produces a protein that produces the desired phenotype on its own, e.g. a soil bacterium, ***Bacillus thuringiensis*** produces a crystal **(Cry)** protein that is toxic to the larvae of certain insects. There are several different types of Cry proteins and each toxic to different group of insects.

The gene encoding Cry protein (*cry* gene) has been isolated and transferred into several crops. The crop expressing a *cry* gene is usually resistant to the group of insects for which the concerned Cry protein is toxic.

In such cases, the transgene product directly produces the phenotype of interest. *cry gene* have been transferred into corn, enabling the corn to produce its own pesticides against insects such as the **European corn borer,** etc.

It may be noted that the symbols for a gene and for its product are similar. However, gene symbol has small letters and is invariably in italics. e.g., *cry*, where, the first letter of the protein symbol is always capital and is always written in Roman letters, e.g. Cry.

(iii) **Modifies an Existing Biosynthetic pathway** It modifies an existing biosynthetic pathway so that a new end-product is obtained. **Transgenic rice** and **transgenic potato** produce higher content of vitamin-A and protein, respectively. Golden rice is a variety of rice produced through genetic engineering to biosynthesis beta-carotene, a precursor of vit-A and protein.

(iv) **Masks the Expression of a Native Gene** It also prevents the expression of an existing native gene, e.g. tomato variety ***'Flavr savr'*** the expression of a native gene, i.e. **polygalacturonase** has been blocked. This can stopped by several different methods.

Fruit softening is promoted by the enzyme **polygalacturonase** which degrades the pectin layer of the cell wall. Hence, fruits of this tomato variety remain fresh and retain their flavour much longer than do the fruits of normal tomato varieties. In addition, these fruits have a superior taste and increased total soluble solids, these are unexpected bonus.

Other examples of GM crops are :

Bt cotton Pest resistant, herbicide tolerant and high yielding plant. It is also resistant to bollworm infestations.

Golden rice Vitamin-A rich rice.

Potato With higher protein content.

Corn, brinjal Insect resistance provided.

Soyabean, maize Herbicide resistance provided.

Single Cell Protein

Conventional agricultural production of cereals, pulses, vegetables, fruits, etc., may not be able to meet the demand of food at the rate at which human and animal population is increasing. The shift from grain to meet diets also creates more demand for the cereals as it takes about 3-10 kg of grain to produce atleast 1 kg of meat by animal farming. Thus, one of the alternate sources of proteins for animal and human nutrition is Single Cell Protein (SCP).

Microbes are being grown on an industrial scale as source of good protein. Microbes like ***Spirulina*** can be grown easily on materials like wastewater from potato processing plants containing starch, straw, molasses, animal manure and even sewage, to produce large quantities of SCP and can serve as food rich in proteins, minerals, fats, carbohydrates and vitamins.

Such utilisation also reduces environmental pollution. It has been estimated that in a day, 250 g of microorganisms like *Methylophilus methylotrophus*, because of its high rate of biomass production and growth, can be expected to produce **25 tonnes of protein.**

Biotechnology in Industry

In Industry, for thousands of years, natural enzymes made by microorganisms have been used to make products such as cheese, bread, wine, beer, etc. This study of industrial enzymes and their uses is called **enzyme technology**.

The microorganisms selected are usually cultured in the large fermentation chembers called **fermenters**. Under controlled condition to maximise the production of enzymes.

Microbes in a fermenter are grown under very controlled conditions. There are various requirements that must be fulfilled in order to grow a suitable microbial product. These are:

Various requirements for the production of microbes in fermenters are given below:

(i) Oxygen is needed for aerobic respiration of (some) microorganisms — others are strict anaerobes and oxygen must be excluded.

(ii) A source of **carbohydrate** is needed as an energy source for respiration to release energy needed for growth.

(iii) A source of nitrogen is needed, for protein synthesis, i.e. **ammonia** (NH_3) and **urea** (NH_2CONH_2) are both widely used as (cheap) sources of usable nitrogen.

Biotechnology in Ethical Issues

With the development of GMOs, (Genetically Modified Organisms) a continuous problem is also growing against them, i.e. the people all over the world feels that use of GMOs is unethical and such manipulations of living being should not continue without regulations. Thus, several societies have started objecting to genetic modifications and regulatory measures has been established to take care of morality on such issues and biological importance of these studies.

Biopatent

A patent is the right granted by the government, to an inventor, to prevent others to make commercial use of one's invention. Different countries have different patent laws, which are changed with the advent of time.

A patent is granted for

(i) An invention (including a product).

(ii) An improvement in an earlier invention.

(iii) The process of generating a product.

(iv) A concept or a design.

Biopatents are awarded for the achievements in the following areas of biotechnology:

(i) Strains of microorganisms
(ii) Cell lines
(iii) Genetically modified strains of plants and animals
(iv) DNA sequences
(v) The proteins encoded by DNA sequences
(vi) Various biotechnological procedures
(vii) Production processes
(viii) Products
(ix) Product applications

Biopiracy

Biopiracy can be defined as an illegal removal of biological materials. Some organisations and multinational companies exploit and/or patent biological resources (bioresources) of other nations without proper authorisation from the countries concerned and this is called **biopiracy.**

Biowar

The war, which is fought by bioweapons (biological weapons) against humans and their crops and animals is called as biowar or **biological war.** Viruses, bacteria and some other harmful organisms can be used as Bioweapons (BW) in biological warfare.

Chapter Exercise

1. Protein engineering is used to study the proteins to compare the catalytic properties of
(a) normal and mutated form of enzyme
(b) normal form of enzyme
(c) mutated form of enzyme
(d) normal and mutated form of proteins

2. Plasmids that carry genes to provide resistance to antibiotics are called
(a) R-plasmids (b) C-plasmids
(c) A-plasmids (d) Ti-plasmids

3. DNA sequencing is
(a) nucleoside sequence of *m*RNA fragments
(b) nucleotide or base sequence of a DNA fragment
(c) nucleotide design of DNA fragment
(d) nucleotide or base sequence of *t*RNA

4. Cells, in which recombinant DNA multiplies is
(a) *E. coli* (b) virus (c) fungi (d) algae

5. In order to express human gene in a bacterium, *c*DNA must be made because bacteria
(a) splice RNA (b) destroy human DNA
(c) cannot remove introns (d) have reverse transcriptase

6. Sticky ends are produced by following restriction enzymes except
(a) *Pst* I (b) *Sma* I
(c) *Hae* II (d) *Bam* HI

7. Primer DNA, *m*RNA hybrids are made by using
(a) reverse transcriptase (b) DNA polymerase
(c) RNA polymerase (d) DNA hybridase

8. Which of the following could be a restriction enzyme recognition site?
(a) ATGCAT (b) ATCATC (c) AAAGGA (d) ATCCTA

9. Which of the following restriction enzymes produce blunt ends?
(a) *Sma* I (b) *Hae* III (c) *Alu* I (d) All of these

10. What is the basis of tissue culture technique that enables the cloning of plants?
(a) Plant cells have a semi-rigid cell wall that increases their ability to divide
(b) Plants have no resistance to pathogens, so they need to be grown in sterile conditions
(c) Plant cells are totipotent
(d) Plant cells are pleuripotent

11. Protoplast fusion between *Nicotiana glauca* and *N. langsdorffii* was obtained by
(a) Robertis (b) Carlson
(c) Kuchko (d) Vasil

12. A 'genomic library' is a term used to describe a
(a) collection of books about genes in DNA technology
(b) collection of known DNA sequences
(c) collection of genes that code for specific protein sequences
(d) culture of bacteria that contain DNA fragments representing the genome of an organism

13. Genetic modification has
(a) made crops more tolerant to abiotic stresses
(b) reduced reliance on chemical pesticides
(c) helped to reduce post harvest losses
(d) All of the above

14. The linking of antibiotic resistance gene with the plasmid vector become possible with the enzyme
(a) DNA ligase (b) RNA polymerase
(c) DNA polymerase (d) All of these

15. A procedure, through which a piece of DNA is introduced in a host bacterium is
(a) transduction (b) transformation
(c) conjugation (d) *r*DNA technology

16. The cutting of DNA at specific locations became possible with the discovery of
(a) cloning (b) restriction endonuclease
(c) restriction enzyme (d) vectors

17. The plasmid vector is cleaved by
(a) exonuclease (b) ribonuclease
(c) endonuclease (d) any nuclease

18. A method of cellular defence that takes place in all eukaryotic organisms is
(a) RNA (b) downstream processing
(c) extension (d) annealing

19. The process by which multiple copies of the gene of interest is synthesised *in vitro* using two of primers is known as
(a) PCR (b) gel electrophoresis
(c) downstream processing (d) genetic printing

20. Now-a-days, the most commonly used matrix in gel electrophoresis is
(a) agarose (b) ethidium bromide
(c) sulphate bromide (d) sulphite bromide

21. When an alien DNA is linked with the origin of replication, so that this alien piece of DNA can replicate and multiply itself in the host organism, this can be called as
(a) gene transfer (b) gel electrophoresis
(c) cloning (d) transformation

22. In recombinant DNA technology, the plasmid vector DNA is cleaved by
(a) ribonuclease (b) exonuclease
(c) endonuclease (d) any nuclease

23. The bacterial resistance to antibiotics is a genetic qualification present in
(a) virus (b) bacterial DNA
(c) bacterial membrane (d) plasmid

24. Plasmid that has the ability of integrating into the bacterial DNA are called
(a) nucleotides (b) hyposomes (c) virion (d) episome

25. A gene carried by recombinant DNA is cloned when
(a) its host bacterium divides by binary fission
(b) it is transcribed
(c) it is fragmented by restriction enzymes
(d) it is hybridised

26. DNA of plasmid is
(a) double-stranded and circular
(b) single-stranded and circular
(c) double-stranded and linear
(d) single-stranded and linear

27. Rennet enzyme obtained from calf's stomach was reported by
(a) A Fleming (b) Christian Gram
(c) Waksman (d) Christian Hansen

28. Retroviruses, in animals have the ability to transform normal cell into
(a) dead cell (b) RNA containing cell
(c) DNA containing cell (d) cancerous cell

29. The Father of DNA fingerprinting is
(a) Sunder Lal Bahuguna (b) Vishwanath
(c) Jeffreys (d) Rockfeller

30. Genetically identical copy of individual cell or gene is known as
(a) species (b) phagemid (c) clone (d) plasmid

31. The technique of breakage of DNA fragment and inserting it into another DNA molecule, is related to
(a) gene splicing (b) gene cloning
(c) gene typing (d) DNA fingerprinting

32. The number of copies of a gene present in a cell is called
(a) gene dosage (b) gene pool
(c) gene amplification (d) gene frequency

33. Type-I restriction enzyme
(a) recognise the specific sequence and cleave DNA at the specific sequence
(b) do not recognise or cleave the specific sequence
(c) recognise specific sequence but cleave non-specific sequence
(d) None of the above

34. Which of the following is not associated with genetic engineering?
(a) Plastids (b) Plasmids (c) ATPase (d) Histones

35. VNTRs are
(a) Variable Number Tandem Repeats
(b) Very Narrow Tandem Repeats
(c) Variable Non-cistronic Transposon Repeats
(d) Valuable Non-cistronic Transposic Regions

36. In a chromosome, there is a specific DNA sequence, which is responsible for initiating replication, known as
(a) vector (b) origin of replication
(c) recognition sequence (d) marker gene

37. The cutting of DNA by restriction endonucleases results in the fragments of DNA. These fragments can be separated by a technique known as
(a) PCR (b) SDS page
(c) Northern blotting (d) Gel electrophoresis

38. You can see bright orange coloured bands of DNA in a ethidium bromide stained gel exposed to
(a) X-rays (b) sun rays (c) UV rays (d) gamma rays

39. What are important products of yeast fermentation?
(a) Ethanol and acetic acid
(b) Butanol, glycerol and pyric acid
(c) Phenylethanol, amylalcohol and caproiacid
(d) All of the above

40. A thermostable DNA polymerase, which remains active during the high temperature induced denaturation of double-stranded DNA is isolated from a bacterium
(a) *E. coli* (b) *Thermus aquaticus*
(c) *Salmonella* (d) *Thermus coccus*

41. Who establishes the scientific basis of vaccination
(a) G Kohler (b) Ceaar Milestein
(c) Louis Pasteur (d) Edward Jenner

42. In the year 1963, the two enzymes responsible for restricting the growth of bacteriophage in *E. coli* were isolated. One of these added methyl groups to DNA, while the other cut DNA. The latter was called
(a) restriction endonuclease (b) methylase
(c) ligases (d) plasmid

43. What are important products of yeast fermentation?
(a) Ethanol and acetic acid
(b) Butanol, glycerol and pyric acid
(c) Phenylethanol, amylalcohol and caproiacid
(d) All of the above

44. In PCR, short primers are added to single-stranded DNA molecules in a test tube and the appropriate enzymes are included to make a copy of the DNA. Which of the following primers is needed for copying the single-stranded DNA sequence 5′TACGGTAGGTC?
(a) 5′ATGCC (b) 5′GACCT
(c) 5′TACGG (d) 5′GGCAT

45. The procedure, in which DNA fragments are separated by gel electrophoresis and then transferred on to a filter for radioactive probing is known as
(a) gene mapping (b) gene cloning
(c) polymerase chain reaction (d) Southern blotting technique

46. Antibiotics are usually associated with
(a) blocking of transcription of prokaryotic dna
(b) blocking of ets
(c) inhibition of atp synthesis
(d) None of the above

47. Megaplasmid is found in
(a) *E. coli* (b) *Agrobacterium tumefaciens*
(c) *Pseudomonas* (d) *Bacillus subtilis*

48. Main phenomenon on which genetic engineering is based is
(a) alteration of genetic constitution of cells
(b) insertion of plasmid in cells
(c) alteration in cell organelles
(d) joining of two cells

49. To prevent recirculation of vector DNA and to increase the frequency of production of recombinant DNA technology
(a) reverse transcriptase are used
(b) alkaline phosphatases are used
(c) acid phosphatases are used
(d) All of the above

50. Synthetic DNA or sDNA is synthesised in
(a) lab without any template
(b) the cell without any template
(c) the lab without any nucleotide
(d) the cell without any nucleotide

51. Molecular cloning is
(a) replication of recombinant DNA molecule inside the host cell
(b) transduction of recombinant DNA
(c) transformation of recombinant DNA
(d) mutation of recombinant DNA molecule

52. The process used for separation of the proteins in polyacrylamide gel is called
(a) Southern blotting (b) Northern blotting
(c) Western blotting (d) Eastern blotting

53. Restriction enzymes were firstly isolated from
(a) *Haemophilus influenzae* (b) *E. coli*
(c) *Pseudomonas putida* (d) *Agrobacterium*

54. Reverse transcriptase enzyme was discovered by
(a) Saffron and Morris (b) Tamin and Baltimore
(c) Singer and Nicholson (d) Meselson and Stahl

55. Which of the following scientists won the Nobel Prize for the development of PCR in 1993
(a) Philip Sharp (b) Kary Mullis
(c) Milstein (d) Nirenberg

56. Restriction enzymes are used in genetic engineering because they
(a) are nucleases that cut DNA at variable sites
(b) are proteolytic enzymes which can degrade harmful proteins
(c) can cut DNA at specific base sequence
(d) can join different DNA fragments

57. It is preferable to use yeasts rather than bacteria as recipient cells for recombination of eukaryotic DNA because yeast can
(a) produce restriction enzymes
(b) excise introns from the RNA transcript
(c) remove methyl groups
(d) reproduce at a faster rate

58. PCR stands for
(a) Polymerase Cyclic Reaction
(b) Polymerase Chain Reaction
(c) Polyethyl Cytosine Reaction
(d) Polymerisation Chain Reaction

59. The role of restriction enzymes in DNA technology is to
(a) provide a vector for the transfer of recombinant DNA
(b) produce cDNA from *m*RNA
(c) produce a cut (usually staggered) at specific recognition sequences on DNA
(d) reseal 'sticky ends' after base-pairing of complementary bases

60. Complementary DNA does not create a complete gene library as the shotgun approach because
(a) it has eliminated introns from the genes
(b) cell produces *m*RNA for only small portion of its genes
(c) the shotgun approach produces more restriction fragments
(d) cDNA can easily integrated into plasmids or phage genomes

61. By the use of biotechnology in which bacteria production of vitamins-B_{12} has been increased to about 20000 times?
(a) *Ashbya gossypii* (c) *Pseudomonas denitrificans*
(b) *E. coli* (d) *Propionibacterium sharmanii*

62. The fruit juice becomes bitter after keeping for sometime. This is due to
(a) fermentation by microorganisms
(b) oxidation
(c) reduction
(d) None of the above

63. Antibiotic used against mycosis disease is
(a) Jawaharene (b) Cephalosporin
(c) Griseofulvin (d) Neomycin

64. Organic acid first produced through fermentation is
(a) propionic acid (b) lactic acid
(c) citric acid (d) oxalic acid

65. Which of the following beverage is obtained from European rye?
(a) Rum (b) Wine (c) Cider (d) Gin

66. α-1 antitrypsin is
(a) an antacid (b) an enzyme
(c) used to treat arthritis (d) used to treat emphysema

67. An artificial seed is raised from
(a) meristem (b) root tip
(c) somatic embryo (d) immature seed

68. In RNAi, genes are silenced using
(a) *ss*DNA (b) *ds*DNA (c) *ds*RNA (d) *ss*RNA

69. The trigger for activation of toxin of *Bacillus thuringiensis* is
(a) acidic pH of stomach
(b) high temperature
(c) alkaline pH of gut
(d) mechanical action in the insect gut

70. Match the following column of bacteria and their commercially important products.

Column I	Column II
A *Aspergillus niger*	1. Lactic acid
B. *Acetobacter aceti*	2. Butyric acid
C. *Clostridium butylicum*	3. Acetic acid
D. *Lactobacillus*	4. Citric acid

Codes

	A	B	C	D		A	B	C	D
(a)	2	3	4	1	(b)	2	4	3	1
(c)	4	3	2	1	(d)	1	2	3	4

71. Source of genes for high vitamin content in Golden rice is
(a) *Daffodalis* (b) *Erwinia*
(c) Both (a) and (b) (d) None of these

72. Which one is known as 'Superbug'?
(a) *Pseudomonas putida* (b) *E. coli*
(c) *Aspergillus niger* (d) *Acetobacter aceti*

73. Which one is called wine yeast?
(a) *Saccharomyces cerevisiae*
(b) *S. ellipsoides*
(c) *S. sake*
(d) *S. pireformis*

74. A bioreactor is
(a) fermentation tank
(b) culture of bacteria
(c) hybridoma
(d) culture for synthesis of new chemicals

75. Which *Bt* crop is recently recommended for cultivation in India?
(a) Cotton (b) Wheat
(c) Soya bean (d) Rice

76. Red biotechnology is applied to
(a) agricultural processes
(b) medical processes
(c) Bath (a) and (b)
(d) industrial processes

77. The slow ripening transgenic tomato was developed in USA by using
(a) antisense RNA technology
(b) ribozyme technology
(c) cosuppression approach
(d) transgene silencing approach

78. An enzyme used in cheese making is
(a) protease (b) rennin
(c) glucoamylase (d) lactase

79. Wakesman and Woodruff (1941) were responsible for the isolation of
(a) streptomycin (b) penicillin
(c) actinomycin (d) neomycin

80. Antibiotics of fungal origin are
(a) penicillin (b) cephalosporin
(c) Both (a) and (b) (d) gentamycin

Answers

1. (a)	**2.** (a)	**3.** (b)	**4.** (a)	**5.** (c)	**6.** (b)	**7.** (a)	**8.** (a)	**9.** (d)	**10.** (c)
11. (b)	**12.** (d)	**13.** (d)	**14.** (a)	**15.** (b)	**16.** (b)	**17.** (c)	**18.** (b)	**19.** (a)	**20.** (a)
21. (c)	**22.** (c)	**23.** (d)	**24.** (d)	**25.** (a)	**26.** (a)	**27.** (b)	**28.** (d)	**29.** (c)	**30.** (c)
31. (b)	**32.** (a)	**33.** (c)	**34.** (c)	**35.** (a)	**36.** (b)	**37.** (d)	**38.** (c)	**39.** (d)	**40.** (b)
41. (a)	**42.** (a)	**43.** (d)	**44.** (b)	**45.** (d)	**46.** (b)	**47.** (a)	**48.** (a)	**49.** (b)	**50.** (a)
51. (a)	**52.** (c)	**53.** (b)	**54.** (c)	**55.** (a)	**56.** (c)	**57.** (b)	**58.** (b)	**59.** (c)	**60.** (b)
61. (a)	**62.** (a)	**63.** (c)	**64.** (b)	**65.** (d)	**66.** (b)	**67.** (c)	**68.** (c)	**69.** (b)	**70.** (c)
71. (c)	**72.** (c)	**73.** (a)	**74.** (a)	**75.** (a)	**76.** (b)	**77.** (a)	**78.** (b)	**79.** (a)	**80.** (c)

41

Basic Concepts of Plant Breeding

Plant breeding is the branch of biology, concerned with the developing varieties superior to the existing ones. The chief objective of plant breeding is to develop such improved varieties of crop plants that will be commercially successful, i.e. they become more desirable, agronomically and economically.

Methods of Plant Breeding

The methods used in crop improvement programmes or plant breeding depends upon the nature of the crop, mode of pollination and mode of reproduction.

Various methods of breeding are used for the crop improvement mechanisms. These are as follows:

1. Plant Introduction

This method consists of taking a genotype or a group of plants into a new environment, where they were not being grown before, even within a country.

Plant introduction may be classified into following two categories:

(i) **Primary introduction** It is the type of plant introduction, in which the introduced variety of plant is well-adapted to the new environment, released for the commercial cultivation without any alteration in the original genotype.

e.g. In India, several varieties has been produced by the method of primary plant introduction such as the introduction of semi-dwarf wheat (*Triticum aestivum*) varieties such as Sonora 64, Lerma Roja and of semi-dwarf rice (*Oryza sativa*) varieties such as Taichung Native 1 (TN1), TR 8, IR 28 and IR 36.

(ii) **Secondary introduction** In this method, the introduced variety may be subjected to the selection in order to isolate a superior variety. Rather, it may be hybridised with the local varieties to transfer one or few characters from this variety to the local ones, e.g. Sonalika and Kalyan Sona are wheat varieties selected from the materials introduced from CIMMYT, Mexico.

2. Selection

It is the oldest method of plant breeding and is the basis of all crop improvement programmes. The selection may be either **natural**, which takes a longer duration to bring about improvement or **artificial**, which may bring about improvement within few years.

A number of artificial selection methods are available for the improvement in self-pollinated, cross-pollinated and asexually propagated crops.

The selection method include the following sub methods:

(i) **Mass selection** It is the process of selecting the best variety out of all and then bulking the seeds till the desired result is achieved. In cross-pollinated crops, mass selection is usually practiced for population improvement purposes, whereas in self-pollinated crops, it has only a limited use of maintaining purity of a variety.

Mass selection is the simplest, easiest and the most oldest method of plant breeding followed by the farmers to improve the cross-pollinated crops mainly. The presence of **heterozygosity** or **variability** is the basis of mass selection.

It is the selection of seeds and number of plants at the time of harvesting, on the basis of **phenotypes** such as, fruit size, number of seeds, size of seeds, etc.

Dharwar American, Dodahatti local, Cambodias, etc., are important cotton varieties. Similarly, TMU 1, TMU 2, AK 10, K 122, etc., are some important groundnut varieties, been resulted from the mass selection method.

(ii) **Pure line selection** Pure line is the progeny of a single, homozygous, self-pollinated plant. In pure line selection, large number of plants are selected from a self-pollinated crops and are harvested individually.

Individual plant progenies from them are then evaluated, and the best progeny is released as a pure line variety.

The term pure line was suggested for the first time by a Danish botanist, WL Johannsen (1903), while working with the Princess variety of French bean (*Phaseolus vulgaris*). It is the only method of improvement of local varieties of self-pollinated crops.

Pure line selection is the favourite method for the improvement of local varieties that have considerable genetic variability. Various varieties have been developed by using this method.

e.g. **NP 4** and **NP 52** wheat (*Triticum aestivum*), NP 11 and NP 12 linseed (*Linum usitatissimum*), T1 cowpea (*Vigna unguiculata*) and T1 moong (*Vigna radiata*). These are pureline selections from the old local varieties.

Some examples of pure line varieties isolated from old pure line are Chafa gram, Jalgaon 781 (from China moong 781), Pusa Baisakhi (from T 44), etc.

Pure line selection was also made in introduced varieties like Kalyan Sona in wheat and Shining moong 1 in moong.

(iii) **Progeny Selection** This method is mainly employed in cross-pollinated crops, It also overcomes some of the demerits of mass selection. It is based on the principle of testing the progenies of selected plants and the final selection is based on this test. The most common method used in maize is **ear-to-row** method, developed by **Hopkins** in the year 1908.

(iv) **Clonal Selection** A clone is a group of plants produced from a single plant through the asexual reproduction. Clonal selection is the selection of desirable clones from the mixed population of **vegetatively** (asexually) **propagated** plants. In asexually propagated crops, progenies derived from a plant resemble in genetic constitution with the parent plant. Therefore, plants of a clone are **genotypically** as well as **phenotypically** alike making selection within a clone futile. As in pure line selection, the importance is given to the phenotypes only.

e.g. this method is used in improving crops such as sugarcane, banana, potato, citrus, mango, grapes, sugarbeet, etc.

A generalised clonal selection involves the basic step :

- The collection of possible clonal variability.
- Critical evaluation of each clone.

Each member of a clone for the yield and quality following artificial inoculation of diseases and pests. The disease free and high yielding clones may be selected, further evaluated and multiplied as a variety.

3. Hybridisation

It is method of crossing of two different crops. It is practiced both in the self and cross pollinated crops. Although it is impossible to produce new characters by this method, but new combinations can be formed easily.

4. Polyploid Breeding

The somatic chromosome number of any individual, whether it is diploid or polyploid, is denoted as $2n$ and the chromosome number of gametes is denoted as n.

An individual having the gametic chromosome number n, is known as **haploid**. The basic chromosome number (x) of an individual, which often corresponds to the gametic number (n) in a true diploid ($2n$) is known as **genome** or **chromosome complement**.

The different chromosomes of a single genome are distinct from each other in morphology and/or gene content and homology. The members of a single genome do not show a tendency of pairing with each other. This phenomenon of having one or more genomes in addition to the normal number for a species is called **polyploidy**.

The polyploid individuals have more than diploid or two sets of chromosomes. Depending upon the number of chromosome sets present, a polyploid is called triploid ($3n$), tetraploid ($4n$), pentaploid ($5n$), hexaploid ($6n$), heptaploid ($7n$), octaploid ($8n$), etc.

Individuals carrying chromosome numbers other than the diploid ($2x$ and not $2n$) numbers, are known as **heteroploids** and the condition is known as **heteroploidy**.

Based on the levels of ploidy or heteroploidy, two groups of plants exist:

(i) **Euploids** whose somatic chromosome number is an exact multiple of the basic number.

(ii) **Aneuploids** whose somatic number is not an exact multiple of the basic number.

5. Mutation Breeding

'Mutation is a sudden heritable change in the characteristics of an organism'. It may be spontaneous or induced in nature.

The mutation which occurs in nature at a very low rate is called **spontaneous mutations**. Its frequency is usually 10^{-6} (i.e. 1 in 10 lacs), while the mutations, which are created artificially by treatment with certain physical or chemical agents are called **induced mutations**.

The agents used to produce such mutations are known as **mutagens**. The utilisation of induced mutations for crop improvement is known as **mutation breeding**.

Mutation breeding involves the production of mutations in plants and then the selection among mutant. Mutations can create genetic variability.

Dr MS Swaminathan is a pioneer in mutation breeding. He is called founder of Radiation Genetics and also known as the Father of Green Revolution in India.

e.g. Sharbati Sonora was produced from Sonora 64 by the mutation breeding using gamma rays.

Some other new crop varieties developed by mutation breeding are NP 386 (wheat), Jagannath (rice), Aruna (castor), Mu 7 and Indore 2 (cotton), Pusa Lal Meeruti (tomato), Primex (white mustard), etc.

About 200 mutant varieties have been released in India till now. Among the seeds propagated, the largest number of varieties have been developed in rice (278) followed by barley (229) and wheat (113).

China has developed the largest number (281) of mutant varieties followed by India (116), Russia (82) and Japan (65).

6. Plant Tissue Culture

Plant breeding by micropropagation of plant parts of high yielding verieties is done by the tissue culture technique.

In strict sense, tissue culture denotes the *in vitro* cultivation of plant cells in an unorganised mass of tissue, e.g. callus culture.

For tissue culture, a good laboratory is required having the following facilities:

- Aseptic condition/environment.
- Nutrient medium for preparation, sterilisation, cleaning and storage of supplied.
- Regulated environmental conditions.
- Observation and evaluation of cultures.
- Recording the observations.

Types of Tissue Culture

The method of plant tissues culture are of ten categorised according to the type of *in vitro* growth. These are as follows:

- Callus and suspension culture
- Embryo culture
- Anther culture
- Protoplast culture
- Somatic hybridisation.

Chapter Exercise

1. In tissue culture medium, the embryoids formed from pollen grains is due to the

(a) cellular totipotency (b) organogenesis
(c) double fertilisation (d) test tube culture

2. Haploid plantlets can be produced by

(a) pollen culture (b) cotyledon culture
(c) embryo culture (d) meristem culture

3. The three topmost major crops of the world in the order of total production are

(a) maize > wheat > rice (b) rice > wheat > maize
(c) wheat > maize > rice (d) rice > maize > wheat

4. Bread wheat is

(a) autopolyploid (b) allopolyploid
(c) aneuploid (d) trisomic

5. Which variety is not a result of induced mutations?

(a) Sharbati Sonara of wheat (b) Jagannath of rice
(c) Bombay green of banana (d) Aruna of castor

6. Most cultivated plants are

(a) autopolyploids (b) allopolyploids
(c) aneuploids (d) haploids

7. Naturally occurring mutation in nature as an example of rice strain is

(a) Padma (b) Jaya
(c) Taichung Native (d) GEB 24

8. Improved dwarf wheat variety with higher percentage of lysine/protein is

(a) Kalyan (b) Sonalika
(c) Lerma Roja (d) Sharbati Sonara

9. Androgenic haploid culture was first performed by

(a) Guha and Maheshwari (b) Steward
(c) Skoog and Miller (d) Halperrin and Wetherell

10. Seedless watermelon is

(a) hexaploid (b) tetraploid
(c) triploid (d) pentaploid

11. Polyploidy leads to rapid formation of new species through the

(a) genetic recombination
(b) isolation
(c) mutation
(d) development of multiple sets of chromosomes

12. Polyploidy gives

(a) hybrid vigour (b) improved varieties
(c) gigas effect (d) All of these

13. Crosses between the plants of the same variety are called

(a) interspecific (b) intervarietal
(c) intravarietal (d) intergeneric

14. Drastic pruning of root system yields Bonsai. It involves the

(a) deficiency of cytokinin
(b) impairment of water absorption
(c) insufficient mineral absorption
(d) deficiency of auxin

15. Bagging is done to

(a) achieve desired pollination
(b) prevent contamination of unwanted pollen
(c) avoid self-pollination
(d) avoid cross-pollination

16. Tomato and potato belongs to
(a) North America (b) South America
(c) Asia Minor (d) Ethiopia

17. The reason why vegetatively reproducing crop plants are best suited for maintaining hybrid vigour is that
(a) these can be easily propagated
(b) once a desired hybrid has been produced there are few chances of losing it
(c) they have a longer life span
(d) they are more resistant to diseases

18. Picking up plants with superior phenotype for further propagation is called
(a) pure line selection (b) mass selection
(c) hybrid vigour (d) introduction

19. Artificial male sterility can be obtained for hybridisation experiments with the help of
(a) NAA (b) 2, 4-D
(c) Maeic hydraside (d) All of these

20. Which method of crop improvement can be practised by a farmer if he is inexperienced?
(a) Clonal selection (b) Mass selection
(c) Pure line selection (d) Hybridisation

21. The first cultivated wheat was
(a) *Triticum tauschii* (b) *T. monococcum*
(c) *T. boeoticum* (d) *T. speltoides*

22. Main technique involved in agricultural biotechnology is called
(a) plant breeding (b) replication
(c) tissue culture (d) hybridisation

23. Emasculation is called
(a) killing pollen grains (b) removal of stamens
(c) removal of anthers (d) All of these

24. Cellular totipotency was demonstrated by
(a) Robert Hooke
(b) FC Steward
(c) Theodore Schwann
(d) Antony von Leeuwenhoek

25. For isolation of protoplasts, one needs
(a) cellulase (b) pectinase
(c) Both 'a' and 'b' (d) chitinase

26. Progeny obtained as a result of repeat self-pollination of a cross-pollinated crop is called
(a) pure line (b) pedigree line
(c) inbreed line (d) heterosis

27. The reason why some mutations, which are harmful do not get eliminated from gene pool is that
(a) they have future survival value
(b) they are recessive and carried by heterozygous individuals
(c) they are dominant and show up more frequently
(d) genetic drift occur because of a small population

28. In tissue culture the nutrients which are thermolabile are sterilise by
(a) autoclave (b) filter sterilisation
(c) steam sterilisation (d) None of these

29. The technique involving the isolation and fusion of protoplasts and culturing the hybrid protoplast to regenerate the whole plant, is known as
(a) para sexual hybridisation
(b) genetic engineering
(c) protoplast culture
(d) androgenic haploids

30. The Father of Green Revolution is
(a) Dr N Borlaug (b) William S Gaud
(c) MS Swaminathan (d) Gurudev S Khush

31. PV 18, Kalyan Sona and S 227 wheat varieties have developed through
(a) gamma ray irradiation (b) multiple crossing
(c) selection (d) genetic engineering

32. The most convenient way of maintaining germplasm is
(a) to raise tissue culture
(b) by developing new varieties
(c) by storing seeds
(d) to treat plant by low temperature

33. In an economically important plant species, embryo mortality is very high. What method do you suggest to check the mortality rate?
(a) Embryo culture (b) Meristem culture
(c) Micropropagation (d) Polyploid mutation

34. The first cultivated wheat was
(a) *Triticum speltoides* (b) *T. monococcum*
(c) *T. boeoticun* (d) *T. tauschi*

35. Triploids are said to be more useful than other polyploids because these are
(a) fruitless and are always propagated by stem cuttings
(b) produced when definitely nucleus fuses with a male gamete
(c) sterile and show increased vigour
(d) fertile and more healthier

36. Polyploidy involves
(a) increase in number of chromosomes
(b) increase in number of genomes
(c) increase in number of plasmons
(d) deletion of part or whole of a chromosome

37. Best varieties of sugarbeet are
(a) tetraploids (b) diploids
(c) triploids (d) allopolyploids

38. Newly introduced plants from other place of country can be tested in
(a) greenhouse (b) quarantine
(c) autoclave (d) All of these

39. Which one of the following statement regarding pomato is correct?
(a) It is a product of somatic hybridisation
(b) It is a product of gene manipulation
(c) Product of sexual hybridisation
(d) Product of cloning

40. IR-36 was developed through breeding of
(a) 6 rice varieties with *Oryza sativa*
(b) 13 rice varieties with *Oryza sativa*
(c) *Oryza sativa* and *Oryza indica*
(d) *Oryza indica* and *Oryza nivara*

41. Root knot disease of brinjal/egg plant is caused by
(a) *Meloidogyne incognita*
(b) *Anguina solanacearum*
(c) *Fusarium udum*
(d) *Pseudomonas rubrilineans*

42. Chick pea/gram is infected with
(a) fungal blight (b) bacterial blight
(c) cucumovirus (d) All of these

43. Purity of seed is guaranteed by
(a) ICAR (b) NBPGR
(c) Nationa Seed Corporation (d) All of these

44. Most cultivated plants are
(a) autopolyploids (b) allopolyploids
(c) aneuploids (d) haploids

45. The main technique involved in agricultural biotechnology is called
(a) tissue culture (b) transformation
(c) plant breeding (d) DNA replication

46. Central Sugarcane Breeding Research Institute is situated at
(a) Lucknow
(b) Delhi
(c) Coimbatore
(d) Bhopal

47. In India the first haploid plant by using anther culture technique was
(a) datura (b) mango
(c) *Atropdia* (d) reserpine

48. Which technique can be useful in over coming hybridisation barrier?
(a) Shoot tip culture (b) Embryo rescue
(c) Protoplast fusion (d) Both (b) and (c)

49. Regeneration in tissue culture represents formation of
(a) root (b) root and shoot
(c) somatic embryos (d) All of these

50. Which is faster method of rapid clonal propagation?
(a) Callus culture (b) Suspension culture
(c) Embryo culture (d) Anther culture

51. Tissue culture technique was first attempted by
(a) Haberlandt (b) Hanning
(c) Hobecourt (d) Gautheret

52. Tissue culture idea was floated by Haberlandt (1902). Who was first to raise and maintain culture of tomato roots for over 30 years?
(a) Muir (b) Skoog and Miller
(c) White (d) Guha and Maheswari

53. Induced fusion of protoplasts is brought by the treatment of
(a) sodium nitrate (b) calcium ions
(c) polyethylene glycol (d) All of these

54. *Raphanobrassica* is an
(a) autopolploid (b) allotetraploid
(c) hexaploid (d) trisomy

55. Seedless watermelon plants are obtained
(a) by inducing parthenogenesis
(b) by applying manure to soil
(c) by raising triploids
(d) by colchicine application

Answers

1. (a)	**2.** (a)	**3.** (a)	**4.** (b)	**5.** (c)	**6.** (b)	**7.** (d)	**8.** (d)	**9.** (a)	**10.** (c)
11. (d)	**12.** (d)	**13.** (c)	**14.** (d)	**15.** (b)	**16.** (b)	**17.** (b)	**18.** (b)	**19.** (d)	**20.** (b)
21. (b)	**22.** (c)	**23.** (d)	**24.** (b)	**25.** (c)	**26.** (c)	**27.** (b)	**28.** (b)	**29.** (c)	**30.** (a)
31. (c)	**32.** (c)	**33.** (a)	**34.** (b)	**35.** (a)	**36.** (a)	**37.** (b)	**38.** (b)	**39.** (a)	**40.** (d)
41. (a)	**42.** (d)	**43.** (c)	**44.** (b)	**45.** (a)	**46.** (c)	**47.** (a)	**48.** (d)	**49.** (d)	**50.** (b)
51. (a)	**52.** (c)	**53.** (d)	**54.** (b)	**55.** (d)					

42

Hybridisation and Heterosis

Hybridisation

It is the method of producing new crop varieties, in which two or more plants of unlike genotype (genetically dissimilar) are crossed. It is practiced both in self-pollinated and cross-pollinated crops.

The process of hybridisation aims to combine breeding, transgressive breeding or hybrid varieties. The main aim of **combination breeding** is the transfer of one or more characters into a single variety from another variety. These characters may be governed by oligogenes or polygenes. **Transgressive breeding** aims at improving yield or its contributing characters through transgressive segregation.

The **transgressive segregation** refers to the appearance of such plants in F_2-generation that are superior to both parents for one or more characters. In most self-pollinated crops, F_1 is more vigorous and have higher yielding than the parents and wherever it is commercially feasible, F_1 may be used directly as a variety. This kind of variety is called **hybrid variety**.

Kinds of Hybridisation

On the basis of taxonomic relationships of the parents involved in a cross, hybridisation may be of following four major types :

(i) **Intravarietal hybridisation** It is a cross between two plants of same variety having different genotypes. This method is useful for the improvement of self-pollinated crops.

(ii) **Intervarietal hybridisation** It is a cross between the plants of different varieties of the same species. These may be of two strains, varieties or races of the same species. It is also known as interspecific hybridisation. This process is employed in the improvement of both self-pollinated and cross-pollinated crops.

(iii) **Interspecific hybridisation** It is a cross made between the two different species of the **same genus**. Hence, also known as **intrageneric hybridisation**. These crosses are generally practiced to produce stress resistance for diseases, pests, drought, etc. The genes for resistance located in the **wild species** are brought into a single hybrid individual.

(iv) **Intergeneric hybridisation** It is a cross made between the plants belonging to the different genera of the same family. This is the most difficult method among all types of crosses. It is generally practiced to create new crop species. The hybrids produced by this method are both scientifically as well as agronomically significant.

The useful varieties developed by this method are as follows:

Triticum aestivum × *Secale cerele* → *Triticale*

Radish × cabbage → Rabbage

Raphanus sativus × *Brassica oleracea* → *Raphanobrassica*

Brinjal × Tomato → Bromato

Sugarcane × *Sorghum* → Sugarcane *Sorghum*

For example, *Triticale* is the first man-made cereal. It is the result of artificial **allopolyploidy** between wheat (*Triticum*) and rye (*Secale*) species. Depending upon whether tetraploid ($2n = 4x = 28$) or hexaploid ($2n = 6x = 42$), wheat is utilised for the synthesis. *Triticale* may be **hexaploid** *Triticale* or **octoploid** *Triticale*. In each case, only diploid rye ($2n = 2x = 14$) is used.

Procedure of Hybridisation

The following steps are involved in hybridisation:

1. Choice of Parents

The choice of parents is based upon the objectives of the breeder. But, atleast one of the parents involved in a cross should be well-adapted and proved in the area for which the new variety is being developed.

First of all, desirable traits, which are required in the new variety are listed after that the parents with these desirable traits are searched. It is always better to select locally available plants as parents because they are already adapted to that climate.

2. Evaluation of Parents

Evaluation is not necessary, if the performance of parents in the area, where breeding is to be done is known. But if their performance is unknown, it should be evaluated particularly for the characters which are expected to contribute for the disease resistance.

3. **Emasculation**

It is the mechanism of removal of stamens or the killing of pollen grains of a flower without affecting the female reproductive organs, in any way is known as emasculation. It is generally done to prevent self-fertilisation. It is essential for bisexual flowers. In case of dioecious plants, male plants are removed while in case of monoecious plants, the male flowers are removed from the plant and female flowers are left.

Types of Emasculation

In bisexual flowers, depending upon the type of flower and flower biology, the method of emasculation may differ. Emasculation is done before the anthers are mature and stigma has become receptive. Here, it can be done by various methods, such as, hand emasculation, suction method, hot water emasculation, alcohol treatment, cold treatment and genetic emasculation.

It can be done in the following ways:

(i) **Hand emasculation** is used for flowers with large anthers. The anthers are carefully removed with the help of fine-tip-forcep generally in the evening (between 4-6 pm).

(ii) **Suction method** is used for the small flowers and in this case, a thin rubber or glass tube attached to a suction hose is used to suck the anthers from the flower. With the suction method, considerable amount of self-pollination (upto 15%) is likely to occur.

4. **Bagging**

After emasculation, the emasculated flower or inflorescence is immediately bagged to avoid pollination by any foreign pollen. For this purpose, emasculation bags made up of butter paper, fine cloth, polythene, etc. That are used depending upon the type of crop to be hybridised. In some cases, the inflorescence for the supply of pollen in the male parent plant is also bagged to avoid contamination of pollen with any foreign pollen.

5. **Tagging**

The emasculated and bagged inflorescence or flowers are tagged. They contain following informations:

- Date of emasculation.
- Date of pollination.
- Name of female and male plants.
- The name of the female parent plant is written first and that of the male parent plant is written later.

6. **Cross Pollination or Artificial Pollination**

The pollination procedure consists of collecting pollen from freshly dehisced anthers of the male parent and dusting this pollen to the stigmas of the emasculated flowers. During pollination a mature, fertile and viable pollen from the male parent should be placed on receptive stigma of emasculated flowers to bring about fertilisation.

7. **Harvesting and Storing the F_1 Seeds**

On maturity, the seeds are develop after crossing, i.e. the F_1 seeds, are then harvested, dried and stored. Certain seeds, such as of sugarcane start losing viability and hence they need to be sown immediately. The seeds from each cross should be kept separately.

Preferably, the seeds should be kept along with the original tags. If the parents involved in a cross differ for genes affecting seedling characteristics, the parent with the dominant form of the characters can be used as the male parent. This helps in the identification of selfed seeds in F_1 - generation.

Heterosis

The term 'heterosis' was first elucidated by Shull in 1914. It was though, first studied by Koelreuter in 1763 followed by Darwin in 1876. The heterosis is lost by inbreeding (self-fertilisation).

It is the phenotypic superiority of the hybrid over either of its parents in one or more traits. Hybridisation between unrelated strains generally leads to an increased vigour and fertility (i.e. hybrid vigour or heterosis).

It is generally expressed as an increase in size and yield, but it may be expressed in several ways, such as earlier or later flowering and maturity, greater adaptability, faster growth rate, etc. It is observed both in cross and self-pollinated species. There are following basis to explain heterosis :

Genetic Basis

There are two genetic hypothesis to explain heterosis, e.g.

(i) **Dominance hypothesis** Which states that heterosis is the result of masking of harmful effects of recessive alleles by their dominant alleles.

(ii) **Overdominance hypothesis** Which states that heterosis is the result of superiority of the heterozygotes.

There is no conclusive evidence to support or to reject either of the two hypothesis. Heterozygotes at certain loci are superior to homozygotes. It is generally believed that heterosis is the result of dominance and epistatic interactions. Some genes may also show overdominance.

Biochemical Basis

At the biochemical level, heterosis may result in one or many ways. Some of the mechanisms known so far are to reduce the amount of a single gene product, to separate gene products by two alleles, to combine gene product or the hybrid substance and effects in two different tissues.

Commercial Utilisation of Heterosis

Heterosis is commercially exploited in the form of hybrid or synthetic varieties. It has been extensively used in many crop species, vegetables, fruit trees and in animals.

A list of some examples of plant and animal species, where heterosis is being commercially exploited is given below:

Commercial Uses of Heterosis

Category	Example
Crop species	Asexually propagated species, cross-pollinated species, e.g. maize, jowar, bajra, sugarbeet, sunflower, forage grasses, forage legumes, cotton and some self-pollinated crops, e.g. rice.
Vegetable crops	Tomato, brinjal (*Solanum melongena*), onion, cucurbits, Brussel's sprouts.
Fruit trees	In almost all fruit trees.
Animals	Silkworm, poultry, cattle, swine.

Chapter Exercise

1. A cybrid is a hybrid carrying
(a) genomes and cytoplast of two different plants
(b) cytoplasm of two different plants
(c) cytoplasm of two different plants but genome of one plant
(d) genome of two different plants

2. Hybrid vigour occurs mainly due to
(a) heterozygosity
(b) superiority of all the genes
(c) homozygosity of pure characters
(d) mixing up of cytoplasm of the male with that of the female exclusively

3. Who is known as the Father of Green Revolution in India?
(a) Prof MS Swaminathan (b) Guha
(c) Maheswari (d) Prof Kashyap

4. The Indian Agricultural Research Institute is located in
(a) New Delhi (b) Calcutta
(c) Odisha (d) Bhopal

5. A progeny that has resulted due to a cross between two genetically unrelated parents is known as
(a) hybrid (b) variety
(c) species (d) None of these

6. The another name for hybrid vigour is
(a) heterosis (b) hybridisation
(c) mutation (d) male sterility

7. Somatic hybridisation can be used for
(a) gene transfer (b) transfer of cytoplasm
(c) formation of allopolyploids (d) All of these

8. Hybridisation process performed between more than two plants is known as
(a) single cross (b) multiple cross
(c) dihybrid cross (d) intergeneric cross

9. Heterosis means
(a) hybrid vigor
(b) hybrids are weak
(c) hybrids are weak as well as vigorous
(d) hybrids are neither weak nor vigorous

10. Two methods of selection in plants are
(a) mass selection (b) pure line selection
(c) hybridisation (d) Both (a) and (b)

11. Hybridoma cells are
(a) fusion product of two somatic cells
(b) fusion product of two sex cells
(c) fusion product of cancer cell and antibody producing cell
(d) fusion product of two intergenic animals

12. The hybrid variety is
(a) genetic (b) F_1-generation
(c) sterile (d) fertile

13. In hybridisation, the haploids combines the advantages of
(a) recombination (b) segregation
(c) fixation (d) All of these

14. Heterosis lost of inbreeding is known as
(a) outbreeding depression (b) inbreeding depression
(c) hybrid vigor (d) None of these

15. Hybridisation performed between two plants belonging to the same variety is known as
(a) intervarietal cross (b) interspecific cross
(c) intergeneric cross (d) Intravarietal

16. Hybrid vigour is best maintained in
(a) emasculation
(b) vegetatively propagated crops
(c) hybridisation
(d) None of the above

17. Somatic hybridisation (protoplast fusion of parasexual hybridisation or cybrid hybridisation) is
(a) protoplast culture (b) cell culture
(c) organ culture (d) None of these

18. Bromato is made from
(a) brinjal (b) tomato
(c) Both (a) and (b) (d) None of these

19. Which of the following is not a part of methods of hybridisation?
(a) Bagging (b) Tagging
(c) Emasculation (d) PCR

20. *Triticale* is
(a) monoploid (b) hexaploid
(c) diploid (d) None of the above

Answers

1. (c)	**2.** (a)	**3.** (a)	**4.** (a)	**5.** (a)	**6.** (a)	**7.** (d)	**8.** (b)	**9.** (a)	**10.** (d)
11. (c)	**12.** (b)	**13.** (d)	**14.** (b)	**15.** (d)	**16.** (b)	**17.** (a)	**18.** (c)	**19.** (d)	**20.** (b)

43

Plant Resource Utilisation

Origin and Centers of Crop Plants

Many plants, their parts and products are eaten as food including cereals, legumes, vegetables, spices, etc. Also they are used for extracting the variety of oils, fats, natural rubber, drugs, sugars, etc.

Tobacco, fibres, fodder, paper are also plant yielding products.

Plants come to man as a gift of God as per oldest belief. Civilisation process and its developments has improved the quality of plants. Apart from this hunting and gathering nature of man also improved plant quality.

It is believed that agriculture originated somewhere in Indus, Tigris and Nile river above 7,000-13,000 years ago. Some evidences were also found which emphasised that agriculture has originated in different parts independently. The old world (Asia, Europe and Africa) and new world (North, South America) is considered as origin center of cereals. The concept of center of origin of crop plants was given by NI Vavilov.

According to him, it is an area where the crop plant has highest number of wild varieties. Further, Vavilov (1926) described following eight main geographical area as an origin and centre of crop/cultivated plants. Further, Chinese Center is consider as the largest and oldest independent center. 136 endemic plants belongs to this place.

Some of the important cultivated plants centre are millet, buck, wheat, soybean, bamboo, etc.

(i) **Indian center** It has following two centers:
   (a) **Main centers** It is in Asom and Burma region. It has 117 endemic plants such as rice, sugarcane, mango, orange, jute, coconut, etc.
   (b) **Indo-Malayan center** It is in Malay region having 55 endemic plants such as banana, coconut, etc.

(ii) **Central Asiatic center** It is in Punjab, Afgan region. The main crop is bean, pea, almond, etc.

(iii) **Near Eastern center** It includes Asia Minor and Iran. It has 83 endemic species such as cherry, pomegranate, etc.

(iv) **Mediterranean center** It includes Mediterranean sea. It has 84 endemic species such as turnip, cabbage, etc.

(v) **Abyssinian center** It includes Ethiopia and Somalia. It has 38 endemic species, such as wheat and barley.

(vi) **South Maxican and Central American center** It include maize, papaya, cotton are main plants.

(vii) **American center** It has following two centres:
   (a) The Peruvian–Ecuadorian–Bolivian center, e.g. tobacco.
   (b) Brazilian paraguayan center, e.g. groundnut.

Concept of Sustainable Development

Sustainable development is defined the process of development that meets the needs of the present generation without compromising the ability of future generation. It involves two main concerns, i.e.,

(i) Carrying capacity of natural systems.

(ii) Social and economic challenges of food by human.

It covers the domains such as economics, ecology, culture and politics.

Some Important Crops in India

1. Cereals

These are the edible seeds of grasses or other seeds having identical characteristics of its grain.

In their natural form, they are rich source of vitamins, minerals, carbohydrates, fats, proteins, etc. Thus, they are grown in greater quantities worldwide and provide more food energy than any other crop. Hence, called **staple crops**.

Brief details about two important cereals are given as under:

(i) **Wheat** It the most widely cultivated crop and used as a staple food in some areas as eight world. Botanical name is *Triticum aestivum*. It is found in major geographical region temperate region. The largest wheat producing country is USA. It is a native plant of West Asia and in India it is considered as a **Rabi crop.** The sowing time of wheat is between October and November.

The important states of India that produce good amount of wheat are UP, Punjab, MP, Maharashtra and the important varieties are Sonora, Sonalika, Lerma, Roja, etc. It majorly contains 65-74% of carbohydrates and used in the form of flour in India.

(ii) **Rice** Botanical name is *Oryza sativa*. It is most important crop of India. China and India are the center of origin. It is mainly cultivated in the coastal states of India. Its major varieties are Pusa, Krishna, Basmati, etc., and the carbohydrate content is 73-82%.

2. **Legumes**

A legumes is a plant belonging to the family–Fabaceae. These are grown primarily for their food grain seed and as a soil-enhancing green manure. Many legumes also have a nitrogen-fixing ability such as (alfa-alfa, clover, peas, beans, soybeans, etc.

These are considered as the best protein source in the plant kingdom. Thus, for vegetarians eating legumes is compared to the nutritional benefits of meat, chicken, etc.

These are the second most important source of food. Mainly used for protein consumption. Some common useful legumes are given as under:

- *Cicer arietinum* (gram or chick pea or Bengal gram) is a native of Southern Europe. Used in roasted form mainly.
- *Cajanus cajan* (pigeon pea or red gram or arhar) is a native of Africa. Mainly used as 'dal' in India.
- *Glycine max* (soybean) is native of South Eastern Asia having maximum protein content (32-48%).
- *Pisum sativum* (pea) is native of Southern Europe. Used as vegetable in India.

Other important legumes and their botanical names are:

Legume	Botanical name
Black gram	*Phaseolus mungo* L
Mat or moth bean	*Phaseolus aconitifolius jacq*
Lentil	*Lens esculenta moench*
Horse gram	*Macrotyloma uniflorum*
Green or golden gram	*Phaseolus aureus roxb*
Chickling vetch or grass pea	*Lathyrus sativus L*
Cow pea or lobia	*Vigna unguiculata (L) Walp*
Cluster bean	*Cyamopsis tetragonoloba (L) taub*

3. **Sugarcane**

Botanical name of sugarcane is *Saccharum officinarum*. Member of family–Graminae. It is major varieties are CO 312, 313, 419, etc. **Bagasse** and **molasses** are the byproduct of sugar industry which are further used for various purposes.

Oil Yielding Plants

Various plants are used for the extraction of various types of oils. They are categorised into two main types, i.e. fixed oils and essential or volatile oils.

(i) **Fixed oils** are those oils which are composed of mainly glycerides and have the tendency of polymerising when exposed to air. These are mainly used in dressing of leather and in textile industries. They are divided further as saturated (containing single bond) and unsaturated (containing double and triple bonds) oils.

Types of oils and their plants sources are as under:

Major oils

(a) Coconut oil, Corn oil, Cotton seed oil
(b) Olive oil, Peanut oil, Rapeseed oil
(c) Safflower oil

Nut oil

(a) Almond oil, Beech nut oil from, *Fagus sylvatica*
(b) Cashew oil, Hagelnut oil, Pine nut oil

Citrus oil

(a) Grape fruit oil, Lemon oil, Orange oil
(b) Oil from melon and ground seed

Other edible oils

(a) Amaranthus oil, apricot oil
Oils used for **biofuel**
(b) Castor oil, Rapeseed oil, *Jatropha* oil, etc.

(ii) **Essential oils** are those that are mainly the derivatives of **terpenes** having a characteristic aroma or taste. These are majorly used informing perfumes, flavouring agents, etc.

Fragrance oil producing plants

Plant material	Botanical name
Bergamot	*Citrus aurantium* sub sp *bergamia* (Risso and Poit)
Cananga	*Cananga odorata* (lamb), Hook fet Thoms.
Capsicum	*Capsicum frutescens* L
Cedarwood	*Cedrus deodara* (Roxb) Loud.
Cinnamon	*Cinnamomum verum* JS Presl, (Syn. C *zeylanicum*)
Citronella	*Cymbopogon winterianus* Jowitt
Clove	*Eugenia caryophyllus* (Spreng) Bullock and Harrison
Jasmine	*Jasminum-officinale* L var. *grandiflorum* Bailey
Lavender	*Lavandula officinalis* Chaix et Kitt.
Lemon	*Citrus limon* (L) Burm. F.
Lemon grass	*Cymbopogon flexuosus* (Nees ex steud) Wats.
Lime	*Citrus aurantifolia* (Christm) Swingle
Pine	*Pinus roxburghii* Sarg.
Mint	*Mentha arvensis* L
Orange (sweet)	*Citrus sinensis* (L) Osbeck
Sandalwood	*Santalum album* L.

Plants Used as Beverages

Beverages are the liquids that are specifically prepared for the human consumption to refresh or to stimulate their body.

Beverages can be **alcoholic**, i.e. containing ethanol (such as beer, wine, whisky, vodka, etc) and can be **non-alcoholic** containing caffeine and other related alkaloids or carbonated water such as juices, soft drinks, tea and coffee.

Tea

Botanical name of tea is *Camellia sinensis.* It belongs to the family–Theaceae and is also known as the **queen of non-alcoholic beverages**. It mainly grows in the sub-tropical and the hilly areas of tropical regions. It grows better in well drained acidic soils. **China** is the longest producer and India is the largest exporter of tea.

Tea is mainly composed of alkaloids, polyphenols (tannins) which are responsible for providing bitterness to tea, ethereal oil, is responsible for providing the aroma and flavour and theine which provides a stimulating and refreshing effect to it.

There are mainly four types of tea, depending upon the method of processing these are

(i) **Black tea** containing 3.5% caffeine.

(ii) **Green tea** containing high amount of tannin and caffeine when compared to the black tea.

(iii) **Oolong tea** category residing between both black and the green tea.

(iv) **Brick tea** prepared from the remains of black and green tea.

Coffee

Botanical name of coffee is *Coffea arabica* which is belong to the family of coffee is Rubiaceae. It is a native plant of *Abyssinia*. It is commonly cultivated in Southern area of India and the soils seeds are used for obtaining coffee.

A volatile oil **caffeol oil** is responsible for providing flavour to the coffee.

Fibres

Fibres are the sclerenchymatus cells of plants mainly found is vascular bundles. They have deposition of both cellulose and lignin on their walls.

There are mainly three types of fibres depending upon the tensile strength :

(i) **Surface fibre or cotton** It is obtained from *Gossypium* plant which belongs to family–Malvaceae and mainly grows in Gujarat and Maharashtra.

(ii) **Soft fibres** Some common soft fibres and their source plants are as under :

- **Flax** is obtained from *Linum usitatissinum.*
- **Jute** is obtained from *Corchorus capsularis*.
- **Sunnhemp** is obtained from *Crotalaria juncea*.
- **Patsan** is obtained from *Cannabis sativa.*
- **Ramie** is obtained from *Boehmeria nivea*.

(iii) **Hard fibres** Some important fibres and their source plant are as under:

- **Manila hemp** is obtained from *Musa*.
- **Bow string hemp** is obtained from *Sansevieria*.
- **Coir** is obtained from *Cocos nucifera*.
- **Sisal hemp** is obtained from *Agave*.

Medicinal Plants

There are over around more than hundred chemical substances that have been derived from plants for the use as drugs and medicines.

A plant medicinal value is due to the presence in its tissues of some chemical substance that produces a physiological action on the body. Most important are the alkaloids compounds of carbon, hydrogen, oxygen and nitrogen.

The drugs can be obtained from the different parts of the plants such as roots, stems, flowers, barks, fruits, seeds, leaves, etc. These are given below :

- **Drugs /medicines from roots**

  Acotine it is obtained from *Acotinum napellus*.

  Belladona it is obtained from *Atropa*.

  Ashwagandha it is obtained from *Withania*.

  Asafoetida it is obtained from *Ferula*.

- **Drugs from stem**

  Turmeric it is obtained from *Curcuma*.

  Ginger it is obtained from *Zingiber*.

- **Drugs from leaves**

  Holy basil it is obtained from *Ocimum*.

  Vasaka it is obtained from *Adhatoda*.

- **Drugs from flowers**

  Saffron it is obtained from *Crocus*.

- **Drugs from bark**

  Quinine it is obtained from *Cinchona*.

  Cinnamon it is obtained from *Cinnamomum*.

- **Drugs from fruits**

  Opium it is obtained from *Papaver*

  Emblica it is obtained from *Emblica*

- **Drugs from seeds**

  Croton it is obtained from *Croton*

Chapter Exercise

1. Which of the following is an example of major oil?
   (a) Grape fruit oil (b) Corn oil
   (c) Almond oil (d) None of these
2. Almond oil is
   (a) major oil (b) nut oil
   (c) citrus oil (d) All of these
3. Which of the following is an example of citrus oil?
   (a) Almond oil (b) Peanut oil
   (c) Lemon oil (d) Rapeseed oil
4. Which part of melon and ground seed are used for oil extractions?
   (a) Seeds (b) Leaves
   (c) Fruits (d) None of these
5. Which of the following plants are used for producing biofuels?
   (a) Castor (b) Rapeseed
   (c) Jatropha (d) All of these
6. *Camellia sinensis* is the botanical name of
   (a) pea (b) coffee
   (c) tea (d) None of these
7. Tea belongs to the family
   (a) Malvaceae (b) Theaceae
   (c) Rutaceae (d) Thalamiflorae
8. Which of the following part of India is used for cultivation of tea?
   (a) Asom (b) Darjeeling
   (c) J & K (d) Both (a) and (b)
9. Botanical name of coffee is
   (a) *Coffea arabica* (b) *Camellia sinensis*
   (c) *Pisum sativum* (d) None of these
10. Which of the following plant belong to family–Rubiaceae?
    (a) *Amaranthus* (b) Pea
    (c) Coffee (d) Banana
11. Which of the following part of plant is used for coffee production?
    (a) Leaves (b) Stems
    (c) Seeds (d) Flowers
12. Cotton is a
    (a) hard fibre (b) surface fibre
    (c) soft fibre (d) None of these
13. Which of the following plant is used for obtaining sunnhemp fibres?
    (a) *Cannabis* (b) *Boehmeria*
    (c) *Crotalaria* (d) *Linum*
14. *Agave* is used for obtaining
    (a) coir (b) sisel hemp
    (c) manila hemp (d) None of these
15. Belladona is obtained from
    (a) *Atropa* (b) *Aconitum*
    (c) *Ferula* (d) *Withania*
16. Which of the following part of holy basil is used for obtaining drugs?
    (a) Root (b) Stem
    (c) Leaves (d) Flowers
17. Quinine is a drug obtained from bark of which of the following plant?
    (a) *Adhatoda* (b) *Crocus*
    (c) *Cinchona* (d) None of these
18. Opium is obtained from
    (a) *Papaver* (b) *Emblica*
    (c) *Cinchona* (d) *Ocimum*
19. Corn oil is a type of
    (a) nut oil (b) citrus oil
    (c) major oil (d) None of these
20. Coffee is cultivated in which part of India?
    (a) Eastern area (b) Western area
    (c) Southern area (d) Northern area
21. Which of the following is not a domain of sustainable development?
    (a) Economics (b) Politics
    (c) Psychology (d) Culture
22. Which of the following is the fundamental domain of sustainable domain?
    (a) Economics (b) Politics
    (c) Culture (d) All of these
23. Politics deals with the .... of good habit.
    (a) proliferation (b) marketing
    (c) Both (a) and (b) (d) None of these
24. *Triticum aestivum* is the botanical name of
    (a) wheat (b) maize
    (c) rice (d) pea
25. Most widely cultivated crop is
    (a) rice (b) pea
    (c) wheat (d) maize
26. Largest wheat producing country is
    (a) China (b) India
    (c) USA (d) South Africa

27. Wheat is sown in which of the following months?
(a) May-June (b) April-May
(c) Oct-Nov (d) Jan-March

28. The carbohydrate % in wheat is
(a) 65-74% (b) 60-65%
(c) 90-95% (d) 100%

29. Botanical name of rice is
(a) *Pisum sativum* (b) *Oryza sativa*
(c) *Triticum aestivum* (d) None of these

30. Which of the following is the centre of origin of rice?
(a) India and China (b) Australia
(c) Brazil (d) Africa

31. Rice is cultivated in which of the following area of India?
(a) Himalayan region (b) Coastral region
(c) Plain region (d) None of these

32. Pusa is a variety of
(a) wheat (b) maize
(c) rice (d) banana

33. Legumes are used for obtaining
(a) carbohydrates (b) oils
(c) protein (d) fibres

34. *Glycine max* is the botanical name of
(a) pea (b) soybean
(c) gram (d) maize

35. *Saccharum officinarum* is the botanical name of
(a) maize (b) barley
(c) sugarcane (d) banana

36. CO 312, 313 and 419 are the varieties of
(a) pea (b) gram
(c) sugarcane (d) maize

37. Bagasse and molasses are byproducts of which of the following industry?
(a) Flour (b) Textile
(c) Sugar (d) Leather

38. The botanical name of pigeon pea is
(a) *Pisum sativum*
(b) *Cicer arietinum*
(c) *Cajanus cajan*
(d) None of these

39. Sonora is a variety of
(a) barley (b) maize
(c) wheat (d) None of these

40. Most important crop of the world is
(a) wheat (b) maize
(c) rice (d) barley

41. Wheat is an important cereal of
(a) tundra region (b) taiga region
(c) temperate region (d) Both (b) and (c)

42. Which of the following crop even highest foreign exchange?
(a) Sugarcane (b) Cotton
(c) Tea (d) Papaya

43. Which of the following crop is considered as gift of new world to old world?
(a) Cotton (b) Tea
(c) Wheat (d) Banana

44. The largest producer of jute and tea is
(a) China (b) USA
(c) Brazil (d) India

45. Which of the following is considered as first man-made cereal plant?
(a) *Zea mays* (b) *Pisum*
(c) *Triticale* (d) *Brassica*

46. Which of the following is situated at Hyderabad?
(a) IARI (b) NBRI
(c) CCMB (d) NPL

47. The chile center is famous for
(a) tomato (b) pea
(c) potato (d) alium

48. The concept of origin and center of crop plants was given by
(a) Vavilov (b) Darwin
(c) Haeckel (d) E Mayer

49. Longest and oldest center is
(a) Mediterranian center (b) Indian center
(c) Chinese center (d) None of these

50. Chinese center have following number of endemic species
(a) 26 (b) 25 (c) 136 (d) 83

Answers

1. (b)	**2.** (b)	**3.** (c)	**4.** (a)	**5.** (d)	**6.** (c)	**7.** (b)	**8.** (d)	**9.** (a)	**10.** (c)
11. (c)	**12.** (b)	**13.** (c)	**14.** (b)	**15.** (a)	**16.** (c)	**17.** (c)	**18.** (a)	**19.** (c)	**20.** (c)
21. (c)	**22.** (a)	**23.** (a)	**24.** (a)	**25.** (c)	**26.** (c)	**27.** (c)	**28.** (a)	**29.** (b)	**30.** (a)
31. (b)	**32.** (c)	**33.** (c)	**34.** (b)	**35.** (c)	**36.** (c)	**37.** (c)	**38.** (c)	**39.** (c)	**40.** (c)
41. (c)	**42.** (c)	**43.** (c)	**44.** (d)	**45.** (c)	**46.** (c)	**47.** (c)	**48.** (a)	**49.** (c)	**50.** (c)

44

Analytical Techniques in Plant Science

Analytical techniques are the methods of determining the concentration of any compound or an element from the mixture of substances. There is a wide variety of techniques been available today that are majority used for the analytical purposes.

The most common techniques are given below:

Microscopy

It is the technical field of using microscopes, an instrument used to view samples and objects that are unable to be seen with an unaided eye (i.e. objects having the resolution less than the normal eye resolution).

There are many type of microscopes, the most common and the first to be invented was optical microscope, that uses light to image the samples.

The other important microscopes also have been developed by different scientists with the rise in the field of analytical chemistry and biological sciences to view sample even less than 0.1 μm. These are light microscope, fluorescence microscopy, confocal microscopy, electron microscopy, etc.

Centrifugation

It is a process used in biological sciences such as molecular biology, plant sciences, etc., to separate two immiscible liquids from each other during experimentations.

In this process, a centrifugal force is used for the sedimentation of heterogenous mixtures (i.e. containing more than two different compounds) with an instrument called centrifuge.

During this technique of separation, the more dense component of a mixture migrates away from the axis of the centrifuge when compared to the less dense component which migrates towards the axis.

Depending upon the densities of separation, centrifugation can be either

(i) **Differential centrifugation** In which the separation is done due to the change in mass/size.

(ii) **Density gradient centrifugation** In which the separation is done in the medium which gradually increase in density from top to bottom.

Spectrophotometry

It is a quantity based study of electromagnetic spectra and the instrument used for this purpose is called **spectrophotometer** which majorly uses the amount of light reflecting from a sample object on the amount of light absorbed by the sample object. Thus, a spectrophotometer consists of two instruments, namely a spectrometer, for producing light of any selected colour (wavelength) and a photometer for measuring the intensity of light.

These instruments are arranged, so that liquid in a cuvette can be placed between the spectrometer beam and the photometer. If development of colour is linked to the concentration of a substance in solution then that concentration can be measured by determining the extent of absorption of light at the appropriate wavelength.

For example, haemoglobin appears red, because the haemoglobin absorbs blue and green light rays much more effectively than red.

Electrophoresis

It is the technique in which the molecules are separated on the basis of their charges when kept in an electric field.

Common types of electrophoresis used are:

- **Gel electrophoresis** in which an agarose or polyacrylamide gel are used for electrophoresis.
- **Discontinuous electrophoresis** in which the gel is divided into sample gel, stacking gel and separating gel.
- **Two dimensional gel electrophoresis** which is performed in two directions for the separation of molecules having minute mass or charge differences.
- **Sodium Dodecyl Sulphate** (SDS) is used for making unique charge on each separating molecules in this electrophoresis.
- **Native page electrophoresis** in which the protein are not degenerated.
- **Immuno-blotting** which uses antibody for affinity purpose.

Chromatography

It is known as the method of physical separation of organic as well as inorganic substances being known by the scientists which utilise the concept of partition or distribution coefficient. So, that they can be analysed and studied. Hence, it is a well known method for observing mixture and solvents.

Following are the types of chromatography:

- **Planer chromatography** Stationary phase is planer.
- **Columner chromatography** Stationary phase is column or tube.
- **Partition chromatography** Relies on distribution of solutes based on solubility differences.
- **Absorption chromatography** Relies on specific interactions between the solutes and the molecules of stationary phase.
- **Size exclusion chromatography** Relies on the basis of size and shape of molecules.
- **Ion exchange chromatography** Separation of charged molecule based on action of opposite ions.
- **Affinity chromatography** Relies on the absorption by the affinity molecule.

Chapter Exercise

1. Separation of molecules on the basis of their partition coefficient is called
(a) electrophoresis
(b) centrifugation
(c) chromatography
(d) All of these

2. Separation of molecules on the basis of difference in charge is called
(a) electrophoresis (b) centrifugation
(c) chromatography (d) None of these

3. Molecular are separated on the basis of their sedimentation is called
(a) chromatography (b) electrophoresis
(c) centrifugation (d) None of these

4. Immunoblotting uses
(a) antigen (b) antibody
(c) Both (a) and (b) (d) None of these

5. Planer chromatography uses following stationary phase
(a) tube (b) column
(c) cube (d) None of these

6. Partition chromatography relies on
(a) man differences (b) charge differences
(c) solubility differences (d) None of these

7. Stationary phase is tubular in
(a) paper chromatography
(b) column chromatography
(c) absorption chromatography
(d) All of the above

8. Size exclusion chromatography separates the molecules on the basis of
(a) size (b) shape
(c) Both (a) and (b) (d) None of these

9. Action of opposite ion is utilised in
(a) affinity chromatography
(b) size exclusion chromatography
(c) ion exchange chromatography
(d) None of these

10. Agarose and polyacrylamide gels are used in
(a) centrifugation (b) chromatography
(c) microscopy (d) None of these

11. Discontinuous gel electrophoresis uses
(a) sample gel (b) stacking gel
(c) separating gel (d) All of these

12. SDS is used in SDS page for
(a) equating the charge (b) equating the mass
(c) Both (a) and (b) (d) None of these

13. Ultracentrifuge has following RPM
(a) 40-60 K (b) 20-40 K
(c) 5-10 K (d) 60-70 K

14. Proteins are used in normal state in which of the following electrophoresis
(a) SDS page (b) Native page
(c) Both (a) and (b) (d) None of these

15. 2D page utilise how many direction
(a) 1 (b) 2 (c) 3 (d) 4

16. The maximum magnification possible with light microscopy is
(a) 400 X (b) 500 X
(c) 1000 X (d) 2000 X

17. The outline of a cell, nucleus, mitochondria and vacuoles are viewed with the help of
(a) bright field microscopy (b) dark field microscopy
(c) fluorescent (d) ultraviolet

18. The size of objects that can be seen with the light microscope is
(a) larger than 0.24
(b) larger than 1 μ
(c) larger than 1nm
(d) larger than 2Å

19. The blank used to calibrate the spectrophotometer for testing the absorbance spectrum of chlorophyll is
(a) water (b) acetone
(c) chlorophyll-*a* (d) cobalt chloride

20. The path followed by the light through a spectrophotometer is
(a) meter, photodetector, filter, sample, light source
(b) meter, sample, filter, light source, photodetector
(c) light source, filter, sample, photodetector, meter
(d) light source, sample, filter, photodetector, meter

21. Microscopes that are will remain nearly focused after the low-power objective lens is changed to the high-power objective lens
(a) monocular (b) parcentered
(c) parfocal (d) properly adjusted

22. Which system consists of a camera and/or a video screen?
(a) Viewing and recording (b) Illuminating
(c) Imaging (d) Contrasting

23. If the ocular of a microscope is 10X and the objective is set at 100X then what is the total magnification of the microscope?
(a) 10X (b) 43X
(c) 1000X (d) 430X

24. Always begin examining microscope slides with which power objective?
(a) High
(b) Low
(c) 100X
(d) It doesn't matter which objective

25. What must be done to a specimen to increase the contrast of the structures viewed?
(a) Illuminated
(b) Stained
(c) Placed under a cover slip
(d) Thinly sliced

26. For which of the following specimens would you use a dissecting scope?
(a) Human skin cells (b) *E.coli*
(c) Insect mouth parts (d) Newspaper print

27. Which of these techniques is often considered a suitable 'polishing' step in a protein purification?
(a) Affinity Chromatography (AC)
(b) Ion-Exchange Chromatography (IEX)
(c) Hydrophobic Interaction Chromatography (HIC)
(d) Size-Exclusion Chromatography (SEC)

28. What properties of a protein does hydrophobic interaction chromatography exploit for purification?
(a) Charged amino acids
(b) Hydrophobic amino acids on the protein surface
(c) Molecular weight
(d) Enzyme activity

29. To elute target proteins from an affinity chromatography matrix, which of the following conditions would be the most appropriate?
(a) Low salt concentrations
(b) High salt concentrations
(c) Adding a soluble ligand which competes with the affinity tagged protein for binding to the column
(d) Just keep washing buffer through the column, isocratic elution

30. Which of the following methods could be used to check the molecular weight of your purified protein?
(a) SDS-PAGE
(b) Mass spectrometry
(c) Analytical SEC
(d) All of the above

31. In gas chromatography, the basis for separation of the components of the volatile material is the difference in
(a) partition coefficients (b) conductivity
(c) molecular weight (c) molarity

32. In reverse phase chromatography, the stationary phase is made
(a) non-polar (b) polar
(c) either non-polar or polar (d) None of these

33. Thin layer chromatography is
(a) partition chromatography
(b) electrical mobility of ionic species
(c) adsorption chromatography
(d) None of the above

34. Ion exchange chromatography is based on the
(a) electrostatic attraction
(b) electrical mobility of ionic species
(c) adsorption chromatography
(d) partition chromatography

35. The field of view of a microscope with a 10X ocular and a 4X objective is 5mm. What is will be the field of view with a 10X objective?
(a) 3.14 mm^2 (b) 20 mm (c) 2 mm^2 (d) 2 mm

36. Which objective provides the greatest depth of field?
(a) High
(b) 100X
(c) Low
(d) This depends on If the specimen is stained

37. Immuno blotting is used for
(a) ELIZA test (b) bacterial infection
(c) viral test (d) genetic disorder

38. At high power, always use which adjustment knob to focus the image?

(a) Fine
(b) Coarse
(c) Diaphragm
(d) It doesn't matter

39. SDS-PAGE, the protein sample is first

(a) treated with a reducing agent and then with anionic detergent followed by fractionation by electrophoresis
(b) fractionated by electrophoresis then treated with an oxidising agent followed by anionic detergent
(c) treated with a oxidising agent and then with anionic detergent followed by fractionation by electrophoresis
(d) None of the above

40. Electrophoresis of histones and myoglobin under non-denaturing conditions (pH = 7.0) results in

(a) both proteins migrate to the anode
(b) histones migrate to the anode and myoglobin migrates to the cathode
(c) histones migrate to the cathode and myoglobin migrates to the anode fans
(d) both proteins migrate to the cathode

41. Proteins can be visualised directly in gels by

(a) staining them with the dye
(b) using electron microscope only
(c) measuring their molecular weight
(d) None of the above

42. In a native PAGE, proteins are separated on the basis of

(a) net negative charge
(b) net charge and size
(c) net positive charges size
(d) net positive charge

43. Proteins are separated in an SDS-PAGE experiment on the basis of their

(a) positively charged side chains
(b) molecular weight
(c) negatively charged side chains
(d) different isoelectric points

44. Which of the following is not a mechanism for the separation of a physically unstable suspension of magnesium hydroxide in water?

(a) Piocculation (b) Aggregation
(c) Ostwald ripening (d) Hydrolysis

45. How might solid sodium carbonate be obtained from sodium carbonate solution?

(a) Centrifugation (b) Filtration
(c) Evaporation (d) It cannot be extracted

46. Which one of the following dispersions does not have liquid continuous phase?

(a) Nano suspension (b) Microemulsion
(c) Gel (d) Foam

47. Which one of the following systems has the smallest sized domains in its dispersed phase?

(a) Nano emulslon (b) Coarse suspension
(c) Coarse emulsion (d) Microemulsion

48. SDS is used in

(a) electrophoresis (b) chromatography
(c) centrifugation (d) All of these

49. Cuvette used in

(a) centrifugation (b) electrophoresis
(c) microscopy (d) spectrophotometry

50. Paper chromatography done of

(a) seperation of pigments (b) seperation of A As
(c) Both 'a' and 'b' (d) None of these

Answers

1. (c)	**2.** (a)	**3.** (c)	**4.** (b)	**5.** (d)	**6.** (c)	**7.** (b)	**8.** (c)	**9.** (c)	**10.** (d)
11. (d)	**12.** (a)	**13.** (c)	**14.** (b)	**15.** (b)	**16.** (c)	**17.** (b)	**18.** (a)	**19.** (b)	**20.** (c)
21. (c)	**22.** (a)	**23.** (c)	**24.** (b)	**25.** (b)	**26.** (c)	**27.** (d)	**28.** (b)	**29.** (c)	**30.** (d)
31. (a)	**32.** (a)	**33.** (a)	**34.** (b)	**35.** (d)	**36.** (c)	**37.** (a)	**38.** (a)	**39.** (a)	**40.** (c)
41. (a)	**42.** (b)	**43.** (b)	**44.** (d)	**45.** (c)	**46.** (c)	**47.** (d)	**48.** (a)	**49.** (d)	**50.** (c)

45

Biostatistics

'Statistics' refers to the subject of scientific activity dealing with the theories and methods of collection, analysis and interpretation of such data and when these tools of statistics are applied to the data been derived from biological sciences it is termed as biostatistics.

Now-a-days, biostatistics has became an elementary tool in the field of biology and its branches because of its importance and need in different aspects such as in physiology and anatomy.

Data

It is a collection of observations expressed in numerical figures. This collection can be done in either of the two ways, i.e. either by complete enumeration or by the simple survey method.

It is to be noted that the data is always in collective sense, i.e. it can never be used singular. The data can be **primary data** collected directly from the field of enquiry or a **secondary data** numerical information that have been already collected by some agency.

Population

It is an entire group of people or study elements-persons, things or measurements having some common fundamental characteristics. It may be either finite or infinite.

Sample

It is a relatively small group of selected number of individuals or objects or cases drawn from a particular population. It is used to throw light on the population characteristics.

Random sample, is a sample chosen in a very specific way every element in the population has an equal opportunity of being included in the sample.

Representation of Data

Statistical data can be represented in the following three main processes :

(i) **Textuals presentation** in which the data is presented in a descriptive form.

(ii) **Tabular representation** in which the data is represented in the form of tables, i.e. in rows and columns.

(iii) **Graphical representation** in which the quantitative data is represented by graphs and charts. Graph includes both charts and diagrams. The major objective of this type of representation is to emphasise on the relative position of different sub-division and not simply record the details.

Measures of Central Tendency

(Representative of Data)

These are following types:

Measure of Central Tendency of Ungrouped Data

Mean The mean or average of a number of observations is the sum of the values of all the observations divided by the total number of observation. Denoted by $\bar{x}$.

Median The median is that value of the given number of observations, which divides it into exactly two parts.

So, when the data is arranged in ascending or descending order, the median of ungrouped data is calculated as follows:

(i) When the number of observations (n) is odd, the median is the value of the $\left\{\frac{n+1}{2}\right\}$ observation.

For example, if $n = 13$ the value of the $\left\{\frac{13+1}{2}\right\}$th, i.e. the 7th observation will be the median.

(ii) When the number of observations (n) is even, the median is the mean of the $\left\{\frac{n}{2}\right\}$th and the $\left\{\frac{n}{2}+1\right\}$th observation.

For example, if $n = 16$th mean of the values of the $\left\{\frac{16}{2}\right\}$th and $\left\{\frac{16+1}{2}\right\}$th observation, i.e. which occur the mean of the values of 8th and 9th observation will be the median.

Mode It is that value of observation which occurs most frequently, i.e. an observation repeated maximum number of times.

Measure of Central Tendency of Grouped Data

Mean It can be done as in the following :

(i) **Direct method** Make the ungrouped data into grouped data form the following table :

Class Interval	No. of Std / Frequency	Class mark $\frac{\text{Upper limit + Lower limit}}{2}$	
			$\Sigma f_1 x_1$
	f_1	x_1	
	Σf_i		$\Sigma f_i x_i$

$$\bar{x} = \frac{\sum f_i x_i}{\sum f_i}$$

(ii) **Mean of deviation** (assumed mean method) Following table will be used

Class Interval	No. of Student or Frequency	Class Marks	$d_i = x_1 - mi$ d value of x	$f_1 d_1$

Mean of deviations, $\bar{d} = \dfrac{\sum f_i d_i}{\sum f_i}$

Relation between $\bar{a}$ and $\bar{d}$

$$\Rightarrow \quad \bar{x} = a + \bar{d}$$

$$\therefore \quad \bar{x} = a + \frac{\sum f_i d_i}{\sum f_i}$$

(iii) **Step deviation method**

Class Interval	f_1	x_1	$d_1 = x_1 - a$	$u = \frac{x-a}{h}$	$f_1 u_1$

$$\bar{u} = \frac{\sum f_i u_i}{\sum f_i}, \quad \bar{x} = a + h\bar{u}$$

$$\Rightarrow \quad \bar{x} = a + h\frac{\sum f_i a_i}{\sum f_i}$$

Mode Modal class is a class having maximum frequency.

$$\text{Mode} = l + \left\{\frac{f_1 - f_0}{2f_1 - f_0 - f_2}\right\} \times h$$

where, l = lower limit of the modal class

h = size of the class interval

f_1 = frequency of the modal class

f_0 = frequency of the class preceeding the modal class

f_2 = frequency of the class succeeding the modal class

Median Firstly median class is calculated by n/2 th observation method and then the value is put Median

$$= l + \left\{\frac{\frac{n}{2} - cf}{f}\right\} h$$

where, l = lower limit

n = number of observation

cf = cumulative frequency before the median class

f = frequency of median class

h = class size

Note 3 Median = Mode + 2 Mean

Different Modes of Data Representation

There are various types of graphs in the form of charts and diagrams. Some of them are:

(i) **Line Diagram or Graph** (representing statistical information)

(ii) **Bar Diagram** (representation of data in the form of bars)

(iii) **Pie chart** representation of data in the form of a circular graph, whose area is sub-divided into sectors by radii.

(iv) **Histogram** diagrammatic representation of grouped frequency distribution of both continuous and discontinuous type.

(v) **Frequency polygon** (area diagram represented is the form of curve obtained by joining the middle points of tops of rectangles)

(vi) **Ogives** (cumulative frequency polygon), a graphical representation of cumulative frequency distribution that are plotted against the corresponding class boundaries.

Measures of Dispersion

The second important characteristic of a distribution is given by dispersion. Two distributions may be identical in respect of its first important characteristic, i.e. central tendency and yet they may differ on account of scatterness.

Dispersion for a given set of observations may be defined as the amount of deviation of the observations, usually, from an appropriate measure of central tendency.

Measures of dispersion may be broadly classified as :

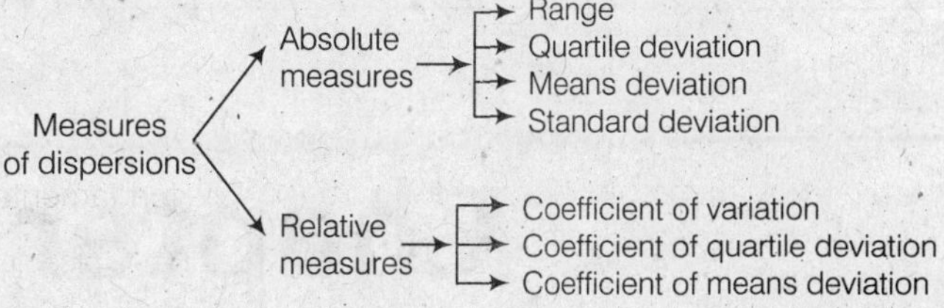

Range

For a given set of observations, range may be defined as the difference between the largest and smallest of observations. Thus, if L and S denote the largest and smallest observations, respectively then we have

$$\boxed{\text{Range} = \text{L} - \text{S}}$$

The corresponding relative measure of dispersion, known as coefficient of range, which is given by

$$\boxed{\text{Coefficient of range} = \frac{\text{L} - \text{S}}{\text{L} + \text{S}} \times 100}$$

Mean Deviation (MD)

It is the average difference between the items in a distribution and the median and mean that series. It is also called average deviation.

Mean deviation can be either

(i) Mean deviation about the mean.

(ii) Mean deviation about the median.

Coefficient of mean deviation is the ratio of mean deviation to its arithmetic mean or median multiplied by 100.

$$\text{Coefficient of MD} = \frac{\text{MD}}{\text{Mean / Median}} \times 100$$

Standard Deviation (SD)

It was introduced by **Paul Pearson** in the year 1893 and is considered as the most widely accepted concept of studying dispersion.

SD is defined as the square-root of the arithmetic mean squares of deviation from arithmetic mean.

In short SD may be defined as 'Root Mean Square Deviation from Mean'. It is usually denoted by the Greek small letter sigma (δ).

Coefficient of standard deviation is ratio of the standard deviation to its arithmetic mean, i.e.

$$\text{Coefficient standard deviation} = \frac{\delta}{\bar{X}}$$

Variance

It is defined as the square of standard deviation is called variance and is denoted by δ^2 Thus,

$$\text{Variance} = (\text{SD})^2 = \delta^2$$

Coefficient of variance is the product of the coefficient of standard deviation ×100

$$\text{Coefficient of variance} = \frac{\text{SD}}{\text{AM}} \times 100, \text{ i.e. } \frac{\text{Sigma}}{\text{A.M}} \times 100$$

Chi-square Test

It is a statistical test that involves the calculation of a quantity which is used to compare an 'observed' ratio with an 'expected' or 'theoretical' ratio and to determine how closely the former fits the latter.

The measure of chi-square enables us to find out the degree of discrepancy between 'observed' frequencies and 'theoretical' or 'expected' frequencies and thus to determine whether the discrepancy between observed and theoretical frequencies is due to the error of sampling or due to the chance.

Thus, X^2 test is statistically written as

$$X^2 = \frac{\sum_{i=1}^{n}\left[|(O-E)| - \frac{1}{2}\right]^2}{E}$$

where, O = Observed frequency, E = Expected frequency

$\frac{1}{2} = 0.5 =$ Yates correction

Test for Goodness of Fit (Pearsonian $- x^2$)

Chi-square test is also applied as a test of "goodness of fit" to determine whether the actual or (observed) numbers or frequencies are similar or in 'good agreement', with the expected (theoretical) number of frequencies.

Thus, the test is called "goodness of fit".

$$x^2 = \Sigma \frac{\left[|(O-E)| - \frac{1}{2}\right]^2}{E}$$

where, O = Observed, E = Expected

$\frac{1}{2}$ = Yates correction

Chapter Exercise

1. Why the data is classified?

(a) To discriminate is similar and dissimilar data
(b) For amusement
(c) Both (a) and (b)
(d) None of the above

2. Which of the following measures the central tendency?

(a) Mean (b) Median
(c) Mode (d) All of these

3. Which of the following is not a part of central tendency?

(a) Mean (b) Median
(c) Integration (d) None of these

4. Frequency distribution was

(a) grouping the data (b) class size
(c) upper and lower limit (d) All of these

5. The sum of the values of all the observation divided by the total number of observation is

(a) mean (b) median
(c) mode (d) All of these

6. Mean is denoted by

(a) x^2 (b) $\bar{x}$
(c) Both (a) and (b) (d) None of these

7. The value of the given number which divides it exactly two parts is

(a) mean (b) median
(c) mode (d) None of these

8. When the number of observation is odd the median is the value of which observation?

(a) $\left(\frac{n+1}{2}\right)$th (b) $\left(\frac{n}{2}\right)$th (c) $\left(\frac{n-1}{2}\right)$th (d) $\left(\frac{n+2}{2}\right)$th

9. When the number of observation is even, the median is the value of which observation?

(a) $\left(\frac{n-1}{2}\right)$

(b) Mean of $\left(\frac{n}{2}\right)$th and $\left(\frac{n+1}{2}\right)$th

(c) $\left(\frac{n-1}{2}\right)$

(d) $\left(\frac{n+2}{2}\right)$th

10. Mode is that value of observation which occurs most

(a) rarely (b) average
(c) frequently (d) None of these

11. Which formula belongs to direct method of mean of grouped data?

(a) $\bar{x} = \frac{\sum f_i x_i}{\sum x_i}$ (b) $\bar{x} = \frac{\sum f_i}{\sum f_i x_i}$
(c) $\bar{x} = \frac{\sum f_i}{\sum x_i}$ (d) None of these

12. $\bar{d} = \frac{\sum f_i d_i}{\sum f_i}$ is

(a) median (b) mode
(c) mean of deviation (d) None of these

13. Stepwise deviation method uses which of the following formula?

(a) $\bar{d} = \frac{\sum f d_i}{\sum f_i}$ (b) $\bar{x} = \frac{\sum f_i x_i}{\sum f_i}$
(c) $\bar{u} = \frac{\sum f_i u_i}{\sum f_i}$ (d) All of these

14. In the formula of mode of grouped data $l + \left\{\frac{f_i - f_o}{2f_1 - f_o - f_2}\right\} h$, h is

(a) lower limit (b) frequency of modal class
(c) size of class interval (d) None of these

15. 3 Median = Mode +... Mean

(a) 1 (b) 3
(c) 2 (d) 4

16. Which one is easier to compute?

(a) Relative measures of dispersion
(b) Absolute measures of dispersion
(c) Both (a) and (b)
(d) Range

17. The absolute measure of dispersion is

(a) range (b) mean deviation
(c) standard deviation (d) All of these

18. Most commonly used measure of dispersion is

(a) coefficient of variation (b) range
(c) quartile deviation (d) standard deviation

19. The range of 15, 12, 10, 9, 17, 20 is

(a) 5 (b) 12
(c) 13 (d) 11

20. For any two numbers SD is always

(a) twice the range (b) half of the range
(c) square of the range (d) None of these

21. If all the observations are multiplied by 2, then

(a) new SD would be also multiplied by 2
(b) new SD would be half of the previous SD
(c) new SD would be increased by 2
(d) new SD would be decreased by 2

22. Ratio of mean deviation to its arithmetic mean or median multiplied by 100 is

(a) coefficient of mean deviation
(b) coefficient of standard deviation
(c) coefficient of variance
(d) None of the above

23. Standard deviation of 10, 16, 10, 16, 10, 10, 16, 16 is

(a) 4 (b) 6
(c) 3 (d) 0

24. Which data provides the detailed information?

(a) Primary data
(b) Secondary data
(c) Both (a) and (b)
(d) None of the above

25. Representation of data in the pure of a circular graph is

(a) histogram (b) pictogram
(c) Pie-chart (d) frequency polygon

Answers

1. (a)	**2.** (d)	**3.** (c)	**4.** (d)	**5.** (a)	**6.** (b)	**7.** (b)	**8.** (a)	**9.** (b)	**10.** (c)
11. (a)	**12.** (c)	**13.** (c)	**14.** (c)	**15.** (c)	**16.** (d)	**17.** (d)	**18.** (d)	**19.** (d)	**20.** (b)
21. (a)	**22.** (a)	**23.** (c)	**24.** (a)	**25.** (c)					

Practice Set 1

1. One chromosome of a homologous pair carries the genes J and K. The other chromosome of the pair carries the genes j and k at corresponding loci. Crossing over results in exchange of chromosome segments and production of gametes with new combinations of genes. A 'recombinant' type gamete resulting from this crossover might contain
   (a) genes J and K (b) genes j and K
   (c) genes J and j (d) genes j and k

2. Which of the following statements is incorrect?
   (a) Fluorescence microscopy may be used with specially stained antibodies to detect specific microorganisms
   (b) Chromogenic stains which consist of a single dye rather than several dyes in combination are known as basic stains
   (c) Transmission electron microscopy has a practical magnification of greater than 100,000 *X*
   (d) Acid fast stains interact with the waxy structures of the cell wall of *Mycobacterium* species

3. Position-specific scoring matrices, consensus sequences, and sequence logo are needed because a transcription factor
   (a) always binds to the same DNA sequence found in many locations in the genome
   (b) always binds to a specific DNA sequence but its affinity varies with the genomic location of that sequence
   (c) often binds to DNA with any one of several very similar sequences
   (d) often binds to other transcription factors

4. Lichens are important in the studies on atmospheric pollution because they
   (a) can grow in polluted atmosphere
   (b) can readily multiply in polluted atmosphere
   (c) efficiently purify the atmosphere
   (d) are very sensitive to pollutants like sulphur dioxide

5. Ecologists would consider all of the following to be natural disturbances except
   (a) droughts (b) floods
   (c) deforestation (d) fires

6. Which of the following is an example of a plant morphological defense?
   (a) Beltian bodies (b) Mustard oil
   (c) Epiphytes (d) Silica

7. The innermost portion of a mature plant cell wall is the
   (a) primary cell wall (b) secondary cell wall
   (c) plasma membrane (d) plasmodesmata

8. Many plant species have adaptations for dealing with the periodic fires typical of a
   (a) chaparral
   (b) temperate grassland
   (c) temperate deciduous forest
   (d) All of the above

9. Which of the following statements is correct?
   (a) A Chi-square value lower than the critical value demonstrates that the hypothesis was correct
   (b) Critical values for the Chi-square test are larger when there are more degrees of freedom because each class of data is expected to show a small amount of variation
   (c) A high Chi-square value always means that our hypothesis is wrong
   (d) A Chi-square value below the critical value means that our hypothesis is correct

10. Plants grow sparsely in arid regions of the world
   (a) only because of high temperature
   (b) because no seeds fall on their soil
   (c) because of several factors combining together
   (d) because the soil is sandy

11. Classification of plants based on chromosome number is
   (a) cytotaxonomy
   (b) numerical taxonomy
   (c) biochemical taxonomy
   (d) All of these

12. Which branch of botany deals with the development of improved varieties of plants?
   (a) Embryology (b) Plant breeding
   (c) Palaeobotany (d) Palynology

13. Seeds are characteristically found in
   (a) pteridophytes (b) bryophytes
   (c) fungi (d) spermatophytes

14. A plant bearing fruit has a column of vascular bundle and a tap root system. This plant is a
   (a) angiosperm and dicot
   (b) gymnosperm and dicot
   (c) gymnosperm plant
   (d) angiosperm and monocot

15. System of classification that was proposed by Bentham and Hooker is
   (a) numerical (b) artificial
   (c) natural (d) phylogenetic

16. According to Hutchinson's classification, the total number of familes in angiosperms are

(a) 202 (b) 252
(c) 431 (d) 411

17. John Hutchinson divided dicots into two divisions

(a) Calciflorae and Corelliflorae
(b) Lignose and Herbaceae
(c) Sympetalae and Monochlamydae
(d) Polypetalae and Gamopetalae

18. The five kingdom classification was proposed by

(a) Copeland (b) Whittaker
(c) Linnaeus (d) John Ray

19. Which of the following is not generally a feature of shifting cultivation?

(a) Relatively small field
(b) Slash and burn forest clearing
(c) Regular pesticide application
(d) Local consumption of crops

20. Monera devoid of cell wall are

(a) Actinomycetes (b) cyanobacteria
(c) mycoplasma cell (d) eubacteria

21. Reserve food in *Euglena* is

(a) true starch (b) paramylum starch
(c) glycogen (d) mannitol

22. In bacterial chromosome, the nucleoid or nucleic acid polymers are

(a) circular DNA molecule
(b) linear DNA molecule
(c) Both types DNA and RNA
(d) linear RNA molecule

23. Virus and bacteriophage are made of

(a) protein (b) nucleoprotein
(c) DNA and RNA (d) lipoprotein

24. The foliose lichens are attached to substratum by means of

(a) rhizinae (b) roots
(c) rhizoids (d) None of these

25. Ergot of rye is caused by

(a) *Claviceps purpurea*
(b) *Sclerospora graminicola*
(c) *Claviceps macrophala*
(d) *Erysipha graminis*

26. The flagella of *Chlamydomonas* are

(a) tinsel type (b) whiplash type
(c) Both (a) and (b) (d) absent

27. The life cycle of *Funaria* is not completed without water because of

(a) *Funaria* is aquatic
(b) *Funaria* is very delicate
(c) zooidogamic fertilisation
(d) branches are not formed without water

28. If a sporangium is developed proma group of cells it is called as

(a) eusporangiate development
(b) heterosporangiate development
(c) leptosporangiate development
(d) None of the above

29. The male cone of *Cycas* is made of a group of

(a) microsporophylls (b) megasporophylls
(c) sporophylls (d) archegonia

30. Highly reduced and disc-like stem develops in

(a) *Ruscus* (b) *Asparagus*
(c) *Smilex* (d) *Allium*

31. Cotton fibres are made up of

(a) cellulose (b) starch
(c) lignin (d) protein

32. The bacterial culture prepared by pure culture method is

(a) inoculum (b) suspension
(c) dilution (d) None of these

33. Amphisarca is found in

(a) *Aegle marmelos* (b) *Michelia*
(c) *Calotropis* (d) *Catharanthus*

34. Which of the following statements is incorrect?

(a) The amount of light available in an ecosystem affects the types and amount of plant life that will grow there
(b) In a species, the organisms offspiring must be able to breed
(c) An ecosystem is composed only biotic factors
(d) The same niche in a habitat may be occupied by more than one species

35. Which of the following plant show guttation?

(a) *Acacia nilotica* (b) *Pisum sativum*
(c) *Ficus bengalensis* (d) *Tropaeolum najas*

36. Characteristics of intracellular receptors that regulate gene transcription include all of the following except

(a) a DNA binding site
(b) an extracellular binding site
(c) a transcription activating domain
(d) may be signaled by lipid soluble molecules

37. The common feature(s) between prokaryotes and eukaryotes is/are

(a) cytoskeleton structure (b) ribosomes
(c) plasmid (d) All of these

38. Trees may be damaged by animals which rub against them, wearing away a strip of bark right around the tree trunk and exposing the xylem. The tree will then

(a) die quickly because the leaves are deprived of food and water
(b) die quickly because fungi enter the trunk through the wounds
(c) continue to grow, because bark always regrows to cover a wound
(d) die slowly because the roots cannot replenish their food reserves

39. What relationship are respiration and photosynthesis?
(a) Reciprocal (b) Reverse
(c) Same (d) No relationship

40. Electron carriers in non-cyclic flow of electrons
(a) pass electrons from a more negative to a more positive reduction potential
(b) pass electrons from higher free energy to lower energy
(c) pass electrons from a more positive to more negative reduction potential
(d) Both (a) and (b)

41. Which of the following phytohormones plays a role in seed dormancy?
(a) Gibberellins (b) ABA
(c) Cytokinin (d) Auxins

42. ABA controls which of the following?
(a) Abscission
(b) Fruit climacteric
(c) Stomatal closure
(d) Cell elongation

43. A population would grow exponentially
(a) if it were limited only by density dependent factors
(b) until it reaches carrying capacity
(c) if there were no limiting factors
(d) if it were a population with an equilibrial life history

44. Coactivators assist in transcriptional activity by
(a) interacting with general transcription factors
(b) modifying the activity of RNA polymerase II directly
(c) changing the chromatin structure to make transcription of certain genes more accessible
(d) All of the above

45. A single layer of phospholipid molecules coats the water in a beaker. Which part of the molecules will face the air?
(a) The phosphate groups
(b) The hydrocarbon tails
(c) Both head and tail because the molecules are amphipathic and will lie sideways
(d) The phospholipids would dissolve in the water and not form a membrane coat

46. To improve the resolution of light microscope up to 1.5 times, we have to use
(a) red light + immersion oil between the specimen and objective
(b) blue light + immersion oil between the specimen and objective
(c) blue light + glycerine between the specimen and objective
(d) None of the above

47. Pair of contrasting characters controlling the same trait is called
(a) factors (b) alleles
(c) allelomorphs (d) paramorphs

48. Which one of the following statements about ionisation in a mass spectrometre is incorrect?
(a) Gaseous atoms are ionised by bombarding them with high energy electrons
(b) Atoms are ionised so they can be accelerated
(c) Atoms are ionised so they can be deflected
(d) It doesn't matter how much energy you use to ionise the atoms

49. If GUG codon is used as a starting codon then it codes for
(a) valine (b) cysteine
(c) methionine (d) leucine

50. Chemoautotrophs
(a) are photosynthetic
(b) use organic molecules for an energy and carbon surface
(c) oxidise inorganic substances for energy and use CO_2 as a carbon source
(d) use light to generate ATP but need organic molecules for a carbon source

51. The system of classification proposed by Linnaeus was artificial because
(a) it considered physiological characters along with morphology
(b) it considered evolutionary relationships of plants
(c) it takes into account similarities and differences in few unifying characters
(d) None of the above

52. A system of classification, which takes into account, all noticeable characters in morphology in nature is
(a) phylogenetic system
(b) natural system
(c) artificial system
(d) cytotaxonomy

53. Advantages of phylogenetic system over artificial system of classification is/are
(a) includes overall similarities
(b) based on evolution any history
(c) high predictive value
(d) All of the above

54. The member of kingdom–Protista differs from members of kingdom–Monera mainly due to
(a) membrane bound nucleus
(b) the presence of cell wall
(c) DNA
(d) RNA

55. What do host cells provide for viruses?
(a) Nucleotides and amino acids
(b) ATP
(c) Ribosomes
(d) All of the above

56. Who discovered synthesis of DNA from RNA in Rous sarcoma virus?
(a) Lous (b) Temin
(c) Smith (d) Mayer

57. Prokaryotes found inhabiting the Great Salt Lake would be the

(a) cyanobacteria (b) extreme halophiles
(c) extremophiles (d) methanogens

58. Bacteria that use light for their energy source and CO_2 for their carbon source are called

(a) photoautotrophs (b) photoheterotrophs
(c) chemoheterotrophs (d) chemoautotrophs

59. The bacteria that cause tetanus can be killed only by prolonged heating at temperature considerably above boiling. Tetanus is caused by

(a) *Agrobacterium tumefaciens*
(b) *Vibrio cholerae*
(c) *E.coli*
(d) *Clostridium tetani*

60. A widely but periodically occurring fungal plant disease will be

(a) epiphytotic (b) epidemic
(c) sporadic (d) endemic

61. When two host species are required for completion of parasitic fungus life cycle, the condition is described as

(a) autoecious (b) heteroecious
(c) autotrophic (d) heterokaryotic

62. A .... is a predatory fungus.

(a) *Puccinia* (b) *Alternaria* (c) *Fusarium* (d) *Arthobotrys*

63. Plasmotomy occurs in case of

(a) uninucleate protists (b) binucleate protists
(c) multinucleate protists (d) All of the above

64. The storage product in *Euglena* (green euglenoids) is

(a) glycogen (b) oil
(c) paramylum (d) chrysolaminarin

65. Photosynthetic protists belong to

(a) Bacillariophyceae
(b) Bacillariophyceae and Euglenophyceae
(c) Bacillariophyceae, Euglenophyceae and Dinophyceae
(d) Zooflagellates

66. In Bentham and Hooker's system of classification, dicots are considered primitive over

(a) monocots (b) gymnosperms
(c) bryophytes (d) pteridophytes

67. The sporogonium of members of class–Bryopsida is differentiated into

(a) foot (b) seta
(c) capsule (d) All of these

68. In lemon, the petiole is

(a) absent
(b) modified into spines and prickles
(c) modified to store proteins and carbohydrates
(d) winged

69. If two vegetative fungal cells fuse, it is

(a) parthenogenesis
(b) somatogamy
(c) spermatisation
(d) plasmogamy

70. Fibrous roots of *Zea mays* (maize) develop from

(a) lower nodes (b) upper nodes
(c) internodes (d) leaves

71. What is the site of enzyme synthesis in cells?

(a) Golgi body
(b) Ribosome
(c) Lysosomes
(d) Smooth endoplasmic reticulum

72. The term 'suicide bag' is applicable to a class of cell organelle called

(a) Golgi apparatus (b) lysosomes
(c) microsomes (d) peroxisomes

73. Most of the hydrolytic enzymes of lysosomes function at

(a) basic pH (b) any pH
(c) neutral pH (d) acidic pH

74. The limit of resolution of an instrument

(a) varies directly with its resolving power
(b) varies inversely with its resolving power
(c) is equal to its resolving power
(d) is not related to its resolving power

75. The technique that counts and identifies cells and chromosomes is

(a) electrophoresis (b) flow cytometry
(c) Both (b) and (c) (d) X-ray crystallography

76. The meristems that develop at the time of emergency are

(a) promeristems
(b) primary meristems
(c) secondary meristems
(d) Both (a) and (c)

77. Vascular cylinder in dicot stems occurs in the form of distinctly separated vascular bundles. But, in some dicot stems, *e.g. Linum, Lonicera*, it appears to be a continuous cylinder because

(a) interfascicular region is not developed at all
(b) the interfascicular region is there but its cells are not differentiable from the cell of vascular bundles
(c) interfascicular region is extremely narrow
(d) continuous vascular cylinder differentiates since, the early stage of development

78. Which of the following statements does not apply to reverse osmosis?

(a) It is used for water purification
(b) In this technique, pressure greater than osmotic pressure is applied to the system
(c) It is a passive process
(d) It is an active process

79. Based on the figure given below which of the following statements is not correct?

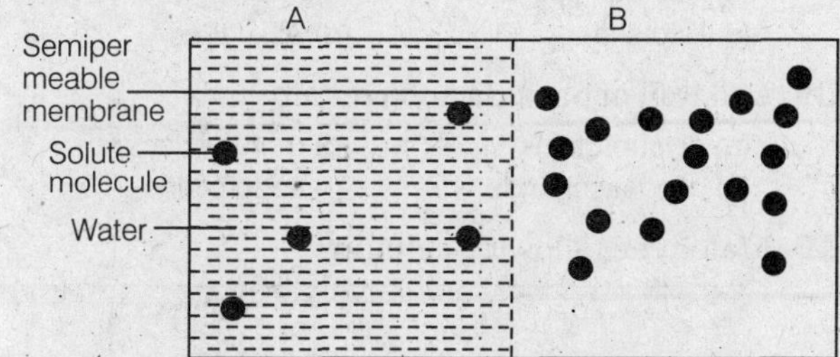

(a) Movement of solvent molecules will take place from chamber A to B.
(b) Movement of solute will take place from A to B.
(c) The presence of a semipermeable is a pre-requisite for this process to occur.
(d) The direction and rate of osmosis depends on both the pressure gradient and concentration gradient.

80. Match the followings and choose the correct option.

A. Leaves	1. Anti-transpirant
B. Seed	2. Transpiration
C. Roots	3. Negative osmotic potential
D. Aspirin	4. Imbibition
E. Plasmolysed cell	5. Absorbtion

Codes

	A	B	C	D		A	B	C	D
(a)	3	4	1	2	(b)	1	2	3	4
(c)	3	2	4	1	(d)	3	2	1	4

81. ATP is injected in cyanide poisoning because it is
(a) necessary for cellular functions
(b) necessary for $Na^+ - K^+$ pump
(c) $Na^+ - K^+$ pump operates at the cell membrane
(d) ATP breakdown cyanide

82. Which of the following product of glycolysis is consumed in alcoholic fermentation?
(a) $NADH_2$ (b) CO_2 (c) ATP (d) Both (a) and (c)

83. Which of the following connects the glycolysis and Krebs' cycle?
(a) Pyruvic acid (b) Glucose
(c) Acetyl Co-A (d) ATP

84. During which stage of meiosis, do centromeres divide?
(a) Prophase-I (b) Prophase-II
(c) Metaphase-I (d) Anaphase-II

85. During mitosis, nuclear membrane disappear at
(a) telophase (b) prophase
(c) metaphase (d) anaphase

86. Identify the correct matched pair.
(a) Anaphase-II of meiosis — Centromeric division
(b) Exchange of genetic material — Diakinesis
(c) Significance of meiosis — Production of genetically similar cells
(d) Segregation — Metaphase-II

87. Which one of the following is an agent orange?
(a) Weedicide with dioxin
(b) Chemical used in luminous paint
(c) Biodegradable insecticide
(d) Colour used in fluorescent lamp

88. The growth in plants is
(a) limited (b) unlimited
(c) unlocalised (d) None of these

89. If the seedlings are grown in darkness they are
(a) similar to those grown in light
(b) taller than those grown in light
(c) of the same size as those grown in light
(d) much healthier than those grown in light

90. The usual shape of growth curve is
(a) linear (b) inverted bell-shaped
(c) sigmoidal (d) *zig-zag*

91. What was the basis for the Griffith experiment?
(a) The virulent strain was transformed into non-virulent strain.
(b) Both the non-virulent and virulent strains of *Pneumococcus* were heat killed.
(c) All the non-virulent bacteria were heat killed
(d) The non-virulent strain was transformed into virulent strain.

92. The substance of *Pneumococcus* used for transformation in the Griffith experiment was known to be?
(a) DNA (b) RNA
(c) Polysaccharide (d) Protein

93. What do you mean by the sapwood?
(a) Outer part of secondary xylem
(b) Inner part of secondary phloem
(c) Both (a) and (b)
(d) None of the above

94. What type of fatty substance does cork cell possess?
(a) Suberin (b) Cutin
(c) Hemicellulose (d) Lignin

95. State the type of vascular bundles found inside a dicot stem
(a) Closed, collateral, endarch
(b) Closed, collateral exarch
(c) Closed, radial, endarch
(d) Open, collateral, endarch

96. All algae possess
(a) Chlorophyll-*a* and carotene
(b) Chlorophyll-*a* and chlorophyll-*b*
(c) Chlorophyll-*b* and carotene
(d) Phycobilins and carotene

97. Identity the well known reserve food found in case of algae
(a) Fat
(b) Cellulose
(c) Glycogen
(d) Starch and oil

98. The fungi having capability of growing at and below the level of freezing point are
(a) coprophilous (b) thermophilic
(c) keratinophilic (d) coenocytic fungi

99. Name the fungus which out of these is edible in nature
(a) *Aspergillus* (b) *Ustilago*
(c) *Morchella* (d) *Polyporus*

100. In which form genetic materials are meant to be preserved in the gene banks.
(a) Vegetative material (b) Seeds
(c) Both 'a' and 'b' (d) None of these

101. Select the odd one for the *ex situ* conservation
(a) Wildlife sanctuaries (b) Botanical garden
(c) Tissue culture repository (d) Field gene bank

102. The Global Plan of Action (GPA) for the conservation and sustainable utilisation of genetic resources includes.
(a) *In* situ conservation
(b) *Ex* situ conservation
(c) Use of plant genetic resources
(d) All of the above

103. Life on other plants is indicated by the presence of
(a) O_2 (b) Water
(c) CH_4 (d) Organic molecules

104. The biopoiesis is related to the
(a) life supporting system
(b) origin of life
(c) evolution of biomolecules
(d) organic evolution

105. Match the following columns.

Column I	Column II
A. Modern theory	1. Richter
B. Theory of panspermia	2. Curvier
C. Theory of spontaneous creation	3. Oparin and Haldane
D. Theory of catastrophism	4. Von Helmont
	5. Pasteur

Codes

	A	B	C	D		A	B	C	D
(a)	4	1	2	3	(b)	4	2	1	3
(c)	5	1	4	2	(d)	5	2	3	4

106. In the absence of air life on earth would be
(a) aerobic
(b) anaerobic
(c) reductive
(d) free from oxidative metabolism

107. Cellular organelles incase of eukaryotes are found to be embedded inside the embedded in
(a) protoplasm (b) nucleoplasm
(c) cytoplasm (d) cytosol

108. The type of prokaryotic cell out of the following is
(a) *Nostoc* (b) *Anabaena*
(c) Bacteria (d) All of these

109. Cell wall of bacteria also contains
(a) glycoproteins (b) muramic acid
(c) oligosaccharide (d) lipoproteins

110. Match the following columns.

Column I	Column II
A. Prokaryotic cell	1. PPLO
B. Eukaryotic cell	2. *Amoeba*
C. Smallest cell	3. Blue-green algae

Codes

	A	B	C		A	B	C
(a)	1	2	3	(b)	3	1	2
(c)	1	2	3	(d)	2	1	3

111. In a unit membrane model of a plasma membrane of eukaryote, its average thickness is estimated to be
(a) 6.5 nm (b) 7.5 nm
(c) 7 nm (d) 8 nm

112. Due to which property the phospholipids are known to be better when compared to other?
(a) Polar and non-polar parts
(b) Glycerol
(c) Proteins bindings sites
(d) None of the above

113. Number of chromatids found in the chromosome at metaphase stage of meiosis-II is
(a) 2 (b) 4 (c) 6 (d) 8

114. What does the hollow spindle fibres determines during cell division?
(a) Spindle fibres are absent in the centre
(b) Complete the absence of chromosomes in the nucleus
(c) The chromosomes are absent in the centre
(d) None of the above

115. Spindle has ………… type(s) of fibres.
(a) 2 (b) 1 (c) 3 (d) 4

116. The division of grown up cells occurs at the stage when
(a) it becomes difficult for the nucleus to manage the organelles
(b) nucleus produce mitogens
(c) karyocytoplasmic ratio gets disturbed
(d) number of organelles double in the cytoplasm

117. Types of cross which does not follow Mendelian rule and ensures the exactly similar ratio for both phenotype and genotype is
(a) incomplete dominance (b) multiple alleles
(c) polygeny (d) pleotropism

118. How many types of gametes produced by aaddcc will be of?
(a) 1 (b) 3 (c) 4 (d) 8

119. Human skin determines the trait called
(a) digenic (b) trigenic
(c) pentagenic (d) hexagenic

120. By the removal of which character Mendel reaches the conclusion of development of pure strain of peas.
(a) Weak plants
(b) Male plants
(c) Female plants
(d) A typical plants in each generation

121. Which statements are true about genotype of an organism?
I. Remains same throughout the life.
II. Can be known through direct observation.
III. Mutations can change genotype.
IV. Is not influenced by phenotype.
(a) I, II and IV (b) III and IV
(c) I, II and III (d) I, III and IV

122. Which of the Mendel's law would not be possible if he had failed to proved the phenomenon of linkage of genes.
(a) Law of independent assortment
(b) Law of segregation
(c) Law of paired factors
(d) Law of dominance

123. One turn of DNA is approximately
(a) 0.34 nm (b) 34 Å (c) 20 Å (d) 2 nm

124. During the process of replication, the type of enzyme responsible for producing nick and the resealing of the DNA strand is called
(a) helicase (b) topoisomerase
(c) unwindase (d) polymerase

125. How many nucleotides does the Okazaki fragments possess?
(a) 2000-3000 nucleotides (b) 100-1000 nucleotides
(c) 100-200 nucleotides (d) 3000-4000 nucleotides

126. The type of attaching site found on the *t*RNA is
(a) anticodon site (b) DHU site
(c) CCA end (d) 5′ end

127. Information for decoding on *m*RNA is
(a) transition (b) translation
(c) transversion (d) transcription

128. Nonsense codons in protein synthesis are meant for
(a) proofread the sequence of amino acid in protein chain
(b) make the protein chain non-functional
(c) terminate ion protein synthesis
(d) initiate protein synthesis

129. If one strand of DNA has 5′-AGCATCGG-3′ the sequence in opposite strand is
(a) AGCATCGG (b) TCGTAGCC
(c) GGCTACGA (d) GTGTAGCT

130. Name the method used for the introduction of the desired genes into
(a) gene therapy
(b) gene modification
(c) gene splicing
(d) gene cloning

131. For the production of which produce the genetically engineered bacteria is commercialised?
(a) Hormones (b) Antibiotics
(c) Vitamins (d) All of these

132. On the atmospheric layer called
(a) mesosphere (b) stratosphere
(c) ionosphere (d) troposphere

133. What do you understand if we say that tolerance range is high?
(a) It has no impact on the distribution of organisms
(b) It depends upon organism to restrict to one region or covers a wider area
(c) Distribution of organisms is restricted
(d) Distribution of organism is wide

134. Match the following terms as per the daily rhythms of animals.

Column I	Column II
A. Active during dawn	1. Vesperal
B. Active during day time	2. Diurnal
C. Active during dusk	3. Auroral
	4. Crepuscular

Codes

	A	B	C		A	B	C
(a)	1	2	3	(b)	3	2	1
(c)	4	2	3	(d)	3	1	4

135. What does the limiting factor determines?
(a) Factor which limits the activity of other over active factors
(b) Factor present in smallest amount
(c) Factor that limits growth, development and reproduction by its deficiency or excess
(d) Factor which causes small growth of organism

136. Ozone layer protects biosphere from
(a) UV rays (b) infra red rays
(c) cosmic rays (d) X rays

137. Embryonic mesoderm forms the
(a) endothelium of blood vessels
(b) exocrine cells
(c) endocrine cells
(d) glial cells

138. What does the inter-related organisms of different species in particular environment forms?
(a) Family (b) Biotic community
(c) Population (d) Ecosystem

139. Why the exotic species are sometimes called the serious pasts?
(a) Better adaptability to new environment
(b) Better environment conditions in new locality
(c) Increase in reproductive potential
(d) The absence of natural competition

140. In what condition, the parasite is known to be successful in nature?
(a) Have high biotic potential
(b) Remains attached to the host for a long
(c) Have minimum demand from the host
(d) Show rapid growth

141. Shown by the female *Anopheles* while visiting the host for sucking blood at intervals is the type of parasitism
(a) hemiparastie
(b) endoparasite
(c) holoparasite
(d) intermittent parasite

142. DDT and BHC are the type of
(a) degradable and qualitative pollutant
(b) non-biodegradable and quantitative pollutant
(c) degradable and quantitative pollutant
(d) non-biodegradable and qualitative pollutant

143. CFC, act as air pollutant although it is used in many areas due to of its multiple uses,
(a) it causes calcification of ligaments
(b) it impair the sensation of smell
(c) it corrods the metals and deteriorate paints
(d) it along with NO_x react with ozone

144. In what term the method of decomposition of organic matter by microbes is measured
(a) biochemical oxygen demand
(b) amoung of primary pollutants produced
(c) amount of secondary pollutants produced
(d) growth of algae

145. Mark the correct statements suitable to the effect of increase in CO_2 concentration
I. Plants will grow more successfully in water scarcity and nutrient poor soil.
II. Rate of transpiration increases
III. Rate of photosynthesis in C_3 plant increases.
IV. Rate of stomatal conductance will decrease.
(a) I and III (b) II and IV
(c) I, II and IV (d) I, III and IV

146. Wildlife Protection Act was passed
(a) 1981 (b) 1989
(c) 1964 (d) 1972

147. Match the following columns.

Column I	Column II
A. Endangered species	1. Depleted species
B. Extinct species	2. Taxon is died out from earth
C. Vulnerable species	3. Taxon facing high risk of extinction in the near future
	4. Threatened species having lower risk of extinction

Codes

	A	B	C		A	B	C
(a)	3	2	1	(b)	2	3	4
(c)	3	2	1	(d)	4	3	2

148. The endangered species out of the following is known to be
(a) Red panda (b) Dodo
(c) Great Indian bustard (d) Snow leopard

149. Number of biosphere reserve been established in India till today are
(a) 7 (b) 10 (c) 14 (d) 18

150. The national animal of India is known to be
(a) *Panthera leo* (b) *Panthera tigris*
(c) *Pavo cristatus* (d) *Moshus moschiferous*

Answers

1. (b)	**2.** (b)	**3.** (c)	**4.** (d)	**5.** (c)	**6.** (d)	**7.** (b)	**8.** (d)	**9.** (b)	**10.** (d)
11. (a)	**12.** (b)	**13.** (a)	**14.** (a)	**15.** (c)	**16.** (d)	**17.** (b)	**18.** (b)	**19.** (c)	**20.** (c)
21. (b)	**22.** (c)	**23.** (b)	**24.** (a)	**25.** (a)	**26.** (b)	**27.** (c)	**28.** (a)	**29.** (a)	**30.** (d)
31. (a)	**32.** (a)	**33.** (a)	**34.** (c)	**35.** (d)	**36.** (b)	**37.** (d)	**38.** (a)	**39.** (b)	**40.** (d)
41. (b)	**42.** (c)	**43.** (c)	**44.** (d)	**45.** (b)	**46.** (b)	**47.** (b)	**48.** (d)	**49.** (c)	**50.** (c)
51. (c)	**52.** (b)	**53.** (c)	**54.** (a)	**55.** (d)	**56.** (b)	**57.** (b)	**58.** (a)	**59.** (d)	**60.** (c)
61. (b)	**62.** (d)	**63.** (c)	**64.** (c)	**65.** (c)	**66.** (a)	**67.** (d)	**68.** (c)	**69.** (b)	**70.** (a)
71. (d)	**72.** (c)	**73.** (d)	**74.** (b)	**75.** (b)	**76.** (c)	**77.** (c)	**78.** (c)	**79.** (b)	**80.** (c)
81. (d)	**82.** (a)	**83.** (b)	**84.** (d)	**85.** (b)	**86.** (a)	**87.** (c)	**88.** (b)	**89.** (b)	**90.** (c)
91. (a)	**92.** (a)	**93.** (a)	**94.** (a)	**95.** (d)	**96.** (a)	**97.** (d)	**98.** (b)	**99.** (a)	**100.** (c)
101. (a)	**102.** (b)	**103.** (b)	**104.** (d)	**105.** (d)	**106.** (a)	**107.** (b)	**108.** (b)	**109.** (c)	**110.** (a)
111. (b)	**112.** (a)	**113.** (b)	**114.** (c)	**115.** (c)	**116.** (a)	**117.** (a)	**118.** (a)	**119.** (b)	**120.** (d)
121. (d)	**122.** (a)	**123.** (b)	**124.** (b)	**125.** (d)	**126.** (c)	**127.** (b)	**128.** (c)	**129.** (b)	**130.** (a)
131. (d)	**132.** (d)	**133.** (d)	**134.** (b)	**135.** (c)	**136.** (a)	**137.** (a)	**138.** (b)	**139.** (d)	**140.** (c)
141. (d)	**142.** (d)	**143.** (d)	**144.** (a)	**145.** (d)	**146.** (d)	**147.** (a)	**148.** (c)	**149.** (c)	**150.** (b)

Practice Set 2

1. The primary ecological factor determining the distribution of deserts is
(a) windiness (b) elevation
(c) moisture (d) temperature

2. The signalling system that operates over the shortest range is
(a) contact-dependent cell signalling, as in the inhibition of neighbours by differentiating nerve cells
(b) paracrine secretion, as in the secretion of histamines by mast cells
(c) neural signalling, as in the transmission of a nerve impulse
(d) None of the above

3. Which of the statement is correct?
(a) Most carbon atoms were formed during the big bang
(b) Humans are about as old as the earth is
(c) There are ten million microns in one metre
(d) Carbon atoms are formed in stars

4. The internal factors that influence the growth of plants are
(a) hormones (b) genes
(c) Both (a) and (b) (d) None of these

5. .... are parts of apical meristem which have become separated from the apex by permanent tissues.
(a) Primary meristem (b) Intercalary meristem
(c) Lateral meristem (d) None of these

6. Which hormone allows seeds to ignore environmental conditions and germinate?
(a) Abscisic acid (b) Cytokinins
(c) Auxins (d) Gibberellins

7. Which of the following statement is correct about nitrate reductase?
(a) FAD, Mo and haeme act as prosthetic groups
(b) Catalyses conversion of NO_3^- to NH_4^+
(c) $FADH_2$ acts as electron donor
(d) Very sensitive to O_2

8. Cell wall of Gram-positive bacteria contain
(a) teichoic acid
(b) protein only
(c) protein and lipid
(d) cellulose and protein

9. Photophosphorylation occurs in
(a) plastids (b) mitochondria
(c) cytoplasm (d) cell membrane

10. The site where the protein coat of virus is synthesised is
(a) mitochondria of the host
(b) plasma membrane of the host
(c) RNA of the virus
(d) ribosome of the host

11. TMV is isolated in crystalline form by
(a) EC Stakmanz (b) KM Smith
(c) D Ivanowsky (d) WH Stanley

12. The parasitic fungi takes nutrition from their hosts with the help of
(a) soredia (b) haustoria
(c) fruiting bodies (d) paraphyses

13. Red rot of sugarcane is caused by
(a) *Helminthosporium* (b) *Puccinia*
(c) *Ustilago* (d) *Colletotricum*

14. Agar-agar, a common gel material used in labs and research work is isolated from
(a) blue-green algae (b) brown algae
(c) red algae (d) septate fungi

15. The rhizoids of *Funaria* are
(a) unicellular
(b) unicellular with septa
(c) multicellular with oblique septa
(d) green coloured

16. Fern prothallus is
(a) homothallic (b) heterothallic
(c) dioecious (d) heterotrophic

17. The seed known by the name of 'chilgoza' that is used as dry fruit belongs to
(a) *Cycas revoluta* (b) *Cedrus*
(c) *Zamia* (d) *Pinus gerardiana*

18. The botanical name of black gram (urd) is
(a) *Phaseolus vulgaris* (b) *Glycine max*
(c) *Cicer arietinum* (d) *Phaseolus mungo*

19. Which of the following group has amphivasal vascular bundle?
(a) *Dracaena* and *Yucca*
(b) *Cycas* and *Dryopteris*
(c) *Helianthus* and *Cucurbita*
(d) Maize and *Aestivum*

20. Bundle sheath is chlorophyllous in
(a) C_3-plants (b) C_4-plants
(c) hydrophytes (d) CAM plants

21. Functional xylem of a dicot tree is
(a) springwood (b) sapwood
(c) autumnwood (d) heartwood

22. Thorn is a modification of
(a) stem (b) leaf
(c) root (d) stipule

23. Dispersal of seeds takes place in a fountain like manner in the
(a) coconut (b) lotus
(c) *Ruellia* (d) *Echallium*

24. Censer mechanism is reported in
(a) *Papaver somniferum* (b) *Antirrhinum*
(c) *Aristolochia* (d) All of these

25. Zygotic meiosis is found in
(a) Fern (b) *Marchantia*
(c) *Chlamydomonas* (d) *Cycas*

26. Cytochrome oxidase and cytochrome-*c* deficiency in mitochondria causes
(a) Menke's disease
(b) Kearns-Says syndrome and Menke's disease
(c) Kearns-Says syndrome
(d) Leber's optic neuropathy

27. Synaptic signalling involves
(a) endocrine signals
(b) paracrine signals
(c) autocrine signals
(d) neurotransmitters

28. A dominantly inherited trait affects a child and his grandmother but neither parent. This best illustrates which of the following principles?
(a) Variable expressivity
(b) New mutation
(c) Somatic mosaicism
(d) Non-penetrance

29. Consider the following three events.
I. Although 18% of the student body is minorities, in a random sample of 20 students, 5 are minorites.
II. In a survey about sexual habits, an embarrassed student deliberately gives the wrong answers.
III. A surveyor mistakenly records answers to one question in the wrong space.
Which of the following correctly characterises the above?
(a) I. sampling error, II. response bias, III. human mistake.
(b) I. sampling error, II. non-response bias, III. hidden error
(c) I. hidden bias, II. voluntary sample bias, III. sampling error
(d) I. undercoverage error, II. voluntary error, III. unintentional error

30. Another very important cycle is the water cycle. All living things need water to live. This cycle has four main processes. The two processes that return water to the earth are
(a) evaporation and condensation
(b) condensation and precipitation
(c) transpiration and condensation
(d) evaporation and transpiration

31. The biodiversity characteristic of fauna and sub-dominant plant forms, is a function of .... factors and the biomass productivity of the vegetation.
(a) abiotic, dominant
(b) biotic, dominant
(c) abiotic, inferior
(d) biotic, inferior

32. At low temperatures, peaks in the NMR spectrum of *p*-nitrophenol, $C_6H_4NO_2OH$, are observed as doublets split by ~100 Hz, because the molecule exists in equilibrium, as both protonated and deprotonated forms. At higher temperatures, the peaks coalesce to produce a single broadened peak with a line width equivalent to the separation between the individual peaks. Estimate the rate constant for proton exchange.
(a) $200\ s^{-1}$ (b) $5\times10^{-3}\ s^{-1}$
(c) $100\ s^{-1}$ (d) $50\ s^{-1}$

33. Which of the following statements about DNA polymerase-I is correct?
(a) It functions as a DNA repair enzyme but is not involved in the DNA replication process.
(b) It requires a template and primer to polymerase deoxyribonucleoside triphosphates
(c) It joins together Okazaki fragments to complete the lagging strand during DNA replication
(d) It produces Okazaki fragments linked to RNA primer chains

34. The newly synthesised discontinuous segments of polynucleotides in replication process, is called as
(a) Kimura fragments
(b) Suzuki fragments
(c) Okazaki fragments
(d) Pinaki fragments

35. All the organisms that occur in the redwood community
(a) make up a single population
(b) use the redwoods for food
(c) have identical geographic distributions
(d) have niches that overlap

36. Which one of the following is a pair of endangered species?
(a) Hornbill and Indian aconite
(b) Indian peacock and carrot grass
(c) Garden lizard and mexican poppy
(d) Rhesus monkey and sal tree

37. Choose the odd item.

(a) Composite rank test (b) Chi-square test
(c) Mann-whitney test (d) Median test

38. Which of the following statement is false regarding MRI scanning technique?

(a) It is based on the magnetic resonance principle
(b) It provides much greater contrast between the different soft tissues of the body than computed tomography
(c) It uses no ionising radiation and radioactive water and uses a powerful magnetic field to align the nuclear-magnetisation of (usually) hydrogen atoms in water in the body
(d) It is especially useful in neurological (brain), musculo-skeletal, cardiovascular and oncological (cancer) imaging

39. In certain fungi and algae, cells undergo mitosis repeatedly without subsequently undergoing cytokinesis. What would result from this?

(a) A decrease in chromosome number
(b) Inability to duplicate DNA
(c) Division of the organism into many cells, most lacking nuclei
(d) Large cells containing many nuclei

40. How many types of BLAST programmes have been developed to support sequence similarity searching using variety of nucleotide and protein sequence queries and databases?

(a) 2 (b) 3 (c) 5 (d) 9

41. 1/3 spiral phyllotaxy (called tristichous) means

(a) three rows of alternate rows
(b) in one circle there are three leaves
(c) angular divergence between two leaves is 120°
(d) All of the above

42. Sugar moves in phloem vessels as

(a) cellulose (b) glucose
(c) starch (d) sucrose

43. .... ions help in photolysis of water

(a) Mn^{2+} (b) Mg^{2+}
(c) Cl^- (d) Both (a) and (c)

44. Ethylene receptor

(a) is a membrane bound receptor dimer of two component system
(b) is a soluble protein in the cytoplasm
(c) has a chromophore similar to phytochrome
(d) it has pterin and flavin as prosthetic groups

45. Which of the following phytochromes produces under water-deficient condition plays an important role in the tolerance response of plants to draughts?

(a) ABA (b) Cytokinin
(c) Ethylene (d) Gibberellins

46. Which of the following statement is correct about diffusion?

(a) It is very rapid over long distance
(b) It requires an expenditure of energy by the cell
(c) It is a passive process
(d) It occurs when molecules move from a region of lower concentration to one of higher concentration

47. Osmosis is a form of diffusion in which

(a) the solute moves freely from a region of higher concentration to lower concentration through a semipermeable membrane
(b) the solvent moves through a semipermeable membrane from region, where a solute is in higher concentration to region of lower concentration
(c) the solvent moves through a semipermeable membrane from higher solvent concentration to lower solvent concentration
(d) solute as well as solvent moves freely from region of higher concentration to lower concentration through a semipermeable membrane

48. Which of the following alkenes would have the largest λ_{max}?

(a) $(H_3C)_2C{=}CH_2$ (b) $(H_3C)_2C{=}CH(CH_3)$
(c) $CH_3(H)C{=}C(H)CH_3$ (d) $(H_3C)_2C{=}C(CH_3)_2$

49. Which of the following method is useful in computing the probability that an given amino acid belongs to a particular secondary structure element when there are no homologous protein sequences or structure data?

(a) PSA method (b) EMBL NN method
(c) Helical wheel method (d) None of these

50. Phenyl Mercuric Acetate (PMA) acts as a

(a) PS-II blocker
(b) PS-I blocker
(c) anti-transpirant
(d) transpiration promoter

51. Cancer cells in any form tends to undergo

(a) Meiosis (b) Amitosis
(c) Binary fission (d) Mitosis

52. Insect-pollinated flowers usually possess

(a) brightly coloured pollen
(b) dry pollens with smooth surface
(c) sticky pollens with rough surface
(d) large quantities of pollen.

53. Process of emasculation is known as?

(a) removal of the anthers from the flowers
(b) pollination between flowers of the same plants
(c) pollination between flowers of different plants
(d) artificial pollination

54. The efficiency of C_4-plant is better when compared to C_3-plants due to

(a) they have photorespiration
(b) CO_2 acceptor is PEP chloroplasts
(c) they lack photorespiration
(d) they have two types of chloroplasts

55. Phenomenon of flowering in pineapple is induced to the condition

(a) GA_3 (b) Cytokinin
(c) Short days (d) IAA

56. How have human activities increased the supply fixed nitrogen available to the primary producers?

(a) Use of nitrogen fertilisers
(b) Increased cultivation of legumes
(c) Deliberate burning of fields
(d) All of the above

57. The destruction of the ozone layer in most directly linked to the

(a) destruction of tropical rainforests
(b) release of chlorofluorocarbons (CFCs) into the atmosphere
(c) decrease in the production of ozone precursors by chemical production plants
(d) global warming

58. Which shows a result of deforestation and the effect it has on the environment?

Result		Effect
(a) Fewer flowering plants	—	Reduced CO_2 in air
(b) Fewer trees	—	Increased humidity of air
(c) More ground cover	—	Wind removes soil
(d) More water drains away	—	Soil washed away

59. What is a long-term effect of cutting down larger areas of rainforests?

(a) Decreased carbon dioxide content in the air
(b) Decreased flooding of low-lying land
(c) Increased rainfall in these areas
(d) Increased rate of soil erosion

60. Great Indian Bustards as a sanctuary national park is meant for

(a) Annamalai sanctuary
(b) Desert National Park
(c) Simlipal sanctuary
(d) Sultanpur bird sanctuary

61. Number of National parks and wildlife that sanctuaries associates with the project Tiger is.

(a) 10 (b) 21
(c) 17 (d) 8

62. Match the following National Parks with the important animals it is famous for

Column I	Column II
A. Kanha National Park	1. Tiger
B. Gir National Park	2. Asiatic lion
C. Bandipur National Park	3. Elephant
	4. Rhinoceros

Codes

	A	B	C		A	B	C
(a)	2	3	4	(b)	3	4	1
(c)	2	1	3	(d)	1	3	2

63. Type of *ex situ* conservation made to conserve white tigers.

(a) Gir forest
(b) Corbett National Park
(c) Kaziranga National Park
(d) Periyar Sanctuary

64. The cases of skin diseases are increasing day-by-day due to

(a) Increase in atmosphere pressure
(b) Increase in air pollution
(c) Excessive use of detergent
(d) Depletion of ozone

65. Minamata disease is related to which of the following?

(a) Oil spills into sea
(b) Release of human organic waste into drinking water
(c) Accumulation of arsenic in water
(d) Release of industrial waste mercury into fishing water

66. Extinction is a naturally occurring phenomenon that has occurred on earth, since life began. Scientists are so concerned about extinction now because

(a) human activities are directly responsible for today's mass extinctions
(b) extinction is occurring faster than ever before and faster than rates of respeciation
(c) large numbers of plant species, which form the base of terrestrial food webs are becoming extinct
(d) All of the above

67. The endangered plant species of India are

(a) *Coptis teeta* (b) *Vanilla piliferae*
(c) *Rauvolfia serpentina* (d) All of these

68. Earth Summit of Rio de Janerio (1992) resulted in

(a) compilation of red list
(b) establishment of biosphere reserves
(c) conventions on biodiversity
(d) IUCN

69. The world summit on sustainable development was held in August, 2002 in

(a) New York (b) Johannesburg
(c) New Delhi (d) London

70. Producers are often termed as transducers due to the fact that?

(a) The cannot work in the absence of sun light
(b) They convert sun energy into chemical energy
(c) They convert chemical energy into mechanical energy
(d) They relase O_2 and consume CO_2

71. What do you mean by key industrial animals.

(a) Phytoplanktons (b) Primary consumers
(c) Secondary consumers (d) Tertiary consumers

72. Which functional unit of environment is comprised of both living and non-living components?

(a) Population (b) Organism
(c) Ecosystem (d) Biosphere

73. Watch the following in terms of interaction between different species

Column I	Column II
A. Both are benefitted and can live independently	1. Protocooperation
B. One is benefitted other remains unaffected	2. Commensalism
C. One is benefitted and other is harmed	3. Parasitism
	4. Mutualism

Codes

	A	B	C		A	B	C
(a)	1	2	3	(b)	2	4	3
(c)	2	1	3	(d)	1	2	3

74. Food derived by killing other organism is
(a) predation (b) protocooperation
(c) parasitism (d) amensalism

75. Parasite living outside the body of the host is
(a) ectophyte (b) endoparasite
(c) mycorrhizae (d) None of these

76. The type of relationship between sea anaemone and hermit crab is
(a) mutualism (b) commensalism
(c) parasitism (d) symbiosis

77. Identify the pair of used commodity in the genetic engineering experiments.
(a) *Escherchia, Agrobacterium*
(b) *Rhizobium, Azotobacter*
(c) *Rhizobium, Diplococcus*
(d) *Nitrosomonas, Nitrosococcus*

78. Transgenic organisms are called the ones that have their
(a) foreign DNA in all cells
(b) foreign RNA in all cells
(c) foreign DNA in germ cells
(d) foreign RNA in germ cells

79. The percentage of DNA one person resembles to the other person is
(a) 21% (b) 0.1%
(c) 99.9% (d) 10%

80. Phenomenon that help to purify the environment with the help of microbes is called
(a) bioindicators (b) bioreactors
(c) bioremediation (d) biocontrollers

81. Fungi can be stained by
(a) cotton blue (b) safranine
(c) glycerin (d) lactophenol

82. Out of 64 codons, the codons coding for amino acids are
(a) 20 (b) 60
(c) 61 (d) 64

83. Identify the pair of codons
(a) UAA, UGA (b) UAG, UGA
(c) UGA, UAA (d) UAA, AUU

84. Number of nucleotides found in the DNA molecule of eukaryotic cell is
(a) 2000 nucleotides (b) 40000 nucleotides
(c) 400,000 nucleotides (d) 100000 nucleotides

85. In prokaryotes an amino acid used as an initiated amino acids in the protein chain of prokaryotes is
(a) methionine (b) formylmethionine
(c) tryptophan (d) formyltyrosine

86. Genetic code consists of
(a) adenine, guanine (b) thymine, adenine
(c) thymine, cytosine (d) All of these

87. The functional unit of synthesis of protein is
(a) glyoxisome (b) peroxisome
(c) polysome (d) lysosome

88. Balbiani rings are structural feature of
(a) Lampbrush chromosomes
(b) polytene chromosomes
(c) allosomes
(d) autosomes

89. Nucleic acid in the stranded form occurs in
(a) *E. coli* (b) λ phage
(c) $\phi \times 174$ (d) T_4 phage

90. In which out of these, the extra chromosomal genes can be traced out
(a) colourblind person
(b) *Paramecium*
(c) sex linked diseased person
(d) sex influenced diseased person

91. Colourblindness is well known
(a) X-linked dominant trait
(b) X-linked recessive trait
(c) Y-linked dominant trait
(d) Y-linked recessive trait

92. On which type of chromosomes, holandric genes are present?
(a) Salivary gland chromosome
(b) X-chromosomes
(c) Y-chromosomes
(d) Lampbrush chromosome

93. What is genophore in accordance of the option given below?
(a) RNA + Histone protein
(b) Single-stranded DNA
(c) Double-stranded DNA
(d) Double-stranded RNA

94. *Drosophila* in which both male and female traits are found is known as?

(a) homomorphic (b) homozygous
(c) gynandromorph (d) hemizygous

95. Which of the following contains an ion carriers?

(a) Cell wall (b) Cell membrane
(c) Intercellular spaces (d) Nucleus

96. Identify the well known barrier found between the protoplasm and an outer cell environment

(a) cell wall (b) cytoplasm
(c) cell membrane (d) tonoplast

97. The function of Golgi complex is

(a) formation of acrosome
(b) recycling of plasma membrane
(c) formation of cell plate during cytokinesis
(d) All of the above

98. Type of protein found in microtubules is

(a) actin (b) myosin
(c) troponin (d) tubulin

99. The arrangement of microtubules shown by the flagellum of eukaryote is

(a) 2 + 9 arrangement
(b) 9 + 2 arrangement
(c) 9 + 0 arrangement
(d) 0 + 9 arrangement

100. The supportive framework of the cell is formed by which of the following?

(a) Microfilament (b) Microtubule
(c) Intermediate filaments (d) All of these

101. Organelle known as the engine of the cell is

(a) ribosome (b) mitochondria
(c) nucleus (d) chloroplast

102. The digestion of wornout cell organelles takes place in

(a) peroxisome (b) lysosome
(c) glyoxisome (d) ribosome

103. The component of the mitochondria acting an ATP synthetase is

(a) F_1 site of oxisome (b) F_0 site of oxisome
(c) F_{5-6} site of oxisome (d) Mitochondrial matrix

104. Nucleus was discovered by

(a) Robert Brown (b) Robert Hooke
(c) Benda (d) Palade

105. Genes are known to be active in the

(a) euchromatin (b) heterochromatin
(c) chromocenter (d) centromere

106. In which of the following plants columella is present in sporangium?

(a) Yeast (b) *Spirogyra*
(c) *Ulothrix* (d) *Rhizopus*

107. Which statement is false regarding the life cycle of *Rhizopus* and *Spirogyra*?

(a) Asexual reproduction takes place in both by spores
(b) When conjugation fails azygospores and parthenospores are formed in both
(d) Zygotic meiosis occurs in both
(d) Sexual reproduction takes place by conjugation in both

108. Which of the following groups contain typically coenocytic forms?

(a) Lichens (b) Phycomycetes
(c) Liverworts (d) Chlorophyceae

109. Which one of the following causes plague?

(a) *Salmonella typhimurium*
(b) *Trichinella spiralis*
(c) *Yersinia/Pasteurella pestis*
(d) *Leishmania donovani*

110. Extra nuclear DNA in *E. coli* is termed as

(a) F^+ factor (b) sex factor
(c) episome (d) All of these

111. Smallest cell/organism is that of

(a) *Vibrio* (b) *Bacillus*
(c) Mycoplasma (d) *Rhizobium*

112. Bacteria were considered more as plants than animals because of the presence of

(a) small nucleus (b) plasma membrane
(c) cell wall (d) spore formation

113. Food poisoining is caused by the infection of

(a) *Bacillus megatherium* (b) *Salmonella typhi*
(c) *E. coli* (d) *Clostridium botulinum*

114. What is the general shape of the bacilli bacteria?

(a) Rod-shaped (b) Spherical
(c) Spiral (d) Comma

115. A primitive filament is formed in protist

(a) *Gonyaulax* (b) *Gymnodinium*
(c) *Melosira* (d) *Pinnularia*

116. *Euglena* is found in

(a) sea (b) freshwater
(c) Both (a) and (b) (d) moist soils

117. Decomposer protists are

(a) sarcodines (b) dinoflagellates
(c) slime moulds (d) diatoms

118. An alga, which is used in space flights for regular supply of oxygen is

(a) *Anabaena* (b) *Nostoc*
(c) *Chlorella* (d) *Chlamydomonas*

119. Which is not the characteristic feature in *Spirogyra* life history?

(a) Isogamious conjugation (b) Fragmentation
(c) Zoospore formation (d) Akinete formation

120. Which example given below is not related to vascular plant?

(a) Lycophytes (b) Pteridophytes
(c) Seed plants (d) Bryophytes

121. Fertilisation in moss occurs when sperms swim from a .... and down the neck of a ....

(a) antheridium; sporangium
(b) sporangium; antheridium
(c) antheridium; archegonium
(d) archegonium; antheridium

122. Bud is the modification of

(a) root (b) shoot
(c) leaf (d) flower

123. Stipulate leaf cannot be found in

(a) *Hibiscus* (b) *Petunia*
(c) *Lathyrus* (d) *Gossypium*

124. Tendrils modified from leaflets are not found in

(a) *Lathyrus aphaca* (b) *Lathyrus odoratus*
(c) *Pisum sativum* (d) Both (b) and (c)

125. The electron microscope differs from the ordinary microscope in having high resolving power due to the use of

(a) electrons, which have a longer wavelength than ordinary light ray, as a source of illumination
(b) ultraviolet light as the source of illumination
(c) fast electrons, which have a shorter wavelength than ordinary light ray, as a source of illumination
(d) gamma rays as the source of illumination

126. During ultracentrifugation of cell homogenate, which fraction would be separated at $10{,}000\ g \times 20$ minutes?

(a) Mitochondria and lysosomes
(b) Ribosome and microsome
(c) Nucleus and mitochondria
(d) Endoplasmic reticulum

127. Which is the suitable technique to obtain correct information regarding the location and synthesis of new DNA?

(a) Electron microscopy
(b) X-ray crystallography
(c) Immunoelectrophoresis
(d) Use of radioactive precursors of DNA and autoradiography

128. The stain used for chromosome

(a) crystal violet (b) aniline blue
(c) toludine blue (d) Both (a) and (c)

129. Collenchyma cells are absent in the stems and leaves of many monocotyledonous plants. It is so because

(a) they do not need any mechanical support
(b) dedifferentiate into secondary meristem in the early stage
(c) their wall thickening get dissolved, hence not easy to be identified
(d) sclerenchyma appear at an early stage of development

130. The cells of tunica undergo anticlinal divisions and give rise to

(a) epidermis (b) cortex
(c) endodermis (d) pericycle

131. The thread-like cytoplasmic strands joining one cell to the other are called

(a) pneumatophores (b) cytodesmata
(c) plasmodesmata (d) plasmophores

132. Collenchyma are present in

(a) woody climbers (b) herbaceous climbers
(c) hydrophytes (d) monocot stems

133. The tissue that brings transport of organic food is

(a) xylem (b) vascular rays
(c) phloem (d) parenchyma

134. The chief function of parenchymatous tissue is to

(a) conduct water and minerals
(b) synthesise food
(c) store food material
(d) provide mechanical support

135. If a plant is cut just above the root, wager seeps out of the cut stem. This is due to the

(a) capillary rise (b) transpiration pull
(c) hydrostatic pressure (d) root pressure

136. The apparatus used to measure osmotic pressure of a solution is known as

(a) osmometer (b) photometer
(c) calorimeter (d) manometer

137. A thin slice of sugarbeet, when placed in a concentrated solution of sodium chloride would

(a) loose water from its cells
(b) become turgid
(c) absorb water from salt solution
(d) neither absorb water nor loose water

138. Decomposers are also called

(a) microconsumers (b) osmotrophes
(c) Both (a) and (b) (d) macroconsumers

139. Which one of the following best describes the base of a pyramid of net production?

(a) Its size depends on the energy available from detritivores
(b) It contains the energy left after the producers have died
(c) It represents the energy available to the producers secondary consumers
(d) It contains the energy captured by photosynthesis

140. Energy and nutrients enter a community by the way of

(a) producer (b) consumer
(c) scavenger (d) detritivores

141. How much of the solar energy that falls on the leaves of a plant is converted to the chemical energy?

(a) 1% (b) 10%
(c) 30% (d) 50%

142. When the carrying capacity is equal to the population size, the net rate of increase in the population equals

(a) the carrying capacity
(b) less than zero
(c) the intrinsic rate of growth
(d) zero

143. To determine the density of a population you would need to know the number of individual and

(a) the factors that limit population growth
(b) their birth rate
(c) the size of the area, in which they live
(d) their population growth rate

144. The role of an organism in ecological system is known as

(a) habitat (b) herbivory
(c) niche (d) interaction

145. The study of inter-relationship between the species/individual and its environment in all stages of its life cycle

(a) synecology (b) forest ecology
(c) autecology (d) ecology

146. Picking up plants with superior phenotype for further propagation is called

(a) pureline selection
(b) mass selection
(c) hybrid vigour
(d) introduction

147. Jinghya 1 is a variety of

(a) cotton (b) maize
(c) rice (d) winter wheat

148. In which of the following plants, pollen tube enters through integuments?

(a) *Casuarina*
(b) *Malva neglecta*
(c) *Cucurbita*
(d) *Beta vulgaris*

149. After penetrating stigmatic and stylar tissue, the pollen tube usually grows downwards towards egg because

(a) no other path to follow
(b) it grows under the control of egg nucleus
(c) attracted by dissimilar electric charge
(d) the filiform apparatus of synergids is believed to attract the pollen tube

150. Even after killing the generative cell with a laser beam the pollen grain of a flowering plant germinate and produce normal pollen tube because

(a) laser beam stimulates pollen germination and pollen tube growth
(b) laser beam does not the region from which pollen tube emerges
(c) the contents of the killed generative cell peri germination and pollen growth
(d) the vegetative cell has not been damaged.

Answers

1. (c)	**2.** (a)	**3.** (d)	**4.** (c)	**5.** (b)	**6.** (d)	**7.** (a)	**8.** (a)	**9.** (a)	**10.** (d)
11. (d)	**12.** (b)	**13.** (d)	**14.** (c)	**15.** (c)	**16.** (a)	**17.** (d)	**18.** (d)	**19.** (a)	**20.** (b)
21. (b)	**22.** (a)	**23.** (d)	**24.** (d)	**25.** (c)	**26.** (b)	**27.** (d)	**28.** (d)	**29.** (a)	**30.** (b)
31. (a)	**32.** (c)	**33.** (b)	**34.** (c)	**35.** (d)	**36.** (a)	**37.** (b)	**38.** (c)	**39.** (d)	**40.** (c)
41. (d)	**42.** (b)	**43.** (d)	**44.** (a)	**45.** (a)	**46.** (c)	**47.** (c)	**48.** (c)	**49.** (a)	**50.** (c)
51. (d)	**52.** (a)	**53.** (a)	**54.** (c)	**55.** (c)	**56.** (d)	**57.** (b)	**58.** (d)	**59.** (d)	**60.** (b)
61. (b)	**62.** (d)	**63.** (b)	**64.** (d)	**65.** (d)	**66.** (b)	**67.** (d)	**68.** (c)	**69.** (b)	**70.** (b)
71. (b)	**72.** (c)	**73.** (a)	**74.** (a)	**75.** (a)	**76.** (b)	**77.** (a)	**78.** (a)	**79.** (c)	**80.** (c)
81. (a)	**82.** (c)	**83.** (a)	**84.** (c)	**85.** (b)	**86.** (d)	**87.** (c)	**88.** (b)	**89.** (c)	**90.** (b)
91. (b)	**92.** (c)	**93.** (b)	**94.** (c)	**95.** (b)	**96.** (c)	**97.** (d)	**98.** (d)	**99.** (b)	**100.** (d)
101. (d)	**102.** (b)	**103.** (a)	**104.** (a)	**105.** (a)	**106.** (d)	**107.** (b)	**108.** (b)	**109.** (c)	**110.** (d)
111. (c)	**112.** (c)	**113.** (d)	**114.** (a)	**115.** (c)	**116.** (b)	**117.** (c)	**118.** (a)	**119.** (c)	**120.** (d)
121. (a)	**122.** (b)	**123.** (b)	**124.** (a)	**125.** (c)	**126.** (a)	**127.** (d)	**128.** (c)	**129.** (d)	**130.** (a)
131. (c)	**132.** (b)	**133.** (c)	**134.** (c)	**135.** (d)	**136.** (a)	**137.** (a)	**138.** (c)	**139.** (d)	**140.** (a)
141. (a)	**142.** (d)	**143.** (c)	**144.** (c)	**145.** (c)	**146.** (b)	**147.** (d)	**148.** (c)	**149.** (d)	**150.** (d)

Practice Set 3

1. **Potato type leaves possess stomata which open during**
(a) Only day time (b) Only night time
(c) Both (a) and (b) (d) None of these

2. **Respiration may take place**
(a) in the presence of oxygen
(b) only in the absence of oxygen
(c) in the presence and absence of oxygen
(d) only in the presence of CO_2

3. **Pyruvic dehydrogenase is used in converting**
(a) pyruvic acid to acetyl coenzyme-A
(b) pyruvate to glucose
(c) pyruvic acid to lactic acid
(d) glucose to pyruvate

4. **The chemical structure of naturally occurring auxin is very similar to the structure of**
(a) adenosine monophosphate
(b) phosphate
(c) tryptophan
(d) pectin

5. **Photoperiodic stimulus is perceived by**
(a) leaves (b) buds
(c) flowers (d) None of these

6. **Most appropriate temperature for vernalisation is**
(a) 1-6°C (b) 7°C (c) –6°C (d) 0°C

7. **Stimulus given in *Mimosa pudica* travels in the form of**
(a) auxin (b) alcohol
(c) hormone (d) turgor changas

8. **Diclinous (unisexual) flowers are**
(a) advanced (b) primitive
(c) intermediate (d) wind pollinated

9. **Unit of chromosomal map is**
(a) picogram (b) Å (c) morgan (d) micron

10. **Experimental observation are confirmed by using**
(a) Chi-square test (b) *t*-test
(c) *F*-test (d) None of these

11. **Inducible enzymes**
(a) are usually involved in anabolic pathways
(b) are produced when a catabolite inactivates the repressor protein
(c) are produced when an activator molecule enhances the attachment of RNA polymerase with the operator
(d) are regulated by inherently inactive repressor molecules

12. **Microtubules are components of all of the following except**
(a) centrioles
(b) the spindle apparatus for separating chromosome in cell division
(c) tracks along which organelles can move using motor molecules
(d) the pinching apart of the cytoplasm in animal cell division

13. **Facultative anaerobes**
(a) can survive with or without oxygen
(b) are poisoned by oxygen
(c) are anaerobic chemoautotrophs
(d) are all able to fix atmospheric nitrogen to make NH_3 which is then available for plant growth

14. **According to Chou and Fasman rule based on a database of protein structures, a set of conformational preference parameters were calculated for each amino acid type based on**
(a) relative frequency of a specific amino acid type within each protein
(b) occurrence of that amino acid in a given type of secondary structure
(c) the fraction of amino acid residues occurring in each type of secondary structure
(d) All of the above

15. **The CMR spectrum of an unknown compound shows 6 absorptions and the PMR spectrum shows 5 absorptions. Which of the following compounds is the unknown compound?**

(a) $CH_3—CH(CH_3)—CH(CH_3)—CH(CH_3)—CH_3$

(b) $CH_2—CH_2(CH_3)—CH_2—CH_2—C(CH_3)(CH_3)—CH_2$

(c) $CH_2—CH_2(CH_3)—CH_2—CH_3$

(d) $CH_3—CH(CH_3)—CH_2—CH_2—CH_3$

16. Prokaryotes have a .... surface area to volume ratio giving then a .... metabolic rate, a ....growth rate and as a consequence, a .... generation time compared to eukaryotes.

(a) larger, higher, higher, larger
(b) larger, smaller, shorter, higher
(c) larger, higher, higher, shorter
(d) shorter, higher, higher, larger

17. The classic work 'Systematics and the Origin of Species' presented in 1942, belong to

(a) Haldane (b) Ernst Mayer
(c) Wright (d) Fisher

18. During secondary growth, xylem rings are produced on the .... of dividing cambium.

(a) inner side
(b) outer side
(c) Both (a) and (b)
(d) None of the above

19. What is the role of stem cells with regard to the functions of adult tissues and organs?

(a) Stem cells are undifferentiated cells that divide asymmetrically, giving rise to one daughter that remains a stem cell and one daughter that will differentiate to replace damaged and worn out cells in the adult tissue or organ
(b) Stem cells are embryonic cells that persist in the adult and can give rise to all of the cell types in the body
(c) Stem cells are determined cells that reside in fully differentiated tissue and can, when needed, differentiate to supply new cells for growth of the tissue
(d) Stem cells are differentiated cells that have yet to express the genes and proteins characteristic of their differentiated state and do so, when needed for repair of tissues and organs

20. During non-cyclic photophosphorylation, plants and cyanobacteria produce

(a) ATP (b) NADPH
(c) NADPH and ATP (d) NADH and ATP

21. In Calvin cycle,

(a) fructose 1, 6-diphosphate undergoes dephosphorylation
(b) ATP is formed during dephosphorylation of fructose
(c) 1, 3 di PGA undergo phosphorylation
(d) None of the above

22. Which of the following is not a part of the triple response of etiolated pea seedlings to ethylene?

(a) Stunted growth
(b) Increased radial expansion of cells of the stem
(c) Lose of gravity perception
(d) Unfolding of the apical hook of the seedling

23. Unlike tropisms, nastic movements are in response to

(a) paratonic stimuli
(b) gravity
(c) non-directional stimuli
(d) directional stimuli

24. Space between cell wall and plasma membrane in a plasmolysed cell is occupied by

(a) pure water (b) air
(c) cell sap (d) plasmolysing solution

25. Blood agar medium is

(a) enrichment medium (b) enriched medium
(c) selective medium (d) differential medium

26. Mark the odd member in each of the following series.

A. (I) sum of squares (II) variance (III) quartile deviation (IV) standard deviation
B. (I) central moment (II) range (III) standard deviation (IV) variance
C. (I) coefficient of variation (II) quartile deviation (III) root-mean square (IV) mean deviation
D. (I)standard deviation (II) absolute measure of dispersion (III) range (IV) coefficient of dispersion

(a) A-III, B-III, C-IV, D-I
(b) A-IV, B-II, C-IV, D-III
(c) A-III, B-II, C-I, D-IV
(d) A-II, B-II, C-III, D-III

27. During initial carboxylation in C_4-plants, the main enzyme required is

(a) pep-carboxylase (b) RuBisCo
(c) phosphorylase (d) decarboxylase

28. Translocation of carbohydrates in flowering plants occurs in form of

(a) glucose (b) sucrose
(c) maltose (d) starch

29. Leaves lack stomata in

(a) epiphytes
(b) xerophytes
(c) submerged hydrophytes
(d) floating leaved hydrophytes

30. RQ is represented as

(a) O_2/CO_2 (b) CO_2/O_2
(c) O_2 taken in (d) $\frac{V_2}{V_2 - V_1}$

31. Flowering in pineapple is promoted by

(a) IA (b) ethylene
(c) GA (d) ABA

32. Opening of flowers involves

(a) epinasty (b) phototaxis
(c) mutation (d) hyponasty

33. If a guanine is lost from DNA due to hydrolysis of the glycosidic bond, this is repaired by

(a) base excision repair
(b) DNA polymerase adding the guanine back by forming a new glycosidic bond
(c) purine salvage pathway
(d) nucleotide excision repair

34. Which of the following is true statement about sampling?

I. Careful analysis of a given sample will indicate whether or not it is random.

II. Sampling error implies an error, possibly very small but still an error, on the part of the surveyor.

III. Data obtained, while conducting a census are always more accurate than data obtained from a sample, no matter how careful the design of the sample.

(a) Only I
(b) Only II
(c) Only III
(d) None of the above

35. In patch clamp setup, the word 'patch' represents

(a) the shape of glass micropipette
(b) the open tip of micropipette
(c) open tip of micropipette with a membrane surface
(d) None of the above

36. Which of the following is an example of a plant chemical defense?

(a) Primary compound (b) Secondary compound
(c) Cactus spine (d) None of the above

37. *In situ* conservation national genetic resources can be achieved by establishing

(a) National Park
(b) Wildlife Sanctuaries
(c) Biosphere Reserve
(d) All of the above

38. Biosphere reserves are being threatened with

(a) population growth
(b) rains
(c) pollution
(d) All of the above

39. Consider the following statements.

I. Absorption of energy from the photon elevates the molecule to an excited electronic state.

II. The disappearance of substrate or appearance of product is measured by spectrophotometery.

III. Circular dichorism measures interaction of polarised light with chiral protein components.

Which of the above statements are correct in reference to biophysical ways?

(a) I and II (b) II and III
(c) I and III (d) I, II and III

40. Calmodulin activates protein kinases in response to a transient increase in

(a) *t*AMP (b) calcium ions
(c) DAG (d) NO

41. Which of the following is not a chemical class of hormones?

(a) Amines (b) Peptides
(c) MAP kinases (d) Steroids

42. Among the following which is longest cell?

(a) Hemp (b) Ramie
(c) Jute (d) Nerve fibre

43. Which of the following statements regarding plasmid is incorrect?

(a) Mostly circular *ds*DNA with multiple copies of pseudogenes
(b) A bacterial cell can harbor more than one type of plasmids
(c) Replication of plasmid may be unidirectional or bidirectional
(d) θ-replication in plasmid is more common form of replication then rolling circle replication

44. Removal of a .... from a community affects community structure significantly.

(a) keystone species (b) niche
(c) competitor (d) predator

45. The ecological time scale of phosphorus cycling involves

(a) the uptake of phosphate from drinking water
(b) the weathering of rock to add O_4^{3-} to the soil
(c) sedimentation to form rocks in the sea bed
(d) the incorporation of phosphorus into organisms that become fossils

46. Cystic fibrosis is a hereditary disease that affects the respiratory and digestive systems. Cystic fibrosis occurs when two recessive genes (cc) are present. A person with one allele for cystic fibrosis is called a carrier (Cc) of the disease. If the mother is a carrier of the disease and the father is homozygous dominant, what are chances that their child will be a carrier of cystic fibrosis?

(a) 25% (b) 50%
(c) 75% (d) 100%

47. A broad-based, pyramid shaped age structure is characteristic of a population that is

(a) growing rapidly
(b) at carrying capacity
(c) stable
(d) limited by density dependent factors

48. Which of the following types of microscope uses UV-rays of higher wavelength?

(a) Ultraviolet microscope
(b) Fluorescence microscope
(c) Polarising microscope
(d) Electron microscope

49. Which is the correct order from least to most complex?

(a) Glucuronate, hyaluronate, proteoglycans, GAG
(b) Hyaluronate, glucuronate, GAG chains, proteoglycans
(c) GAG chains, hyaluronate, glucuronate, proteoglycans
(d) Glucuronate, hyaluronate, GAG chains, proteoglycans

50. Sites where mutations occur at rates higher than normal are known as

(a) suppressor sites (b) hot spots
(c) mutator sites (d) cistrons

51. Young pollen is
(a) partially developed male gametophyte
(b) a sexual microspore
(c) a partially developed embryo
(d) partially developed male gamete

52. To avoid cross-pollination, the flowers, are closed. It is
(a) homogamy (b) cleistogamy
(c) protogamy (d) allogamy

53. In angiosperms, each pollen grain produces two male cells. What do these sperms do in the fertilisation of a flower?
(a) Each one fertilises a separate egg cell to give rise to two seeds
(b) One fertilises an egg cell and the other fertilises another cell that gives rise to the tissue of the fruit
(c) Both fertilises a single egg cell
(d) One fertilise an egg cell and the other fertilises the cell that forms food reserve tissue

54. Maize is
(a) cleistogamous
(b) anemophilous
(c) entomophilous
(d) hydrophilous

55. In 'dicot type' development of anther wall, the middle layers arise from
(a) outer-secondary parietal layer
(b) inner-secondary parietal layer
(c) Both (a) and (b)
(d) None of the above

56. The phenomenon of polyteny is associated with, which wall layer of the anther?
(a) Epidermis (b) Tapetum
(c) Endothecium (d) Middle layers

57. Identify the correct statement.
(a) Meiosis in plants occur when there is a change from sporophyte to gametophyte
(b) Meiosis produces genetically identical daughter cell
(c) Mitosis ensures genetic continuity from generation to generation
(d) After meiosis-I, the cells are still diploid

58. Which of the following serves as a mitotic spindle poison?
(a) Ca^{2+} (b) Mg^{2+}
(c) Tubulin (d) Colchicine

59. Which is the shortest phase in the cell cycle?
(a) G_1 (b) G_2
(c) S (d) M

60. The spindle fibres are composed of
(a) 3% RNA and 97% tubulin
(b) 10% RNA and 90% tubulin
(c) 15% RNA and 85% tubulin
(d) 22% RNA and 78% tubulin

61. Cellular structures, which always disappear during mitosis are
(a) plastids and mitochondria
(b) nuclear envelope and nucleolus
(c) cell wall and plasmalemma
(d) cell wall and nucleolus

62. Numerous new leaves and shoots in plants are produced
(a) when the old ones fall of
(b) when the rainy season sets in
(c) when the weather cools down as in winter
(d) during spring

63. A new cell immediately undergoes
(a) intussusception (b) apposition
(c) plasmatic growth (d) elongation

64. A growing shoot apex follows a helical path and this phenomenon corresponds to
(a) diurnal growth rhythm
(b) circumnutation
(c) allometric growth
(d) a biological clock

65. Which amongst the following is a natural growth regulator?
(a) NAA (b) Ethylene
(c) 2, 4-D (d) Benzaldehyde

66. Growth rhythms are set in a plant when it is
(a) CO_2 environment (b) light
(c) low temperature (d) UV radiation

67. The genetic information of DNA is encoded in
(a) the major groove (b) the backbone
(c) base pairs (d) sequences of bases

68. In DNA one nucleotide is linked with another nucleotide through a covalent bond, which can be found between.
(a) Sugar and phosphates
(b) Sugar and nitrogenous bases
(c) Phosphates and nitrogenous bases
(d) Nitrogenous bases

69. Match correctly between column I and column II.

Column I	Column II
A. Complementary ratio	1. 9 : 7
B. Supplementary ratio	2. 9 : 3 : 4
C. Epistatic ratio	3. 12 : 3 : 1
D. Inhibitory ratio	4. 13 : 3

Codes

	A	B	C	D		A	B	C	D
(a)	4	1	3	2	(b)	1	2	3	4
(c)	4	1	2	3	(d)	1	3	2	4

70. The sequence of bases in one strands of DNA is AAATGGCCCTT, then the complementary sequence of bases on the other side of the strand of DNA would be.

(a) TTTACCGGGAA (b) TTTTGGCCAAA
(c) TTTTGGCCCAA (d) ATATGGCCCCA

71. Which of the following sentences is correct?

(a) The relative amounts of nitrogenous bases vary within the members of the same species
(b) The base composition of all species is the same
(c) In all DNA, the amount of guanine is always equal to the amount of cytosine
(d) Combined pureness are not equal to combined pyrimidines

72. Callose is found in

(a) tracheids (b) sclerenchyma
(c) companion cells (d) old sieve tube

73. Silica forms component of the cell wall of

(a) bacteira (b) fungi
(c) diatoms (d) None of these

74. Elongated tube-like dead cells with tapering ends is feature of one of the following

(a) tracheae (b) tracheids
(c) sieve tubes (d) companion cell

75. Intercalary meristem are found at

(a) lateral side of stem (b) stem apex
(c) root apex (d) base of inter node

76. Organism commonly known as ponds silk is

(a) *Ulothrix* (b) *Vaucheria*
(c) *Volvox* (d) *Spirogyra*

77. Number of pyrenoids found in each cell of *Spirogyra* is

(a) one (b) two
(c) two to five (d) numerous

78. Organism deviod of performing sexual reproduction is

(a) *Nostoc* (b) *Ulothrix*
(b) *Volvox* (d) *Spirogyra*

79. Non-flagellate, thin-walled spores are termed as

(a) aplanospores (b) zygospores
(c) zoospores (d) hypnospores

80. Which out of these is a type of a biodegradable pollutant?

(a) Mercury (b) Sewage
(c) Plastic (d) Asbestos

81. Identify the Marsh gas

(a) H_2S (b) CO
(c) SO_2 (d) CH_4

82. Which gases out of these were responsible for the threat to Taj Mahal from the Mathura refinery

(a) CO_2, CH_4, NO
(b) H_2O, SO_2, NO_x
(c) Fluorine, Freon, SO_2
(d) PAN, NO, H_2SO_4

83. The source of chlorofluorocarbons/freon is

(a) refregerators
(b) propellents
(c) solid plastic foam
(d) All of the above

84. The term 'Stone leprosy' is commonly known because of

(a) lichens (b) PAN
(c) acid rain (d) smoke

85. National environment planning engineering organisation is located at

(a) Kanpur (b) Nagpur
(c) Delhi (d) Mumbai

86. Ozone destroying substances include

(a) CFC (b) N_2O
(c) CH_4 (d) All of these

87. Decomposition of organic matter by microbes is measured in terms of

(a) biochemical oxygen demand
(b) amount of primary pollutants produced
(c) amount of secondary pollutants produced
(d) growth of algae

88. Electron microscope is used for

(a) cell division study
(b) whole mount study
(c) structure of the pollen grain
(d) viewing structure of the cell

89. Homogenisation and isolation of organelles is called

(a) centrifugation (b) fractionation
(c) Both (a) and (b) (d) microtomy

90. In centrifugation, which of the following settles as the first sediment?

(a) Ribosome (b) Mitochondria
(c) Nucleus (d) None of these

91. The resolution of microscope can be increased by

(a) shortening the λ of visible source
(b) increasing NA
(c) Both (a) and (b)
(d) None of the above

92. 'Equilibrium sedimentation' is another name of

(a) equal density centrifugation
(b) density gradient centrifugation
(c) buoyant density centrifugation
(d) None of the above

93. Which of the following developments was most instrumental in enabling us to determine the function of mitochondria?

(a) Technique of cell disruption and ultracentrifugation
(b) Technique of culturing bacteria
(c) The phase contrast microscope
(d) The electron microscope

94. The wall appears in consumer-decomoser protists during

(a) somatic phase (b) gametangium
(c) spore stage (d) Both (b) and (c)

95. A naked sporangium occurs in

(a) cellular slime moulds
(b) acellular slime moulds
(c) all the slime moulds
(d) commonly acellular slime moulds but at time cellular slime moulds

96. Communal slime moulds are

(a) cellular slime moulds
(b) acellular slime moulds
(c) slime moulds feeding on bacteria
(d) slime moulds feeding on plants

97. The presence of capillitium is the characteristic feature of

(a) slime moulds (b) chrysophyta
(c) euglenophyta (d) pyrophyta

98. Match the following columns.

Column I	Column II
A. Capillitia	1. Multinucleate adult stage
B. Swarm cells	2. Mating type
C. Zygote	3. Diploid stage of slime mould
D. *Plasmodium*	4. Calcareous thread

Codes

	A	B	C	D		A	B	C	D
(a)	4	1	3	2	(b)	2	1	3	4
(c)	3	1	4	2	(d)	4	2	3	1

99. Much of the research on gene expression has been done with *Escherichia coli*, which inhabits the human intestine. This organism is a

(a) plasmid (b) virus
(c) bacterium (d) protozoan

100. Non-photosynthetic, unicellular, cell wall bound spherical or rod-shaped microorganisms usually smaller than 5 micron in size which lack an organised nucleus, are called

(a) viruses (b) bacteria
(c) algae (d) mycoplasma

101. *Escherichia coli* is a bacterium, which is a common inhabitant of

(a) human intestine (b) human muscle
(c) Both (a) and (b) (d) human milk

102. Conidia bearing structures of *Penicillium* are

(a) metulae (b) rami
(c) sterigmata (d) asci

103. Yellow rust of Gramineae is caused by

(a) *Cystopus* (b) *Albugo*
(c) *Cythium* (d) *Puccinia*

104. In what form, the food is stored in the mycelium of *Cystopus*?

(a) Starch and protein
(b) Glycogen and oil drops
(c) Protein and steroids
(d) Oil and sugar

105. In a virus, nucleic acid is complexed with

(a) capsid (b) envelope
(c) glycolipid (d) cytoplasm

106. Average size range of virus is

(a) 10-30 nm (b) 20-14000 nm
(c) 100-1000 nm (d) 1-10 nm

107. Each capsomere of TMV contains amino acids whose number is

(a) 185 (b) 158 (c) 518 (d) 815

108. TMV is a

(a) helical virus (b) polyhedral virus
(c) spherical virus (d) binal virus

109. Virus having enzyme lysozyme is

(a) retrovirus (b) influenza virus
(c) bacteriophage (d) vaccinia virus

110. Bacteria are grouped in which one of the following kingdom?

(a) Protista (b) Plantae
(c) Animalia (d) Monera

111. Match the following columns.

Column I	Column II
A. Animalia	1. Unicellular prokaryotes
B. Monera	2. Multicellular and photosynthetic
C. Protista	3. Unicellular eukaryotes
D. Plantae	4. Multicellular ingestive heterotrophs

Codes

	A	B	C	D		A	B	C	D
(a)	4	3	1	2	(b)	4	1	3	2
(c)	1	2	4	3	(d)	2	3	1	4

112. Which of the following have chitinous cell wall?

(a) Monerans (b) Protists
(c) Fungi (d) Animals

113. The 70 S ribosomes in monerans are arranged in the form of

(a) monosome (b) polysome
(c) mesosome (d) Both (b) and (c)

114. Endosperm in angiosperms results after ferilisation from

(a) zygote
(b) secondary nucleus
(c) antipodals
(d) synergids

115. The larger nucleus in a pollen grain is known as
(a) generative nucleus (b) tube nucleus
(c) vegetative nucleus (d) None of these

116. An anther is mainly formed of
(a) three sporangia (b) two sporangia
(c) four sporangia (d) only one sporangium

117. Collar-like outgrowth arising from the base of ovule and forming a sort of third integument is known as
(a) coma (b) caruncle
(c) aril (d) operculum

118. Leaves of many grasses are capable of folding and unfolding because they
(a) have parallel vascular bundles
(b) are very thin
(c) are isobilateral
(d) have specialised bulliform cells

119. Endogenous rhythmic changes that occur in an organism with periodicity of twenty four hours is known as
(a) diurnal rhythm (b) circadian rhythm
(c) Both (a) and (b) (d) None of these

120. Tertiary roots are
(a) plageotropic (b) ageotropic
(c) heliotropic (d) phototropic

121. Which of the following is a group of essential elements
(a) Zn, Co, B, Mo (b) K, Mg, Na, Cl
(c) N, S, P, Si (d) C, H, O, P

122. Which of the following is a group of elements have balancing function in plants
(c) Cu, Fe, Mn (b) N, P, S
(c) Ca, K, Mg (d) Cu, Fe, Zn

123. Root hairs occur in the zone of
(a) cell elongation (b) cell division
(c) cell maturation (d) None of these

124. The name of *Ginkgo* tree has now been removed from the list of endangered plants as a result of
(a) *Ex situ* conservation
(b) *In situ* conservation
(c) Off-site collection
(d) None of the above

125. The collection of which of the following plants from its natural habitat has been banned?
(a) Orchids (b) Neem
(c) Melons (d) Rice cultivars

126. The *Triticale* was developed by a cross between wheat and rye. It is not suitable for making bread because of its
(a) gluten content
(b) cold-tolerant nature
(c) hybrid nature
(d) palatability

127. Which of the following is not true about jojuba?
(a) Its seeds contain about 50% liquid wax
(b) Its liquid wax is comparable to sperm whale oil
(c) Its oil is used as high-performance lubricant
(d) This plant occurs in tropical rainforests

128. Gamete contains
(a) one allele of a gene
(b) two alleles of a gene
(c) all alleles of a gene
(d) many alleles of a gene

129. Gene pool is the sum total of genes present in
(a) cell (b) organism
(c) population (d) ecosystem

130. Intragenic non-Mendelian inheritance include
(a) Dominance, codominance, incomplete dominance
(b) Codominance, multiple allele, incomplete dominance
(c) Epistasis, multiple allele, dominance
(d) Polygene, pleotropism, codominance

131. When pure tall plant is crossed with pure dwarf plant, all the F_1 plants are tall because of
(a) dominance
(b) disappearance of factor for dwarfness
(c) segregation of factors
(d) incomplete dominance

132. An individual with AbBbCc will produce
(a) 8 types of gametes
(b) 16 types of gametes
(c) 2 types of gametes
(d) All similar types of gametes

133. A pond is a
(a) man-made ecosysterm (b) natural ecosystem
(c) biome (d) macro ecosystem

134. The source of energy in the ecosystem is
(a) Sun (b) DNA
(c) Glucose (d) ATP

135. The % of sunlight trapped by producers is
(a) 1-5% (b) 5-10%
(c) 10-15% (d) 15-20%

136. The process which helps in nutrient conservation
(a) mineralisation (b) leaching
(c) nutrification (d) immobilisation

137. The transfer of energy in different trophic level
(a) reversible (b) bidirectional
(c) multidirectional (d) unidirectional

138. Biological diversity day is
(a) 5th June
(b) 21st March
(c) 14th January
(d) 29th December

139. The number of botanical gardens and arboreta worldwide is
(a) 1500 (b) 1200 (c) 1050 (d) 850

140. Rare endangered and endemic taxa can be found intact and flourishing in
(a) sacred grooves (b) buffer zone
(c) tropical forests (d) temperate forests

141. Bold eagle has been at high risk of extinction due to its
(a) large size
(b) low reproduction potential
(c) high trophic level
(d) change in migrating route

142. The number of known species of organisms on this earth is about
(a) 5.0 million (b) 3.2 million
(c) 10.9 million (d) 13.2 million

143. Nitrogen fixation is performed by
(a) green alga and fungi
(b) mycorrhiza
(c) mycoplasma
(d) blue-green algae and bacteria

144. *Nitrosomonas* is a
(a) photoautotroph
(b) chemoautotroph
(c) chemoheterotroph
(d) photoheterotroph

145. Which one of the following is non-symbiotic biofertiliser?
(a) VAM (b) *Azotobacter*
(c) *Anabaena* (d) *Rhizobium*

146. Thuricide is a proteinaceous toxin, which is obtained from
(a) biofertiliser (b) bacterial origin
(c) green manure (d) farmyard manure

147. Formation of tropical forests needs mean annual temperature and mean annual precipitation
(a) 18-25°C and 150-400 cm
(b) 5-15°C and 50-100 cm
(c) 30-50°C and 100-150 cm
(d) 5-15°C and 100-200 cm

148. Relation between viceroy butterfly and monarch butterfly is that of
(a) Camouflage (b) Mullerian mimicry
(c) Batesian mimicry (d) Warning colouration

149. A plant devoid of roots is
(a) *Wolffia/Ceratophyllum* (b) *Vallisneria*
(c) *Hydrilla* (d) *Azolla*

150. Which character is not present in the xerophytes?
(a) Thick cuticle
(b) Well-developed mechanical tissues
(c) Well-developed conducting tissues
(d) Spongy parenchyma

Answers

1. (c)	**2.** (c)	**3.** (a)	**4.** (b)	**5.** (a)	**6.** (b)	**7.** (d)	**8.** (b)	**9.** (c)	**10.** (c)
11. (b)	**12.** (d)	**13.** (a)	**14.** (d)	**15.** (b)	**16.** (c)	**17.** (b)	**18.** (a)	**19.** (a)	**20.** (c)
21. (a)	**22.** (c)	**23.** (c)	**24.** (d)	**25.** (b)	**26.** (c)	**27.** (a)	**28.** (b)	**29.** (c)	**30.** (b)
31. (b)	**32.** (a)	**33.** (a)	**34.** (d)	**35.** (c)	**36.** (b)	**37.** (d)	**38.** (c)	**39.** (d)	**40.** (b)
41. (c)	**42.** (d)	**43.** (a)	**44.** (a)	**45.** (b)	**46.** (b)	**47.** (a)	**48.** (b)	**49.** (d)	**50.** (b)
51. (b)	**52.** (b)	**53.** (d)	**54.** (b)	**55.** (a)	**56.** (b)	**57.** (a)	**58.** (d)	**59.** (d)	**60.** (a)
61. (b)	**62.** (d)	**63.** (c)	**64.** (b)	**65.** (b)	**66.** (b)	**67.** (d)	**68.** (a)	**69.** (b)	**70.** (a)
71. (c)	**72.** (c)	**73.** (c)	**74.** (b)	**75.** (d)	**76.** (d)	**77.** (d)	**78.** (a)	**79.** (a)	**80.** (b)
81. (d)	**82.** (b)	**83.** (d)	**84.** (c)	**85.** (b)	**86.** (d)	**87.** (a)	**88.** (d)	**89.** (b)	**90.** (c)
91. (b)	**92.** (c)	**93.** (a)	**94.** (c)	**95.** (c)	**96.** (a)	**97.** (d)	**98.** (c)	**99.** (c)	**100.** (b)
101. (a)	**102.** (a)	**103.** (d)	**104.** (b)	**105.** (a)	**106.** (b)	**107.** (b)	**108.** (a)	**109.** (c)	**110.** (d)
111. (b)	**112.** (c)	**113.** (b)	**114.** (b)	**115.** (a)	**116.** (c)	**117.** (c)	**118.** (d)	**119.** (b)	**120.** (a)
121. (d)	**122.** (c)	**123.** (c)	**124.** (a)	**125.** (a)	**126.** (a)	**127.** (d)	**128.** (a)	**129.** (c)	**130.** (b)
131. (a)	**132.** (a)	**133.** (b)	**134.** (a)	**135.** (a)	**136.** (d)	**137.** (d)	**138.** (d)	**139.** (a)	**140.** (a)
141. (c)	**142.** (c)	**143.** (d)	**144.** (b)	**145.** (b)	**146.** (b)	**147.** (a)	**148.** (c)	**149.** (a)	**150.** (b)

Practice Set 4

1. Which of the following are included in Tracheophyta?
(a) Algae and fungi
(b) Gymnosperm and angiosperm
(c) Algae and bryophytes
(d) Pteridophyta, gymnosperms and angiosperm

2. Ovary is absent in
(a) *Cycas* (b) neem (c) wheat (d) mango

3. *Amoeba* is placed in the phylum—Protozoa because of
(a) the presence of cell wall
(b) acellular body
(c) phagocytic mode of nutrition
(d) the presence of contractile vacuole

4. Parasexual cycle is found in
(a) deuteromycetous fungi (b) bacteria
(c) cyanobacteria (d) All of these

5. Which statement is wrong in regard to bacteria?
(a) Bacteria carry out neither mitosis nor meiosis
(b) Bacteria reproduce by the form of cell divisioin called binary fission
(c) Bacteria have nucleoid
(d) Bacteria are eukaryotes

6. RNA particles causing symptoms like that of a virus disease are known as
(a) viroids (b) virion (c) HIV (d) None of these

7. Longest known virus is
(a) $\phi \times 174$ (b) *Ti* phase
(c) TMV (d) *Citrus tristeza*

8. Lichen show
(a) very fast growth (b) very slow growth
(c) normal growth (d) fast growth

9. The fungi that are used as a food source are
(a) slime moulds (b) mushrooms
(c) rust and smut (d) mildews

10. A gill of *Agaricus*
(a) bears spores which help in reproduction
(b) is meant for respiration
(c) is equivalent to the gills of fishes
(d) has no specific function

11. The double diffusion method allows for disease identification, diagnosis and an application of what immunological technique?
(a) Agglutination (b) Precipitation
(c) Immunoassay (d) Antigenesis

12. An .... is made up of both living things and the physical environment.
(a) ecosystem (b) habitat
(c) niche (d) None of these

13. Living things in an ecosystem are ....
(a) abiotic (b) physical
(c) limiting (d) biotic

14. Which one of the following statements about retrotransposons is not correct?
(a) They are mobile genetic elements residing in the genomes of all classes of organisms
(b) They are transposons that behave much like retroviruses and have a retroviruses and have a retrovirus like structure
(c) The reverse transcriptase make a cDNA copy of the retrotransposon
(d) RNA of retrotransposons is packed into virus particles

15. The proton pump in plant cells is the functional equivalent of an animal cells
(a) cotransport mechanism
(b) sodium-potassium pump
(c) contractile vacuole for osmoregulation
(d) receptor-mediated endocytosis of cholesterol

16. To determine the ultrastructure of a cell organelle, the most likely method to be used would be
(a) electron microscopy
(b) microdissection
(c) autoradiography
(d) phase contrast microscopy

17. The resolution power of a microscope means the capacity to
(a) magnify the image
(b) distinguish between organelles
(c) distinguish between two objects
(d) distinguish between two points

18. In wild populations, individuals most often shown a .... pattern of dispersion.
(a) random (b) density dependent
(c) equilibrial (d) clumped

19. Which of the following is not an example of autosomal dominant trait?
(a) Brachydactyly
(b) Marfan syndrome
(c) Familial hyper cholesterolemia
(d) Cystic fibrosis

20. Acid rain is caused by recent reports of acid in some industrial cities are due to the effect of atmospheric pollution by

(a) excessive release of CO_2 by burning of fuels like wood and charcoal, cutting of forests and increased animal population
(b) excessive release of NO_2 and SO_2 in atmosphere by burning of fossil fuel
(c) excessive release of NH_3 by industrial plants and coal gas
(d) excessive release of CO in atmosphere by incomplete combustion of coke, charcoal and other carbonaceous fuel in paucity of oxygen

21. In the widely accepted nucleosome model of chromosome the following are found

(a) histone core with DNA molecule wrapped around it
(b) four types of basic histone, one acidic histone and DNA molecule loosely linked
(c) one acidic histone forming the core with four histone and DNA molecule present in separate whorls
(d) DNA core surrounded by five histone protein

22. 'Zinc fingers' are important in cellular function because they are

(a) the catalytic site of many metabolic enzymes
(b) structural motifs in many DNA-binding proteins
(c) structures with high redox potential
(d) characteristic of palindromic stretches of unique DNA sequence

23. A researcher planning a survey of heads of households in a particular state has census lists for each of the 23 counties in that state. The procedure will be to obtain a random sample of heads of households from each of the counties rather than grouping all the census lists together and obtaining a sample from the entire group. Which of the following is a true statement about the resulting stratified sample?

I. It is not a simple random sample.
II. It is easier and less costly to obtain than a simple random sample.
III. It gives comparative information that a simple random sample would not give.

(a) Only I
(b) I and II
(c) I and III
(d) I, II and III

24. Circular dichroism (CD) is observed only when molecule is

(a) optically active
(b) planar
(c) in helix form
(d) in sheet form

25. The difference between prokaryotes and eukaryotes is that

(a) eukaryotes do have 'true' nuclei containing their DNA
(b) eukaryotic organisms are always multicellular
(c) eukaryotes possess cytoskeleton
(d) eukaryotes contain large RNA/protein structures called ribosomes

26. The growth in diameter is called

(a) primary growth
(b) secondary growth
(c) apical growth
(d) intercalary growth

27. Which of the following statement correctly describes the relationship between the Krebs' cycle and electron transport pathways?

(a) The Krebs' cycle releases H^+ used by the electron transport
(b) The electron transport pathway obtains electrons from the CO_2 produced by the Krebs' cycle
(c) ATP produced by the Krebs' cycle derives electron transport and formation of a H^+ gradient
(d) NADH produced by the Krebs' cycle is used to make ATP by electron transport.

28. What is phytochrome?

(a) A pigment that absorbs light and affects photosynthesis
(b) A pigment that absorbs light and affects seed germination
(c) A pigment that senses gravity and affects seed germination
(d) A macronutrient

29. Which of the following wavelengths of light would stimulate germination?

(a) 450 nm
(b) 540 nm
(c) 660 nm
(d) 730 nm

30. Which of the following observations shows that ion uptake in plants is energy dependent?

(a) Ion uptake shows saturation kinetics
(b) Ion uptake rates are different for different ions
(c) Some ions accumulate against a concentration gradient
(d) Some ions enter the symplast before reaching the endodermis

31. *Parthenium* is

(a) a *K*-selected species
(b) a *r*-selected species
(c) Both (a) and (b)
(d) None of these

32. When chromatin is treated with non-specific nucleases, what is the length of the resulting pieces of DNA?

(a) Random numbers of base pairs
(b) About 60 base pairs
(c) About 8 base pairs
(d) About 200 base pairs

33. To stimulate transcription, enhancer sequences

(a) must be within a few base pairs of the gene they enhance
(b) must be within a few hundred base pairs of the gene they enhance
(c) can be ten thousands of base pairs away from the genes they enhance
(d) will not function if they are moved experimentally

34. Which of the following chemical mutagen is likely to cause GC .... AT transitions?

(a) 5-bromouracil
(b) 2-aminopurine
(c) acridine orange
(d) hydroxylamine

35. Glycoproteins and glycolipids are important for
(a) facilitated diffusion
(b) active transport
(c) cell-cell recognition
(d) signal-transduction pathways

36. When birth rate equals to death rate?
(a) A population grows rapidly
(b) The size of a population remains constant
(c) Density-dependent limiting factors do not affect the population
(d) A population goes through up and down cycles

37. In an electron microscope higher magnification is due to the use of
(a) higher wavelengths of light
(b) high velocity electrons
(c) achromatic lenses
(d) magnetic system

38. In a cross section of a tree or log, one of the ribbons of tissue extending radially from the pith form
(a) cortical rays (b) necrotic rays
(c) medullary rays (d) All of these

39. Which of the following illustrates pleiotropy?
(a) In furit flies, the genes for scarlet eyes and hairy body are located on the same chromosome
(b) Matings between earless sheep and long eared sheep always result in short eared offspring
(c) Wheat kernels can range from white to red in colour, a trait controlled by several genes
(d) The human cystic fibrosis gene causes many symptoms, from respiratory distress to digestive problems

40. Productivity in terrestrial ecosystems is affected by
(a) temperature (b) light intensity
(c) availability of nutrients (d) All of these

41. Which of the following statements is correct?
(a) Bacteria chromosome has many origin of replication
(b) Bacterial chromosome is present mostly in positively supercoiled structure
(c) Bacterial chromosome is a circular *ss*DNA
(d) Bacterial chromosome has no centromere and telomere

42. A plasmid is a
(a) fragment of a chromosome that is transferred during bacterial conjugation
(b) piece of DNA carrying a multiple cloning site
(c) circular *ds*DNA which can replicate independently
(d) piece of DNA carrying gene essential for bacteria

43. If two populations of bacteria cultured in a U-shaped tube are separated by a membrane filter (which does not allow phage particles to pass), but recombination takes place anyway, the mechanism of genetic exchange is
(a) specialised transduction
(b) site-specific recombination
(c) conjugation
(d) transformation

44. The line-intercept method of assessing plant communities measures the distance from randomly choosen points to from a .... to the nearest plant.
(a) plot (b) quadrant
(c) transect (d) allelopathy

45. Proteins are transported around the cell in membranous vesicles. These vesicles
I. form when a section of membrane protrudes and buds off.
II. gave a layer of coat protein around the inside of the vesicle.
III. use their protein coat to find their target membrane.
IV. are uncoated after they reach their target.
(a) I and II (b) I and III
(c) I, III and IV (d) I, II, III and IV

46. Highly reduced and disc-like stem takes place in
(a) *Ruscus* (b) *Smilex*
(c) *Allium* (d) *Asparagus*

47. The leaves of *Funaria* are
(a) stalked (b) sessile
(c) petiolate (d) None of these

48. 10 molecules of fructose 1, 6-diphosphate when oxidised through aerobic cycle yield a total of
(a) 200 ATP (b) 100 ATP
(c) 300 ATP (d) 400 ATP

49. The growth rate is highest in
(a) red light (b) blue light
(c) green light (d) white light

50. A lake rich in nutrients and species is classified as
(a) dystrophic (b) oligotrophic
(c) eutrophic (d) ectotrophic

51. The leaves of desert plants are not torn away by high wind velocity because of
(a) spines
(b) bending of leaves towards opposite side
(c) sclerenchymatous tissue that provides extra mechanical support
(d) tearing of corners, while the middle remains saved

52. In an aqueous environment, microscopic animals and plants are collectively known as
(a) plankton (b) herbivore
(c) commensals (d) flora and fauna

53. Xeric environment is characterised by
(a) precipitation
(b) low atmospheric humidity
(c) extremes of temperature
(d) high rate of vapourisation

54. The type of plants having adaptations to check transpiration is
(a) xerophytes (b) lithophytes
(c) halophytes (d) epiphytes

55. Black rust of wheat is caused by
(a) *Rhizopus* (b) Yeast
(c) *Penicillium* (d) *Puccinia*

56. Hydrophyte with both hydrophytic and xerophytic traits is
(a) *Agave* (b) *Nerium*
(c) *Vallisneria* (d) None of these

57. Penicillin was discovered by Alexander Fleming in the year
(a) 1901-02 (b) 1919-20
(c) 1928-29 (d) 1899-1900

58. Which one of the following species do not have the ability to fix atmospheric nitrogen?
(a) *Anabaena* (b) *Azotobacter*
(c) *Nostoc* (d) *Spirogyra*

59. Match the following columns.

Column I	Column II
A. *Aspergillus niger*	1. Lactic acid
B. *Acetobacter aceti*	2. Butyric acid
C. *Clostridium butylicum*	3. Acetic acid
D. *Lactobacillus*	4. Citric acid

Codes

	A	B	C	D		A	B	C	D
(a)	2	3	4	1	(b)	2	4	3	1
(c)	4	3	2	1	(d)	4	1	3	2

60. The bacterium which helps in the degradation of man-made and natural organic pollutants is
(a) *Pseudomonas* (b) *Clostridium*
(c) *Leuconostoc* (d) None of these

61. Match the following columns.

Column I	Column II
A. *B. propioni*	1. Gallic acid
B. *A. spergillus sp.*	2. Propionic acid
C. *E. coli*	3. Butyric acid
D. *C. acetobutyricum*	4. Oxalic acid

Codes

	A	B	C	D		A	B	C	D
(a)	1	2	4	4	(b)	4	1	2	3
(c)	2	4	1	3	(d)	2	1	4	3

62. An antifungal drug is obtained from
(a) *P. griseofulvum* (b) *P. camemberti*
(c) *P. roqueforti* (d) *P. notatum*

63. The main nitrogen reservoir in the biosphere is the
(a) atmosphere (b) ocean
(c) organism (d) rocks

64. Amount of light is used in the production process by plants?
(a) 1-5% (b) 5-10%
(c) 20-40% (d) More than 70%

65. The efficiency of any ecosystem is best depicted by the pyramid of
(a) energy (b) number
(c) biomass (d) volume

66. The first trophic level in any kind of food chain is always a green plant because
(a) of wide distribution
(b) they alone have the capacity to fix atmospheric N_2
(c) their richness is chlorophyll
(d) their capacity to convert light energy to chemical energy

67. In an ecosystem, the population of
(a) primary producer is larger than primary consumers
(b) secondary consumer is largest
(c) primary consumers are more than primary producers
(d) primary consumers are least dependent upon primary producers

68. Dry seeds can tolerate higher temperature than the germinating seeds due to the reason
(a) dry seeds have more reserve food
(b) hydration makes the enzymes more sensitive to temperature
(c) seeds are meant for perennation
(d) None of the above

69. Respiration is different from the combustion in that
(a) energy from carbohydrates is released in one step
(b) energy from carbohydrates is released in different steps
(c) efficiency is very low
(d) All of the above

70. How much energy is released by the aerobic oxidation of 1 mole of glucose?
(a) 68,60,000 cal (b) 6,86,000 cal
(c) 68,600 cal (d) 6,860 cal

71. Alternate name of Krebs' cycle is
(a) TCA cycle (b) citric acid cycle
(c) Both (a) and (b) (d) None of these

72. The electron acceptors in ETS are arranged according to
(a) decreasing positive potential
(b) increasing positive potential
(c) increasing negative potential
(d) None of the above

73. Hybridisation through protoplast fusion is called
(a) crossing over (b) emasculation
(c) asexual hybridisation (d) parasexual hybridisation

74. The method most useful for improving potato plant is
(a) pureline (b) inbreed line
(c) mass selection (d) clonal selection

75. Cryopreservation refers to
(a) breeding with wild varieties
(b) tissue culture
(c) low temperature treatment
(d) reduction in moisture

76. Improved dwarf wheat variety with higher percentage of lysine/protein is
(a) Kalyan (b) Sonalika
(c) Lerma Roja (d) Sharbati Sonara

77. The three topmost major crops of the world in the order of total production are
(a) wheat > rice > maize (b) rice > wheat > maize
(c) wheat > maize > rice (d) rice > maize > wheat

78. Which pathogen can be eliminated through flooding of soil?
(a) Soft rot bacterium of sugarcane
(b) Wilt fungus of pigeon pea
(c) Nematodes of root knots and earcockle
(d) Blight fungus of chick pea

79. Out of two spiny stipules of *Zizyphus*, both are
(a) equal with one straight and other curved
(b) equal and curved
(c) unequal and straight
(d) unequal with one straight other curved

80. When a large number of sessile flowers grow from a suppressed rachis giving rise to a globose structure, the inflorescence is called
(a) catkin (b) corymb
(c) capitate (d) umbel

81. Spike inflorescence with fleshy axis enclosed by one or more large bracts are found in
(a) mulberry (b) *Colocasia*
(c) *Achyranthes* (d) None of these

82. The inflorescence of poplar and willow are
(a) amentum only
(b) catkin and spikes, respectively
(c) spadix and catkin, respectively
(d) spike and catkin, respectively

83. The female inflorescence of *Zea mays* is best described as
(a) spadix (b) spike of spikelets
(c) spadix of spikes (d) spadix of spikelets

84. In mulberry, the inflorescence is
(a) catkin (b) capitulum
(c) spadix (d) cyathium

85. In sunflower (capitulum inflorescence), ray florets are
(a) bisexual (b) unisexual
(c) asexual (d) None of these

86. How are gametes produced by bryophytes?
(a) By mitosis of gametophyte cells
(b) By meiosis of gametophyte cells
(c) By meiosis of sporophyte cells
(d) By mitosis of spores

87. In *Funaria*, the haploid structure is a
(a) protonema (b) capsule
(c) columella (d) seta

88. A bryophyte differs from pteridophytes in
(a) archegonia
(b) lack of vascular tissue
(c) swimming antherozoids
(d) independent gametophytes

89. The lower bryophytes are called
(a) mosses (b) ferns
(c) liverworts (d) None of these

90. Why are the bryophytes not included under Pteridophyta?
(a) The presence of vascular tissue
(b) Sporophyte being dominant in their life cycle
(c) The presence of roots
(d) Gametophyte being dominant in their life cycle

91. Which of the following type of paraphysis are found in moss?
(a) Multicelled and club-shaped
(b) Multicelled and bottle-shaped branches
(c) Unicelled and club-shaped
(d) Unicelled and bottle-shaped

92. *Sphagnum* is used as a packing material for transporting living plants because of its
(a) acidic nature, it does not undergo decay
(b) creeping capacity
(c) water absorbing capacity
(d) Both (a) and (c)

93. In moss, branch bearing female reproductive organ is called
(a) gynophore (b) carpophore
(c) archegoniophore (d) antheridiophore

94. Which is the characteristic of *Funaria*?
(a) Scalariform xylem (b) Peristome
(c) Both (a) and (b) (d) None of these

95. The columella in *Funaria* acts as a
(a) nourishing tissue (b) conducting tissue
(c) protecting tissue (d) fertile tissue

96. Heterosporous plants produce
(a) megaspores that develop into female gametophytes and microspores that develop into male gametophytes
(b) megaspores that develop into male gametophytes and microspores that develop into female gametophytes
(c) megaspores that bear antheridia and microspores that bear archegonia
(d) spores that produce both archegonia and antheridia

97. Process of selective transmission of a liquid through semipermeable membrane is called
(a) diffusion (b) osmosis
(c) plasmolysis (d) transmission

98. The water potential can be calculated by
(a) π + TP (b) π + WP
(c) $\psi_s + \psi_p$ (d) ψ + WP

99. Osmotic pressure is responsible for the turgidity of plant cells which
(a) causes cell elongation
(b) causes opening of stomata
(c) prevents wilting of leaves
(d) causes all the three above

100. Imbibition process involves
(a) only diffusion (b) only capillary
(c) Both (a) and (b) (d) None of these

101. Mark the correct relationship
(a) $\psi_w = \psi_p - (\psi_\pi + \psi_m)$
(b) $\psi_w = \psi_p + \psi_\pi + \psi_m$
(c) $\psi_w = \psi_p + \psi_\pi - \psi_m$
(d) None of these

102. If we separate the cell organelles of a living cell, which part should be alive?
(a) Endoplasmic reticulum (b) Chloroplast
(c) Cell wall (d) Ribosomes

103. Non-living substances of a cell are collectively known as
(a) cytoplasm (b) mesoplasm
(c) deutoplasm (d) plasmagel

104. Which of the following is stored by aleurone grains?
(a) Glycogen (b) Fats
(c) Starch (d) Proteins

105. Match the following columns.

Column I	Column II
A. Dictyosomes	1. Storage
B. Mitochondria	2. Photosynthesis
C. Vacuoles	3. Transport
D. Grana	4. Secretion
	5. Respiration

Codes

	A	B	C	D		A	B	C	D
(a)	4	5	1	2	(b)	3	4	2	1
(c)	4	5	3	2	(d)	4	3	1	2

106. The site of detoxification reaction in liver is
(a) SER (b) free ribosomes
(c) RER (d) hydrophobic interaction

107. The mature sclerenchyma fibre cells are
(a) live and active (b) live but inactive
(c) dead (d) All of these

108. In *Helianthus annuus* (sunflower) stem, the pericycle is composed of thin-walled and thick-walled cells. The thick-walled pericycle lies
(a) above the primary phloem of vascular bundles
(b) between the two adjacent vascular bundles
(c) randomly
(d) above the thin-walled pericycle

109. A layer of meristematic cells present in between the primary phloem and the primary xylem is absent in
(a) vascular cryptogams
(b) monocot stems
(c) monocot roots
(d) All of these

110. Fusiform initials and ray initials are the constituents of
(a) xylem (b) phloem
(c) phellem (d) cambium

111. When the vascular cambium is present between the xylem and phloem, the vascular bundle is called
(a) closed (b) open
(c) endarch (d) exarch

112. Root hairs originate from
(a) pericycle (b) hypodermis
(c) endodermis (d) exodermis

113. Undifferentiated ground tissue characteristically occurs in
(a) *Zea mays* stem (b) *Cucurbita* stem
(c) *Helianthus* stem (d) *Helianthus* root

114. What do you mean by an orthodox seed?
(a) Tolerate dehydration and low temperature
(b) Tolerated low temperature but not dehydration
(c) Tolerate dehydration but not low temperature
(d) None of the above

115. The central Sugarcane Breeding Research Institute is located at
(a) Lucknow (b) Coimbatore
(c) Delhi (d) Bhopal

116. Which fungus attack is reponsible for causing the potato famine of Ireland.
(a) *Phytophthora palmivora*
(b) *Plasmophora viticola*
(c) *Puccinia graminis*
(d) *Phytophthora infestans*

117. Which part of coconut produces coir of commerce?
(a) Mesocarp (b) Epicarp
(c) Seed coat (d) Pericarp

118. Exogenously produced nonmotile, asexual spores which are formed terminally are called.
(a) zoospores (b) conidia
(c) sporangiospores (d) chlamydospores

119. In *Mucor*, zygospores do not germinate immediately after their formation and undergo a resting period. Meiosis in zygospores occurs
(a) before the zygospore becomes dormant
(b) at the time of zygospore germination
(c) at the time of zygospore fusion
(d) in the germ sporangium

120. The devastating famine in Ireland was caused by Late blight or potato in 1845 was caused by a fungus. Which of the following was the causal organism?

(a) *Alternaria solani*
(b) *Albugo Candida*
(c) *Phytophthora infestans*
(d) *Fusariurn*

121. Sporangia in *Mucor* are borned terminally on sporangiophores. Which of the following is true about sporangiophores?

(a) Sporangiophores grow vertically
(b) Sporangiophores mycelium is geotropic
(c) Sporangiophores mycelium is hydrotropic
(d) All the above

122. Arrangement of conidia of *Albugo* is

(a) interalary (b) acropetally
(c) irregulary (d) basipetally

123. The synchrony in meiosis within an anther lobe is maintained through the

(a) tapetal cytoplasm (b) linker DNA
(c) channels (d) All of these

124. What is meant by double fertilisation?

(a) Fusion between egg and male gamete
(b) Fusion between secondary nucleus and male gamete
(c) Both (a) and (b)
(d) Formation of two zygotes

125. The triple fusion in plants was discovered by

(a) Winkler (b) Theophrastus
(c) Nemec (d) Nawaschin

126. Pollen grains of a flower fall on the stigma of another flowers of same plant. The phenomenon is

(a) genetically autogamy and ecologically allogamy
(b) genetically autogamy
(c) ecologically allogamy
(d) genetically allogamy and ecologically autogamy

127. .... is favoured by heterostyly in *Primula*.

(a) Autogamy (b) Allogamy
(c) Homogamy (d) All of these

128. Consider the following statements regarding angiosperms and choose the combination of correct statements from among them.

I. As a rule, only one pollen tube is formed from one pollen grain.
II. The protoplast of the pollen grain migrates out and position itself at the tip of pollen tube.
III. Backward flow of cytoplasm from the tip of the pollen tube towards the microspore is prevented by the formation of callose plugs at intervals.

(a) I, II and III (b) I and III
(c) II and III (d) I and II

129. Cross-pollination offers certain advantages as well as disadvantages to plants, which one of the following statements is not true about cross-pollination.

(a) It can fail to take place due to distance barrier
(b) It is less economic because plants have to produce large number of pollen grains
(c) It often gives high yield of crop
(d) It takes place only in monoecious flowers

130. In 1957 Meselson and Stahl did density gradient centrifugation of *E.coil* DNA after culturing *E.coil* on N^{15} containing medium. They concluded from their studies that.

(a) DNA replicated conservatively
(b) DNA replicates dispersively
(c) DNA replicates semi-conservatively
(d) DNA dose not replicate

131. The N^{15} DNA in the Meselson experiment when replicates for first round in N^{14} medium shows that

(a) A single band in CsCl (b) Two bands is CsCl
(c) Three bands in CsCl (d) Four bands is CsCl

132. Replication of DNA is a

(a) bidirectional
(b) unidirectional
(c) faster in eukaryotes process
(d) without any direction

133. DNA polymerase picks up the complementary nucleoid triphosphate and join the nucleoid monophosphate to

(a) 3′– OH group of the terminal deoxyribose sugar
(b) 2′–OH group of the terminal deoxyribose sugar
(c) 5th carbon of the sugar
(d) 1st carbon of the sugar

134. Proofreading of a DNA molecule during replication indicates

(a) correction of errors during replication
(b) transcription
(c) removal of RNA primer
(d) removal of pyrophosphate

135. Transcription starts when the enzyme RNA polymerase binds to the

(a) enhancer (b) promoter (c) operator (d) regulator

136. Codon and anticodon are matched due to the

(a) base pairing (b) covalent bonding
(c) hydrophobic interaction (d) ionic interaction

137. A hormone used for inducing morphogenesis in plant tissue culture is

(a) abscisic acid (b) gibberellins
(c) cytokinins (d) ethylene

138. In decapitated plants, axillary buds are activated because

(a) auxin is made available to them
(b) more food material is available to them
(c) amount of cytokinin increase in them
(d) they get more light

139. Hormones differ from enzymes in that hormones are
(a) found only in animals
(b) completely used up during metabolism
(c) found only in plants
(d) not used up at all in the metabolism

140. Auxin inhibits the growth of
(a) apical buds
(b) lateral axillary buds
(c) roots on stem cuttings
(d) parthenocarpic development of fruits

141. The synthetic hormone used as a weedicide is
(a) indole 3-acetic acid
(b) gibberellic acid
(c) 2,4-dichlorophenoxyacetic acid
(d) malic hydrazide

142. When a green plant is exposed to light only on one side, it bends toward the source of light as it grows because
(a) green plants need light for carrying on photosynthesis
(b) green plants seek light because they are phototrophic
(c) light stimulates plant cells on the lighted side to grow faster
(d) auxin accumulates on the shaded side stimulating greater cell elongation on that side

143. During mitosis, metaphase differ from anaphase in having
(a) same number of chromosomes and half number of chromatids
(b) half number of chromosomes and half number of chromatids
(c) half number of chromosomes and same number of chromatids
(d) same number of chromosomes and same number of chromatids

144. The cell, which undergoes reduction division, is called
(a) androcyte (b) meiocyte
(c) zygote (d) zoospore

145. Dyad is a pair of
(a) sister chromatids
(b) non-sister chromatids
(c) homologous chromosomes
(d) non-homologous chromosomes

146. Chiasmata was first discovered by
(a) Morgan in *Drosophila*
(b) Beadle in *Neurospora*
(c) Johnsan in amphibians
(d) Hammerling in *Acetabularia*

147. The thread-like structures, which begin radiating from each centromere by the end of prophase are
(a) kinetochore microtubules
(b) polar microtubules
(c) aster microtubules
(d) spindle microtubules

148. In a chromosome, on the sides of centromere disc-shaped structures are present, called
(a) telomere (b) centronema
(c) kinetochore (d) chromocentre

149. Good soil is the one which
(a) allows water to percolate slowly
(b) allows water to pass quickly
(c) allows limited amount of water into it
(d) holds whole of water entering it

150. Water held tightly by soil particles as a thin film over their surface is
(a) hydroscopic water (b) capillary water
(c) chemical water (d) gravitational water

Answers

1. (d)	**2.** (a)	**3.** (b)	**4.** (d)	**5.** (d)	**6.** (a)	**7.** (c)	**8.** (b)	**9.** (b)	**10.** (a)
11. (b)	**12.** (a)	**13.** (d)	**14.** (d)	**15.** (b)	**16.** (a)	**17.** (c)	**18.** (d)	**19.** (d)	**20.** (b)
21. (a)	**22.** (b)	**23.** (d)	**24.** (a)	**25.** (a)	**26.** (b)	**27.** (d)	**28.** (b)	**29.** (c)	**30.** (c)
31. (b)	**32.** (a)	**33.** (c)	**34.** (d)	**35.** (c)	**36.** (b)	**37.** (d)	**38.** (c)	**39.** (d)	**40.** (d)
41. (d)	**42.** (c)	**43.** (d)	**44.** (c)	**45.** (c)	**46.** (c)	**47.** (b)	**48.** (d)	**49.** (a)	**50.** (c)
51. (c)	**52.** (a)	**53.** (b)	**54.** (a)	**55.** (d)	**56.** (d)	**57.** (c)	**58.** (d)	**59.** (c)	**60.** (a)
61. (c)	**62.** (a)	**63.** (a)	**64.** (a)	**65.** (a)	**66.** (d)	**67.** (a)	**68.** (c)	**69.** (d)	**70.** (d)
71. (c)	**72.** (c)	**73.** (d)	**74.** (d)	**75.** (c)	**76.** (d)	**77.** (a)	**78.** (c)	**79.** (a)	**80.** (c)
81. (b)	**82.** (a)	**83.** (d)	**84.** (a)	**85.** (b)	**86.** (a)	**87.** (a)	**88.** (b)	**89.** (c)	**90.** (d)
91. (a)	**92.** (d)	**93.** (c)	**94.** (b)	**95.** (b)	**96.** (a)	**97.** (b)	**98.** (c)	**99.** (d)	**100.** (c)
101. (d)	**102.** (b)	**103.** (c)	**104.** (a)	**105.** (b)	**106.** (d)	**107.** (c)	**108.** (c)	**109.** (a)	**110.** (d)
111. (c)	**112.** (d)	**113.** (b)	**114.** (a)	**115.** (b)	**116.** (d)	**117.** (a)	**118.** (b)	**119.** (a)	**120.** (c)
121. (d)	**122.** (d)	**123.** (c)	**124.** (c)	**125.** (d)	**126.** (a)	**127.** (b)	**128.** (b)	**129.** (d)	**130.** (c)
131. (a)	**132.** (a)	**133.** (a)	**134.** (a)	**135.** (b)	**136.** (a)	**137.** (c)	**138.** (a)	**139.** (b)	**140.** (b)
141. (c)	**142.** (d)	**143.** (c)	**144.** (b)	**145.** (a)	**146.** (c)	**147.** (a)	**148.** (c)	**149.** (a)	**150.** (a)

Practice Set 5

1. Mycelium is the name given to the plant body of
   (a) Algae (b) Bryophyta
   (c) Fungi (d) Pteridophyta

2. In herpes virus DNA is
   (a) single-stranded
   (b) double-stranded
   (c) may be single or double
   (d) None of the above

3. The root nodules of leguminous plant contain
   (a) *Streptococci* (b) *Staphylococci*
   (c) *Rhizobium* (d) *Bacillus anthracis*

4. The smallest unit of classification is
   (a) order (b) species (c) genus (d) family

5. The chloroplast in *Ulothrix* is
   (a) spiral-shaped (b) ladder-shaped
   (c) stellate (d) girdle-shaped

6. Which is a gametophytic tissue?
   (a) Peristome (b) Apophysis
   (c) Calyptra (d) Operculum

7. The sporangia of *Selaginella* are attached
   (a) to the axis of strobilus
   (b) between the axis and sporophyll
   (c) to the sporophyll
   (d) None of the above

8. Winged pollen grains are found in
   (a) *Pinus* (b) *Cycas*
   (c) *Cycas* and *Pinus* (d) Absent in (b) and (c)

9. Highly reduced and disc-like stem develop in
   (a) *Smilex* (b) *Asparagus*
   (c) *Allium* (d) *Ruscus*

10. The major source of sugar in world is
    (a) watermelon (b) sugarcane
    (c) *Beta vulgaris* (d) date plam

11. Sieve tube cells lack
    (a) protoplasm
    (b) organic solutes
    (c) streaming movement in protoplasm
    (d) None of the above

12. The conditions required for autoclave
    (a) 121°C temperature and 15 lbs pressure for 20 min
    (b) 120°C temperature and 20 lbs pressure for 30 min
    (c) 150°C temperature for 1 h
    (d) 130°C temperature for 2 h

13. Which of the following statements is correct?
    (a) Mendel had extensive training in biology and heredity prior to his ground breaking work
    (b) Fusion of gametes is called fertilisation which creates a diploid individual
    (c) Cross fertilisation is the typical mode of reproduction for pea plants
    (d) The probability that two heterozygotes produce an affected and a heterozygous offspring is 3/4. Keep in mind that each offspring is considered a separate event

14. Secondary productivity
    (a) is measured by the standing crop
    (b) is the rate of biomass production in consumers
    (c) is 10 per cent less than primary productivity
    (d) is the gross primary productivity minus the energy used for respiration

15. Using a transect of 10 m, an ecologist found 157 individuals of a species of bluebonnet. According to her measurements, what is the density of bluebonnets?
    (a) 1570 bluebonnets per metre
    (b) 15.7 bluebonnets per metre
    (c) 1570 bluebonnets per m^2
    (d) 0.06 per m^2

16. Both chloroplasts and mitochondria
    (a) are found within the nucleus
    (b) have linear DNA
    (c) carry extranuclear genes
    (d) are inherited from both parents

17. Which of the following statements is incorrect?
    (a) Osmosis is the net diffusion of water through a semipermeable membrane, so pure water has a lower osmotic pressure than any solution.
    (b) Osmosis is the movement of water through a semipermeable membrane from the side with a more concentrated solution to the side with a more dilute solution.
    (c) A given osmolality will lower the freezing point of a solution the same amount no matter what solutes are in the solution.
    (d) Oxygen and carbon dioxide gas molecules can move in and out of cells by simple diffusion.

18. Which of the following cell junction can be seen through light microscope?
    (a) Tight junction
    (b) Hemidesmosomes
    (c) Desmosomal junctions
    (d) Adhesion plaques

19. Which compound has a molecular ion at m/z =58, an infrared absorption at 1650 cm^{-1} and just one singlet in its nmr spectrum?

(a) Butane (b) CH_3COCH_3
(c) CH_3CH_2CHO (d) 2-methylpropane

20. One of the *ex situ* conservation methods for endangered species is

(a) wildlife sanctuaries (b) biosphere reserves
(c) cryopreservation (d) national park

21. Air pollution causing photochemical oxidants production include

(a) carbon monoxide and sulphur dioxide
(b) nitrous oxide, nitric acid fumes and nitric oxide
(c) ozone, peroxyacetyl nitrate and aldehydes
(d) oxygen, chlorine and fuming nitric acid

22. DNA methylation of cytosine residues

(a) can be induced by drugs that reactive genes
(b) may contribute to long term gene inactivation
(c) produces the promoter region that specifically bind RNA polymerase
(d) makes satellite DNA a different density so it can be separated ultracentrifugation

23. How many absorptions will the following compound have in its carbon NMR spectrum?

$$(CH_3)(CH_3H_2C)C=C(CH_3)(CH_3)$$

(a) 3 (b) 4
(c) 5 (d) 6

24. Match column I (theories of evolution/inheritance) with column II (proposed by) and select the correct answer using the codes given below the columns.

Column I	Column II
A. Catastrophism	1. Cuvier
B. Inheritance of acquired characters	2. Lamarck
C. Performation	3. Bonner
D. Pangenesis	4. Darwin

Codes

	A	B	C	D		A	B	C	D
(a)	1	2	3	4	(b)	4	3	2	1
(c)	1	3	2	4	(d)	4	2	3	1

25. Root cap is a specialised protective structure found on the tips of the terrestrial vascular plants only. The cells of this root cap are

(a) simple and usually forms a homogenous parenchymatous tissue of short lived cells
(b) simple and usually forms a homogenous collenchymatous tissue
(c) complex and consists of outer collenchyma and inner sclerenchyma
(d) complex and consists of outer suberised cells and inner parenchyma

26. The quiescent centre in root meristem serves as a

(a) site for storage of food which is utilised during maturation
(b) reservior of growth hormones
(c) reserve for replenishment of damaged cells of the meristem
(d) region for absorption of water

27. The two photosystem-I and II during non-cycle photophosphorylation acts in

(a) parallel (b) series
(c) Both (a) and (b) (d) None of these

28. Assimilatory power produced during photosynthesis are

(a) RuBP and RuBisCO (b) H_2O and O_2
(c) ATP and NADPH (d) $C_6H_{12}O_6$ and PGAL

29. The accumulation of one of the following causes seed dormancy

(a) cytokinin (b) auxin
(c) ABA (d) gibberellins

30. Which of the following properties of water is most directly related to its ability to rise in the capillary spaces of plants?

(a) Neutral pH (b) High density
(c) Low compressibility (d) High surface tension

31. Which of the following is not recycled in an ecosystem?

(a) Water (b) Carbon (c) Energy (d) Nitrogen

32. Which of the following is not a character of *r*-selected species?

(a) Small adult size
(b) Long generation time
(c) A high allocation to reproduction
(d) Maximum population growth

33. Eukaryotes regulate gene expression at all of the following levels except

(a) transcription (b) translation
(c) RNA processing (d) All of these

34. Mutation is the change in

(a) gene frequency
(b) base pairs in the DNA
(c) genetic drift
(d) environmental mechanism of evolution

35. DIG based technique is used for the detection of certain known DNA in an animal cell it is preferred over radio labelling techniques because of

I. Quick result II. Safety
III. Stability IV. Easy handling

Choose the correct answer.

(a) I, III and IV (b) III and II
(c) I, II, III and IV (d) I, II and III

36. The niche of an animal is the

(a) number of individuals of the species the environment will support
(b) same as its habitat
(c) way the animal fits into its environment
(d) specific place in the habitat where the animal lives

37. An oak tree produces thousands of acorns but very few grow into mature oak trees. The oak tree exhibits a .... survivorship curve.
(a) type I (b) type II
(c) type III (d) type I or II

38. Consider the following statements regarding linkage.
I. The linked genes are located on the same chromosome.
II. Crossing over between linked genes is rare.
III. Linked genes are always inherited together.
IV. Linkage affects the percentage of homozygosity following hybridisation.
(a) I and III are correct (b) I and II are correct
(c) II, III and IV are correct (d) I, III and IV are correct

39. Upwellings in the ocean
(a) are locations of reef communities
(b) occur over deep sea hydrothermal vents
(c) are responsible for ocean currents
(d) bring nutrients rich water to the surface

40. In India temperate evergreen vegetation occurs mostly in
(a) Rajasthan and South Punjab
(b) Eastern and Western Himalayas less than 3500 m
(c) Western Himalayas above 3500 m
(d) Western Ghats and Asom

41. In an $F^+ \times F^-$ cross the
(a) F^+ cell becomes an Hfr cell
(b) F^- cell becomes an Hfr cell
(c) F^+ cell becomes an F^- cell
(d) F^- cell becomes an F^+ cell

42. To determine an organism's niche, all of the following must be determined except
(a) how it is classified
(b) what it eats
(c) where it lives
(d) what relationships it has with other organisms

43. Petroleum products which contain carbon are burned and the carbon escapes into the atmosphere as carbon dioxide but how does it get into the petroleum in the first place?
(a) Refineries (b) Plant respiration
(c) Decomposing plankton (d) Photosynthesis in plants

44. The chromatin is made up of repetitive units known as
(a) chromosomes (b) chromonemata
(c) nucleosomes (d) nucleotides

45. Which of the following cell junction link cell together (intercellular junctions)?
(a) Focal contacts (b) Adhesion plaques
(c) Hemidesmosomes (d) Adherens

46. .... is the careful and responsible management of something for which you are responsible.
(a) Stewardship (b) Enhancement
(c) Restoration (d) Sustainability

47. Growing points possess
(a) meristematic cells (b) mature cells
(c) dead cells (d) slowly enlarging cells

48. Pollinators are of no need in the flower known as
(a) anemophilous (b) cleistogamous
(c) herkogamous (d) chasmogamous

49. Humus is mainly composed of
(a) nucleoprotein (b) lipoproteins
(c) lignin proteins (d) None of these

50. A close physical association between organisms, such as the association of mycorrhizal fungi with the roots of plants in forests which is beneficial to both species is known as which of the following?
(a) Parasitism (b) Competition
(c) Pollination (d) Symbiosis

51. In *Aspergillus*, the ascogonium incase of *Aspergillus* after the processs of plasmogamy developed into a fruiting body known as
(a) ascocarp (b) germ sporangium
(c) pollinadium (d) antheridium

52. The aspergilloses of lungs, resembling tuberculosis, in animals and human beings. Which of the following species cause this disease?
(a) *Aspergillus famigatus* (b) *Aspergillus niger*
(c) *Aspergillus flavus* (d) All of these

53. The species of *Penicillium* commonly used for cheese making is
(a) *Penicillium notatum* (b) *P. chrysogenum*
(c) *P. grisefulvum* (d) *P. roquefortii*

54. Identify the mismatched pair out of the following
(a) *Penicillium italicum*–Blue mould of citrus fruits
(b) *P. grisefulvum*–Antibiotic griseofulvin
(c) *P. camemberi*–Roquefort cheese
(d) *P. expansum*–Spoilage of apple during storage

55. In which of the following the pseudomycelium is formed?
(a) *Aspergillus* (b) *Penicillium*
(c) *Saccharomyces* (d) *Helminthosporium*

56. Which one is not a mode of reproduction in yeast?
(a) Budding (b) Plasmogamy
(c) Ascospore formation (d) *Oogamy*

57. The mingling of protoplasm without nuclear fusion is called
(a) pseudogamy (b) polygamy
(c) plasmogamy (d) karyogamy

58. Which out of these is the property of a dikaryotic cell?
(a) Two dissimilar haploid nuclei
(b) Two similar nuclei
(c) Diploid zygotes
(d) Two haploid nuclei

59. When a cell is kept in 0.5 M solution of sucrose its volume does not alter. If the same cell is placed in 0.5 M solution of sodium chloride, the volume of the cell will
(a) increase (b) decrease
(c) cell will be plasmolysed (d) will not show any change

60. The most abundant solute in phloem sap is
(a) potassium ions (b) glucose
(c) sucrose (d) starch

61. Swelling of gums on being placed in moistened environment confirms
(a) diffusion (b) imbibition
(c) endosmosis (d) exosmosis

62. The phenomenon of permeability has been explained by traube through the following concept
(a) solution theory (b) mosaic theory
(c) colloidal theory (d) retention pressure theory

63. The sequence of events by which a cell duplicates its genome, synthesises the other constituents of the cell and eventually divides is termed
(a) cell division (b) cell cycle
(c) cell synthesis (d) cell lysis

64. The phase between two successive M-phases, is called
(a) G_0-phase (b) interphase
(c) S-phase (d) intermediate phase

65. In mitosis, the number of chromosomes in the parent and progeny cells is same, thus also called as
(a) equational division (b) reductional division
(c) may be (a) or (b) (d) None of the above

66. Anaphase stage is characterised by
(a) centromeres split and chromatids separate
(b) chromatids move to opposite poles
(c) Both (a) and (b)
(d) None of the above

67. In the following reaction, what is expressed by A and B?

$$P_r \underset{B}{\overset{A}{\rightleftharpoons}} P_{fr}$$

(a) A–Light B–Dark
(b) A–Dark B–Light
(c) A–Far-red light B–Red light
(d) A–Red light B–Far-red light

68. Indole-3-acetic acid called auxin, was first isolated from
(a) human urine (b) corn germ oil
(c) *Fusarium* (d) *Rhizopus*

69. Plants bend towards light because
(a) they need light for photosynthesis
(b) they need light for respiration
(c) light attracts them
(d) cells on the shaded side elongate more

70. The auxin 2, 4, 5-T is
(a) phenoxy acid (b) benzoic acid
(c) indole auxin (d) naphthalenic acid

71. Gibberellic acid has been successfully employed to induce flowering in
(a) short-day plants under long-day conditions
(b) long-day plants under short-day conditions
(c) the day-neutral plants
(d) few plants

72. The impermeability to oxygen by the seed coat can be found in
(a) Leguminious seeds (b) *Brassica* seeds
(c) *Xanthium* seeds (d) *Pisum* seeds

73. Ribosome found in a cluster is called
(a) oligosome (b) polyribosome
(c) megasome (d) microsome

74. Point mutation is DNA may leads to
(a) termination of protein synthesis
(b) addition of wrong amino acid during protein synthesis
(c) neither termination of protein synthesis nor substitution or wrong amino acid
(d) All of above

75. During *m*RNA processing, the exons are joined make a mature *m*RNA. This process is called
(a) polymerisation (b) hydrolysis
(c) splicing (d) addition

76. Identify the incorrect statement about *t*RNA from the following.
(a) It has anticodon
(b) It binds to amino acid
(c) It does not bind to ribosome
(d) It binds with *t*RNA synthetase

77. What is meant by the statement that the genetic code is universal?
(a) All organisms have the same code
(b) All organisms have nucleus
(c) All organisms have DNA
(d) All organisms synthesis proteins

78. A double-stranded DNA is radioactive in both. It is allowed to replicate twice in non-radioactive medium. The result would be
(a) all double strands have radioactive
(b) half the double strands have radioactive
(c) three double strands have radioactive
(d) radioactivity does not appear is any double strand

79. Genetic code is not
(a) redundancy (b) degeneracy
(c) overlapping (d) universality

80. What is the basis of chromosomal theory of inheritance?
(a) Genes are in mutable
(b) Chromosomes are de-condensed in telophase
(c) Genetic codes are found in chromosomes
(d) Chromosomes carry alleles which are segregated as chromosomes are pulled apart during cells division

81. Adansonia (boabab tree) is
(a) malacophilous (b) ornithophilous
(c) chiropterophilous (d) anemophilous

82. In pin-eyed flowers of *Primula*
(a) stamens are long
(b) style is long
(c) there are two series of long stamens and one short style
(d) there are two series of long styles and one series of short stamens

83. Caruncle is formed by
(a) peduncle (b) cotyledons
(c) integument (d) None of these

84. Ategmic ovules are found in
(a) *Utricularia* (b) *Nepenthes*
(c) *Olax* (d) *Rafflesia*

85. The ovule curvature is more pronounced and the embryo sac becomes horse-shoe-shaped in the following ovule
(a) amphitropous (b) campylotropous
(c) anatropous (d) orthotropous

86. Hemianatropous ovules occur in
(a) *Sida* (b) *Ranunculus*
(c) *Brassica* (d) *Aloe*

87. Perisperm is the
(a) outer part of embryo sac
(b) degenerated secondary nucleus
(c) degenerated synergid
(d) remains of nucellus

88. A middle integument has been observed in
(a) *Annona* (b) *Magnolia*
(c) *Ranunculus* (d) *Cananga*

89. Red rust of tea is caused by
(a) red algae (b) a fungus
(c) a green algae (d) All of these

90. Algae known as sea lettuce is
(a) *Laminaria* (b) *Ulva*
(c) *Sargassum* (d) *Porphyra*

91. Carrageenin is jelly-like substance is derived from marine algae known as
(a) *Fucus* (b) *Gelidium*
(c) *Gracilaria* (d) *Chondrus crispus*

92. The sporophytic generation is represented by the zygote only
(a) Pines (b) *Funaria*
(c) *Chlamydomonas* (d) *Selaginella*

93. Cell wall of fungi are made up of
(a) cellulose (b) lipids
(c) proteins (d) chitin and hemicellulose

94. Who among the following is famous Indian mycologist
(a) P Maheshwari (b) AK Sharma
(c) MOP Iyengar (d) Sadasivan

95. Chickenpox is caused by
(a) bacteria (b) bacteriophage
(c) varicella virus (d) pox virus

96. The parasite that is non-living outside a living cell, but becomes living on entering in it.
(a) Bacterium (b) Bacteriophage
(c) *Plasmodium* (d) Yeast

97. A virus that can reproduce without killing its host is called a
(a) lytic virus (b) retroactive virus
(c) temperate virus (d) virion

98. Viroids enter plant cells only through
(a) roots (b) wounds
(c) stomata (d) lenticels

99. Retroviruses and reverse transcriptase were discovered by
(a) Sinsheimer (b) Stanley
(c) Temin and Baltimore (d) Enders

100. The insect transmitting a virus is known as
(a) vector (b) transformer
(c) collateral host (d) None of these

101. Asexual spores of *Mucor* are
(a) uninucleate (b) binucleate
(c) multinucleate (d) multicellular

102. When a moist bread is kept exposed in air, it becomes mouldy and black because
(a) spores are present in the water
(b) spores are present in the bread
(c) spores are present in the air
(d) the bread decomposes

103. The closed ascocarp is
(a) apothecium (b) amphithecium
(c) endothecium (d) cleistothecium

104. Fungal filament system is known as mycelium. Pseudomycelium is formed in
(a) *Synchytrium*
(b) *Rhizopus*
(c) yeast
(d) *Rhizopus* during spore germination

105. Uredosorus is formed by *Puccinia* on
(a) wheat (b) barberry
(c) Both (a) and (b) (d) *Dalbergia* and wheat

106. Syphilis causing *Treponema pallidum* belongs to the group
(a) Rickettsiae (b) *Bacillus*
(c) Actinomycetes (d) *Spirillum*

107. Halophiles can comfortably live in
(a) Dead sea (b) Dal lake
(c) Arabian sea (d) Godavari

108. Which is not a component of the bacterial cell wall?
(a) Fats (b) Mucopeptide
(c) Cellulose (d) Non-cellulose polysaccharide

109. Botulism is a
(a) type of food poisoning due to the saprophytic bacterium
(b) disease in man due to the parasitic bacterium
(c) disease in various organisms
(d) disease of plants due to the viruses

110. Cellulose is stained blue with
(a) eosin (b) phloroglucinol
(c) chlor-zinc iodine (d) methylene blue

111. The compound microscope is based upon the principle that
(a) the image formed by one optical element works as the object for second element
(b) a single lens system magnifies all diameters of the object being viewed
(c) Both (a) and (b)
(d) None of the above

112. The use of oil immersion can
(a) increase the value of refractive index
(b) decrease the value of refractive index
(c) increase the value of numerical aperture
(d) decrease the value of numerical aperture

113. Dark field microscopy is used to observe bacteria. It makes use of
(a) no light (b) visible light
(c) invisible infrared (d) invisible UV-light

114. Colourblind person fails to distinguish between
(a) all colours (b) red and green colours
(c) green colour (d) blue colour

115. What is the number of sex chromosomes found in human males?
(a) 2 (b) 22 (c) 23 paris (d) 2 paris

116. The genetic information is carried by long chain molecules of
(a) chromosomes (b) enzymes
(c) nucleotides (d) amino acids

117. Character inherited only through female gamete is known to be
(a) Gynacomastia (b) Cytoplasmic inheritance
(c) Incomplete dominance (d) Gynander

118. During metaphase-I of meiosis, Y-chromosome always pairs with
(a) SAT chromosomes (b) X-chromosome
(c) Y-chromosome (d) Autosome

119. The bacterium which is used in the preparation of vitamin riboflavin is
(a) *Leuconostoc* (b) *Clostridium butylicum*
(c) *Lactobacillus* (d) None of these

120. During alcoholic fermentation, conversion of sugar into alcohol is due to direct action of
(a) temperature
(b) pH
(c) concentrate of sugar solution
(d) zymase enzyme

121. Sausages are fermented
(a) vegetables (b) meats (c) milk (d) sauce

122. Which group is responsible for the formation and flavour of yoghurt?
(a) *Bacillus megathermus* and *Xanthomonas* sp.
(b) *Bacillus subtilis* and *E. coli*
(c) *Rhizobium meliloti* and *Azotobacter*
(d) *Lactobacillus casei* and *Streptococcus thermophilus*

123. Which is false about antibiotics?
(a) The term was coined by Waksman in 1942
(b) Antibiotics are capable of curing any disease
(c) Some persons develop allergy to antibiotics
(d) Antibiotics are produced by microorganisms

124. Citric acid is got from
(a) *Aspergillus niger* (b) *Rhizobium nigricans*
(c) *Penicillium citrinum* (d) *Lactobacillus vulgaricus*

125. Match the following columns.

Column I	Column II
A. Neomycin	1. *Streptomyces fradiae*
B. Terramycin	2. *Penicillium notatum*
C. Viridin	3. *Streptomyces ramosus*
D. Penicillin	4. *Gliocladium virens*

Codes
A B C D A B C D
(a) 3 1 4 2 (b) 3 3 1 2
(c) 1 3 4 2 (d) 4 1 3 3

126. The most important use of fungi in recent times is as
(a) food (b) fermenting agents
(c) biological assay (d) antibiotics

127. The most important property of yeast is
(a) distillation (b) fermentation
(c) oxidation (d) All of these

128. Which air pollutant could cause rains to be acidic?
(a) Dust particles from cement factories
(b) Insecticides from crop sprays
(c) Smoke from wood fires
(d) Sulphur dioxide from coal-fired power stations

129. Which kind of power station produces sulphur dioxide, causing air pollution and acid rain?
(a) Coal-fired (b) Nuclear
(c) Hydroelectric (d) Wind-powered

130. Acid rain occurs where atmosphere is heavily polluted with
(a) CO, CO_2 (b) smoke particles
(c) O_3 (d) SO_2

131. The material generally used for sound proofing of rooms like a recording studio and auditorium, etc., is
(a) cotton (b) coir (c) wood (d) styro foam

132. The trace gas which is produced in rice field and is associated with global warming is
(a) CH_4 (b) N_2O (c) CO_2 (d) H_2S

133. Best arrangement of an energy system consists of hawks, snakes, mice and grasses is
(a) grasses → mice → snakes → hawks
(b) grasses → snakes → mice → hawks
(c) grass → mice → hawks → snakes
(d) hawks → snakes → mice → grasses

134. Aquatic ecosystem consists of
(a) biotic factor (b) biotic and abiotic factors
(c) consumers only (d) producers only

135. Match the following columns.

Columns I	Columns II
A. Mobius	1. Biocoenosis
B. Forbes	2. Microcosm
C. Sukachey	3. Biogeocoenosis

Codes
A B C A B C
(a) 3 2 1 (b) 3 2 1
(c) 1 2 3 (d) 1 3 2

136. Detritus food chain in comparison to grazing food chain is generally
(a) longer (b) equal (c) shorter (d) None of these

137. Ecological pyramids were given by
(a) Elton (b) Haeckel
(c) Odum (d) Reiter

138. Rate of storage of organic matter by heterotrophs is termed as
(a) net productivity
(b) net primary productivity
(c) gross primary productivity
(d) secondary productivity

139. Piliferous layer refers to
(a) epiblema (b) exodermis
(c) epidermis (d) pericycle

140. Sclerenchymatous hypodermis is found in
(a) monocot roots (b) dicot roots
(c) dicot stems (d) monocot stems

141. Sclerenchymatous sheath surrounding the vascular bundles in a monocot stem, is called
(a) vascular sheath (b) circular sheath
(c) bundle sheath (d) starch sheath

142. Protoxylem cavity of lacuna (water containing cavity) is .... in origin.
(a) schizogenous (b) lysigenous
(c) schizo-lysigenous (d) None of these

143. Vascular bundles are arranged in two rings in the dicot stems of
(a) *Helianthus* (b) *Ricinus*
(c) *Cucurbita* (d) *Aristolochia*

144. The most mutable virus is
(a) chickenpox virus (b) HIV
(c) influenza virus (d) dengue virus

145. Sedimentation coefficient or Svedberg unit is equal to
(a) 1×10^{-13} s (b) 1×10^{-2} s
(c) 1×10^{-10} s (d) 1×10^{-5} s

146. Microfilaments are long thin fibres 40-60Å in diameter forming clusters or networks at various places in the cell, these fibres are composed of a protein called
(a) tubulin
(b) actin
(c) myosin
(d) haemoglobin

147. The pigment molecules of a chloroplast are located within
(a) its thylakoid membranes
(b) the space between the inner and outer membranes
(c) the inner membrane
(d) intrathylakoid spaces

148. Inclusion bodies such as sulphur granules are bounded in a bacterial cell by
(a) unit membrane
(b) membrane but without unit membrane structure
(c) no membrane
(d) None of the above

149. The correct match is
(a) Binucleate cell — Nucleus
(b) Peripheral nucleus — Prokaryotic cell
(c) Nucleonema — Nucleolus
(d) Chromonema — Phycoerythrin

150. A cell membrane consisting of a phospholipid bilayer with a single continuous layer of protein molecules on either surface is called as
(a) Sandwich model
(b) Lamellar theory
(c) Danielli and Davson model
(d) All of the above

Answers

1. (c)	**2.** (b)	**3.** (c)	**4.** (b)	**5.** (d)	**6.** (c)	**7.** (b)	**8.** (a)	**9.** (a)	**10.** (c)
11. (d)	**12.** (c)	**13.** (b)	**14.** (b)	**15.** (b)	**16.** (c)	**17.** (b)	**18.** (d)	**19.** (b)	**20.** (c)
21. (c)	**22.** (b)	**23.** (d)	**24.** (a)	**25.** (a)	**26.** (c)	**27.** (b)	**28.** (c)	**29.** (c)	**30.** (d)
31. (c)	**32.** (b)	**33.** (d)	**34.** (b)	**35.** (c)	**36.** (c)	**37.** (c)	**38.** (d)	**39.** (d)	**40.** (b)
41. (d)	**42.** (a)	**43.** (d)	**44.** (c)	**45.** (d)	**46.** (a)	**47.** (a)	**48.** (b)	**49.** (c)	**50.** (d)
51. (a)	**52.** (d)	**53.** (a)	**54.** (c)	**55.** (c)	**56.** (d)	**57.** (c)	**58.** (a)	**59.** (b)	**60.** (c)
61. (b)	**62.** (d)	**63.** (b)	**64.** (b)	**65.** (a)	**66.** (c)	**67.** (d)	**68.** (a)	**69.** (d)	**70.** (a)
71. (b)	**72.** (c)	**73.** (b)	**74.** (c)	**75.** (c)	**76.** (c)	**77.** (b)	**78.** (b)	**79.** (c)	**80.** (d)
81. (c)	**82.** (b)	**83.** (c)	**84.** (c)	**85.** (a)	**86.** (b)	**87.** (d)	**88.** (d)	**89.** (c)	**90.** (b)
91. (d)	**92.** (d)	**93.** (d)	**94.** (d)	**95.** (c)	**96.** (b)	**97.** (c)	**98.** (b)	**99.** (c)	**100.** (a)
101. (c)	**102.** (c)	**103.** (d)	**104.** (c)	**105.** (a)	**106.** (d)	**107.** (a)	**108.** (c)	**109.** (a)	**110.** (c)
111. (a)	**112.** (c)	**113.** (b)	**114.** (b)	**115.** (a)	**116.** (c)	**117.** (b)	**118.** (b)	**119.** (b)	**120.** (d)
121. (b)	**122.** (d)	**123.** (b)	**124.** (a)	**125.** (b)	**126.** (d)	**127.** (b)	**128.** (d)	**129.** (a)	**130.** (d)
131. (d)	**132.** (a)	**133.** (a)	**134.** (b)	**135.** (c)	**136.** (c)	**137.** (a)	**138.** (d)	**139.** (d)	**140.** (a)
141. (c)	**142.** (b)	**143.** (a)	**144.** (b)	**145.** (a)	**146.** (b)	**147.** (a)	**148.** (b)	**149.** (c)	**150.** (d)